Homogeneous Hydrogenation

HOMOGENEOUS HYDROGENATION

BRIAN R. JAMES
The University of British Columbia

M · Þ · K

A WILEY-INTERSCIENCE PUBLICATION

JOHN WILEY & SONS, New York · London · Sydney · Toronto

Copyright © 1973, by John Wiley & Sons, Inc.

Library of Congress Cataloging in Publication Data:

James, Brian R.   1936–
   Homogeneous hydrogenation.

"A Wiley-Interscience publication"
Includes bibliographical references.
1. Hydrogenation. 2. Catalysis. I. Title.

QD281.H8J36   547′.23   72-11804
ISBN 0-471-43915-0

Printed in the United States of America

10–9 8 7 6 5 4 3 2 1

TO JANE

Jennifer
Peter
Sarah
Andrew

Preface

Over the last decade interest and expansion in the field of homogeneous catalysis have been phenomenal. Studies on the activation of a variety of small gas molecules by metal complexes in solution have revealed remarkable chemistry, the potential of which remains to be fully realized. Studies on activation of molecular hydrogen are the best documented from a historical point of view, and the general impetus in the homogeneous catalysis area has advanced the knowledge of some hydrogenation reactions to such a sophisticated level that highly selective and efficient catalysts have been developed for the reduction of a variety of unsaturated systems at ambient conditions.

Despite the appearance of many reviews there has been no definitive treatise on the vast subject of homogeneous hydrogenation. I have attempted to cover the literature exhaustively up to the end of 1970, and in a discriminating manner. I hope that there are few serious omissions and I offer my apologies if I have slighted work that readers may regard as particularly significant. An Appendix has been added to incorporate the more relevant references of 1971 and part of early 1972.

Classification of material proved a difficult matter; metal triads have been used primarily as a basis with cross-references given where necessary (e.g., Chapters I to XIII), although separate chapters (XIV to XVI) are devoted to certain topics such as hydrogenation of unsaturated fats and Ziegler catalysts.

This book is especially intended for research workers in the area of homogeneous hydrogenation and also for researchers in the more general area of homogeneous catalysis. The currently developing technique of heterogenizing homogeneous catalysts, besides offering greater practicability for homogeneous catalysts, should contribute significantly to further understanding of more complex heterogeneous and biological catalysis.

Some attempt has been made in the Summary (Chapter XVII) to offer help to the synthetic organic chemist by listing catalytic systems that have been found effective for the hydrogen reduction of various unsaturated bonds. The index also lists organic compounds by classifications such as steroids, terpenes, etc., which have been hydrogenated using homogeneous systems. It is hoped that this book will stimulate further interest in a field in which there still remains great scope for more systematic investigation.

Most of this book was written during a sabbatical year in the Chemistry Department at the University of Sussex in England and the Department of Chemical Engineering at the University of Waterloo in Ontario, Canada. The facilities and cooperation of these two departments were very much appreciated. I thank the Royal Society and Nuffield Foundation for a Commonwealth Bursary that I held at the University of Sussex. The assistance of I. Wender and C. Zahn of the Pittsburgh Energy Research Center, U.S. Bureau of Mines, especially in the editing of the final manuscript, was much. appreciated. Several colleagues at the University of British Columbia offered their translation services: F. Aubke (German), E. Ochiai (Japanese), C. Reid (Russian), and J. Serreqi (Italian).

I am grateful to Jack Halpern, at the University of Chicago, for first arousing my interest in the subject of homogeneous hydrogenation and to Irving Wender who initially encouraged me to undertake the task of reviewing the subject. The National Research Council of Canada and the University of British Columbia have given generous financial support for the studies carried out by our group.

Finally, I thank my wife both for her help in compiling the list of references and for her unfailing patience, love, and understanding.

BRIAN R. JAMES

Vancouver, Canada
November 1972

Contents

Homogeneous Hydrogenation

I

Introduction

A. HISTORICAL AND GENERAL CONSIDERATIONS

Molecular hydrogen is relatively unreactive chemically, particularly at low temperatures (below 100°). The dissociation of gaseous hydrogen into atoms is endothermic to the extent of 104 kcal mole^{-1}, which partly accounts for its low reactivity. However, many reactions involving molecular hydrogen have been recognized which involve catalysis by various substances. Until about a decade ago this hydrogen activation was largely confined to heterogeneous systems involving metals, metal oxides, and some salts, but since then a wide range of metal ions and complexes, particularly of the platinum metals, have been found to catalyze hydrogenation reactions homogeneously in solution.

We shall consider the detailed mechanisms of hydrogenation reactions later, but they must inevitably involve formation of intermediate hydride species. It is of interest to estimate the energies required for the dissociation of hydrogen in solution for the two feasible processes involving homolytic splitting into atoms and heterolytic splitting into ions:

$$H_2(aq) \qquad 2H(aq) \qquad\qquad (1)$$

$$H_2(aq) \longrightarrow H^+(aq) + H^-(aq) \qquad\qquad (2)$$

The data required are the heats of hydration per mole for the proton, -260 kcal (1), the hydride ion, -104 kcal (2), the hydrogen molecule, -1.0 kcal (3), the hydrogen atom, -1.0 kcal (4), as well as the established values for the ionization potential (313 kcal) and electron affinity (-17 kcal) of the hydrogen atom, and the hydrogen bond dissociation energy. Simple calculations by means of Born-Haber cycles show that splitting into free atoms requires about 100 kcal mole^{-1}, whereas the ionic splitting requires about 37 kcal. Such considerations are clearly important among others in determining the mode of hydrogen activation and do suggest that a reasonably favorable energetic path through hydride formation is available through the heterolytic mechanism, particularly in polar solvents.

The many advances in organometallic chemistry have provided methods for preparing a wide variety of metal complexes and knowledge of their ligand substitution and oxidation-reduction reactions in solution. Homogeneous catalytic systems are inherently simpler chemically and kinetically than heterogeneous systems and are much more amenable to detailed study. Heterogeneous systems remain the major synthetic tool for the organic chemist employing catalytic hydrogenation, but an increasing number of workers are turning to homogeneous systems for more selective and controlled hydrogenations. Industrially important catalytic hydrogenations employ heterogeneous systems extensively, but the increasing list of patent applications involving homogeneous systems indicates that they will become increasingly important. A 1968 report by a chemistry committee of the Science Research Council in the United Kingdom (5) indicates that in 12 to 20 years possibly half of the catalytic systems used generally in the heavy organochemical industry will be homogeneous. On the other hand, later articles (e.g., 5a, 5b) suggest that the idea of such a boom in homogeneous catalysis is unlikely.

This book devotes itself to homogeneous hydrogenation systems in solution and attempts to cover the literature up to the end of 1970. Patent literature, which involves data not covered in the journals, is also included. Moreover, several volumes and articles dealing with heterogeneous hydrogenation have appeared (e.g., Refs. 6–13b). Relationships between homogeneous and heterogeneous hydrogenation will not be discussed here in any detail (see Chapter XVI).

Before considering the many individual systems studied, it is worthwhile to review briefly the apparently more significant contributions to the development of homogeneous hydrogenation, particularly the reports that have stimulated other researchers and have brought the field to its fairly advanced present stage.

Historically the first documented example of a homogeneous catalytic hydrogenation was reported in 1938 by Calvin (14, 15), who discovered that in quinoline solution at about 100° cuprous acetate catalyzed the reduction by dissolved hydrogen ($\sim$1 atm) of cupric acetate or benzoquinone. Ipatieff and Werchowsky (16) in 1909 had observed that aqueous solutions of cupric acetate were reduced under quite mild conditions to cuprous oxide, but it was not until 1954 that this system was studied kinetically and shown to be homogeneous (17, 18).

Iguchi (19) in 1939 first reported the activation of molecular hydrogen by rhodium(III) complexes; in aqueous acetate solution at 25° catalysts such as $[Rh(NH_3)_5H_2O]Cl_3$, $[Rh(NH_3)_4Cl_2]Cl$, and $RhCl_3$ were found to be active for the reduction of a number of inorganic and organic substrates, such as quinone, fumaric acid, and sodium nitrite, while the saturated amine complexes $[Rh(NH_3)_6]Cl_3$ and $[Rh(en)_3]Cl_3$ were inactive. The kinetics of these reactions were not extensively studied, however, and possible mechanisms were not discussed.

The ready absorption of hydrogen by cobaltous cyanide solutions at room temperature in an amount corresponding to almost one hydrogen atom per cobalt atom was accidently discovered by Iguchi (20) in 1942 while he was investigating the reducing action of metal complexes using hydrogen as an inert nonoxidizing atmosphere. This finding has led to the most widely studied homogeneous hydrogenation catalyst, the pentacyanocobaltate(II) anion. Such systems have been reviewed by Kwiatek and Seyler (21).

Considerable progress in the understanding of hydrogen activation by metal ions was made between 1955 and 1960 by Halpern and his coworkers. These studies dealt mainly with the reduction of inorganic substrates, and several reviews appeared during this period (22–26a).

In 1961 Halpern's group (27) reported that under mild conditions aqueous solutions of chlororuthenate(II) were effective catalysts for the hydrogen reduction of a number of activated olefins. Kinetic, spectroscopic, and deuterium tracer studies indicated a mechanism involving reaction of hydrogen with a ruthenium(II)-olefin complex. When further reported in 1962 (28, 29), this system became the best understood process involving olefin hydrogenation catalyzed by a platinum metal complex. About the same time, workers at the Du Pont Company (30) reported that a platinum-tin chloride complex in methanol solution was the first effective catalyst for the reduction of ethylene at room temperature and 1 atm hydrogen pressure. In 1961 Köster and coworkers (31) and DeWitt and coworkers (32), using knowledge of the hydroboration reaction developed by Brown (33), reported that boranes catalyze the hydrogenation of olefins; however, quite severe conditions (200°, >70 atm hydrogen) were required. This appears to be one of the few systems reported that involves an intermediate hydride of an essentially nonmetallic element (Sections XVI-D, XVI-G, and XVI-H).

In 1962 Vaska and Diluzio (34) discovered that a synthetic complex of iridium(I) (35) reacted with hydrogen to give a reversible but isolable and stable adduct, formally an iridium(III) complex.

$$trans - IrCl(CO)(PPh_3)_2 + H_2 \rightleftharpoons IrH_2Cl(CO)(PPh_3)_2 \qquad (3)$$

This is now the prototype of the so-called oxidative addition reaction of square planar d^8 complexes. The importance of these complexes to hydrogenation and homogeneous catalysis in particular is clear; such systems offer the opportunity of directly observing the electronic and stereochemical properties of a metal-gas-activated adduct and of obtaining an understanding of the factors that determine the reversible activation of small covalent gas molecules by transition metal complexes. Reviews that stress the importance of these oxidative-addition reactions of homogeneous catalysis generally, including hydrogenation, have been written by Vaska (36), Halpern (37–37b), Collman (38, 39a) Collman and Roper (39), Carrà and Ugo (40), Cramer (41), Strohmeier (41a), and Wilkinson (42).

Reports of the more active hydrogenation catalysts in 1960–1962 did involve platinum metal complexes; it is in this area that subsequent intense activity has resulted. The literature on such systems is now substantial, as we shall see, but very significant advances have been made by Wilkinson's group in the isolation and detailed study of the extremely active (at room temperature, 1 atm hydrogen) rhodium(I) complex, $RhCl(PPh_3)_3$ (43, 43a), and the ruthenium(II) complex, $HRuCl(PPh_3)_3$ (44), both originally reported in 1965. The activity of the rhodium complex was also discovered independently by Coffey (45).

Besides the extensively studied pentacyanocobaltate(II) systems, the catalytic activity of other complexes of the first-row transition metals for hydrogenation has also been studied, particularly by Tulupov and his group, for example, the hydrogenation of cyclohexene in alcohol solutions containing chromic stearate (46). These systems generally seem more complex than those involving platinum metal complexes and often require more severe conditions of temperature and hydrogen pressure for effective hydrogenation, making them less amenable for a detailed study and understanding of kinetics. Similar limitations apply to studies of catalyzed hydrogenation by metal carbonyls of the first-row transition metals, such as $Fe(CO)_5$ (47, 48) and $Co_2(CO)_8$ (49, 50), which were undoubtedly stimulated by reported studies of the hydroformylation(Oxo) reaction which was often accompanied by reduction processes, for example, aldehydes to alcohols. The hydroformylation reaction itself has been extensively reviewed (51–55); it will not be studied in this book, although hydrogenation under these conditions will be considered.

Finally, note should be taken of the discovery by Sloan and coworkers in 1963 (56) of soluble Ziegler-type catalysts, such as $R_3Al-Cr(acac)_3$, for hydrogenation of organic molecules, the systems again requiring a few atmospheres of hydrogen pressure. Such systems have since been investigated extensively.

A significant number of general review articles have appeared in journals and specialized texts on the subject of homogeneous hydrogenation; they are entitled essentially as such (57–57h, 58–58h, 81) or are considered in a particular subsection of more general articles that have discussed homogeneous catalysis (36–41a, 59–62a, 64–68e); indeed, it is now common to find brief sections on homogeneous hydrogenation in senior undergraduate texts on inorganic and organometallic chemistry (69–74b). A monograph (13) on the more industrial aspects of catalysis, for example, also includes a brief section on homogeneous hydrogenation. Several articles that include information on the activity of platinum metal systems have also been published (42, 63, 75–79g, 149–151).

Homogeneous hydrogenation was the subject of a Faraday Society meeting (80), and this process has also been discussed at meetings of the New York Academy of Sciences (80a, 80b).

Several reviews specifically on hydrogenation of unsaturated fats have appeared (Chapter XIV). An article on asymmetric synthesis, including a brief discussion of homogeneous hydrogenation, has also appeared (79h).

In Chapters II to XIII the data reported on the activity of metal ions and their complexes for homogeneous hydrogenation will take, in turn, the Group B triads from IB to VIIB and then the Group VIII triads, as classified below in the three rows of transition metals.

IIIB	IVB	VB	VIB	VIIB		VIII		IB	IIB
Sc	Ti	V	Cr	Mn	Fe	Co	Ni	Cu	Zn
Y	Zr	Nb	Mo	Tc	Ru	Rh	Pd	Ag	Cd
La	Hf	Ta	W	Re	Os	Ir	Pt	Au	Hg

This manner of presentation has merit in that it follows to some extent the chronological development of the subject.

Separate sections are devoted to the hydrogenation of unsaturated fats (Chapter XIV) and Ziegler-type catalysts (Chapter XV). Chapter XVI is devoted to miscellaneous systems, including hydrogenase and nonmetal catalyzed hydrogenation. A few general conclusions are presented in the final chapter (XVII) of this book.

Isomerization reactions, particularly double-bond migration, are frequently encountered during studies in homogeneous hydrogenation. Review articles on this topic include those by Orchin (152), Bird (53), Davies (153), Cramer (41), and Hubert and Reimlinger (154). Three mechanisms have been presented for such isomerization as follows:

(*a*) Metal hydride addition-elimination

$$RCH_2CH{=}CH_2 \rightleftharpoons RCH_2\underset{\underset{M}{|}}{CH}{-}CH_3 \rightleftharpoons RCH{=}\underset{\underset{M}{|}}{CH}{-}CH_3$$
$$\underset{\underset{MH}{|}}{\ }$$

(*b*) π-Allyl hydride formation

$$\underset{\underset{\textstyle M}{|}}{RCH_2CH}{=}CH_2 \rightleftharpoons \underset{\underset{\textstyle MH}{|}}{RCH'}\overset{\textstyle CH}{\underset{\textstyle \diagdown}{\diagup}}\underset{}{CH_2} \rightleftharpoons \underset{\underset{\textstyle M}{|}}{RCH}{=}CHCH_3$$

(*c*) Carbene formation

$$\underset{\underset{\textstyle M}{|}}{RCH_2CH}{=}CH_2 \rightleftharpoons \underset{\underset{\textstyle M}{\|}}{RCH_2CCH_3} \rightleftharpoons \underset{\underset{\textstyle M}{|}}{RCH}{=}CHCH_3$$

Details of isomerization studies will not be considered as such, although they are often important in elucidating hydrogenation mechanisms; however, the isomerizations will often be mentioned.

Further, the chemistry of alkyl complexes will not be considered in any detail as such, although these are intimately involved in many catalytic hydrogenation reactions. Papers by Sneedon and Zeiss (155) and Wilkinson and coworkers (156) refer to important advances in this area where at least four fundamental processes have been recognized:

(*a*) α-Metal hydride elimination

$$RCH_2CD_2M \rightarrow RCH{=}CDH + DM$$
$$RCD_2CH_2M \rightarrow RCD{=}CHD + HM$$

(*b*) β-Metal hydride elimination

$$RCD_2CH_2M \rightarrow RCD{=}CH_2 + DM$$

(*c*) Homolysis, attack on solvent

$$RCD_2CH_2M + \text{solvent} \rightarrow RCD_2CH_3 + M$$

(*d*) Homolysis, hydrogen transfer between β-position of one alkyl to the α-position of another

$$2RCD_2CH_2M \rightarrow RCD_2CH_2D + RCD = CH_2 + M$$

The reverse of processes shown in (*a*) and (*b*), the hydrometallation of olefins, will be encountered frequently. Processes such as (*c*) and (*d*) are always possible steps leading to formation of saturated product.

B. FORMULATION AND NOMENCLATURE OF METAL HYDRIDES

Three ways have thus far been recognized in which a metal species may form a hydride complex, and these have been briefly referred to in Section

I-A (eqs. 1 to 3). Cleavage of molecular hydrogen homolytically has been formulated by various workers in three ways, for example, for a divalent metal complex:

$$2M^{II} \text{ or } (M^{II})_2 + H\text{:}H \longrightarrow \begin{cases} 2M^{II}(\cdot H) \\ 2M^{I}(H) \\ 2M^{III}(\text{:}H) \end{cases} \tag{4}$$

The three products differ only in the position of the electron originally associated with the hydrogen atom.

In a transition metal hydride the hydrogen is thought to be present as an anionic hydride ligand ($\text{:}H^-$) occupying one or more of the normal coordination positions around the central metal atom and can be treated as, for example, halide ligands with ligand substitution properties and occupancy of a position in the spectrochemical series (68c, 82–84).

This book will use this formulation, and hence both homolytic splitting and dihydride formation will result in formal oxidation of the metal center, the former by one unit, the latter by two units; for example:

$$2M^{I} + H_2 \rightarrow 2M^{II}H \tag{5}$$

$$M^{I} + H_2 \rightarrow M^{III}H_2 \tag{6}$$

Heterolytic splitting of the hydrogen molecule with resulting hydride production results in no change in the formal oxidation state of the metal:

$$M^{I} + H_2 \rightarrow M^{I}H + H^+ \tag{7}$$

Such formulations have not always been used in the literature and the terminology used in discussing mechanisms can be confusing. In the formulation above, for example, it should be realized that heterolytic activation of hydrogen by copper(II) and a homolytic activation by copper(I) produce a hydride that is formally a copper(II) derivative, namely $Cu^{II}H$ (or CuH^+).

The classification of heterolytic and homolytic splitting of hydrogen can clearly arise from consideration of the hydride products. Mechanistically such a distinction may not always be meaningful. As we shall see, both types of splitting may occur by prior *dihydride* formation.

C. LIST OF ABBREVIATIONS

The following abbreviations have generally been used in the book, unless stated otherwise.

acac, acetylacetonate

Ar, aryl grouping

atm, atmosphere

β_n, overall stability constant for formation of ML_n

Bu, butyl

^{i}Bu or Bu^i, isobutyl

^{n}Bu or Bu^n, n-butyl

(Me, Et, and Pr used similarly for simple alkyl groups)

bipy or dip, 2,2'-bipyridyl

CDT, cyclododecatriene

COD, cyclooctadiene

Cp, π-C$_5$H$_5$

dien, diethylenetriamine

diphos, 1,2-bis(diphenylphosphino)-ethane

DMA, N,N'-dimethylacetamide

DMF, N,N'-dimethylformamide

DMGH$_2$, dimethylglyoxime

DMSO, dimethylsulfoxide

en, ethylenediamine

e.s.r., electron spin resonance

H$_4$EDTA, ethylenediaminetetra-acetic acid

i.r., infrared

L, unless specified, is an undetermined ligand

M mole liter^{-1}

MA, maleic acid

n.m.r., nuclear magnetic resonance

OAc, acetate

OBu, butyrate

OHp, heptanoate

OPr, propionate

Ph, phenyl

phen, o-phenanthroline

pn, propylenediamine

py, pyridine

R, alkyl

R.T., room temperature

THF, tetrahydrofuran

trien, triethylenetetramine

u.v., ultraviolet

μ, ionic strength

[ ]$_T$ denotes *total* concentration of a species present in the system

II

Group I Metal Ions and Complexes

A. CUPROUS SPECIES

As mentioned in the Introduction, cuprous species were involved in the first well documented example of a homogeneously catalyzed hydrogenation. Calvin (14, 15) reported that quinoline solutions of cupric acetate or salicylaldehyde absorbed hydrogen in an autocatalytic reaction at temperatures around 100° and hydrogen pressures up to 1 atm. The uptake essentially ceased at a stoichiometry corresponding to complete reduction of Cu^{II} to Cu^{I}, although a much slower subsequent uptake was observed due to the slow reduction of Cu^{I} to the metal. The rate of hydrogen uptake was proportional to the amount of Cu^{II} reduced, and hence the cuprous salt produced is the catalytic species responsible for activating hydrogen. Later magnetic measurements by Wilmarth and coworkers (85) confirmed that in the autocatalytic region a paramagnetic species (Cu^{II}) was being reduced to a diamagnetic one (Cu^{I}). Calvin (14) originally reported that these autocatalytic plots indicated a first-order dependence on cuprous. Other substrates such as p-benzoquinone were then shown to be hydrogenated in the presence of dissolved cuprous acetate. Such a system gave rise to linear uptake plots from which the reaction was reported (15) to be first-order in hydrogen, between first- and second-order in cuprous, and essentially independent of substrate concentration. This substrate independence is also indicated by the linear uptake plots, which in the case of benzoquinone essentially ceased at a stage corresponding to half reduction, that is, reduction to quinhydrone. From the kinetic data Calvin suggested that a dimeric cuprous acetate

9

quinoline complex was the active catalyst:

$$2Cu^IQ \rightleftharpoons (Cu^IQ)_2 \tag{8}$$

$$(Cu^IQ)_2 + H_2 \underset{k_{-1}}{\overset{k_1}{\rightleftharpoons}} (Cu^IQ)_2H_2 \tag{9}$$

$$(Cu^IQ)_2H_2 + \text{substrate} \overset{k_2}{\longrightarrow} (Cu^IQ)_2 + \text{product} \tag{10}$$

$$(Cu^IQ)_2H_2 \overset{k_3}{\longrightarrow} 2Cu^0 + 2Q + 2H^+ \tag{11}$$

(Q represents quinoline.)

The data require that $k_2 > k_{-1} > k_1 > k_3$; an activation energy of about 13 kcal mole^{-1} was measured for k_1.

Calvin (14) has also observed that para-rich hydrogen was converted to the ortho-para equilibrium mixture but only after reduction of the substrate (Cu^{II}, in this study) was complete. Such a conversion could readily take place through an equilibrium such as eq. 9, but it will not occur effectively until all the substrate has been consumed by the relatively fast reaction of eq. 10. Wilmarth and Barsh (86) later studied the ortho-parahydrogen conversion by quinoline solutions of cuprous acetate and concluded from kinetics and rate data that the same activated intermediate (eq. 9) was involved for this process and catalytic hydrogenation. The same workers (86) also observed an exchange of deuterium gas with cuprous acetate quinoline solutions at a rate of the same order of magnitude as the parahydrogen conversion, and they suggested that it might be an integral part of the conversion mechanism. However, Weller and Mills (87) showed that this deuterium exchange results from aniline impurities in the solvent; they also showed that there was little exchange prior to complete reduction of the substrate, confirming Calvin's result. The basis of the exchange reaction was written as in eq. 12:

$$(Cu^I)_2D_2 + RH \text{ (solution impurity)} \rightleftharpoons (Cu^I)_2HD + RD \text{ (solution)} \tag{12}$$

and was considered important because it suggested that the hydrogen (deuterium) is dissociated on activation, and that eq. 9 should be written in the form

$$(Cu^IQ)_2 + H_2 \rightleftharpoons (Cu^IQ)_2 \cdot 2H \tag{13}$$

A detailed study by Weller and Mills (87) of Calvin's systems (both quinone and Cu^{II} reduction) supported the original mechanism shown in eqs. 8 to 11, and an equilibrium constant for the dimerization process (eq. 8) was also estimated from boiling-point elevation data.

In a later study on the reduction of cupric salts, Calvin and Wilmarth (88) suggested that when due account is taken of the inhibitory effects of water impurity and acetic acid formed in the cupric acetate system, an exact second-order dependence on cuprous concentration is observed. They questioned the dimeric nature of the catalyst and they preferred a single termolecular rate-determining step, involving a homolytic splitting of the hydrogen molecule:

$$2Cu^I + H_2 \underset{k_{-1}}{\overset{k_1}{\rightleftharpoons}} 2Cu^{II}H \tag{14}$$

followed by

$$Cu^{II}H + Cu^{II} \xrightarrow{k_2} 2Cu^I + H^+ \tag{15}$$

$$(k_2 \gg k_{-1} \gg k_1)$$

Wilmarth and Barsh (89) then reinvestigated the cupric acetate reduction and obtained an activation energy for k_1 in good agreement with Calvin's value for the cuprous acetate catalyzed hydrogenation of benzoquinone (15).

The activity of a number of cuprous salts has been investigated in quinoline solution (88); activities of the acetate, salicylaldehyde, and 4-hydroxy-salicylaldehyde were similar and somewhat greater than those of the stearate and benzoate. Nitrobenzoates, nitrosalicylaldehydes, and complexes of certain Schiff bases were inactive. These results were generally interpreted as indicating increasing catalytic activity with increasing basicity of the anion, but steric effects were also thought to be important in the more heavily substituted Schiff bases. For one system, that of cuprous salicylaldehyde anil, experiments involving parahydrogen conversion established that the catalytic inactivity resulted from failure to form an intermediate hydride rather than the inability of the substrate to react with such a hydride. Wright and Weller (90) reported that both ethylenediamine and EDTA inhibit the cuprous acetate catalyzed, hydrogen reduction of benzoquinone and cupric acetate; steric interference in the reduction step of the substrate rather than prevention of hydrogen activation was suggested, but coordination of these ligands with both Cu^I and Cu^{II} species prevented any definite conclusions.

Weller and Mills (87) also made a cursory study of the effect of solvents on the activity of cuprous acetate. In quinoline, pyridine, several of their alkylated derivatives, and dodecylamine the catalytic activity was maintained. No activity was observed in a number of other solvents including indole, formamide, dibutylphthalate, dimethylaniline, dipentylamine, diethanolamine, and 8-hydroxyquinoline. The results suggested that for effective hydrogen activation the solvent should be a nitrogen base, not necessarily heterocyclic, and free of complicating features such as steric factors or strong chelating tendency; there was some indication that activity increased with solvent basicity.

Wright and coworkers (91) showed that the kinetics of hydrogen activation by cuprous acetate in pyridine or dodecylamine were first-order in cuprous, in contrast to the data in quinoline solution. These kinetics were obtained from autocatalytic uptake plots for the cupric acetate system and again the rate was essentially independent of the cupric concentration. Molecular-weight measurements showed that cuprous acetate was unassociated in pyridine, so that the effective catalyst was a monomeric cuprous species. The reason for the differences in kinetics and mechanism in the pyridine solvent and the quinoline solvent system (which Wright and coworkers thought involved a cuprous dimer) was attributed to steric effects that involved differences in the size of the solvent molecules (91). In retrospect, the difference would lie in the mechanism of a reaction such as that shown in eq. 14 involving two cuprous species per hydrogen molecule in quinoline and a reaction in the pyridine system involving one cuprous center. For the latter, both heterolytic and homolytic splitting of the hydrogen molecule were considered and the subsequent fate of the hydrogen atoms or ions was speculated upon, but no definite conclusions were drawn (91).

Chalk and Halpern studied the hydrogen reduction of cupric heptanoate in heptanoic acid solution (92) and also the reduction of other cupric carboxylate salts in heptanoic acid, diphenyl, and octadecane (93). The rate law for the reduction of cupric heptanoate in all these solvents was

$$\frac{-d[H_2]}{dt} = k_1[H_2][Cu^{II}] + k_2[H_2][Cu^{I}] \tag{16}$$

Again the rate curves exhibited autocatalytic behavior but, in contrast to the solvent systems studied previously, a finite initial rate was observed, which was attributed to activation of hydrogen by Cu^{II} (the k_1 term in eq. 16). The first-order dependence on cuprous fell to more nearly half-order with increasing concentration of the cuprous salt, which was thought to be due to dimerization to an inactive form. In all these systems the reactivity of Cu^{I} toward hydrogen was greater than that of Cu^{II} by factors of 20 to 40. Addition of 2,2'-biquinoline, which forms a very stable chelate complex with cuprous ions, completely inhibited the autocatalysis (92).

Heterolytic splitting of the hydrogen molecule was thought to be involved for both the cupric state (Section II-B) and the cuprous species, and for the latter (eq. 17), this was confirmed by isotope-exchange experiments (92). HD

$$Cu^{I} + H_2 \rightleftharpoons Cu^{I}H + H^{+} \tag{17}$$

is formed by exchange of D_2 with heptanoic acid in the absence of the reducible substrate Cu^{II}, through the reversal of reaction 17. Substrate reacts rapidly with the $Cu^{I}H$ intermediate and prevents exchange; the results are analogous to those reported for the cupric acetate system in quinoline (87). Some kinetic data (93), concerning the effect of solvent and anions on the activity of cuprous complexes, are given together with data for the cupric complexes in Table 2 in the following section and are discussed there.

In 1960, Parris and Williams (94), in an attempt to understand the different rate expressions found in the quinoline and pyridine solvents, performed a more detailed study of the hydrogen reduction of cupric ions catalyzed by cuprous salts in these two solvents. Working at much lower concentrations of Cu^I and Cu^{II}, they observed for both solvent systems a rate law of the form

$$-\frac{d[H_2]}{dt} = \frac{k[Cu^{II}][Cu^I][H_2]}{k' + k''[Cu^{II}]} \tag{18}$$

which agreed with that found earlier (91) in pyridine at high cupric concentrations. The variation of rate with different cuprous salts in pyridine was found to follow closely that reported by Calvin and Wilmarth (88) in quinoline solution, but Parris and Williams questioned the difference in mechanism previously proposed (88, 91) in the two solvents. In contrast to earlier work (87) no reduction was observed in 4-methylquinoline solution; also, in 3- or 4-cyanopyridines a faster rate than that in the pyridine system was observed and a second-order dependence in the cuprous salt was indicated. The following scheme was suggested (94) to account for all these data:

$$Cu^IX^- + H_2 \underset{k_{-1}}{\overset{k_1}{\rightleftharpoons}} Cu^IX^-(H_2) \tag{19}$$

$$Cu^IX^-(H_2) + Cu^{II} \overset{k_2}{\longrightarrow} Cu^{II}H + Cu^I + HX \tag{20}$$

$$Cu^IX^-(H_2) + Cu^I \overset{k_3}{\longrightarrow} 2Cu^{II}H + X^- \tag{21}$$

$$Cu^{II}H + Cu^{II} \overset{k_4}{\longrightarrow} 2Cu^I + H^+ \tag{22}$$

A rough correlation between the reaction rate and basicity of the anion X^- suggested that reaction 19 involved essentially heterolytic splitting of the hydrogen molecule [which had been meantime more fully discussed by Halpern (25, 26)], with the H^- more associated with the coordinated pyridine

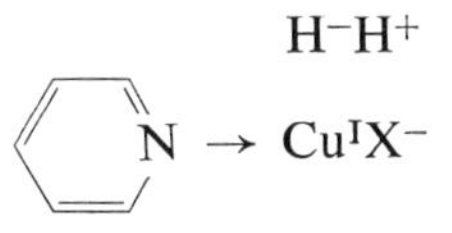

$$H^-H^+$$

$$N \rightarrow Cu^IX^-$$

and the H^+ with the anion X^-. The other intermediate hydride, $Cu^{II}H$, is that postulated previously but produced then by homolytic splitting according to eq. 14. Both hydrides react with the Cu^{II} substrate; assuming steady-state concentrations of the hydrides gave the rate law

$$-\frac{d[H_2]}{dt} = \frac{k_1[Cu^I][H_2]\{k_3[Cu^I] + k_2[Cu^{II}]\}}{k_{-1} + k_2[Cu^{II}] + k_3[Cu^I]} \tag{23}$$

The various rate expressions that had been reported were considered to be special limiting cases of expression 23, depending on the magnitude of k_{-1}, k_2, k_3, and the concentrations of Cu^I and Cu^{II} used. For example, in cyano-pyridines if $k_{-1} > k_3[Cu^I] > k_2[Cu^{II}]$, the observed second-order dependence in Cu^I is explained. Similarly in pyridine and quinoline, $k_3[Cu^I]$ must be $< k_2[Cu^{II}]$ to give rise to rate law 18.

Dunning and Potter (95), while investigating the hydrogen reduction of cupric sulfate in aqueous sulfuric acid solution ($\sim150°$, 1 to 4 atm H_2), observed an induction period preceding an accelerating autocatalytic range; addition of metallic copper removed this induction period because of the production of cuprous through the equilibrium

$$Cu^0 + Cu^{II} \rightleftharpoons 2Cu^I \tag{24}$$

Evidence for hydrogen activation by both Cu^{II} and Cu^I species was obtained and, depending on the temperature, the Cu^I species was up to 100 times more active than the Cu^{II} species. Hahn and Peters (96) have subsequently studied this system in more detail and obtained the rate law

$$\frac{-d[H_2]}{dt} = \frac{k_1[Cu^{II}]^2[H_2]}{\dfrac{k_{-1}}{k_2}[H^+] + [Cu^{II}]} + \frac{k_3[Cu^I][Cu^{II}]^2[H_2]}{\left(\dfrac{k_{-1}}{k_2}[H^+] + [Cu^{II}]\right)\left(\dfrac{k_{-3}}{k_4}[H^+] + [Cu^{II}]\right)} \tag{25}$$

The first term represents activation by Cu^{II} and the second term, activation by Cu^I. The rate law of Dunning and Potter (95) was similar, but the second term was linear, rather than quadratic, in both $[Cu^{II}]$ and $[H^+]$ in the denominator. Both groups of workers supported the mechanisms involving heterolytic splitting of the hydrogen by Cu^{II} and Cu^I as postulated by Chalk and Halpern for the corresponding heptanoic acid system (92), whose rate law (eq. 16) is a limiting form of eq. 25 at high $[Cu^{II}]$ and low $[H^+]$:

$$Cu^{II} + H_2 \underset{k_{-1}}{\overset{k_1}{\rightleftharpoons}} Cu^{II}H + H^+ \tag{26}$$

$$Cu^{II}H + Cu^{II} \overset{k_2}{\longrightarrow} 2Cu^I + H^+ \tag{27}$$

$$Cu^I + H_2 \underset{k_{-3}}{\overset{k_3}{\rightleftharpoons}} Cu^I H + H^+ \tag{28}$$

$$Cu^I H + Cu^{II} \overset{k_4}{\longrightarrow} Cu^{II}H + Cu^I \tag{29}$$

The inverse acid dependences observed gave good evidence for a heterolytic-type mechanism. A cuprous hydride, $Cu^I H$, has been prepared in the solid state (97), and thermochemical data of the gaseous species have been published (98, 99). Its formation in aqueous solution via reaction 28 is probable on energetic grounds (100).

A similar intermediate, Cu^IH, was originally thought to be active for an apparently homogeneous hydrogen reduction of unsaturated fatty acids to unsaturated fatty alcohols catalyzed by copper salts (101, 102). Temperatures around 250°, high pressures ($\sim$150 atm), and a cocatalyst, notably cadmium oleate, were required. The reaction was first-order with respect to copper, hydrogen, and cadmium, although the effect of the cocatalyst was not explained. Some later work (143), however, has concluded that this system is heterogeneous, the active catalyst being colloidal copper stabilized by the cadmium oleate; a prepared sample of cuprous hydride was inactive under the same conditions.

As discussed in the next section, the catalytic activity of Cu^{II} toward hydrogen has been investigated in considerable detail, particularly in perchlorate solutions (25, 26, 103–107). The data were interpreted in terms of activation by Cu^{II} only. Subsequent work (100, 108) has revealed that up to 20% of the reduction rates could be attributed to catalysis by Cu^I. The mechanism and rate laws mentioned above for the sulfate system (95, 96) are probably valid for the perchlorate system; however, perchlorate reduction by Cu^I is a complication and the matter has not been resolved (96).

Kinetic data for the activation of hydrogen by some copper(I) species is summarized in Table 4 and will be considered briefly from a general point of view in Section II-C.

Although catalyzed hydrogenation of olefins has not been reported for simple copper(I) [or copper(II)] complexes, the chemistry of copper alkyls, which would be likely intermediates, has been studied to some extent. Whitesides and coworkers (1373) have, for example, attributed the thermal dissociation of n-butyl(tri-n-butylphosphine) copper(I) into but-1-ene and butane to reactions 29a and 29b:

$$C_4H_7Cu(PBu_3) \longrightarrow CH_3CH_2CH=CH_2 + HCu(PBu_3) \qquad (29a)$$

$$C_4H_7Cu(PBu_3) + HCu(PBu_3) \longrightarrow C_4H_8 + 2Cu^0 + 2PBu_3 \qquad (29b)$$

Equation 29a involves β-elimination of a copper(I) hydride, a reaction well substantiated for other metal alkyls (see Sections I-A and XIII-C-3a and Ref. 41), and eq. 29b has frequently been postulated as the final step in catalyzed hydrogenation reactions (see Section X-B-2a and Chapter XV). The hydridophosphine complex, which is formed when aluminum hydrides are used, has not been isolated, but it has been studied in solution and it does react with copper(I) alkyls, vinyls, or aryls to give saturated hydrocarbons (1374). Dilts and Shriver (1375) have also studied copper(I) hydride-phosphine solutions.

B.   CUPRIC SPECIES

Ipatieff and Werchowsky (16) first reported that aqueous solutions of cupric acetate could be reduced by hydrogen to metallic copper or cuprous oxide, depending on the conditions. Halpern and Dakers (17, 18) later investigated the kinetics of this reaction by a sampling technique at about $100°$ and hydrogen pressures of 7 to 35 atm and then analyzing for unreacted Cu^{II} spectrophotometrically. The reaction (eq. 30) was shown to be homogeneous and followed the rate law of eq. 31, with k being independent of pH between pH 3.5 to 5.0.

$$2Cu(OAc)_2 + H_2 + H_2O \longrightarrow Cu_2O + 4HOAc \tag{30}$$

$$\frac{-d[H_2]}{dt} = \frac{-d[Cu(OAc)_2]}{2\,dt} = k[H_2][Cu(OAc)_2] \tag{31}$$

The catalytic species was thought to be undissociated cupric acetate and the mechanism was written as follows:

$$Cu(OAc)_2 + H_2 \xrightarrow[\text{slow}]{k} Cu(OAc)_2H_2 \tag{32}$$

$$Cu(OAc)_2H_2 + Cu(OAc)_2 + H_2O \xrightarrow[\text{fast}]{} Cu_2O + 4HOAc \tag{33}$$

Peters and Halpern (103) then extended these studies by using the more readily reducible substrate dichromate under the same reaction conditions. Solution samples were withdrawn periodically and analyzed for dichromate (spectrophotometrically) and copper (electrolytically). The same kinetics and rate data were obtained, showing that reaction 32 is again rate determining. Dichromate was then reduced in a fast step according to the stoichiometry:

$$3Cu(OAc)_2H_2 + Cr_2O_7^{2-} + 8H^+ \longrightarrow 2Cr^{3+} + 7H_2O + 3Cu(OAc)_2 \tag{34}$$

The concentration of Cu^{II} thus remains constant during this stage and since the hydrogen pressure is effectively constant, the dichromate is removed at a constant rate. The Cu^{II} is then finally reduced according to reactions 32 and 33.

In order to determine the activity of the uncomplexed Cu^{2+} ion, Peters and Halpern (104) then investigated the hydrogen reduction of several oxidants (dichromate, ceric, and iodate) catalyzed by cupric perchlorate in perchloric acid solutions. The various oxidants were again consumed at a constant rate independent of the oxidant used, with kinetics giving first-order in both Cu^{2+} and hydrogen. The same mechanism, involving a rate-determining hydrogen activation followed by a subsequent fast reaction of the intermediate hydride with substrate, was postulated. The activity of the Cu^{2+} ion was about 100

Table 1   Relative Catalytic Activities of Various Cupric Complexes in Aqueous Solution (25, 109)

Complex	"Mean" Formation Constant[a]	Catalytic Activity[b]
$Cu(OBu)_2$	—	150
$Cu(OPr)_2$	—	150
$Cu(OAc)_2$	30	120
$CuSO_4$	1×10^2	6.5
$CuCl_4^{2-}$	~ 1	2.5
Cu^{2+}	—	1
$Cu(glycinate)_2$	5×10^7	<0.5
$Cu(en)_2^{2+}$	1×10^{10}	0.1

[a] $\sqrt[n]{\beta_n}$.
[b] Relative to that of Cu^{2+}.

times less than that of the $Cu(OAc)_2$ complex under corresponding conditions. This was the first demonstration of homogeneous activation of hydrogen by an "uncomplexed" simple metal ion (i.e., hydrated). The same workers (109) further investigated the effects of complexing of the Cu^{2+} ion by using the dichromate reduction system in aqueous solution. The relative activity of various cupric complexes is summarized in Table 1. The determined second-order rate constants for the sulfate and chloride systems increased with increasing anion concentration, but they approached a limiting value

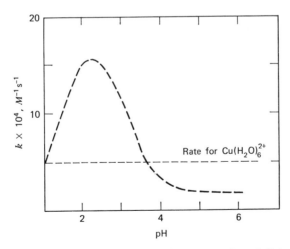

Fig. 1   Effect of pH on the copper(II)-catalyzed hydrogenation of dichromate; 0.05 M $Cu(ClO_4)_2$, 0.2 M glycine, 130°, 20 atm H_2 (109).

in each case; these values were attributed to the fully formed complexes $CuSO_4$ and $CuCl_4^{2-}$, respectively. In the ethylenediamine system the rate constant varied with pH, the rate falling with increase of pH in the region where complex formation started, and leveling off at a low constant value when the complex $Cu(en)_2^{2+}$ was fully formed. The variation of rate constant with pH for the glycine system is shown in Fig. 1. This was interpreted in terms of the following pH-dependent equilibria:

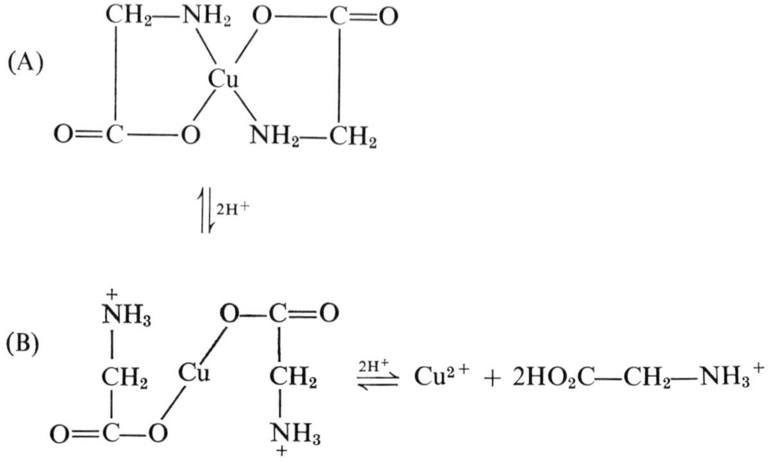

(A)

$$
\begin{array}{ccc}
CH_2\!-\!NH_2 & O\!\!-\!\!-\!\!-C\!=\!O \\
| & \diagdown\;\diagup & | \\
& Cu & \\
| & \diagup\;\diagdown & | \\
O\!=\!C\!-\!-\!-O & NH_2\!-\!CH_2
\end{array}
$$

$$\Updownarrow {\scriptstyle 2H^+}$$

(35)

(B)

$$
\begin{array}{ccc}
\overset{+}{N}H_3 & O\!-\!C\!=\!O \\
| & \diagup\quad | \\
CH_2 & Cu & CH_2 \\
| & \diagup\quad | \\
O\!=\!C\!-\!O & NH_3 \\
& \underset{+}{} \\
\end{array}
\quad \overset{2H^+}{\rightleftharpoons}\ Cu^{2+} + 2HO_2C\!-\!CH_2\!-\!NH_3{}^+
$$

In the intermediate pH range ($\sim$2) the activity occurs owing to the carboxylate complex(B), which is more reactive than either the chelate complex(A) or the uncomplexed ion, predominating at higher and lower pH, respectively. The behavior is reminiscent of that observed for many enzyme-catalyzed reactions whose rates pass through a maximum as the pH is varied (110; see also Section XVI-I). Beck (109a) considers that at pH $\sim$2 a monoglycinate complex exists and that maximum activity occurs because of this species.

The significance of these complexing effects was more fully discussed after Halpern and his group (105) investigated the Cu^{2+} catalyzed reduction of dichromate at higher perchloric acid concentrations. They observed an inverse acid dependence and accordingly presented the following rate law:

$$\frac{-d[H_2]}{dt} = \frac{k_1 k_2 [Cu^{2+}]^2 [H_2]}{k_{-1}[H^+] + k_2 [Cu^{2+}]} \tag{36}$$

This becomes equivalent to eq. 31 at low acid concentration; the dependence on Cu^{2+} changed from first- to second-order with increasing $[H^+]$ concentration. This rate law agrees with the mechanism

$$Cu^{2+} + H_2 \underset{k_{-1}}{\overset{k_1}{\rightleftharpoons}} CuH^+ + H^+ \tag{37}$$

$$CuH^+ + Cu^{2+} \overset{k_2}{\longrightarrow} 2Cu^+ + H^+ \tag{38}$$

$$2Cu^+ + substrate \overset{fast}{\longrightarrow} products + 2Cu^{2+} \tag{39}$$

The reversal of reaction 37 competes with the subsequent reaction of CuH^+, and at $110°$ the value of k_{-1}/k_2 was found to be 0.25. The zero-order dependence on dichromate concentration, even at high acidity, suggested that the hydride did not react directly with dichromate, and this led to the more detailed investigation of the cupric dependence. This paper is significant because it presented the first substantial evidence for the heterolytic splitting of the hydrogen molecule in the activation step.

The promoting influence of certain anions (109) was then explained in terms of their stabilization of the released proton (105):

$$CuX^+ + H_2 \longrightarrow CuH^+ + HX \tag{40}$$

A plausible representation of the transition state in such a reaction (26), for a complex $CuX_n{}^{z+}$, was

$$\begin{array}{ccc} X_{(n-1)} & Cu^{2+} & X \\ | & & | \\ H^- & \text{---} & H^+ \end{array}$$

The catalytic activity decreased with decreasing basicity of the anion (Table 1), that is, butyrate, propionate > acetate > sulfate > chloride > perchlorate. With increasing metal-ligand bond strength, removal of the ligand or its replacement by hydride becomes increasingly difficult, hence very stable complexes such as the glycinate and ethylenediamine complex exhibit low catalytic activity. These effects of complexing have been fully discussed by Halpern (25, 26).

The intermediate hydride is similar to that postulated (88) for the activation of hydrogen by cuprous salts in quinoline solution through homolytic splitting (eq. 14, Section II-A); in aqueous solution, at least, the hydride seemed the most plausible intermediate on energetic grounds (22, 24), considering the observed (104) activation energy of 26 kcal mole^{-1} for reaction 37. Further evidence for reaction 37 has been provided by later studies of the Cu^{II}-catalyzed isotopic-exchange reaction using D_2 and $HClO_4$ solutions (100); HD is produced from the reverse of reaction 37,

$$D_2 + H_2O \longrightarrow HD + DHO \tag{41}$$

Halpern and MacGregor (107) examined the hydrogen reduction of the Cu^{2+} ion in perchlorate solutions in the absence of added substrate. Copper was precipitated and the initial reduction kinetics followed the rate law given in eq. 36. A disproportionation process replaces reaction 39 in the overall mechanism

$$2Cu^+ \longrightarrow Cu^{2+} + Cu^0 \tag{42}$$

There was no significant activity attributed to cuprous species, although it was shown later (100, 108) that up to 20% of the reduction rates were due to this effect.

Of interest here is a related reaction that is catalyzed by Cu^{2+} ions, which involves the recombination of hydrogen and oxygen in nuclear reactors. McDuffie and coworkers (111) studied the system at 250° and found the reaction to be first-order in Cu^{2+} and hydrogen, and independent of acidity and oxygen pressure. They observed that this was inconsistent with the mechanism of eqs. 37 to 39 and the rate law of eq. 36, unless the value of k_{-1}/k_2 was much smaller than 0.25 as estimated by Halpern's group (105) from data at 110°, because the value of $[Cu^{2+}]$ in their own work was very small. This apparent anomaly was resolved by Hahn and Peters (106) who studied the Cu^{2+}-catalyzed hydrogen reduction of dichromate in perchlorate solution by using higher temperatures and high ratios of dichromate to Cu^{2+}. The dichromate was postulated to react directly with the CuH^+ intermediate as well as with the Cu^+ species. This incorporated a further step in the mechanism of eqs. 37 to 39, namely

$$CuH^+ + Cr^{VI} \xrightarrow{k_3} Cu^{2+} + products \tag{43}$$

This mechanism gave rate law 44 and is consistent with the observed data.

$$\frac{-d[H_2]}{dt} = \frac{k_1[Cu^{2+}][H_2]\{(k_2/k_{-1})[Cu^{2+}] + (k_3/k_{-1})[Cr^{VI}]\}}{[H^+] + (k_2/k_{-1})[Cu^{2+}] + (k_3/k_{-1})[Cr^{VI}]} \tag{44}$$

Such a scheme and rate law (with Cr^{VI} replaced by O_2) then become consistent with the data of McDuffie's group (111), if the third term of the denominator predominates, that is, the attack of CuH^+ by oxygen is dominant.

Chalk and Halpern (92, 93) studied the hydrogen reduction of cupric salts in heptanoic acid, diphenyl, and octadecane. The reduction process was autocatalytic owing to the production of the cuprous state, which was a much more effective hydrogenation catalyst than the cupric state in these nonaqueous media (see Section II-A). The initial rates, however, were attributed to activity of Cu^{II} and gave rise to the first term of rate law 16, a heterolytic splitting of the hydrogen molecule being postulated. Further evidence that Cu^{II} was activating hydrogen was provided by its ability to catalyze the hydrogenation of substrates such as benzoquinone or ferric at a constant rate approximating the initial rate in the absence of substrate (92). The activity of cupric heptanoate, $Cu(OHp)_2$, in heptanoic acid decreased on addition of heptanoate owing to the formation of higher complexes. The kinetic data from Halpern's group summarizing the effect of solvent and anions on the activation of hydrogen by both cupric and cuprous species are given in Table 2 (93). The data indicate that a marked decrease occurs in k_1 (per-

Table 2 Comparison of Reactivities of Cupric and Cuprous Salts in Various Solvents at $145°$ (93)

Anion	Solvent	pK_a [a]	$k_1 \times 10^3$ $M^{-1}s^{-1}$	$k_2 \times 10^2$ $M^{-1}s^{-1}$	k_2/k_1
Heptanoate	Heptanoic acid	4.9	2.3	9.0	40
Heptanoate	Octadecane	4.9	2.4	7.5	31
Heptanoate	Biphenyl	4.9	6.2	16.0	26
Propionate	Water	4.9	200	Not detected	$\ll 1$
(Perchlorate)	Water	—	2.0	Not detected	$\ll 1$
o-Toluate	Biphenyl	3.9	8.0	13.8	17
m-Chloro-benzoate	Biphenyl	3.8	10.0	8.5	8.5
Naphthalene-2-sulfonate	Heptanoic acid	~ 1	4.8	2.1	4.4

[a] pK_a of the corresponding acid of the anion at $25°$ in aqueous solution.

taining to Cu^{II}) and a correspondingly large increase in the ratio k_2/k_1 in going from water to less polar solvents, and that possibly k_1 decreases and k_2 increases with basicity of the anion in the nonaqueous media. In the case of k_1 this latter trend is the reverse of that observed in aqueous solution (see Table 1). As previously discussed for the heterolytic splitting process (eq. 40), an inverse dependence of the rate on the Cu—X bond strength and a direct dependence on the H—X bond strength has been postulated (25, 26, 105). Both of these increase with basicity of X^- and, depending on which predominates, the effect of the latter on the rate may be in either direction. The strength of the Cu—X bond and its variation with X was thought (93) to be greater for Cu^{II} than for Cu^I; it has also been proposed that the charge separation in the transition state would be greater for Cu^{II} than for Cu^I, suggesting that k_1 should decrease more than k_2 in going from a polar to nonpolar solvent.

The work of Dunning and Potter (95) and Hahn and Peters (96) on the hydrogen reduction of aqueous cupric sulfate solutions, when hydrogen activation by both Cu^{II} and Cu^I was observed, has already been considered (Section II-A).

Methanolic solutions of cupric acetylacetonate have been used by Frankel's group (144) for the catalytic hydrogenation of linseed methyl esters, which contain an undesirable flavor-instability ingredient as 9,12,15-octadeca-trienoates. Under the conditions used (150 to 180°, ~70 atm H_2), the catalyst decomposes to an inactive black precipitate. The catalyst showed little selectivity and was less active than most other first-row transition metal acetylacetonates for the hydrogenation of unsaturated fats. This work has recently been extended by Japanese workers (145) to other alcohol systems (see Section XIV-E).

In the presence of triethylaluminum in *n*-octane solution, cupric salts such as acetylacetonate give a soluble Ziegler-Natta-type hydrogenation catalyst (Chapter XV), and these systems have been used, for example, to reduce benzene to cyclohexane at 100 to 145° and 5 to 15 atm H_2 pressure (e.g., 147, 164, 175).

Tulupov and Gagarina (146) have reported that a saturated ethanolic solution of cupric stearate slowly catalyzes the hydrogenation of cyclohexene at 30° and 1 atm H_2. In a constant-pressure apparatus, a linear uptake plot was observed; the rate can become independent of both the catalyst and substrate concentration to give the rate law

$$\frac{-d[H_2]}{dt} = k[H_2] \tag{45}$$

No reduction of copper(II) was observed. The mechanism was written as follows:

$$H_2 + C \underset{k_{-1}}{\overset{k_1}{\rightleftharpoons}} H_2\text{*} \tag{46}$$

$$Cu^{II} + C_6H_{10} \overset{K}{\rightleftharpoons} [Cu^{II}{\cdot}C_6H_{10}] \tag{47}$$

$$[Cu^{II}{\cdot}C_6H_{10}] + H_2\text{*} \overset{k_2}{\longrightarrow} [Cu^{II}{\cdot}H_2\text{*}{\cdot}C_6H_{10}] \longrightarrow C_6H_{12} + Cu^{II} \tag{48}$$

$$(k_2 \gg k_{-1}, K \text{ large})$$

C may be the wall or any other colliding species and is in the rate-determining step which involves activation of the H_2 molecule (eq. 46); k is simply $k_1[C]$.

At higher pressures (~70 atm) in the absence of ethanol solvent, more complex behavior was observed due to the production of copper(I) and metal; this was considered to be due to reaction of H_2 with copper(II) (eq. 37) competing with reaction 47. A complex of composition $Cu(II){\cdot}16C_6H_{10}$ was isolated from experiments at the higher pressures.

This cupric system is one of a series studied by Tulupov and his group (57b), which used stearate complexes of the first-row transition metals; the same general mechanism of eqs. 46 to 48, which seems to be unique for hydrogen activation, has been postulated, and this work will be considered separately and in more detail in Section XVI-A.

C. GENERAL CONCLUSIONS ON ACTIVITY OF COPPER SPECIES

Tables 3 and 4 summarize some kinetic data that have been obtained for catalytic activation of hydrogen by copper(II) and copper(I) species, respectively. All the cupric systems listed undoubtedly involve heterolytic splitting of the hydrogen molecule according to a reaction such as 37, and the activation parameters reported refer to the rate constant (k_1) for the forward step. The entropy values are generally in the normal range for a simple bimolecular reaction in solution (112). The effects of complexing have been satisfactorily explained in terms of opposing effects of ligand basicity and complex stability (25, 26), and the decrease in reactivity in going from water to solvents of lower polarity has been attributed to the effect of charge separation in the transition state (25, 26, 93).

No such generalizations seem applicable to the copper(I) systems. In quinoline solution the rate law is second-order in cuprous (88), suggesting that the hydrogen is split homolytically to give $Cu^{II}H$ as an intermediate. Reported data in pyridine (91) indicate a first-order dependence and heterolytic splitting of hydrogen to give a $Cu^{I}H$ intermediate, but apparently a second-order dependence has also been observed (113). These anomalies could be explained by the complete rate law of eq. 23 (94), which can give rise to both dependences, subject to conditions, but is derived assuming that the rate-determining step basically involves heterolytic splitting. There is a general trend of increasing activity of the cuprous salt with increasing basicity of the anion in heptanoic acid (93), diphenyl (93), pyridine (94), quinoline (88), and possibly in aqueous solutions, because cuprous is active in sulfate solutions (95, 96) and relatively inactive in perchlorate solutions (96, 100, 108) under similar conditions, and sulfate is a stronger base than H_2O. The trend in quinoline solution has been interpreted (88) as a decrease in the energy required to promote electrons from the full 3d shell; this must be necessary in a homolytic splitting of hydrogen to form $Cu^{II}H$. Clearly, however, stabilization of the released proton by a basic anion in a heterolytic splitting mechanism would be an equally valid interpretation. However, the only report giving evidence against a homolytic splitting in quinoline solution is that of Parris and Williams (94), and their mechanism *does* involve the formation of $Cu^{II}H$, but it is produced according to reactions 20 and 21 (Section II-A) and not directly by homolytic splitting.

In the other solvent systems, heterolytic splitting of hydrogen by cuprous complexes seems well substantiated.

Table 3 Kinetic Data for Activation of Hydrogen by Copper(II) Species: $Cu^{2+} + H_2 \underset{}{\overset{k_1}{\rightleftharpoons}} CuH^+ + H^+$; $-d[H_2]/dt = k_1[H_2][Cu^{II}]$

Catalyst	Solvent	Reaction Studied	Temperature (°C)	$\Delta H_1^{\ddagger}$ (kcal mole⁻¹)	$\Delta S_1^{\ddagger}$ (eu)[a]	Reference
Cu(OAc)₂	Aqueous HOAc	$Cu^{II} \rightarrow Cu^{I}$	80–140	24.2	−6.5	17, 18
Cu(OAc)₂	Aqueous HOAc	$Cr_2O_7^{2-} \rightarrow Cr^{3+}$	80–140	24.6	−6.5	103
Cu²⁺	Aqueous HClO₄	$Cr_2O_7^{2-} \rightarrow Cr^{3+}$	80–140	26.6	−9.6	104
Cu²⁺	Aqueous HClO₄	$Cr_2O_7^{2-} \rightarrow Cr^{3+}$	160–200	25.8	(−10)	106
Cu(OPr)₂	Aqueous HOPr	$Cr_2O_7^{2-} \rightarrow Cr^{3+}$	80–120	24.8	−4.7	109
CuSO₄	Aqueous H₂SO₄	$2H_2 + O_2 \rightarrow 2H_2O$	190–295	24.1	(−12)	111
CuSO₄	Aqueous H₂SO₄	$Cu^{II} \rightarrow Cu^{I} \rightarrow Cu^{0}$	120–180	22.4	−21	96
CuSO₄	Aqueous H₂SO₄	$Cu^{II} \rightarrow Cu^{I} \rightarrow Cu^{0}$	126–205	23.5	−10.4	95
Cu(OHp)₂	Heptanoic acid	$Cu^{II} \rightarrow Cu^{I}$	125–155	29.4	−1	92

[a] $\Delta S^{\ddagger}$ values given in parentheses have been calculated from data given in the reference.

Table 4 Kinetic Data for Hydrogen Activation by Copper(I) Species

Catalyst	Solvent	Reaction Studied	Temperature (°C)	Rate Law	$\Delta H^{\ddagger}$ (kcal mole^{-1})	$\Delta S^{\ddagger}$ (eu)[a]	Reference
CuOAc	Quinoline	Quinone reduction	100–120	$k[H_2][Cu^I]^2$	13	—	15
CuOAc	Quinoline	Para-H_2 conversion	60–100	$k[H_2][Cu^I]^2$	16	(−5)	86
CuOAc	Quinoline	Cu^{II} reduction	25–100	$k[H_2][Cu^I]^2$	14	(−20)	89
CuOAc	Quinoline	Quinone and Cu^{II} reduction	~100	$k[H_2][Cu^I]^2$	~15	—	87
CuOAc	Pyridine	$Cu^{II} \rightarrow Cu^I$	100	$k[H_2][Cu^I]$	—	—	91
CuOHp	Heptanoic acid	$Cu^{II} \rightarrow Cu^I$	125–155	$k[H_2][CuOHp]$	20.2	−15	92
CuOAc	Quinoline, pyridine	$Cu^{II} \rightarrow Cu^I$	100	Equation 18	—	—	94
Cu^I	Aqueous H_2SO_4	$Cu^{II} \rightarrow Cu^I \rightarrow Cu^0$	126–205	$k[H_2][Cu^I]$	15.9	−35	95
Cu^I	Aqueous H_2SO_4	$Cu^{II} \rightarrow Cu^I \rightarrow Cu^0$	120–180	$k[H_2][Cu^I]$	15.3	−31	96
Cu^I, Cd^{II} oleates[b]	Decane	Fatty acids → alcohols	245–270	$k[H_2][Cu^I][Cd]$	26.7	—	101

[a] $\Delta S^{\ddagger}$ values given in parentheses have been calculated from data given in the reference.
[b] Shown recently to be heterogeneous (see Section II-A).

25

Table 5 Summary of Kinetic Data for Activation of Hydrogen by Some Silver(I) Species

Catalyst	Solvent	Reaction Studied	Temperature (°C)	Rate Law	$\Delta H^{\ddagger}$ (kcal mole^{-1})	$\Delta S^{\ddagger}$ (eu)[a]	Reference
AgOAc	Pyridine	$Ag^{I} \rightarrow Ag^{0}$	78–100	$k[Ag^{I}][H_2]$	16	(−25)	91
AgOAc	Pyridine	$Ag^{I} \rightarrow Ag^{0}$	25–65	$k[Ag^{I}][H_2]$	12.5	(−25)	118
Ag^+	Aqueous HClO$_4$	$Cr_2O_7^{2-} \rightarrow Cr^{3+}$	50–70	$k[Ag^+]^2[H_2]$	15.8	−22	119
Ag^+	Aqueous HClO$_4$	$Cr_2O_7^{2-} \rightarrow Cr^{3+}$	30–120	(1) $k[Ag^+]^2[H_2]$ (2) $k[Ag^+][H_2]$	14.7 24.0	−25 −6	120 120
AgOHp	Heptanoic acid	$Ag^{I} \rightarrow Ag^{0}$	86–108	$k[Ag^{I}][H_2]$	18.0	−10	123
$Ag^+ + MnO_4^-$	Aqueous HClO$_4$	$MnO_4^- \rightarrow MnO_2$	30–60	$k[Ag^+][MnO_4][H_2]$	9.3	−26	124

[a] $\Delta S^{\ddagger}$ values given in parentheses have been calculated from data given in the reference.

D. SILVER SPECIES

It was recorded over a century ago that metallic silver could be formed by hydrogen treatment of aqueous solutions of silver salts (114). Silver permanganate solutions (115, 116) and quinoline solutions of silver acetate (117) also absorbed hydrogen with production of metal. Wilmarth's observation (117) led to kinetic investigation by his own group (118) and by Wright and coworkers (91) in 1955 of hydrogen activation by silver complexes, following the detailed studies of the copper systems.

Both groups studied the reduction of silver acetate by hydrogen in pyridine solution to metallic silver:

$$2AgOAc + H_2 \longrightarrow 2Ag + 2HOAc \tag{48a}$$

Wright and coworkers (91) studied the reaction between 78 and 100° at hydrogen pressures up to 1 atm, and by following hydrogen absorption found a first-order dependence on both silver and hydrogen; the reaction products, silver and acetic acid, had little effect on the hydrogenation rate. Using dodecylamine as solvent gave similar rate data; neither glacial acetic acid, nor an aqueous solution of the acetate, nor a pyridine solution of silver chloride gave any measurable reaction at 100°. Wilmarth and Kapauan (118) studied reaction 48a at somewhat lower temperatures of 25 to 65° and obtained quite comparable rate data (Table 5), although in contrast to the other workers (91) they observed a marked inverse dependence on acetic acid and some enhancement of the rate by metallic silver; their data were thus obtained from initial rates. The reactivity varied with the nature of the silver salt (118): the fluoride reacted about five times as rapidly as the sulfate and acetate, while solutions of trifluoroacetate, chloride, perchlorate, and nitrate were inactive. These variations were ascribed to differences in basicity of the anions and provided some support for a postulated mechanism involving heterolytic splitting of hydrogen:

$$AgOAc + H_2 \xrightarrow{k} AgH + HOAc \tag{48b}$$

$$AgH + AgOAc \xrightarrow[\text{fast}]{} 2Ag + HOAc \tag{48c}$$

The formation of AgH as an intermediate seemed plausible and energetically consistent (118) with the observed activation energy of about 14 kcal mole^{-1}. For the silver nitrate in pyridine system the inactivity extended to para-hydrogen conversion, indicating that the hydrogen-activation step is prevented (118). Using deuterium gas, no exchange with hydrogen donors in solution was observed either during or after reduction of silver acetate in pyridine; this is consistent with the above mechanism if metallic silver does not activate the hydrogen (117). The discrepancy on this last point has not yet been resolved; nevertheless, the reported kinetic data undoubtedly refer to the homogeneous reaction. Surprisingly, no studies of the presence of reducible substrates (such as benzoquinone) in pyridine solutions have been reported, even though this could well eliminate the early production of the metal.

Aqueous solutions of silver acetate had been reported to be relatively unreactive toward 1 atm hydrogen at 100° (117, 118). Halpern and Webster (119), however, reported that aqueous silver perchlorate solutions were reduced to the metal quite readily by hydrogen (50°, 1 atm) and also were found to be effective catalysts for the hydrogen reduction of other substrates such as dichromate and ceric. The kinetics of these substrate reductions were followed spectrophotometrically by the rate of loss of the substrate; the reactions were first-order in hydrogen, second-order in silver, and independent of substrate concentration and perchloric acid concentration from 0.1 to 1.0 M. The Ag^+ concentration remained constant and no silver metal was produced until after complete reduction of the substrate, so possible complications from heterogeneous reactions were eliminated. The rate-determining termolecular step, coupled with energetic considerations (22, 24, 25), suggested the following mechanism involving homolytic splitting of the hydrogen:

$$2Ag^+ + H_2 \xrightarrow{\ k\ } 2AgH^+ \tag{48d}$$

$$2AgH^+ + \text{substrate} \xrightarrow[\text{fast}]{} \text{products} + 2Ag^+ \tag{49}$$

Silver nitrate solutions gave similar results with rates about 10% lower than those for comparable perchlorate solutions. A slight deviation from the second-order dependence on silver concentration, which was reflected in the fact that the linear plots of rate against $[Ag^+]^2$ did not quite extrapolate to the origin, was significant; this deviation suggested that an alternative catalytic path involving only one Ag^+ ion might be possible, particularly at lower silver concentrations where the termolecular path would be less favorable. Use of higher hydrogen pressures (a few atm) and higher temperatures (up to 120°) enabled rates to be measured at these lower catalyst concentrations (120). The rate law determined is shown in eq. 50:

$$\frac{-d[H_2]}{dt} = k[H_2][Ag^+]^2 + \frac{k_1 k_2 [H_2][Ag^+]^2}{k_{-1}[H^+] + k_2[Ag^+]} \tag{50}$$

The first term, which is pH independent, refers to the mechanism, outlined in eqs. 48d and 49, which predominates at the lower temperatures and higher silver concentrations used in the earlier study (119). The second term was identical to that obtained for the Cu^{2+} catalyzed reduction of dichromate by hydrogen in perchloric acid solution (eq. 36) and similarly refers to a mechanism involving heterolytic splitting of hydrogen

$$Ag^+ + H_2 \underset{k_{-1}}{\overset{k_1}{\rightleftharpoons}} AgH + H^+ \tag{51}$$

$$AgH + Ag^+ \overset{k_2}{\longrightarrow} \text{intermediates} \underset{Cr^{VI}}{\overset{\text{fast}}{\longrightarrow}} \text{products} \tag{52}$$

Measurements using D_2O-enriched water confirmed the formation of HD which the mechanism predicts by means of the reverse of reaction 51. The activation energy of 14.7 kcal mole^{-1} for the termolecular path compared with 24 kcal mole^{-1} for the bimolecular path (Table 5) shows that the former may predominate at lower temperatures. Schematic potential energy diagrams depicting the two paths of activation of hydrogen by Ag^+ have been presented (120, 25). The homolytic splitting mechanism can also compete with the heterolytic one at high acidities, when the competing back-reaction of eq. 51 suppresses the bimolecular path. The more basic solvent pyridine would be expected to favor the bimolecular path, an observation consistent with the reported kinetic data for the silver acetate system in this solvent (118, 119).

The reaction rates at higher temperatures did increase with dichromate concentration but approached a limiting value, and rate law 50 was obtained from these limiting rates (120). The later work of Hahn and Peters (106) on the Cu^{2+}-catalyzed dichromate reduction (eqs. 43, 44) suggests that a direct reaction of dichromate with the AgH intermediate might also be involved. Webster and Halpern (120) obtained some complex kinetic data using a very low dichromate concentration that showed an order between 3 and 4 for the Ag^+ ion (depending on the acidity), although the reaction was still first-order in hydrogen and inversely dependent on acid. An aggregate of a number of silver ions into an intermediate was suggested:

$$nAg^+ + H_2 \rightleftharpoons Ag_nH^{(n-1)+} + H^+ \tag{53}$$

$$Ag_nH^{(n-1)+} + Ag^+ \longrightarrow Ag_{n+1}H^{n+} \overset{\text{fast}}{\underset{Cr^{VI}}{\longrightarrow}} \text{products} \tag{54}$$

It is of interest that Blues and Bryce-Smith (121) have recently isolated the first silver cluster complexes that are also stable in solution.

Webster and Halpern (122) extended their studies by investigating the effect of complexing on the silver ion in aqueous solutions. Dichromate could not be used as substrate because of the insolubility of silver chromate at pH's where the complexes (e.g., the acetate) are stable, and hence metal precipitation was followed. Temperatures around 100° and 4 atm hydrogen pressure were used. The rate data indicated that in perchlorate solution both the precipitation reaction and dichromate reduction proceeded by the same mechanism that involved heterolytic splitting of hydrogen under these conditions (eqs. 51, 52). In acetate and ethylenediamine solutions, the determined rate law was

$$\frac{-d[H_2]}{dt} = k_1[Ag^I][H_2] \tag{55}$$

The basic ligands enhance the heterolytic mechanism by stabilization of the proton and inhibit the reverse of eq. 51. The $Ag(CN)_2^-$ complex was not reduced under these conditions. Table 6 shows the relative activity of these

Table 6   Relative Catalytic Activities of Various Silver(I) Complexes in Aqueous Solution (122)

Complex	"Mean" Formation Constant[a]	Relative Activity
AgOAc	~3	80
$Ag(en)_2^+$	7×10^3	25
Ag^+	—	1
$Ag(CN)_2^-$	2.4×10^9	Inactive

$a \;\; \sqrt[n]{\beta_n}.$

silver species. As for the cupric complexes (Table 1), the very stable complexes show inactivity because of the strong metal-ligand bond. The $Ag(en)_2^+$ system showed a marked increase in rate with increasing hydroxide concentration and this was attributed to such equilibria as

$$Ag(en_2^+) + OH^- \rightleftharpoons (en)Ag(NHCH_2CH_2NH_2) + H_2O \tag{56}$$

or

$$Ag(en)_2^+ + OH^- \longrightarrow Ag(en)_2^+OH^- \tag{57}$$

The complexes formed have a more basic group than (en) or H_2O (i.e., $NHCH_2CH_2NH_2^-$ or OH^-), which can serve as an acceptor for the proton released in the reaction with H_2.

Halpern and his group (123) later studied the hydrogen reduction (~100°, 1 atm) of silver heptanoate to the metal in heptanoic acid solution. Rate law 55 was observed with no autocatalysis from the metal production. The value of the rate constant at 70° was very similar to those observed for the reduction of the acetate in aqueous solution (122) and the acetate in pyridine solution (91, 118), suggesting that solvent polarity and hence charge separation in the transition state are much less important for the lower-charge silver(I) complexes than, for example, for the cupric complexes (123).

Silver salts had been known for some time to increase greatly the rate of hydrogen absorption by neutral and alkaline permanganate solutions

(115, 116). Webster and Halpern (124) studied both the uncatalyzed reaction (see Chapter VIII) and the silver perchlorate catalyzed reaction. The latter was studied in perchloric acid solution to avoid complications from the reported heterogeneous catalysis by an $Ag_2Mn_2O_5$ product formed at higher pH's (116). The overall stoichiometry showed

$$2MnO_4^- + 3H_2 + 2H^+ \longrightarrow 2MnO_2 + 4H_2O \tag{58}$$

No silver was detected in the MnO_2 product, and the kinetics of the catalyzed reaction (followed by loss of permanganate), which was homogeneous during the initial stages, followed the rate law

$$\frac{-d[H_2]}{dt} = k[H_2][MnO_4^-][Ag^+] \tag{59}$$

The rate-determining step was thought to be

$$AgMnO_4 \text{ (or } Ag^+ + MnO_4^-) + H_2 \longrightarrow AgH^+ + MnO_4^{2-} + H^+ \tag{60}$$

followed by fast reactions of AgH^+ and MnO_4^{2-} to yield MnO_2 and regenerate the Ag^+ catalyst. The role of MnO_4^- in reaction 60 corresponds to that of one of the Ag^+ ions in the homolytic splitting mechanism of eq. 48d, and its effectiveness is thought to be a result of its high electron affinity. The activation energy of about 9 kcal (Table 5) is one of the lowest yet observed for a homogeneous hydrogenation.

Halpern and Milne (125), in an effort to study the effect of a more systematic variation of ligand properties, investigated the hydrogen reduction of a range of silver amine complexes to the metal in aqueous solution:

$$Ag(NR_3)_2^+ + H_2 \longrightarrow Ag^0 + 2R_3NH^+ \tag{61}$$

Depending on the amine used, the conditions for a measurable rate varied from 1 atm H_2 at 25° to 25 atm H_2 at 100°. By following the loss of Ag^I in solution, the kinetics followed eq. 55, where under the conditions used Ag^I was identified with the bis complex, for example, $Ag(NR_3)_2^+$. The heterolytic mechanism (eq. 51) is favored because of the basic ligands and the high pH of the solutions. Table 7 summarizes the data for the various amines. The order of reactivity without exception is tertiary amines > secondary amines > primary amines > ammonia, and, for the alkylamines, this order corresponds to an inverse dependence of the reactivity on the stability of the complex measured either as K_1 or β_2. The customary heterolytic-type mech-

$$Ag(NR_3)_2^+ + H_2 \longrightarrow \begin{bmatrix} R_3N-Ag^+\cdots NR_3 \\ \vdots \quad\quad \vdots \\ H^-\cdots H^+ \end{bmatrix} \longrightarrow R_3N + AgH + R_3NH^+ \tag{62}$$

anism, shown in eq. 62, was thought to be somewhat less likely, because the configuration allowing simultaneous dissociation and formation of bonds

Table 7 Kinetic Data for Hydrogen Activation by Silver(I) Species in Aqueous Solution (125): $Ag^I + H_2 \underset{k_{-1}}{\overset{k_1}{\rightleftharpoons}} Ag^IH + H^+$; $d[H_2]/dt = k_1[Ag^I][H_2]$

Ligand	Log k_1 (70°) (M^{-1} s^{-1})	$\Delta H^{\ddagger}$ (kcal mole^{-1})	$\Delta S^{\ddagger}$ (eu)	pK_a†	Stability Constants		
					Log K_1	Log K_2	Log β_2
H_2O	−3.20	23.3	−6	—	—	—	—
F^- (dark)^a	0.79	18.3	−3	3.4	−0.2		
F^- (diffuse daylight)^a	0.88	20.3	+4	3.4	−0.2		
Monoamines							
NH_3	−3.48	28.3	+8	9.25	3.31	3.92	7.23
CH_3NH_2	−1.91	22.3	−3	10.72	3.15	3.53	6.68
$C_2H_5NH_2$	−2.42	21.6	−7	10.81	3.37	3.93	7.30
$n\text{-}C_3H_7NH_2$	−2.26	23.3	−1	10.58			7.39
$(C_2H_5)_2NH$	−1.33	20.5	−5	10.96	3.06	3.30	6.36
$(C_2H_5)_3N$	−0.13	13.8	−5	10.77	2.6	2.1	4.76
Diamines							
$NH_2CH_2CH_2NH_2$	−1.80	20.8	−7	10.18	4.62	2.92	7.54
$NH_2CH_2CH_2CH_2NH_2$	−2.00	18.9	−13	10.64	5.77		
$NH(CH_2CH_2)_2NH$	−1.32	16.8	−16	9.81	3.32		
$CH_3N(CH_2CH_2)_2NCH_3$	−0.07	17.9	−7	10.29			
$(CH_3)_2NCH_2CH_2N(CH_3)_2$	0.17	15.1	−14	9.30			
$N(CH_2CH_2)_3N$	0.26	17.1	−8	8.19	1.65		
$NH_2CH_2CH_2N(CH_3)_2$	−0.49	17.1	−11	9.53			
Heterocyclics							
Pyridine	−2.09	19.3	−12	5.45	2.04	2.18	4.22
Amine Derivatives							
$N(CH_2CH_2OH)_3$	−0.36	17.5	−10	7.90	2.30	1.34	3.64
$NH_2CH_2COO^-$	−1.33	21.5	−3	9.78	3.51	3.38	6.89
$NH_2CH_2CH_2COO^-$	−1.63	18.3	−13	10.19			
$NH_2CH(CH_3)COO^-$	−1.61	16.2	−19	9.87	3.64	3.54	7.18

^{a} Reference 132.

† Acid dissociation constants at 25°.

would be difficult to achieve for amine ligands having only a single co-ordinative position. A modified mechanism was postulated in which proton transfer from H_2 to the amine occurred through an intervening water molecule

$$Ag(NR_3)_2^+ + H_2 + H_2O \longrightarrow \left[\begin{array}{c} H \\ R_3N{-}Ag^+{\cdots}O{\cdots}H \longrightarrow NR_3 \\ H^-{\cdots}H^+ \end{array} \right] \longrightarrow$$

$$R_3N + AgH + HNR_3 + H_2O \quad (63)$$

The difficulty of replacing an amine ligand by a water molecule gave rise to the inverse correlation between reactivity and stability. Some support for such a mechanism was provided by n.m.r. studies on the kinetics of proton exchange between R_3NH^+ and R_3N in aqueous solution by Meiboom and his coworkers (126, 127), where both direct proton transfer and transfer through an intervening water molecule were distinguished. For methylamines this latter path was favored, but for ammonia itself the direct path was operative. Halpern and Milne (125) suggested that the very low rate observed for $Ag(NH_3)_2^+$ (Table 7) was consistent with the direct mechanism (eq. 62) for this system, and the abnormally high $\Delta H^\ddagger$ and $\Delta S^\ddagger$ resulted from the non-binding of a water molecule.

Nagy and Simandi (128) later studied the hydrogen reduction of the silver ion in perchloric acid solutions, and they proposed a mechanism based on an analogy with heterogeneous hydrogenation systems, particularly on the electrolytic oxidation of hydrogen on transition-metal surfaces (Pt, Pd). They pointed out that all the observed active catalytic ions for hydrogenation to that date (1963) had positive standard reduction potentials relative to the zero standard hydrogen potential (i.e., capable of oxidizing H_2 to H^+) and that the catalytic activity increased with more positive reduction potentials. In agreement with Webster and Halpern (120), the existence of a homolytic and a heterolytic path for hydrogen fission was confirmed. However, Nagy and Simandi proposed that these two paths arose from a common step that involved the formation of an AgH_2^+ intermediate

$$H_2 + Ag^+ \underset{}{\overset{K}{\rightleftharpoons}} AgH_2^+ \begin{array}{c} \overset{Ag^+}{\nearrow} 2AgH^+ \\ \underset{H_2O}{\searrow} AgH + H_3O^+ \end{array} \quad (64)$$

If the first step is a rapid equilibrium (K), then the rate laws derived become equivalent in form to those of Halpern and Webster (120), shown in eq. 50, but incorporating the equilibrium constant K; such a mechanism requires no termolecular step involving two positively charged species for the production of AgH^+.

Although not pointed out by Nagy and Simandi (128), such a scheme corresponds closely to that of Parris and Williams (94) for the cuprous catalyzed system in quinoline and pyridine, where the production of $Cu^{II}H$ was thought to arise from reaction of $Cu^{I}H_2$ with Cu^{II} (eq. 20), rather than by direct homolytic splitting of hydrogen. Nagy and Simandi (128) also suggested that those species finally reacting with a substrate such as dichromate in reactions such as 49 or 52 are likely to be silver atoms (produced by decomposition of the hydrides by water), since it was observed that freshly precipitated silver quickly reduces dichromate solution.

Nagy and Simandi concluded that for a mechanism such as eq. 62 or 63 some correlation between rate constants and complex stability or ligand basicity should exist for a series of related reactions, and they proposed the linear free-energy relationship (129–131)

$$\Delta G^{\ddagger} = A + B\,\Delta G^{\circ}_{rds} \tag{65}$$

where $\Delta G^{\ddagger}$ is the free energy of activation, ΔG°_{rds} the free energy change for rate-determining step 62 or 63, and A and B are constants. They applied this relationship to the data of Halpern and Milne (125) (Table 7) for the series of aliphatic monoamines. ΔG°_{rds} was determined from a linear combination of known ΔG° values; the ΔG° for the formation of AgH appears for each system and may be incorporated into the constant A. This yielded the relationship

$$\log k_1 = a + b(2\log K_a - \log \beta_2) \tag{66}$$

where a and b are constants. Figure 2 shows this relationship and indicates good linearity over an interval of four orders of magnitude in k_1. These data show nicely that reactivity does increase with increasing ligand basicity,

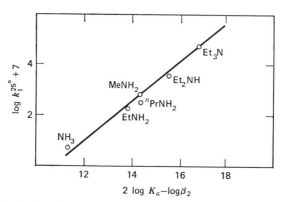

Fig. 2 Correlation between rates and complex stability and ligand basicity for hydrogen reduction of silver amine complexes (eqs. 55 and 66; Ref. 130).

indicating the importance of the ligand as the proton acceptor, and decreases with increasing complex stability as anticipated if the amine and hydride ions are competing nucleophiles.

The very large enhancement by fluoride ion (Table 7) of the reactivity of the silver ion toward hydrogen, observed by Beck and coworkers (132, 133), is an approach to optimum conditions, because although hydrogen fluoride is a rather weak acid, the fluoride ion forms only a very weak monofluorosilver(I) complex. The linear structure of silver(I) complexes made it necessary to again postulate proton transfer through a water molecule:

$$F^- \!-\! Ag \text{----} H^-$$

$$
\begin{array}{ccc}
H & & H^+ \\
& O & \\
& | & \\
& H &
\end{array}
$$

Light had a small but definite enhancing effect on this fluoride system.

The data of Table 7 also showed somewhat higher activities for complexes of diamines relative to those of corresponding monoamines, for example, $NH_2CH_2CH_2NH_2 > C_2H_5NH_2$. Although the predominant species in solution were thought to contain monocoordinated ligands, the higher activity could result from some tendency toward chelate formation (which would assist removal of the second amine ligand in a path such as 62 or 63) or, alternatively, a free amino group of a monocoordinated diamine could act as the proton acceptor. The lower $\Delta H^\ddagger$ and more negative $\Delta S^\ddagger$ values found for the diamines were consistent with binding of an additional amino group in the activated complex. Similar possibilities involving concerted participation of the amino and carboxylate were thought to account for the higher activities of the amino-acid complexes relative to those of other primary amines.

In conclusion, some quite detailed kinetic studies and data have been provided by Halpern's group for activation of hydrogen by silver(I) complexes, particularly in aqueous solution. Both homolytic and heterolytic fission of molecular hydrogen can arise; if the former involves a direct termolecular rate-determining step, it would appear to become significant at lower temperatures because of a lower activation energy than that for the heterolytic process. The systematic data and linear free-energy relationship established for the silver amine systems are of considerable value in understanding the role of the central ion and of the ligand in the heterolytic fission of hydrogen, and they offer substantial evidence for its existence.

The existence of the AgH^+ species at low temperatures has been postulated by Eachus and Symons (131a, 1391) from e.s.r. studies on irradiated solutions of silver sulfate in sulfuric acid; the species could be formed from reaction of the silver(I) ion with radiation produced hydrogen atoms (see also Section XVI-C). The species Ag_2H^{2+} with a probable $(AgHAg)^{2+}$ structure was also detected.

E.   GOLD SPECIES

Some thoughts on the relation of catalytic activity to the configuration of ions, especially electron-acceptor properties (25, 26, 128) (see Chapter XVII), have predicted a high activity for the hydrated Au^+ ion which, however, is unstable and disproportionates to the auric state and metal. Complexes of the type $Au(NH_3)_2^+$ and $[Au^IL_2]^{n-}$ (where $L = S_2O_3^{2-}$, Cl^-, Br^-) are known (134), but, using arguments presented from data for complexes of Cu^I and Ag^I, the gold complexes are likely to be thermodynamically too stable for reactivity.

Halpern and Harrod (135) have reported that chloro-complexes of gold(III) are inactive, although the reaction conditions were not given. Fasman and Ikhsanov (135a), however, have noted that perchloric acid solutions of such complexes can be hydrogenated to give metal (Section XVI-B).

III

Group II Metal Ions and
Complexes

A. ZINC AND CADMIUM SPECIES (AND MAGNESIUM)

Rittenberg (136) has reported that the acetates of magnesium, zinc, and cadmium activate hydrogen catalytically in aqueous solution, but no details were given. Cadmium heptanoate showed no reaction with hydrogen in heptanoic acid solutions as high as 150° (92); magnesium, zinc, and cadmium perchlorates were similarly unreactive in aqueous perchlorate solutions (25, 137). Ethanolic solutions of zinc stearate catalyze the hydrogen reduction of cyclohexene at ∼30° and 1 atm pressure (148) (see Section XVI-A). Zinc acetylacetonate-alcoxyalanate mixtures have been used for hydrogenation (1187; Chapter XV).

B. MERCURY SPECIES

Halpern and coworkers (137) found that mercuric acetate was reduced to the mercurous complex by molecular hydrogen in aqueous solution. The reaction was first-order in both reactants, suggesting that the hydrogen was being activated; however, interpretation of the kinetics was difficult because of uncertainties in the extent of complexing and in the possible influence of the mercurous acetate product. The studies were thus extended to aqueous perchlorate solutions where homogeneous hydrogen reduction was demonstrated at 60 to 100° and hydrogen pressures of 1 to 4 atm (138). The reaction proceeded in two stages, the first corresponding to the reduction of mercuric to mercurous,

$$2Hg^{2+} + H_2 \longrightarrow Hg_2^{2+} + 2H^+ \tag{67}$$

and the second to reduction of mercurous to the metal,

$$Hg_2^{2+} + H_2 \longrightarrow 2Hg + 2H^+ \tag{68}$$

The kinetics were studied by analyzing for Hg_2^{2+}, Hg^{2+}, and total mercury by standard methods. The transition between the two stages corresponded to the first appearance of metallic mercury, and this is governed by the readily reversible equilibrium 69,

$$Hg + Hg_2^+ \rightleftharpoons Hg_2^{2+} \tag{69}$$

whose equilibrium constant, K, is about 70 M^{-1} at these temperatures. Both Hg^{2+} and Hg_2^{2+} contribute to the activation of hydrogen; the kinetics at any stage were given by the following equation:

$$\frac{-d[H_2]}{dt} = k_1[Hg^{2+}][H_2] + k_2[Hg_2^{2+}][H_2] \tag{70}$$

No pH dependence was observed in the range 0.025 to 0.1 M [H^+].

Some doubt remains about the mechanism of hydrogen activation by these ions. The participation of hydrido-metal intermediates similar to CuH^+, CuH, and AgH receives no support on either kinetic or thermodynamic grounds. The observation of no increase in the solubility of mercury in water with increasing acid concentration (139) has been interpreted (59) by the suggestion that HgH^+ is unstable with respect to Hg and H^+. However, HgH^+ could still be the initial product through heterolytic splitting of hydrogen by Hg^{2+} or homolytic splitting by Hg_2^{2+}. On energetic grounds (24) the most likely intermediates are mercury atoms, and these could be formed directly in a rate-determining step involving a two-electron reduction of the metal ion coupled with proton transfer to the solvent (25)

$$Hg^{2+} + H_2 \longrightarrow Hg + 2H^+ \tag{71}$$

$$Hg_2^{2+} + H_2 \longrightarrow 2Hg \text{ (or } Hg_2) + 2H^+ \tag{72}$$

The reduction of Hg^{2+} and Hg_2^{2+} by formate in solution was known to involve a rate law corresponding to 70, and it was concluded that the rate-determining process involved electron transfer from formate to Hg^{2+} or Hg_2^{2+} (140). Halpern and coworkers (138, 141) had initially proposed a similar mechanism to explain the activation of hydrogen by the Cu^{2+} and Ag^+ ions. But later, on kinetic evidence (105, 120) and energetic grounds (24)—the latter ruling out reactions such as 71 and 72 in the cases of Cu^{2+},

Cu^+, and Ag^+—they concluded that hydride intermediates were involved in these systems.

Mercuric perchlorate was also found to catalyze the hydrogen reduction of permanganate in acid solutions (124). The mechanism suggested was the rate-determining reaction 71 followed by the faster steps shown in eq. 73 (see Chapter VIII also):

$$Hg + MnO_4^- \longrightarrow Hg^{2+} + MnO_4^{3-} \qquad (73)$$

$$\downarrow H^+$$

$$MnO_2$$

Korinek and Halpern (142) investigated the effects of complexing on the hydrogen reduction of mercuric salts. The nitrate and sulfate systems behaved similarly to the perchlorate system in their kinetics (eq. 70) and stoichiometry, because of further hydrogen activation by a soluble mercurous salt. The acetate, propionate, chloride, and bromide systems produced inactive (carboxylate) or insoluble (halide) mercurous complexes, and the kinetics corresponded to activation by the mercuric salts only (the k_1 term of eq. 70). The effect of ethylenediamine was examined in basic solution where the mercuric *bis* complex is stable; the mercurous salt disproportionates to metallic mercury and again only the k_1 term is obtained in the rate law. The effect of complexing on the activity of various mercury species is shown in Table 8.

The effects of complexing have been interpreted (25, 26), as for cupric complexes (Table 1; Section II-B) and silver complexes (Tables 6 and 7;

Table 8   Complexing Effects on Activity of Mercury Species
(25, 138, 141)

Complex	Mean Formation Constant[a]	Relative Activity[b]
$HgSO_4$	22	1.8
Hg^{2+}	—	1
Hg_2^{2+}	2×10^8	1.1×10^{-1}
$Hg(OAc)_2$	1.6×10^4	4×10^{-2}
$Hg(OPr)_2$	—	4×10^{-2}
$HgCl_4^{2-}$	6×10^3	3.2×10^{-3}
$HgCl_2$	4×10^6	2.5×10^{-3}
$HgBr_2$	3.5×10^7	1.7×10^{-3}
$Hg(en)_2^{2+}$	5.1×10^{11}	1×10^{-3}

[a] $\sqrt[n]{\beta_n}$.

[b] The values are approximate, since they are based on measurements at 90 to 123°.

Table 9 Kinetic Data for Hydrogen Activation by Mercury Species

Catalyst	Solvent	Reaction Studied	Temperature (°C)	k^a (M^{-1} s^{-1})	$\Delta H^{\ddagger}$ (kcal mole^{-1})	$\Delta S^{\ddagger}$ (eu)	References
Hg_2^{2+}	Aqueous $HClO_4$	$Hg_2^{2+} \rightarrow Hg^0$	65–100	5×10^{-1}	20	−10	138
Hg^{2+}	Aqueous $HClO_4$	$Hg^{2+} \rightarrow Hg_2^{2+}$	65–100	3.3	18	−12	138
$Hg(OAc)_2$	Water	$Hg^{II} \rightarrow Hg^{I}$	69–115	1.4×10^{-1}	19	−15	142
$Hg(OHp)_2$	Heptanoic acid	$Hg^{II} \rightarrow Hg^0$	120	3.0×10^{-3}	—	—	123
$Hg(OHp)_2$	Biphenyl	$Hg^{II} \rightarrow Hg^0$	160–190	1.9×10^{-4}	21	−23	123

^{a} k refers to the bimolecular rate constants for reaction of H_2 with the mercury species (eq. 70); data at 120°.

Section II-D), in terms of proton affinity and complex-forming tendency of the ligands, although the order of activity of the various complexes differs for the three metals because of the relative stabilities of the complexes. (The lower activity of Hg_2^{2+} compared to Hg^{2+} was interpreted by considering Hg_2^{2+} as a highly stable complex of Hg^{2+} containing a covalently bonded Hg atom as a ligand.) This general interpretation follows for a mechanism involving formation of intermediate hydrides. A similar rationale was given (142) for the mechanism that involved direct electron transfer from the hydrogen to the mercuric species: if the unfilled low-lying orbitals of Hg^{2+} were used in forming covalent bonds with ligands and increasingly so with increasing stability, then the electron affinity of the mercuric species would decrease and give reduced reactivity toward hydrogen.

The addition of basic anions such as acetate, carbonate, and hydroxide increased, in the order listed, the rate at which the fully formed $Hg(en)^{2+}$ complex reacted with hydrogen and this development was not because of replacement of the ethylenediamine by the anions (142). The interpretation suggested (142) was that a basic anion stabilized the protons released in the rate-determining step (heterolytic splitting of hydrogen or electron transfer) or, alternatively, that the anion formed a polarizable bridge between the $Hg(en)^{2+}$ and the H_2 molecule and facilitated electron transfer between them. Halpern also suggested later (25, 26) that the enhanced reactivity could be due to formation of an ion-pair complex, $Hg(en)_2^{2+} \cdot X^-$, in which the anion replaces an outer-sphere water molecule. A possible explanation of the effect of hydroxide is an acid-base equilibrium

$$(-Hg-NH_2R)^{2+} + OH^- \rightleftharpoons (-Hg-NHR)^+ + H_2O \qquad (74)$$

The enhanced reactivity is attributed (25, 26) to replacement of ethylenediamine (NH_2R) by a more basic ligand ($-NHR$). A similar effect was observed (122) for the ethylenediamine complex of silver (eq. 56) but not of copper(II) (109).

Halpern and coworkers (123) later studied the hydrogen reduction (1 atm H_2, 120 to 190°) of mercuric heptanoate to metal in both heptanoic acid and biphenyl solution

$$Hg(OHp)_2 + H_2 \longrightarrow Hg + 2HOHp \qquad (75)$$

Kinetic measurements were confined to the initial 5% of the reaction, caused by the complicating autoreduction of the mercuric salt by heptanoate ion or the solvent itself; the rate law gave first-order in both hydrogen and mercuric. The kinetic data obtained are summarized in Table 9 together with that for some aqueous systems.

The reactivity of the mercuric carboxylates falls off markedly with solvent polarity (water > heptanoic acid > biphenyl). Differences in both $\Delta H^{\ddagger}$ and

$\Delta S^{\ddagger}$ contribute to this trend, which resembles that for cupric complexes (Table 2) but differs from that for silver (Table 5). The trend in mercury, as for the cupric species (see Section II-B), has been interpreted in terms of the increasing energy required for charge separation in the transition state for strong metal-ligand bonds with solvents of decreasing polarity. The smaller charge separation for the silver complexes gives rise to little solvent dependence.

IV

Group III Metal Ions and Complexes (Scandium, Cerium)

The only report concerning homogeneous hydrogenation by a Group III metal species is that of Tulupov (161) on the reduction of cyclohexene by ethanolic solutions of scandium(III) stearate at about 30° and 1 atm hydrogen pressure (Section XVI-A).

Ceric heptanoate is unreactive toward hydrogen in heptanoic acid solutions at temperatures up to 150° (92) and the perchlorate was similarly unreactive in aqueous solutions (25), but some activity has been noted for ceric sulfate solutions (135a; see Section XVI-B).

V

Group IV Metal Ions and Complexes

A. TITANIUM SPECIES

Breslow and Newburg (162) in 1959 had reported the first study on a soluble Ziegler-Natta catalyst for the catalytic polymerization of ethylene; the complex used was bis(cyclopentadienyl)-titanium dichloride-triethyl-aluminum. Further studies investigated the possible activation of hydrogen by such complexes, and Breslow's group (56) first reported the successful hydrogenation of olefins and phenylacetylene by such soluble catalysts in 1963 at about 30° and 3 to 4 atm of hydrogen. The titanium systems studied were $(C_2H_5)_3Al-(C_5H_5)_2TiCl_2$, $^nC_4H_9Li-Ti(O-^iC_3H_7)_4$, and $R_3Al-Ti(OR)_4$. A number of other systems employing analogous Ziegler-type catalysts containing titanium have been reported for the reduction of olefins (e.g., 147, 163–168, 195; Chapter XV). The active catalysts are thought to be titanium (IV) species.

Hagihara and Sonogashira (169) have reported that benzene or n-heptane solutions of dicyclopentadienyltitaniumdicarbonyl, $(C_5H_5)_2Ti(CO)_2$, catalyze the hydrogen reduction of acetylenes at about 50° and 50 atm hydrogen. Styrene was only slowly reduced while $trans$-stilbene and dienes were unaffected. Table 10 summarizes the experimental data. The relative activating effects of the substituents of the acetylenes $RC{\equiv}CH$ and $R'C{\equiv}CR'$ toward hydrogenation are seen to be R = Ph > t-butyl > n-alkyl and R' = Ph > n-alkyl. The system may be fairly selective for the hydrogenation of terminal acetylenes. Hydrogenation of phenylacetylene was stated to involve a two-step process by styrene production, but this does not seem too consistent with

Table 10   Hydrogenation of Olefins and Acetylenes Catalyzed by $(C_5H_5)_2Ti(CO)_2$ (169)^a

Substrate	Product (Yield, %)
Styrene	Ethylbenzene (15)
trans-Stilbene	Not reduced
Butadiene	Not reduced
1,3-Cyclooctadiene	Not reduced
Acetylene	Not reduced
1-Pentyne	1-Pentene (95)
1-Hexyne	1-Hexene (90)
3,3-Dimethyl-1-butyne	3,3-Dimethyl-1-butene (40)
	2,2-Dimethylbutane (60)
3-Heptyne	Not reduced
Phenylacetylene	Ethylbenzene (95)
Diphenylacetylene	Dibenzyl (90)

^{a} 0.025 M catalyst, 10 M substrate, 50 to 65°, 50 atm H_2, 30 min reaction time; solvents were benzene, n-heptane, or xylene.

slow styrene reduction. The intermediates, $(C_5H_5)_2Ti(C_6H_5C{\equiv}CR)_2$, R = C_6H_5 or H, of probable structure **1a**, were isolated but these were not active

hydrogenation catalysts. A precursor of **1a**, such as structure **1b**, was thought to be a possible catalyst since an analogous vanadium complex was known (170).

Shikata and coworkers (171) have reported that reduction of $(\pi\text{-}C_5H_5)_2TiCl_2$ by sodium in tetrahydrofuran under a hydrogen atmosphere yields solutions (after filtering) which are active for the catalyzed hydrogenation of styrene and cyclohexene under mild conditions. Hydrogen was absorbed during the reduction process and only those solutions containing less than 1.0 mole H_2 per mole of titanium complex were active for hydrogenation. This system appears to be a homogeneous one, involving a biscyclopentadienyl complex. Reduction of the dichloride using Grignard reagents is reported by Brintzinger (1335) to give the hydride intermediate $[(C_5H_5)_2TiH]_2$ (see Section XV-A).

Very relevant to these cyclopentadiene catalysts are the studies of Clauss and Bestian (1164) who showed that the titanium-alkyl bond in $(C_5H_5)_2TiMe_2$ is readily cleaved by molecular hydrogen in solution:

$$2(C_5H_5)_2TiMe_2 + 2H_2 \longrightarrow \text{``}[(C_5H_5)_2Ti]_2\text{''} + 4CH_4 \qquad (76)$$

A reaction such as 76 is a likely final step in the catalytic hydrogenations. The "titanocene" product has, in fact, been shown recently to be a hydrido species $[(C_5H_5)(C_5H_4)TiH]_2$, containing a cyclopentadienylidene moiety (1371).

Martin and coworkers (1314, 1315) find that the π-allyl-dicyclopentadienyl complex $(C_5H_5)_2Ti(C_3H_5)$ also reacts with hydrogen in cyclohexane at 20° and 1 atm, but 1.9 moles of gas are absorbed per titanium and the soluble product is believed to be a dihydride, which is considered a likely intermediate in reactions such as 76. The 1-methyl- and 1,1-dimethylallyl analogs formed similar hydride solutions which were found to be very active for catalytic hydrogenation of unsaturated hydrocarbons at the mild conditions; the solutions were also effective isomerization catalysts in the absence of hydrogen, and under deuterium they catalyzed exchange with hydrogen including that of the cyclopentadienyl rings.

Of related interest is the finding of O'Brien and coworkers (1334) that $(C_5H_5)_2TiPh_2$ is thermally decomposed in solution to yield benzene by hydrogen abstraction by one phenyl group from the other.

Giannini and coworkers (193, 194) have briefly reported that tetrabenzyltitanium, $Ti(CH_2Ph)_4$, is an effective catalyst for the hydrogenation of terminal olefins at 0° and 1 atm hydrogen. Little detail is given, but the reports indicate that the reactions are homogeneous when carried out in benzene solution. The catalyst itself, unlike the corresponding zirconium one (Section V-B) shows no reaction with hydrogen at the hydrogenation conditions. These compounds were primarily studied for use as polymerization catalysts. The mechanism for these hydrogenations and those above using the allyls must be similar to but somewhat simpler than that of the soluble Ziegler-type catalysts (see Chapter XV).

Abalyaeva and Khidekel (1313) have reduced titanium(IV) complexes of formula R_nTiCl_{4-n}, where $R =$ acetate, β-phenyl- or β-(1-naphthyl)-acetate, heterogeneously to titanium(III) species, which in DMF solution are effective catalysts for the hydrogenation of carbon double and triple bonds, nitrobenzene to aniline, and fluorenone to fluorenol.

A suspension of titanium dihydride in THF, which may contain some dissolved species at higher temperatures, is inactive for the reduction of 1,3-pentadiene at 225° and up to 100 atm hydrogen (172).

Khrushch and Shilov (1476) have mentioned the use of $TiCl_4$-$SnCl_2$ mixtures in polar organic solvents for the catalytic hydrogenation of terminal olefins (see Section VII-B).

The reduction of olefinic bonds by aqueous solutions of low-valence transition-metal salts in noncatalytic reactions in the absence of molecular hydrogen is well known (see Section VII-A), and Karrer and coworkers (173a) have used aqueous-ethanol solutions of ammoniacal titanous chloride to hydrogenate cinnamic acid, cinnamic acid amide, benzalacetone, and certain flavones.

B. ZIRCONIUM AND HAFNIUM SPECIES

The biscyclopentadienylzirconium(IV)dichloride-triisobutylaluminum system has been used as a soluble Ziegler catalyst for the hydrogenation of cyclohexene (56) but it is not very efficient; $(C_5H_5)_2ZrCl_2$ with butyllithium or phenylmagnesium bromide is inactive for the hydrogenation of butadiene in benzene solution at 40° and 4 atm hydrogen (167) (see Chapter XV). Hydride derivatives of zirconium and hafnium, such as $[(C_5H_5)_2MH_2]_n$ and $(C_5H_5)_2MH(BH_4)$, have been prepared, however (e.g., 1336, 1337).

Tetrabenzylzirconium has recently been reported (194) to be a good catalyst for terminal olefin hydrogenation at 0° and 1 atm hydrogen in benzene solution. The $Zr(CH_2Ph)_4$ complex also reacted with hydrogen at 50° to produce toluene and uncharacterized low-valence zirconium products that are themselves active for hydrogenation of aromatic rings at this temperature. The reaction with hydrogen is typical of that postulated in the hydrogenation mechanism by Ziegler catalysts (Section XV-A, eq. 469). The orthohydrogens of the aromatic ring of the tetrabenzyl complex interact with the metal atom; such interactions will be considered in more detail in discussing some ruthenium systems (Section IX-B-2b).

There are no reports apparently of studies involving hafnium species.

VI

Group V Metal Ions and Complexes (Vanadium, Niobium, Tantalum)

Kanai and Miyake (172a) have patented the use of THF solutions of bis-cyclopentadienylvanadium for the selective hydrogenation of conjugated polyenes at 100° and 60 atm hydrogen. Isoprene gave almost exclusively 2-methylbut-2-ene by 1,4-addition (see also Sections VII-A, VII-B, and Table 12); polyolefins conjugated with aromatic rings were also said to be hydrogenated.

Other studies reported on hydrogen activation by vanadium(IV) species have involved reduction of monoenes or dienes by benzene solutions of biscyclopentadienylvanadium dichloride with butyllithium or phenyl-magnesium bromide (167), and n-octane solutions of vanadyl acetylacetonate with trialkylaluminum (147, 163, 164, 195). Some vanadium(V) systems, comprising vanadyl trialkyls and aluminum trialkyls in heptane solution, have been used to reduce cyclohexene and 1-octene (56). These Ziegler-type systems are considered in Chapter XV. Aqueous solutions of metavanadate are reported to be unreactive toward hydrogen at up to 150° (25).

The use of VCl_4-$SnCl_2$ solutions for hydrogenation of terminal olefins has been mentioned (1476; Section VII-B).

It has been known for a long time (e.g., Ref. 1282) that aqueous acid solution of vanadous salts can reduce unsaturated organic compounds in the absence of hydrogen. Vrachnou-Astra and coworkers (1298, 1299) have studied the reduction of fumaric, maleic, chloromaleic, and citraconic acids (and some esters) using vanadium(II) perchlorate solutions. These reactions occur according to reaction 77 (where S = substrate),

$$2V^{2+} + S + 2H^+ \longrightarrow 2V^{3+} + SH_2 \qquad (77)$$

and appear very similar to those involving the more extensively studied chromium(II) systems considered in more detail in Section VII-A. Accompanying *cis-trans* isomerization has been observed. The earlier work of Conant and Cutter (1282) involved reduction of α-, β-unsaturated aldehydes and ketones, which was also accompanied by the formation of reduced coupling products.

There have been no reports of homogeneous systems involving niobium or tantalum complexes. The hydrogenation of olefins using catalysts prepared by the reaction of niobium and tantalum pentachlorides with phenylmagnesium bromide is almost certainly purely heterogeneous involving the metal hydrides (see, e.g., Refs. 58b and 173).

Parshall and coworkers (1361) have reported that the trihydridobis(cyclopentadienyl) complex, $(C_5H_5)_2TaH_3$, catalyzes exchange between deuterium and benzene (or H_2-C_6D_6). The hydride slowly loses hydrogen at 80° and also reacts with triethylphosphine to give a monohydride; the data were interpreted in terms of the following reactions:

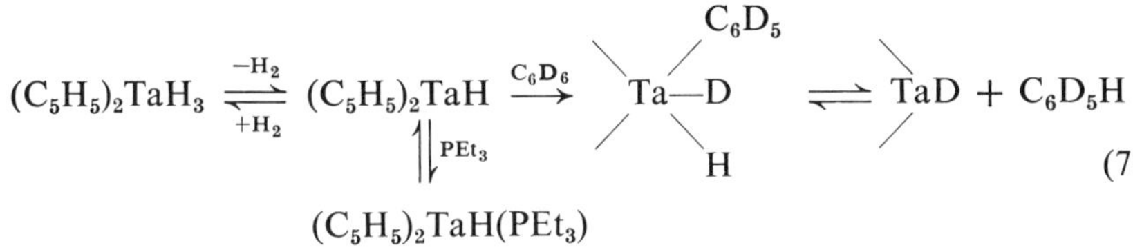

$$(C_5H_5)_2TaH_3 \underset{+H_2}{\overset{-H_2}{\rightleftharpoons}} (C_5H_5)_2TaH \overset{C_6D_6}{\longrightarrow} \overset{C_6D_5}{\underset{H}{Ta-D}} \rightleftharpoons TaD + C_6D_5H$$

$$\Big\Uparrow PEt_3$$

$$(C_5H_5)_2TaH(PEt_3)$$

$$(78)$$

The important step is the oxidative addition of the C—H bond of benzene to the coordinatively unsaturated monohydride intermediate.

VII

Group VI Metal Ions and Complexes

A. CHROMIUM SPECIES

Tulupov (174) first reported activation of hydrogen by a chromium species in 1962. Chromic stearate was one of a number of transition-metal stearates that catalyzed the reduction of cyclohexene in ethanolic solution under mild conditions (see Section XVI-A).

Chromic acetylacetonate has been used in conjunction with aluminum alkyls (56, 147, 163, 164, 166, 175–177, 195) to give soluble Ziegler catalysts in hydrocarbon solution for the reduction of unsaturated organic compounds (usually monoenes) at a few atmospheres hydrogen pressure (Chapter XV).

The use of various chromium tricarbonyl complexes for activation of molecular hydrogen has been reported. Cais and coworkers (178–180) have used arenechromium tricarbonyl complexes in cyclohexane solution ($\sim$150° and 50 atm H_2) to reduce methyl sorbate (*trans*-2, *trans*-4-hexadienoate) selectively to methyl 3-hexenoate in a few hours. The arene ligands used were benzene, toluene, ethylbenzene, anisole, methylbenzoate, chlorobenzene, 1,4-diphenylbutadiene, 3-carbomethoxyanisole, stilbene, diphenyl, diphenylmethane, diphenylethane, benzophenone, phenanthrene, and cycloheptatriene. Electron-withdrawing groups in the benzene moiety and the use of more polar solvents (acetone, methylene chloride) accelerated the reaction rate. In acetone, 5 atm pressure at 100° was effective. The cycloheptatriene derivative was the most active catalyst, but unlike the other complexes, it rapidly decomposed during the reduction. Complexes containing two $Cr(CO)_3$ moieties per molecule, such as **2a** and **2b** and other diphenyl derivatives, were much more efficient as catalysts and could be used under the milder conditions.

Cais (179, 180) suggested the following mechanism for the monochromium complexes:

$$Cr + substrate \underset{k_{-1}}{\overset{k_1}{\rightleftharpoons}} [Cr\text{-substrate}] \tag{79}$$

$$[Cr\text{-substrate}] + H_2 \underset{k_{-2}}{\overset{k_2}{\rightleftharpoons}} [H_2 \cdot Cr\text{-substrate}] \tag{80}$$

$$[H_2 \cdot Cr\text{-substrate}] \overset{k_3}{\longrightarrow} Cr + product \tag{81}$$

[where $Cr = (arene) Cr(CO)_3$]. The mechanism is based on kinetic data (not published) and on comparison with studies involving olefin/iron carbonyl complexes (Section IX-A). An induction period observed in the chromium systems was thought to involve slow formation of the [Cr-substrate] adduct; this could be formed similarly under a nitrogen atmosphere with subsequent immediate reduction under hydrogen.

Frankel and coworkers (181, 1136), however, have suggested a quite different mechanism from some deuterium-tracer studies on the methyl-benzoate-chromium tricarbonyl system. Deuterium reduction yielded exclusively methyl 2,5-dideuterio-cis-3-hexenoate. In the absence of substrate, H_2-D_2 exchange was observed, showing that the chromium complex alone is effective for H_2 activation and H_2-D_2 exchange, whereas, in the presence of substrate, hydrogen (deuterium) addition predominates over exchange. A few percent of free arene was detected, and a mechanism consistent with these data was presented:

$$PhCO_2Me - Cr(CO)_3 \rightleftharpoons PhCO_2Me + \underset{\mathbf{3}}{[Cr(CO)_3]} \tag{82}$$

$$\mathbf{3} + D_2 \rightleftharpoons \underset{\mathbf{4}}{[D_2Cr(CO)_3]} \underset{D_2}{\overset{H_2}{\rightleftharpoons}} [H_2Cr(CO)_3] \rightleftharpoons H_2 + \mathbf{3} \tag{83}$$

The observed induction period was attributed to the rate-determining dissociation of the arene complex; addition of free arene was said to interfere with the hydrogenation. The 1,4 deuterium (hydrogen) addition indicated that the dideuteride (dihydride) species **4** was involved. During reduction the methyl sorbate underwent no exchange and this, together with the stereochemistry of **5**, accounted for the high selectivity of this catalyst system. A

Table 11 Hydrogenation of Dienes Using Methyl Benzoate—Cr(CO)$_3$ (1135, 1136)[a]

Substrate (% Conversion)[b]	k (hr^{-1})[c]	Major Product (%)
Hexa-1,3-diene (100)	0.9	cis-Hex-2-ene (66)
Hexa-2,4-diene (100)	2.1	cis-Hex-3-ene (90)
Hexa-1,4-diene (77)	0.22[d]	cis-Hex-3-ene (50), hex-2-enes (50)
Hexa-1,5-diene (2)	0.004[d]	Hex-1-ene (50), hex-2-enes (50)
Me / —Me (100)	0.69	Me / Me —Me (75)
Me Me (100)	0.61	Me Me / Me Me (82)
—Me / Me (100)	1.1	Me / —Me Me (66)
Me— —Me Me Me (5)	0.007	Me— —Me Me Me (100)
Cyclohexa-1,3-diene (100)	0.50	Cyclohexene (98)
Cyclohexa-1,4-diene (100)	0.42	Cyclohexene (93)
Cycloocta-1,5-diene (100)	0.09	1,3-Diene (64), cyclooctene (35)
	0.19[e]	1,3-Diene (28), cyclooctene (70)
Cycloocta-1,3-diene (25)	0.05	Cyclooctene (100)
Methyl trans, trans-9,11-octadecadienoate	—	Methyl 10-octadecenoates
Methyl cis, trans-9,11-octadecadienoate	—	Methyl 9- and 10-octadecenoate

[a] 10^{-2} M catalyst; 30 atm H$_2$, 160° in pentane or hexane.
[b] 0.19 M.
[c] Pseudo first-order constant for loss of diene, usually after an induction period.
[d] 175°.
[e] 170°.

small inverse isotopic effect (k_H/k_D 0.9) was observed for the reduction with either H_2 or D_2 alone. Hydrogen addition is predominantly molecular since with D_2 monoene-d_2 is the main product, and with $H_2 + D_2$ it is a mixture of monoene-d_0 and monoene-d_2.

Frankel and Little (182, 183) have also reported that the activity of the complexes (X-benzene) $Cr(CO)_3$ for the methyl sorbate reduction decreases in the order X = Cl > acetyl > CH_3 > $(CH_3)_3$ > $(CH_3)_6$ and this is thought to be due to electron donors increasing the stability of the π-arene complex (see eq. 82).

Frankel and coworkers (1135, 1136) have extended the use of these arene–$Cr(CO)_3$ complexes for the hydrogenation and deuteration of other 1,3- and 1,4-dienes to mainly cis-monoenes in hydrocarbon solution (n-pentane, n-hexane, cyclohexane) at 100 to 180° and 30 to 50 atm hydrogen. Table 11 summarizes these reductions using the methyl benzoate complex. The data were rationalized in terms of 1,4 addition of hydrogen by means of a cisoid complexed diene system (eqs. 82 to 84); the most favorable stereochemistry for the addition is with a trans,trans diene. In the cis,trans- and cis,cis-dienes, one and two substituents interfere sterically with the concerted addition. For the hydrogenation of a mixture of isomeric 9,11-octadecadienoates (see also Section XIV-D), the relative rates were 1.0 for cis,cis-, 8.0 for cis,trans-, and 25 for trans,trans-diene (1135). As in the sorbate reduction, deuteration of cyclohexa-1,3-diene and 9,11- and 10,12-octadecadienoates yielded monoenes almost exclusively deuterated at the α-methylene carbons, —CHD—CH=CH—CHD—. No deuterium exchange occurred with any conjugated dienes, thus indicating that the addition step in eq. 84 is not reversible (1136).

Conjugated dienes are readily reduced. Methyl substituents on C-2 and C-3 of some branched 1,3-dienes did not interfere with the 1,4 addition, although substitution on C-1 and C-4 severely inhibited reduction as shown by the 2,5-dimethyl-2,4-hexadiene system. The hydrogenation of 4-methyl-1,3-pentadiene was accounted for in terms of isomerization by a 1,5-hydrogen shift followed by 1,4 addition:

$$\text{(84a)}$$

The unexpected production of 9-monoene from the cis,trans-9,11-octadecadienoate was similarly explained by the closeness of the C-8 and C-12 in this diene. The case of hydrogenation of cyclohexa-1,3-diene compared with cycloocta-1,3-diene may be related to a more favorable conformation in the former for complex formation (1135).

Unconjugated dienes require conjugation prior to hydrogenation. 1,4-Dienes are reduced when somewhat higher temperatures are used to allow isomerization that is likely to be rate-determining in these hydrogenations (see also Section XIV-D). The more stable methyl benzoate and benzene complexes are thus suitable for these hydrogenations (183, 1135). The deuteration of cyclohexa-1,4-diene to give cyclohexene-d_2 with 90% of the deuterium on the α-methylene carbon atoms provides good evidence for the

$$\text{⬡} \longrightarrow \text{⬡} \longrightarrow D-\text{⬡}-D \qquad (84b)$$

preliminary isomerization. As observed, preferential hydrogenation of the 1,3-diene in a mixture with the 1,4-diene under conditions when both are individually readily hydrogenated (Table 11) is probably because of favored coordination of the conjugated diene. The selectivity to form monoenes suggests that they have a much lower affinity for the catalyst (see also Section XIV-D). Hexa-1,5-diene is not isomerized and thus is not reduced effectively. Cycloocta-1,5-diene is reduced with difficulty because of the thermodynamic stability of the 1,3-diene (1135).

Miyake and Kondo (184, 1432) have reported similar selective hydrogenation of polyenes using bis(tricarbonylcyclopentadienylchromium) as a catalyst under similar conditions (70°, >50 atm H_2) in benzene solution. Table 12 summarizes their data. Hydrogenation also occurs slowly at room temperature or 1 atm of hydrogen. Except for sterically hindered dienes, the hydrogen adds preferentially at the ends of the conjugated system; double bonds do not migrate, and isolated double bonds are not reduced. The

Table 12 Hydrogenation of Polyenes Catalyzed by $[C_5H_5(CO)_3Cr]_2$ (184, 1432)[a]

Substrate	Product (Yield, %)
Isoprene	2-Methyl-2-butene (95)
1,3-Pentadiene	2-Pentene (100)
4-Methyl-1,3-pentadiene	2-Methyl-2-pentene (78)
	4-Methyl-2-pentene (22)
1,3-Cyclohexadiene	Cyclohexene (100)
Cyclooctatetrene[b]	Cyclooctene, cyclooctadienes

[a] 2.0 g substrate, 0.15 g catalyst, 10 ml benzene, 70°, 90 atm H_2, 5 hr reaction time.

[b] 85°, 140 atm, 15 hr; $C_5H_5Cr(CO)_3$ (halide) was also used (1432).

reactions are thought to involve tricarbonylcyclopentadienylhydridochromium as an intermediate:

$$2C_5H_5(CO)_3CrH + CH_2{=}\underset{\underset{\displaystyle CH_3}{|}}{C}{-}CH{=}CH_2 \longrightarrow$$

$$[C_5H_5(CO)_3Cr]_2 + CH_3{-}\underset{\underset{\displaystyle CH_3}{|}}{C}{=}CH{-}CH_3 \quad (85)$$

$$[C_5H_5(CO)_3Cr]_2 + H_2 \longrightarrow 2C_5H_5(CO)_3CrH \quad (86)$$

Reaction 85 occurred quantitatively at room temperature as shown for isoprene (184) and also for 2,4-hexadiene (185); the latter, however, gave a mixture of 2- and 3-hexenes. Reaction 86 was known to occur at high pressures ($\sim$150 atm) of hydrogen (186). Mechanistic details were not considered, but the suggested monohydride intermediate indicates a hydrogen transfer process quite different from that postulated by Frankel's group (181); such a mechanism (eqs. 82 to 84), however, cannot be ruled out. The corresponding molybdenum and tungsten carbonyls $[C_5H_5(CO)_3M]_2$ are not active catalysts (185) because of the more stable metal-metal bonds in these species (Section VII-B).

Of interest here is an earlier report by Brown and coworkers (187) who found that ethanolic hydrochloric acid solutions of arene chromium tricarbonyls reduce azobenzene stoichiometrically:

$$PhN{=}NPh + 2H^+ \xrightarrow[\text{(arene) Cr(CO)}_3]{} H_2N{\cdot}C_6H_4{\cdot}C_6H_4NH_2 \quad (87)$$

Reduction is thought to occur by protonation at the chromium atom followed by hydrogen transfer to the π-bonded substrate, with resulting oxidation of the chromium. The reaction was facilitated in this case by electron-donor groups, indicating that the protonation step is probably rate-determining. Although this system does not involve molecular hydrogen it does involve homogeneous hydrogenation, and such studies are important for elucidating the nature of the transfer step of hydrogen to substrate.

The effect of the chromium tricarbonyl and iron tricarbonyl groups on the π-electron distribution of coordinated aromatic ligands has been studied using n.m.r. methods (e.g., Ref. 1322).

Rejoan and Cais (179) have shown that $Cr(CO)_6$ also reduces methyl sorbate catalytically and selectively to the 3-hexenoate in acetone at 165° with 50 atm hydrogen. The reaction is thought to involve initial activation of the substrate (eqs. 79 to 81); initial hydrogen activation (see eq. 83) may hinder the process.

Brown and coworkers (187) have used moist hydrocarbon solutions of $Cr(CO)_6$ under nitrogen to reduce conjugated ketones in noncatalytic reactions—cyclopentadienones gave cyclopent-2-enones; p-benzoquinone gave quinol, and benzil gave benzoin. Tetraphenylcyclopentadienone formed a π-complex with bonding to a $Cr(CO)_3$ moiety by a phenyl substituent. The mechanism was thought to be essentially the same as that described for reaction 87.

A number of binuclear complexes containing chromium carbonyl moieties have been used by Thompson and coworkers (1002, 1007, 1010) as hydrogenation catalysts for olefin/acetylene mixtures in benzene at 175° and 100 atm; the complexes include $(OC)_4Cr(PMe_2)_2Mo(CO)_4$ and $(OC)_4Cr(PPh_2)_2$-$Fe(CO)_3$, with bridging phosphine ligands, and $[(OC)_5Cr(PPh_2)ClPd]_2$; the latter is considered in Section XIII-B.

Arenechromium tricarbonyl complexes (178, 183, 183a, 1135, 1136, 1473) and chromic acetylacetonate (145) have been used for the catalytic hydrogen reduction of unsaturated fats (Sections XIV-D, -E). Aqueous solutions of chromic salts or chromate are unreactive toward hydrogen at 150° (25), and chromic salts are inactive as catalysts for the hydrogen reduction of dichromate at temperatures up to 270° (100).

Closely related to the "protonation-type" hydrogenations described by Brown and coworkers (see eq. 87) are those involving the reduction of unsaturated compounds, while using aqueous or mixed-solvent solutions of chromous salts under nitrogen. Such reductions by metal ions were discovered by Berthelot (1280); those involving chromium(II) have been extensively studied by many workers including Traube and Passarge (1281), Conant and Cutter (1282), Patterson and du Vigneaud (1283), and in more recent years by Bottei and coworkers (1284–1286), Castro and coworkers (1287–1289), Kopple (1290), Malliaris and Katakis (1291), and Shilov and coworkers (1292–1294). Table 13 gives a representative list of the types of reductions reported. Azo compounds, saturated aldehydes, and ketones have also been reduced (1285, 1290).

Acetylenes and olefins, particularly with electronegative substituents, are generally hydrogenated stoichiometrically according to reaction 87a (S = substrate), although reduced coupling products have been observed—especi-

$$2Cr^{2+} + S + 2H^+ \longrightarrow 2Cr^{3+} + SH_2 \tag{87a}$$

ally with the unsaturated aldehyde and ketone substrates (e.g., Ref. 1282). Detailed kinetics have been studied for a number of the systems (1286, 1288, 1294), and the rate laws obtained are essentially consistent with the overall mechanism shown below:

$$Cr^{II} + S \rightleftharpoons Cr^{II} \cdot S \tag{87b}$$

$$Cr^{II}S + Cr^{II} \xrightarrow{2H^+} 2Cr^{III} + SH_2 \tag{87c}$$

At lower substrate concentrations, equilibrium 87b lies to the left and the reaction becomes second order in chromium. The detailed mechanism of

Table 13 Stoichiometric Reductions of Olefins and Acetylenes Using Chromium(II) Systems[a]

Catalyst	Substrate	Reference
$CrCl_2$ in HCl or HCl/EtOH	Maleic, fumaric, cinnamic acids, α,β-unsaturated aldehydes and ketones; $RC{\equiv}CR$ (R=H, CO_2H)	1281–1285
Cr^{II} salts in aqueous NH_3	Cinnamic acid, mesityl oxide, acetylene	1280, 1290
$CrSO_4$ in H_2O or H_2O/DMF	Maleic, fumaric, cinnamic acids and esters, dimethylmaleic anhydride, acrylonitrile, 2,3-diphenylfumaronitrile; $HC{\equiv}CR$ (R=CH_2OH, Ph), $RC{\equiv}C\ CO_2H$ (R=CO_2H, Ph), $RC{\equiv}CCH_2OH$ (R=Me, CH_2OH). 2-carboxydiphenylacetylene, hex-1-yne	1287–1289
$Cr^{II}/EDTA$ in H_2O	Propargyl alcohol ($CH{\equiv}C\ CH_2OH$)	1287
$Cr(H_2O)_6{}^{2+}$ in $HClO_4$	Maleic, fumaric acids and esters, $R^1C{\equiv}CR^2$ ($R^1=R^2=H$ or CO_2H)	1286, 1291–1294

[a] Under nitrogen at 5 to 100° (eq. 87a).

reaction 87c then involves two metal centers. The proton dependence observed depends on the substrate; an inverse dependence is observed with acetylenedicarboxylic acid, since the bonded substrate is in an anionic form (1286). With acetylene itself (1294) the rate law is $k[Cr^{2+}]^2[C_2H_2][H^+]$, for which the following mechanism was suggested:

$$2Cr^{2+} + C_2H_2 \underset{}{\overset{K}{\rightleftharpoons}} \left[Cr\overset{C_2H_2}{\cdots\cdots} Cr \right]^{4+} \xrightarrow[H^+]{k}$$

(87d)

The equilibrium reaction (K) was observed at low temperatures, and thermodynamic parameters $\Delta H°$ and $\Delta S°$ of magnitude 4 kcal mole^{-1} and 4 eu, respectively, were measured (1293). The two chromium atoms can contribute one electron each to the acetylene molecule; and the metal-metal bond could be significant in the electron-transfer process.

In the presence of a strong reducing agent (Zn/HCl, electrochemical, etc.), these hydrogenations can become catalytic in chromium(II) (1294). Castro and coworkers (1295–1297) have also used chromous sulfate solutions for the hydrogenolysis of organic halides.

B. MOLYBDENUM AND TUNGSTEN SPECIES

Frankel and Little (183) have found that a number of molybdenum and tungsten carbonyls are effective for the hydrogen reduction of methyl sorbate to methyl hexenoates in cyclohexane solution at 100 to 180° and 40 atm H_2. The complexes studied were mesitylene-Mo(CO)$_3$, cycloheptatriene-Mo(CO)$_3$, bicyclo(2,2,1)hepta-2,5-diene-Mo(CO)$_4$ and mesitylene-W(CO)$_3$. The molybdenum complexes were generally more active (measured as sorbate consumption over a time interval) at lower temperatures than were the arene-chromium tricarbonyls (Section VII-A). For the mesitylene complexes the relative order of activity was Mo > W > Cr, although the selectivity for production of the 3-hexenoate varied in the order Cr > Mo > W. Unlike the chromium system no induction period was observed in the molybdenum and tungsten mesitylene reactions, which probably accounts for their greater activity. This activity may be due to the greater thermal lability of the molybdenum and tungsten complexes which, unlike the chromium complex, decompose during hydrogenation to other carbonyl species. The nature of the product composition–time curve suggested that the mode of action of the tungsten complex was quite different from that of the comparable molybdenum and chromium systems (see eqs. 82–84), because it probably involves a monohydride intermediate, as has been postulated for some comparable iron carbonyl systems (Section IX-A). The dimer of cyclopentadienyl-Mo(CO)$_3$ is thermally very stable and was found unreactive (183). Miyake and Kondo (185, 185a) have also reported that this complex and the corresponding tungsten one are inactive compared with the chromium complex with its more fissile metal-metal bond (Section VII-A; eq. 86). The tricarbonyl-cyclopentadienylhydrido-molybdenum and tungsten complexes, however, do hydrogenate polyenes stoichiometrically in the absence of hydrogen at 20 to 60°, according to reactions such as 85. Polyenes can be reduced in steps by such reactions, which, however, generally do not seem particularly selective. 2,4-Hexadiene is reduced to a mixture of 2- and 3-hexenes; 4-methyl-1,3-pentadiene gives a mixture of 2-methyl-2-pentene and 4-methyl-2-pentene; 2,4,6-octatriene gives 3,5-octadiene and 2,4-octadiene. However,

1,3,5-hexatriene gives only 2,4-hexadiene; 1,3,5,7-octatetrene gives 90% 2,4,6-octatriene; and isoprene gives 88% 2-methyl-2-butene (cf. Table 12).

The complexes $(OC)_4M(PPh_2)_2Fe(CO)_3$, where M = Mo, W, have been used as hydrogenation catalysts in the same manner as $(OC)_4Cr(PMe_2)_2$-$Mo(CO)_4$ (Section VII-A; 1002).

Otsuka and coworkers (1338) have reported that the molybdenum dihydride $(C_5H_5)_2MoH_2$, but not the tungsten analog, reacts with diphenylacetylene and azobenzene in boiling toluene or THF to give a stoichiometric hydrogenation

$$(C_5H_5)_2MoH_2 + 2PhC\!\equiv\!CPh \longrightarrow (C_5H_5)_2Mo\overset{\displaystyle CPh}{\underset{\displaystyle CPh}{\big|\big|}} + PhCH\!=\!CHPh \tag{87e}$$

Stronger π-acids (acetylenedicarboxylate, perfluorobut-2-yne) give the more usual product from metal-hydride addition across the triple bond. Green and Knowles (1360) have similarly reduced isoprene to isomeric pentenes by using the tungsten dihydride in benzene or toluene at 120°; the tungsten product is a phenyl hydride formed by addition of benzene to the intermediate tungstenocene (see eq. 78 in Chapter VI).

$$(C_5H_5)_2WH_2 + C_5H_8 \longrightarrow [(C_5H_5)_2W] + C_5H_{10} \tag{87f}$$

$$[(C_5H_5)_2W] + HC_6H_4R \longrightarrow (C_5H_5)_2WH(C_6H_4\!-\!R) \qquad R = H \text{ or } Me \tag{87g}$$

Khidekel and coworkers (188 by 57d) have reported that a molybdenum catalyst stabilized by quinonoid and aromatic ligands is effective in DMF solution under mild conditions for the reduction of alkenes (see Section XI-H for a corresponding rhodium system).

Wilkinson's group (187a) have reported that solutions containing the binuclear species Mo_2^{4+}, prepared by protonation of molybdenum(II) acetate using 40% fluoroboric acid in methanol, become effective for the hydrogenation of 1-hexene (35°, <1 atm H_2) on adding triphenylphosphine in the ratio 2:1, ligand:metal. The molybdenum(II) solutions are extremely air-sensitive and are several hundred times less efficient than comparable solutions containing Rh_2^{4+} and Ru_2^{4+} species (Sections XI-I, IX-B-2b).

Khrushch and Shilov (1476, 1477) have described the use of $MoCl_5$-$SnCl_2$ mixtures in THF or methanol for the reduction of ethylene (up to 16 atm) at hydrogen pressures of 30 to 100 atm and 60 to 90°. The reaction rate is first-order in molybdenum, ethylene, and hydrogen, and is a maximum at a Sn/Mo ratio of 2. The diamagnetism of the catalyst solution in THF was explained by formation of a dimeric molybdenum(III) complex:

$$MoCl_5 + 3C_4H_8O \longrightarrow MoOCl_3 \cdot 2C_4H_8O + C_4H_8Cl_2 \tag{87h}$$

$$2MoOCl_3 + 4SnCl_2 \longrightarrow (MoOSnCl_3)_2 + 2SnCl_4 \tag{87i}$$

The reaction rate decreased on addition of acid but increased on addition of water to the point where insoluble hydroxides started to precipitate. The mechanism, analogous to a $PtCl_4^{2-}$-$SnCl_2$ system (see Section XIII-C-2), was written as follows:

$$Mo_2O_2(SnCl_3)_2 + H_2 \underset{K_1}{\rightleftharpoons} H^+ + HMo_2O_2(SnCl_3)_2^-$$

$$+H^+ \uparrow \; -C_2H_6 \qquad\qquad K_2 \updownarrow C_2H_4 \qquad\qquad (87j)$$

$$C_2H_5Mo_2O_2(SnCl_3)_2^- \xleftarrow{k} HMo_2O_2(SnCl_3)_2C_2H_4^-$$

This scheme accounts for the rate law if the molybdenum is essentially all present as $Mo_2O_2(SnCl_3)_2$. The K_1 equilibrium, which was written as involving a molybdenum(V) dihydride intermediate (see Section IX-B-2e, eq. 148), accounts for the acid dependence and the enhancement by water by an increase in medium polarity. Ethane production occurs by hydrolytic decomposition of the ethyl complex. Using deuterium, the product composition of $C_2H_4D_2$, C_2H_5D, and C_2H_6 depended on the extent of hydrogen exchange of D^+ and the hydride, hence on the concentration of water in the system. An activation energy of about 18 kcal mole^{-1} was reported. Other terminal olefins could be reduced but only with accompanying isomerization. The use of tungsten tetrachloride was also referred to (1476).

Some molybdenum and tungsten carbonyl complexes, such as $MCl_2(CO)_3$-$(PPh_3)_2$, have been used in the presence of stannous chloride for the hydrogenation of fats; a binuclear molybdenum carbonyl species has also been used (1139, 1475; Section XIV-G).

In conjunction with triisobutylaluminum, the dioxide acetylacetonate complex, $MoO_2(acac)_2$, furnishes a soluble Ziegler hydrogenation catalyst (see Chapter XV), for example, for the reduction of cyclohexene and 1-octene (56).

Tulupov (57b) notes that he and his coworkers have found that the molybdate ion, MoO_4^{2-}, is an active hydrogenation catalyst for the reduction of dichromate, but no data or references were given.

Denisov and coworkers (188a) have reported the reduction of phenyl-hydrazine by hydrochloric acid solutions of molybdenum(V) in the absence of molecular hydrogen:

$$Ph-NH-NH_2 + 2Mo^V \longrightarrow PhNH_2 + NH_3 \qquad (87k)$$

The reaction is second order in molybdenum and likely to correspond to the more extensively studied chromium(II) systems (Section VII-A).

Moss and Shaw (1308) have evolved a multihydride complex, $WH_6(PMe_2Ph)_3$, but whether it undergoes exchange with acidic hydrogens, as with some related rhenium and osmium complexes (see Sections VIII-B and IX-C), was not reported.

VIII

Group VII Metal Ions and Complexes

A. MANGANESE SPECIES

Aqueous solutions of permanganate were known for many years to be reduced by hydrogen (189–191), and the earliest report by Just and Kauko (189) involved a kinetic study in neutral solutions. Halpern and Webster (124) extended this work and studied the reaction in perchlorate solutions from 1.0 M in acid to 0.3 M in alkali, at 30 to 70° and 1 atm pressure (the reaction was followed by loss of permanganate). The reactions observed in acid, neutral, and basic solutions are as follows:

$$2MnO_4^- + 3H_2 + 2H^+ \longrightarrow 2MnO_2 + 4H_2O \tag{88}$$

$$2MnO_4^- + 3H_2 \longrightarrow 2MnO_2 + 2H_2O + 2OH^- \tag{89}$$

$$2MnO_4^- + H_2 + OH^- \longrightarrow 2MnO_4^{2-} + 2H_2O \tag{90}$$

The kinetics were essentially independent of pH and followed the simple rate law

$$\frac{-d[H_2]}{dt} = k[H_2][MnO_4^-] \tag{91}$$

A bimolecular reaction must give an intermediate manganese species that undergoes further rapid reaction to products. The observed activation energy was 15 kcal mole^{-1}, and on energetic grounds (24) manganese(V), as suggested originally (189), was considered most likely. Its formation could involve transfer of hydride, or two electrons from H_2 to MnO_4^-:

$$MnO_4^- + H_2 \longrightarrow HMnO_4^{2-} + H^+ \tag{92}$$

$$MnO_4^- + H_2 \longrightarrow MnO_4^{3-} + 2H^+ \tag{93}$$

or of an oxygen atom from MnO_4^- to H_2

$$MnO_4^- + H_2 \longrightarrow MnO_3^- + H_2O \tag{94}$$

Either reaction 92 or reaction 94 was preferred because of the large negative entropy of activation (-17 eu), although oxygen transfer was demonstrated in the permanganate oxidation of aldehydes and formic acid (192).

The production of a Mn(VI) intermediate by the reaction

$$MnO_4^- + H_2 \longrightarrow MnO_4^{2-} + H^+ + H \tag{95}$$

was ruled out on energetic grounds (24), but this intermediate was favored for the silver-catalyzed hydrogen reduction of permanganate (124) where the Ag^+ ion can stabilize the hydrogen atom released (eq. 60, Section II-D). The silver-catalyzed path becomes effective because of its lower activation energy (9 vs. 15 kcal mole^{-1}).

Since no autocatalysis was observed during the reduction of basic permanganate solutions, the manganate ion, MnO_4^{2-}, does not activate hydrogen (124). Manganous salts are also inactive in aqueous solutions up to $150°$ (25).

Tulupov (148) later studied the hydrogen reduction of permanganate in sulfuric acid solutions and observed an increasing rate with increasing acidity over the range 1.8 to 12.6 M H_2SO_4. Rate law 91 was not obeyed and at a constant hydrogen pressure the first-order log plots for permanganate fell away from linearity throughout. The reaction apparently initially produced manganate ion (k_1 step) which then disproportionates by a nonrapid reaction such as

$$2MnO_4^{2-} + 4H^+ \xrightarrow{k_2} MnO_2 + MnO_4^- + H_2O \tag{96}$$

A kinetic equation for the loss of permanganate, by two consecutive reactions involving regeneration of the starting permanganate, was derived and written in the form

$$\Delta = a\left(\frac{1 - k_1}{k_1 + k_2} \{1 - \exp\left[-(k_1 + k_2)t\right]\}\right) \tag{96a}$$

where a was the initial $[MnO_4^-]_o$, and $\Delta = [MnO_4^-]_o + [MnO_4^-] - [MnO_4^{2-}]$. It is not clear how this Δ term may be estimated unless the $[MnO_2]$ is determined; also, no equation was given for the k_1 step (see eq. 95).

Tulupov (193a) has studied the hydrogenation of cyclopentene using manganous stearate dissolved in liquid paraffin (Section XVI-A).

Manganous acetylacetonate in conjunction with trialkylaluminum compounds has been used as a soluble Ziegler catalyst (see Chapter XV) for the reduction of monoenes and benzene (56, 147, 163).

Dimanganesedecacarbonyl, $Mn_2(CO)_{10}$, is reported by Sternberg (208) to catalyze the hydrogenation of oct-1-ene at 80 to 150° and 200 atm hydrogen; the carbonyl has also been used for the catalytic reduction of unsaturated fatty esters (Section XIV-C).

Alkyl and acyl manganese pentacarbonyls are claimed in patent literature (1317, 1318) to be more effective than cobalt carbonyls for hydroformylation. In hexane solution at 300° and high carbon monoxide/hydrogen pressure, ethylene is converted into propionaldehyde and propan-1-ol; the alcohol product shows that the system is capable of hydrogenating saturated aldehydes (Section X-E-2).

Hieber and Kruck (1320) have reported that the hydridopentacarbonyl may be formed from the reaction of water with the cationic hexacarbonyl:

$$Mn(CO)_6^+ + H_2O \longrightarrow HMn(CO)_5 + CO_2 + H^+ \qquad (96b)$$

The reaction probably involves a carboxyl intermediate (see Section XIII-C-3a, eq. 449).

B. RHENIUM SPECIES

Despite the upsurge in the chemistry of rhenium and the preparation of a wide range of its organometallic derivatives, only one report has so far appeared on homogeneous hydrogenation by rhenium complexes, although several multihydro-complexes of the type ReH_5L_3, ReH_7L_2, and ReH_8L^- (L = tertiary phosphine or tertiary arsine) are known (1339, 1340) and hydrogen-deuterium exchange reactions with alcohols, hydrogen, and benzene were reported for the $ReH_7(PPh_3)_2$ complex by Chatt and Coffey (1339) (see eq. 78 in Chapter VI). Khrushch and Shilov (1476) have briefly mentioned the use of $ReCl_5$-$SnCl_2$ mixtures in polar organic solvents for the hydrogenation of terminal olefins (Section VII-B).

It should be mentioned that rhenium in a formal -1 valence state should have a strong tendency to undergo oxidative addition reactions (e.g., 39).

Dirheniumdecacarbonyl, $Re_2(CO)_{10}$, is reported by Selin (1318) to be inactive as a homogeneous hydrogenation catalyst up to 200°; decomposition to metal, a heterogeneous catalyst, occurs above this temperature.

No reports of activity by the more limited range of technetium complexes have appeared.

IX

Group VIII Metal Ions and Complexes

THE IRON TRIAD

A. IRON SPECIES

Some work on the hydrogenation of organic compounds by iron carbonyl and carbonyl hydrides has been reviewed by Wender and Sternberg (413, 1119) and Bird (53). The hydride $H_2Fe(CO)_4$ readily decomposes with loss of hydrogen to a mixture of carbonyls (1120). In the presence of an acceptor, stoichiometric hydrogenation may occur, and Sternberg and coworkers (1121) reduced acetone to propan-2-ol at 20° in this manner; hex-1-ene was completely isomerized, however. Iron pentacarbonyl is not a very effective hydroformylation catalyst; an accompanying reaction common under these conditions generally is catalytic hydrogenation, which will be discussed in detail for the better-studied cobalt systems (Section X-E). Gresham and Brooks (1124) noted the reduction product n-propanol as well as the expected propionaldehyde from the hydroformylation of ethylene in cyclohexane solution, when using $Fe(CO)_5$ at 130 to 200° and 400 to 800 atm carbon monoxide/hydrogen. Uchida and Bando (1125) found only hydrogenated product (propionate) from methyl acrylate; the reaction was retarded by carbon monoxide and could be carried out under a purely hydrogen atmosphere.

Reductions have been more commonly carried out by using the pentacarbonyl under carbon monoxide pressure at elevated temperatures and by using alkaline water/methanol as the hydrogen source; hydroxide and amines have been used as base. Such reductions have been recorded by Reppe and Vetter (1123), Sternberg and coworkers (1121), Kutepow and Kindler (426), and Reed and Lenel (1128). Table 14 summarizes some of the data; the

Table 14   Catalytic Reductions Using Alkaline Solutions of Fe(CO)$_5$^{a}

Substrate	Product	Reference
Ethylene	Propanol, propionic acid	1123
Propylene	n- and iso-Butanols	1123, 1126, 1128
But-1-ene	n- and iso-Pentanols	1123
Oct-1-ene	Nonanol, oct-2- and -3-enes	1121
Isooctene	Nonanols	1123
Butadiene	Butenes, butane	1123
Penta-1,3-diene	Hexanols	1123
Hexa-1,3-diene	Heptanols	1123
Methylvinylether	Ethanol, methanol	1123
Allyl alcohol	Propanol	1123
But-2-ene-1,4-diol	Butan-1,4-diol	1123
Dihydrofuran	Butan-1,4-diol	1123
Cyclopentene	Cyclopentylcarbinol	1121

^{a} At elevated temperatures under CO pressure.

reactions usually give a range of products and the table lists the reduction products in particular. Alcohols arise from reduction of the expected saturated aldehyde product, while other products result from hydrogenation of olefinic bonds. Under less strongly alkaline conditions, the products are mainly aldehydes (1121, 1127). The alkaline-carbonyl solutions have also been used at ambient conditions for stoichiometric reductions of nitrobenzene to aniline (1129, 1130), benzil to benzoin (1129), benzoquinone to quinol (1129, 1130), benzaldehyde to benzyl alcohol (1121, 1127), and acetylene to ethylene (1131).

The chemistry of the carbonyl solutions has been studied intensively (e.g., see Ref. 1121). In alkaline solution, the pentacarbonyl forms $Fe(CO)_4^{2-}$ and $HFe(CO)_4^-$ (1122)

$$Fe(CO)_5 + 3OH^- \longrightarrow HFe(CO)_4^- + CO_3^{2-} + H_2O$$
$$\text{OH}^- \downarrow\uparrow$$
$$Fe(CO)_4^{2-} + H_2O \qquad\qquad (97)$$

The reducing properties may be attributed to electron transfer to the substrate (eq. 97a) followed by proton addition from the solvent (1121). The binuclear anion $Fe_2(CO)_8^{2-}$ has been isolated (1120)

$$2Fe(CO)_4^{2-} \longrightarrow Fe_2(CO)_8^{2-} + 2e \qquad (97a)$$

Alternatively, reduction could be achieved by hydrogen transfer, possibly by a dihydride intermediate as suggested by Sternberg and coworkers (1119, 1121, 1127)

$$2HFe(CO)_4^- \longrightarrow [H_2Fe_2(CO)_8]^{2-} \rightarrow H_2 + Fe_2(CO)_8^{2-} \qquad (97b)$$

In this context the same group (413) found that in the preparation of $(C_5H_5)_2Fe_2(CO)_4$ according to eq. 97c, the hydrogen expected was not evolved but was transferred to excess cyclopentadiene to give cyclopentene and cyclopentane. These workers also proposed the formation of a hydride (eq. 97d), followed by hydrogen transfer

$$2Fe(CO)_5 + 2C_5H_6 \longrightarrow (C_5H_5)_2Fe_2(CO)_4 + 6CO + H_2 \qquad (97c)$$

$$Fe(CO)_5 + C_5H_6 \longrightarrow HFe(CO)_2(C_5H_5) + 3CO \qquad (97d)$$

The quantitative reduction of azobenzene to benzidine by an ethanolic hydrochloric acid solution of the carbonyl dimer was explained (413) in terms of protonation at the iron atom, followed by hydrogen transfer with resulting oxidation of the iron (see eq. 87 in Section VII-A).

The hydroformylation reactions by the pentacarbonyl were also thought to occur by the dihydride

$$H_2Fe(CO)_8^{2-} + C_2H_4 \xrightarrow{-CO} \left[\begin{array}{c} CH_2\!\!-\!\!CH_2 \\ (CO)_3FeH \quad HFe(CO)_3 \\ C \\ O \end{array} \right]^{2-} \rightarrow \begin{array}{c} CH_3CH_2CHO \\ + \\ [Fe(CO)_3]_2^{2-} \end{array}$$

$$(97e)$$

The catalyst was thought to be regenerated from the products of eqs. 97b and 97e by reaction of the iron carbonyl with carbon monoxide and water. Other hydrides such as $HFe_2(CO)_8^-$ and $HFe_3(CO)_{11}^-$ are known (1120, 1121, 1323), and Kutepow and Kindler (426) have presented a scheme involving the latter species for the formation of butanol from propylene:

$$HFe_2(CO)_{11}^- + C_3H_6 \longrightarrow C_3H_7Fe_3(CO)_{11}^- \longrightarrow$$

$$C_3H_7\overset{\delta-}{CO}\cdots Fe \xrightarrow{H_2O} C_3H_7CHO + OH^- + 'Fe' \qquad (97f)$$

$$HFe_3(CO)_{11}^- + C_3H_7CHO \longrightarrow C_4H_9OH + 'Fe' \qquad (97g)$$

The mechanisms of these reactions are not well established. $HFe(CO)_4^-$ is isoelectronic with the well-studied $HCo(CO)_4$ analog (see Section X-E), and the mechanisms of hydroformylation and reduction could follow more closely those of the cobalt system (Sections X-E-1 to E-3). Steps there involving reaction with molecular hydrogen could here be involved instead with a carbonyl hydride (see eqs. 250, 260a, 278, etc.) or a proton (eq. 97f).

Hashimoto and Shiina (949), Misono and coworkers (48, 201), and Frankel's group (198) reported that iron pentacarbonyl was effective for the hydrogen reduction of polyunsaturated fats. Stable diene and triene-iron tricarbonyl complexes were isolated later during related studies, and these were also effective hydrogenation catalysts (47, 48, 199, 200, 202–205). Ferric acetylacetonate (144, 145) and Et_3Al-ferric systems (206, 207) also have been used for the hydrogenation of fats, a topic that is considered separately in Section XIV-E.

The diene- and triene-$Fe(CO)_3$ complexes are good model systems for detailed studies of catalytic hydrogenation, and more recently, Frankel and Cais and their coworkers (81, 180, 209, 210) have reported studies on these complexes, monoene-iron tetracarbonyl complexes, and iron pentacarbonyl for the hydrogenation of methyl sorbate (*trans*-2, *trans*-4-hexadienoate). The reactions are carried out in benzene or toluene solution at about 175° and 50 atm hydrogen for catalyst concentrations $\sim 10^{-2}$ M and substrate concentrations of ~ 0.2 M. The reductions are not selective; a mixture of methyl 2-, 3-, and 4-hexenoates as well as methyl hexanoate (the major product at longer reaction times) is obtained (Fig. 3). Table 15 shows the activity of the complexes expressed as the half-life of the substrate relative to that of butadiene-$Fe(CO)_3$.

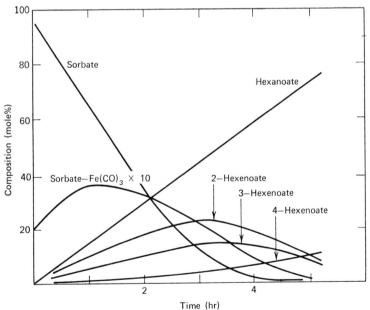

Fig. 3 Hydrogenation of methyl sorbate using methyl sorbate–$Fe(CO)_3$ catalyst; 175°, 50 atm H_2 (209).

Table 15   Activity of Iron Carbonyl Complexes for Methyl
Sorbate Hydrogenation (209)

Complex	Activity[a]
Sorbaldehyde—Fe(CO)$_3$	0.60
Sorbyl methyl ketone—Fe(CO)$_3$	0.60
Dimethyl muconate—Fe(CO)$_3$	0.60
1,4-Diphenylbutadiene—Fe(CO)$_3$	0.90
Butadiene—Fe(CO)$_3$	1.0
Methyl sorbate—Fe(CO)$_3$	1.0
Sorbamide—Fe(CO)$_3$	1.0
Maleic anhydride—Fe(CO)$_4$	1.0
Dimethyl fumarate—Fe(CO)$_4$	1.0
Norbornadiene—Fe(CO)$_3$	1.3
2,4-Hexadiene—Fe(CO)$_3$	1.6
2,4-Dimethyl-2,4-hexadiene—Fe(CO)$_3$	1.9
Sorbyl alcohol—Fe(CO)$_3$	1.9
Sorbic acid—Fe(CO)$_3$	3.2

[a] Activity expressed as half-life of substrate relative to that
of the butadiene—Fe(CO)$_3$ system ($\sim$1.5 hr).

The same product distribution was obtained for catalysis by Fe(CO)$_5$ or by
norbornadiene-Fe(CO)$_3$, and infrared analysis, in the carbonyl stretch region,
detected the formation of methyl sorbate–Fe(CO)$_3$ as an intermediate in
both systems. These ligand-exchange reactions, shown in eqs. 98 and 99, are
an important feature of the hydrogenation; they have been studied separately
under N$_2$ as well as under hydrogenation conditions. L = conjugated diene

$$L \cdot Fe(CO)_3 + D \longrightarrow D \cdot Fe(CO)_3 + L \qquad (98)$$
$$L' Fe(CO)_4 + D \longrightarrow D \cdot Fe(CO)_3 + CO + L' \qquad (99)$$

ligand; L$'$ = monoene ligand; D = diene substrate.

Reports (180, 210) indicate that reaction 98 is first order in both complex
and substrate (data are given for L = methyl sorbate, D = diphenyl-
butadiene), while reaction 99 is first order in complex and independent of
substrate (data are given for L$'$ = dimethyl fumarate, D = methyl sorbate).
An earlier report (209), however, states that reaction 98 for D = methyl
sorbate is second order for some systems (e.g., L = sorbyl alcohol) and first
order for others (e.g., L = norbornadiene). The data reported in the abstracted reports are insufficient for any definite conclusions to be drawn; both
eqs. 98 and 99 probably involve predissociation reactions followed by subsequent reaction with D. The dependence on D will be first-order at low concentrations and decreasing to zero order at higher concentrations of D.

Table 15 shows that in a series of substituted butadiene–$Fe(CO)_3$ complexes, CH_3—CH=CH—CH=CH—$R \cdot Fe(CO)_3$, the activity varies with R in the order $COCH_3 \simeq CHO > CONH_2 \simeq COOCH_3 > CH_3 > CH_2OH > COOH$. Except for hydroxyl and carboxylic acid groups, electron-withdrawing substituents increased activity, while the electron-repelling methyl substituent decreased it. Electron-withdrawing groups facilitate dissociation of the $Fe(CO)_3$ moiety by weakening the π-bond with the diene; the methyl group would have the opposite effect. Other data from studies on the hydrogenation of fats, using iron carbonyl complexes (Section XIV-B), and on the sorbate reduction, using similar chromium complexes (181, 183; Section VII-A), had shown such a dissociation reaction to be important. The —OH and —COOH groups are thought to retard dissociation of a diene–$Fe(CO)_3$ complex through intramolecular hydrogen bonding with the metal (81, 209).

When methyl sorbate–$Fe(CO)_3$ was used as catalyst for the reduction of pure hexenoate isomers, the relative rates were in the order (81) of 2-hexenoate $\gg$ 3-hexenoate > 4-hexenoate; their rates of isomerization were in the opposite order. The hexanoate is produced principally from the 2-hexenoate, and the product distributions (Fig. 3) indicate that the 3- and 4-hexenoates first isomerize to 2-hexenoate before further reduction. A scheme in which hydrogen reduction of a sorbate–$Fe(CO)_3$ complex gives mainly 2-hexenoate and smaller amounts of the 3- and 4-isomers is said to account for the kinetic data of Fig. 3. The overall mechanism postulated is shown in eqs. 100, 101 and Scheme 1 (81, 209):

$$\text{olefin–}Fe(CO)_3 \rightleftharpoons Fe(CO)_3 + \text{olefin} \qquad (100)$$

$$Fe(CO)_3 + \text{sorbate} \longrightarrow \text{sorbate–}Fe(CO)_3 \qquad (101)$$
$$\mathbf{6}$$

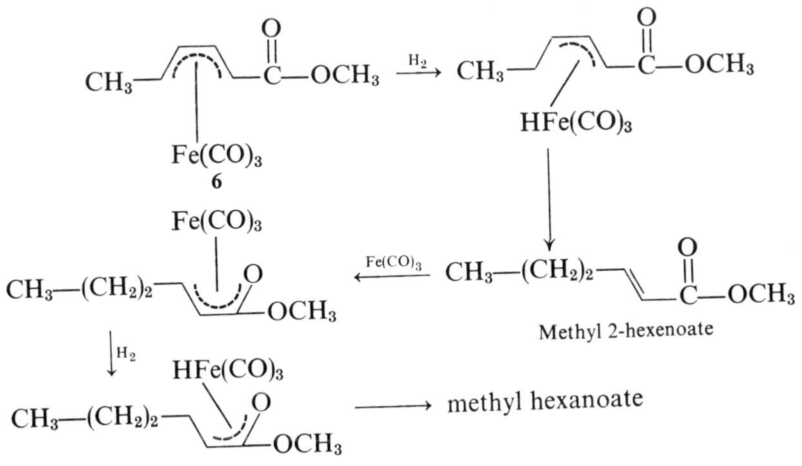

Scheme 1

The reduction of sorbate–Fe(CO)$_3$ is thought to occur by a π-allyl HFe(CO)$_3$ species which is stabilized by the carbonyl group to give 2-hexenoate as preferred product. Further reduction proceeds by an allyl hydride complex and involves the ester carbonyl. The isomerization would involve reversal of the reaction producing this π-allyl hydride complex; the isomerization occurs at a lower rate under purely nitrogen atmospheres (180). Von Rosenberg and coworkers (1321) have reported n.m.r. evidence for π-allylhydridoirontricarbonyl intermediates in some related isomerization reactions.

These iron carbonyl systems contrast with those involving arene-Cr(CO)$_3$ complexes that give rise to selective reduction of methyl sorbate to methyl 3-hexenoate (178–183; Section VII-A). Frankel accounts for the difference in terms of the given mechanisms (81); with the iron systems, activation of substrate occurs before activation of hydrogen, while the reverse seems true for the chromium systems (eqs. 82 to 84). Cais (180), however, suggests that both occur through prior activation of substrate (eqs. 79 to 81) and that the selectivity of the chromium systems results from the negligible rates of isomerization observed with these catalysts (Section XIV-D). These two general types of mechanisms for catalytic hydrogenation have been considered by a number of workers and will be discussed again, particularly in Section XI-B-1.

Iron pentacarbonyl has been used by Levering (211) for the catalytic hydrogenation of nitriles to a mixture of primary and secondary amines at 200° and ∼30 atm hydrogen. The nitriles themselves were used as the solvent media—acetonitrile was converted in about 10 hr to ethylamine (66%) and diethylamine (22%); benzonitrile gave benzylamine (51%) and dibenzylamine (45%); propionitrile gave propylamine (53%) and dipropylamine (41%); and adiponitrile gave hexamethylenediamine and secondary amines of it.

In a search for catalysts for the oligomerization of vinyl monomers, Misono and coworkers (212) have observed that acrylonitrile is converted into propionitrile and hydrodimers, adiponitrile and α-methylglutaronitrile, by a catalyst system consisting of iron pentacarbonyl with sodium borohydride or sodium hydroxide–methanol mixtures under either nitrogen or hydrogen atmospheres:

$$CH_2{=}CHCN \xrightarrow[\text{N}_2 \text{ or H}_2]{\text{catalyst}} CH_3CH_2CN,\ NC(CH_2)_4CN,\ NC(CH_2)_2CH(CH_3)CN$$
$$(102)$$

Dicobalt octacarbonyl gave less effective conversions, whether used in the same manner or on its own under hydrogen.

The reactions were carried out for 4 to 7 hr in benzene solution at 110 to 150° and about 100 atm pressure in the presence of hydroquinone as a polymerization inhibitor. Under hydrogen atmospheres, up to 50% of the acrylonitrile was reduced to propionitrile and up to 10% was converted to the hydrodimers; under nitrogen atmospheres the yields of the hydrodimers were similar but that of the propionitrile was greatly reduced ($\sim$5%). The production of propionitrile then apparently involves catalytic activation of molecular hydrogen with iron carbonyl hydride intermediates. Under nitrogen the hydride comes from borohydride, hydroxide, or solvent; for example:

$$Fe(CO)_5 + 3OH^- \longrightarrow HFe(CO)_4^- + CO_3^{2-} + H_2O \qquad (103)$$

Some iron carbonyl hydride complexes were prepared and reacted with acrylonitrile in methanol (50°, 24 hr), and up to 10% yields of propionitrile and the hydrodimers were obtained. The mononuclear species $HFe(CO)_4^-$ gave the branched hydrodimer while the polynuclear species $HFe_2(CO)_8^-$ and $HFe_3(CO)_{11}^-$ formed the linear one. The following mechanism was proposed:

$$-FeH + CH_2 = CHCN \longrightarrow -FeCH_2CH_2CN \longrightarrow CH_3CH_2CN$$

$$7 \quad \bigg\downarrow CH_2{=}CHCN \qquad\qquad\qquad\qquad (104)$$

$$-FeCHCH_2CH_2CH_2CN \longrightarrow NC(CH_2)_4CN$$
$$\underset{\textstyle CN}{|}$$

$$\mathbf{8}$$

The cyanoethyl complex **7** is formed by addition of acrylonitrile to a hydride complex; further addition of acrylonitrile to **7** gives a dicyanobutyl complex, **8**. The mode of fission of the iron/carbon σ bond to give the products was not discussed.

A similar hydrodimerization process has recently been reported by Trost and Bright (213), although a different mechanism was used. Reaction of pyracylene, **9**, with diiron nonacarbonyl in ether solution at 25° for 15 hr under nitrogen or argon produced 1,2-dihydropyracylene (**10**), the hydrodimer (**11**), and a $C_{14}H_8Fe_2(CO)_7$ complex; the yields were about 40, 50, and 8% respectively. This reaction is a further example of homogeneous hydrogenation not involving molecular hydrogen. No reduction products were obtained when pentane was used as the solvent medium; the ether is thought to be the source of hydrogen. Labile complexes such as $C_{14}H_8 \cdot Fe(CO)_4$ and

$$C_{14}H_8 \cdot Fe_2(CO)_7$$
$$+$$

9   **10**   **11**   **12**

$C_{14}H_8 \cdot Fe_2(CO)_6$ are thought to give rise to **10** and **11**; perhaps these complexes are thermally excited to a triplet state with electron transfer occurring from the highest occupied orbital to the lowest vacant orbital. Hydrogen abstraction from the ether by this triplet species would produce a radical such as **12**, which is probably complexed to the iron. Further hydrogen abstraction would produce **10**, while attack on a pyracylene molecule followed by hydrogen abstraction from solvent would give the dimer **11**.

The use of some binuclear complexes $(CO)_4M(PPh_2)_2Fe(CO)_3$, where M = Cr, Mo, W, as hydrogenation catalysts (1002), was mentioned in Sections VII-A and VII-B; the catalyst $[(OC)_4Fe(PPh_2)ClPd]_2$ is considered in Section XIII-B.

The hydride complexes $H_2Fe[PR(OEt)_2]_4$, where R = OEt or Ph, have been isolated when borohydride is used as the hydrogen source (991, 1345).

An anticipated $Fe(diphos)_2$ complex was isolated by Hata and Miyake (1324) as the hydride $HFe(C_6H_4PPh\text{-}CH_2CH_2PPh_2)(diphos)$ with the iron atom inserted into the C—H bond of a phenyl ligand (see Section IX-B-2b, Scheme 3). Reaction of the complex with hydrogen chloride yielded *trans*-HFeCl(diphos)_2. Other monohydrides are also known (e.g., Ref. 511).

A soluble Ziegler catalyst containing iron, $(C_4H_9)_3Al\text{-}Fe(acac)_3$, was first reported in 1963 by Breslow's group (56) for the reduction of cyclohexene. Since then a number of workers (147, 163, 164, 167, 168, 175, 177, 195) have used such catalysts (usually incorporating the acetylacetonate complex) for the reduction of olefinic bonds. These will be considered in Chapter XV.

Tulupov (196) has described qualitative studies on the hydrogen reduction of substrates such as 2-pentene, oleic acid, cyclohexanone, and benzene in ethanol solutions containing ferric stearate. A more detailed study of the reduction of cyclohexene was reported later (197) and will be considered in Section XVI-A.

Aqueous solutions of ferric salts and heptanoic acid solutions of ferric heptanoate are unreactive toward hydrogen up to 150° (25, 92).

B.   RUTHENIUM SPECIES

Reviews of certain aspects of ruthenium chemistry, in particular homogeneous catalysts, have been published by James (79e) and Schleitzer (79f).

1.   Reduction of Inorganic Substrates.   The first reports of activation of molecular hydrogen by ruthenium complexes appeared in 1961. Halpern's group (214) showed that ruthenium(III) chloride in aqueous hydrochloric acid catalyzed the hydrogen reduction of iron(III) and ruthenium(IV) substrates at up to 1 atm and around 80° (see also Section XVI-B). In a constant-pressure apparatus, linear uptake plots were observed for the iron system; the gas uptake terminated abruptly at stoichiometries corresponding to reduction to iron(II). Ruthenium(IV) is reduced stoichiometrically to ruthenium(III), and the uptake plot is autocatalytic. The rate law observed for both systems was of the form

$$\frac{-d\,[H_2]}{dt} = k_1[H_2][Ru^{III}] \tag{106}$$

where the measured k_1 was the same for both systems. The mechanism was written as a rate-determining step that involved heterolytic splitting of hydrogen to produce an intermediate hydride, which subsequently reacted with the substrate in a faster step:

$$Ru^{III} + H_2 \underset{k_{-1}}{\overset{k_1}{\rightleftharpoons}} Ru^{III}H + H^+ \tag{107}$$

$$Ru^{III}H + 2Fe^{III} \xrightarrow{k_2} Ru^{III} + 2Fe^{II} + H^+ \tag{108}$$

or

$$Ru^{III}H + Ru^{IV} \xrightarrow{k_3} 2Ru^{III} + H^+ \tag{109}$$

The solution at the completion of these reductions contained the initial ruthenium(III) species, which was practically unreactive toward hydrogen itself over the reaction times required for substrate reduction, that is, $k_2 > k_{-1} > k_1$. $\Delta H^{\ddagger}$ and $\Delta S^{\ddagger}$ were estimated to be 23.1 kcal mole^{-1} and $+6$ eu, respectively, for the k_1 step in 3 M HCl. At this acid concentration, a mixture of $Ru(H_2O)_2Cl_4^-$, $Ru(H_2O)Cl_5^{2-}$, and $RuCl_6^{3-}$ is present in solution (215); the reaction rate was independent of the acid concentration between 3 M and 6 M, suggesting that the species had similar reactivities. An observed lower rate at lower acidities was possibly caused by the presence of some hydrolyzed species of lower reactivity.

In the absence of a substrate, the reversibility of reaction 107 has been demonstrated by Halpern and James (29, 216) and Schindewolf (217, 218), who used isotopic exchange between deuterium and water. The exchange followed rate law 106, and when using D_2 gas the total exchange rate $(d[HD]/dt + d[H_2]/dt)$ agreed well with the rate of deuterium reduction of ferric under the same conditions (216, 217). The HD and H_2 species are produced by the following reactions (written for $RuCl_6^{3-}$):

$$RuCl_6^{3-} + D_2 \xrightarrow{k_1} DRuCl_5^{3-} + D^+ + Cl^- \tag{110}$$

$$DRuCl_5^{3-} + H^+ + Cl^- \xrightarrow{k_{-1}} RuCl_6^{3-} + HD \tag{111}$$

$$DRuCl_5^{3-} \xrightarrow[H_2O]{k_4} HRuCl_5^{3-} + HDO \tag{112}$$

$$HRuCl_5^{3-} + H^+ + Cl^- \xrightarrow{k_{-1}} RuCl_6^{3-} + H_2 \tag{113}$$

Isotope exchange between D_2 and H_2O occurs through reversal of step 110 (i.e., through steps 111 and 113). The D^+ released in step 110 undergoes complete exchange to give H^+, whereas the $DRuCl_5^{3-}$ complex exchanges more slowly by step 112. The total exchange rate is determined by step 110 while the ratio of primary exchange products is determined by competition between reactions 111 and 112,

$$\frac{[HD]}{[H_2]} = \frac{k_{-1}[H^+][Cl^-]}{k_4} \tag{114}$$

and this product ratio does increase with the concentrations of acid and chloride (216). The exchange was also studied by using gaseous hydrogen containing 0.2% HD and then measuring the loss of HD in the gas phase (218). Exchange studies were carried out between 75 to 115° in 4 M HCl to yield rate constants and activation parameters (217, 218) $\Delta H^{\ddagger} = 24$ kcal mole^{-1} and $\Delta S^{\ddagger} = +7$ eu, which were in excellent agreement with the data from the ferric reduction. In the presence of substrate the hydride (deuteride) species are removed rapidly and in a manner similar to a reaction such as 108; no exchange is observed.

The rate of parahydrogen conversion catalyzed by chlororuthenate(III) was 1.5 times faster than the deuterium exchange (217), while a kinetic isotope effect of ~1.1 $[k_1(H_2)/k_1(D_2)]$ was reported for the catalyzed ferric reduction (216).

Aqueous acid solutions of chlororuthenate(III) complexes do react with hydrogen at 80° to give blue solutions containing chlororuthenate(II) species but with uptake rates much slower than those of ferric reduction or exchange (217, 219). No kinetic data are reported for this reaction, but it likely proceeds by reaction 107 followed by reaction 115 (214):

$$Ru^{III}H + Ru^{III} \xrightarrow{k_4} 2Ru^{II} + H^+ \tag{115}$$

where $k_{-1} > k_4$.

James and Hui (220–222) have extended these studies to the activation of hydrogen by ruthenium chlorides in some nonaqueous solvents. In dimethylacetamide at room temperature and 1 atm hydrogen, autocatalytic reduction of ruthenium(IV) to ruthenium(III) occurred rapidly and was followed by a slow stoichiometric reduction to ruthenium(II). Both stages are first-order in ruthenium(III) and hydrogen and, as in the aqueous acid system, an intermediate $Ru^{III}H$ is thought to be formed through heterolytic splitting of hydrogen (reaction 107), followed by reaction 109 or 115, consecutively. The more basic solvent system promotes reaction 115 by stabilization of the released proton. At higher temperatures (~60°) a further "base-promoted" reaction occurred in which the blue ruthenium(II) solutions absorbed 0.5 mole H_2 per mole Ru^{II}, the process again being first-order in hydrogen and ruthenium(II). There was no evidence for the presence of a metal-hydrogen bond in the resulting brown solutions; they are thought to contain ruthenium(I) species produced by the following reactions

$$Ru^{II} + H_2 \rightleftharpoons Ru^{II}H + H^+ \qquad (116)$$

$$Ru^{II}H + Ru^{II} \longrightarrow 2Ru^I + H^+ \qquad (117)$$

Support for this mechanism comes from the observation that in the presence of triphenylphosphine, a ruthenium(II) solution absorbs 1.0 mole of H_2 per mole Ru^{II} to produce solutions containing the species $HRuCl(PPh_3)_3$; the intermediate hydride is stabilized and no reduction to ruthenium(I) occurs. The kinetics were again first-order in hydrogen and ruthenium(II) with a measured rate constant in good agreement (Table 16) with that for the reaction producing ruthenium(I), thus indicating a common rate-determining step (reaction 116).

Further data indicating the existence of the ruthenium(I) species in solution resulted from stoichiometric oxidation with oxygen to ruthenium(III) and reaction with carbon monoxide to give a $Ru^I(CO)_2$ species which can be oxidized stoichiometrically to $Ru^{II}(CO)_2$. Spectrophotometric data on the brown solutions and the isolation of some incompletely characterized complexes (some with phosphine ligands) indicate that the ruthenium(I) species are dimeric (see also Section B-2c). Addition of 2,2'-dipyridyl to the brown solutions yielded ruthenium metal and ruthenium(II) dipyridyl complexes in amounts according to the disproportionation shown in eq. 118 (223).

$$2Ru^I \longrightarrow Ru^0 + Ru^{II} \qquad (118)$$

Besides the hydrogen reduction of ruthenium(IV) and ruthenium(III), ruthenium(III) chloride in DMA also catalyzed the reduction of iron(III) to iron(II), and molecular oxygen to water. A second-order rate law such as 106 was again observed (221). In the less basic solvent ethanol, reduction of ruthenium(III) chloride by hydrogen proceeded no further than the divalent stage (221).

The complete rate law for a mechanism involving reaction 107 followed by reaction 119

$$Ru^{III}H + substrate \xrightarrow{k} products + H^+ \qquad (119)$$

is shown in eq. 120:

$$\frac{-d[H_2]}{dt} = k_1[H_2][Ru^{III}]\left\{1 - \frac{k_{-1}[H^+]}{k_{-1}[H^+] + k[substrate]}\right\} \qquad (120)$$

The second term in the bracket represents the competition of the hydride to react with the reducible substrate (reaction 119) or reproduce H_2 from the

Table 16 Summary of Kinetic Data for Activation of Hydrogen by Ruthenium Chlorides: $Ru^{n+} + H_2 \rightleftharpoons RuH^{(n-1)+} + H^+$; $d[H_2]/dt = k_{obs}[H_2][Ru^{n+}]$

Catalyst	Solvent	Reaction Studied	Temperature (°C)	k_{obs} ($M^{-1}\,s^{-1}$)	$\Delta H^\ddagger$ (kcal mole^{-1})	$\Delta S^\ddagger$ (eu)	Reference
Ru^{III}	Aqueous 3 M HCl	$Fe^{III} \rightarrow Fe^{II}$	65–85	1.00(80°)	23.1	+6	214
Ru^{III}	Aqueous 3 M HCl	$Ru^{IV} \rightarrow Ru^{III}$	80	0.98	—	—	214
Ru^{III}	Aqueous 4 M HCl	D_2 exchange	75–115	1.43(85°)	24	+7	217, 218
Ru^{III}	Aqueous 3 M HCl	$Ru^{III} \rightarrow Ru^{II}$	80	<0.05	—	—	214, 219
Ru^{III}	DMA	$Ru^{IV} \rightarrow Ru^{III}$	30–55	1.60(55°)	19.9	+3	221
Ru^{III}	DMA	$O_2 \rightarrow H_2O$	30–70	2.20(55°)	19.3	+2	221
Ru^{III}	DMA	$Ru^{III} \rightarrow Ru^{II}$	65–80	0.23(80°)	15.5	−16	221
Ru^{II}	DMA	$Ru^{II} \rightarrow Ru^{I}$	60–80	0.04(60°)	16.6	−13	221
Ru^{II}	DMA	$Ru^{II} \rightarrow Ru^{II}H$	60	0.055	—	—	221
$Ru^{II}(CO)$	Aqueous 5 M HCl	$Ru^{II}(CO) \rightarrow Ru^{II}(CO)$	80	0.22	—	—	225
$Ru^{II}(CO)$	Aqueous 5 M HCl	D_2 exchange	80	0.18	—	—	225
$Ru^{II}(CO)_2$	Aqueous 5 M HCl	D_2 exchange	80	no exchange	—	—	225

back-reaction of 107. Some kinetic data for the ruthenium chloride–catalyzed systems are summarized in Table 16. The ruthenium(IV) and iron(III) systems in aqueous acid and the ruthenium(IV) and oxygen systems in DMA give essentially the same rate constants and activation parameters, suggesting that these substrates oxidize the $Ru^{III}H$ rapidly; thus k is large and rate law 120 reduces to its simple form of 106, and the kinetic data do refer to the k_1 constant of reaction 107. The observed second-order rate constants for the ruthenium(III) reduction in both solvent systems are very much lower, and this must result from the reversal of step 107 competing effectively with step 119; compared with ruthenium(IV), iron(III), and oxygen, ruthenium(III) is a weaker oxidizing agent and in fact ruthenium(II) is a strong reductant (224). The kinetic data imply that the k[substrate] term is small compared with the $k_{-1}[H^+]$ term for the ruthenium(III) reduction, and hence the first-order dependence on ruthenium(III) is still observed. An inverse acid dependency observed for the ruthenium(III) reduction in DMA (221) is consistent with this reasoning.

The promotion of the reaction of hydrogen with ruthenium chlorides in DMA results from a lower activation energy (Table 16). The dielectric constants for DMA and 3 M HCl are both in the range 30 to 40, and a more important factor is likely to be the greater basic strength (coordinating ability) of DMA compared with water. Such an effect could be realized in stabilization of the released proton and/or assistance in the dissociation of the initial ruthenium(III) complex in reaction 107.

Aqueous acid solutions containing the ruthenium(III) anion, $Ru(CO)Cl_5^{2-}$, are reduced by hydrogen (80°, 1 atm) in an autocatalytic manner to a ruthenium(II) chlorocarbonyl complex, according to eq. 121 (225).

$$2Ru(CO)Cl_5^{2-} + H_2 + 2H_2O \longrightarrow 2Ru(CO)(H_2O)Cl_4^{2-} + 2H^+ + 2Cl^- \quad (121)$$

The autocatalysis is ascribed to the activation of hydrogen by the ruthenium(II) product:

$$Ru^{II}(CO) + H_2 \overset{k}{\rightleftharpoons} Ru^{II}(CO)H + H^+ \quad (122)$$

$$Ru^{II}(CO)H + 2Ru^{III}(CO) \overset{fast}{\longrightarrow} 3Ru^{II}(CO) + H^+ \quad (123)$$

As for the ruthenium(III) chloride system in aqueous acid, the $Ru(CO)(H_2O)Cl_4^{2-}$ complex catalyzes the isotopic exchange of deuterium with water (eq. 122). The value of k estimated from the exchange rate (0.18 M^{-1} s^{-1}) agrees with that calculated from analysis of the autocatalytic hydrogen uptake plots (0.22 M^{-1} s^{-1}) when allowance is made for a small isotopic effect. Solutions containing $Ru(CO)_2Cl_4^{2-}$ did not catalyze the D_2-H_2O exchange (225).

Unlike the $RuCl_5(H_2O)^{2-}/RuCl_6^{3-}$ complexes, the $RuCl_4(bipy)^-$ complex did not react with hydrogen or catalyze the hydrogen reduction of ferric in aqueous acid solutions at 80° and 1 atm pressure (221), and this is attributed to the lower relative lability of chloride in the bipyridyl complex which is required for hydride formation (cf. eq. 110). The bipyridyl complex of ruthenium(II), $RuCl_4(bipy)^{2-}$, was however reduced by hydrogen to the metal under similar conditions; the chloride is more labile because of the higher negative charge and hydride formation is possible:

$$RuCl_4(bipy)^{2-} + H_2 \longrightarrow HRuCl_3(bipy)^{2-} + H^+ + Cl^- \qquad (124)$$

The hydride intermediate, a catalyst for olefin hydrogenation (see Section IX-B-2), decomposes to metal in the absence of substrate (221, 230a)

$$HRuCl_3(bipy)^{2-} \longrightarrow Ru^0 + H^+ + bipy + 3Cl^- \qquad (125)$$

2. Reduction of Organic Substrates. *a. Chlororuthenate(II) and Tetrachloro(bipyridyl)ruthenate(II) Systems.*

Aqueous hydrochloric acid solutions containing chlororuthenate(II) complexes (thought to be $RuCl_4^{2-}$) were reported by Halpern and coworkers to be active for the hydrogenation of maleic, fumaric, and acrylic acids at 65 to 90° and hydrogen pressures up to 1 atm (27, 219). The initial blue solutions, produced by titanous reduction of ruthenium(III, IV) solutions, rapidly turned yellow in the presence of excess unsaturated organic acids, and absorbed hydrogen according to a simple rate law

$$\frac{-d[H_2]}{dt} = k[H_2][Ru^{II}]_T \qquad (126)$$

Spectrophotometric studies showed that the yellow solutions contained 1:1 ruthenium-olefin complexes of moderate stability (Table 17). Activation of the double bond by the presence of an adjacent carboxyl group seemed essential for hydrogenation, since a number of other olefins (such as ethylene, propylene, and 5-norbornene-2,3-dicarboxylic anhydride) formed 1:1 complexes but were not hydrogenated. The two reaction steps 127 and 128

$$Ru^{II} + olefin \overset{K}{\rightleftharpoons} Ru^{II}(olefin) \qquad (127)$$

$$Ru^{II}(olefin) + H_2 \overset{k}{\longrightarrow} Ru^{II} + saturated\ product \qquad (128)$$

accounted for the kinetics of eq. 126. Linear uptake plots were measured in a constant-pressure apparatus, the rate falling off near complete reduction because of the remaining low olefin concentration and production of uncomplexed blue chlororuthenate(II) species. The total hydrogen uptake was stoichiometric for olefin reduction. There was an accompanying but very much slower isomerization of maleic to fumaric acid during the hydrogenation of the former. Hydrogenation of fumaric or maleic acid with D_2 in H_2O solution yielded undeuterated succinic acid while H_2 (or D_2) in D_2O solutions

Table 17 Catalytic Hydrogenation of Monoenes Using Chlororuthenate(II) Complexes in 3 M HCl: $-d[H_2]/dt = k[H_2][Ru^{II}]$

Reducible Substrates	Reaction Observed	Ru^{II}-Olefin Complex[a]	k at 80° ($M^{-1}\,s^{-1}$)	$\Delta H^\ddagger$ (kcal mole^{-1})	$\Delta S^\ddagger$ (eu)	Reference
Fumaric acid	Red. to succinic acid	$K_1 = 2 \times 10^3\,M^{-1}$	3.6	17	−8	219
Maleic acid	Red. to succinic acid	$K_1 = 5 \times 10^3\,M^{-1}$	2.3	14	−17	219
Acrylamide	Red. to propionamide	$K_1 = 6 \times 10^3\,M^{-1}$	7.2	10	−26	228
Acrylic acid	Red. to propionic acid	$K_1 = 3 \times 10^4\,M^{-1}$	5.9	15	−12	229
Crotonic acid	Red. to butyric acid	$K_1 = 2 \times 10^3\,M^{-1}$	3.7	18	−3	229

Nonreducible Substrates	Reaction Observed (No Hydrogenation)	Complex Formation[a]	Reference
Ethylene	Complex + $H_2/C_2H_4 \rightarrow Ru^0$	$K_1 < 5 \times 10^3$	219, 230
Propylene	Complex + $H_2/C_3H_6 \rightarrow Ru^0$	Observed	219
Isobutene	Hydration and/or polymerization in absence of Ru^{II}	Not observed	219
Butadiene	Hydration and/or polymerization in absence of Ru^{II}	Observed	219
Dimethylmaleic anhydride	Complex + $H_2 \rightarrow Ru^0$	Weak complexing	219
5-Norbornene-2,3-dicarb- oxylic anhydride	Catalyzed D_2—H_2O exchange; $k^{80°} = 1.9\,M^{-1}\,s^{-1}$	$K_1 = 2 \times 10^2\,M^{-1}$	219
Vinyl fluoride	Catalytic hydration $\rightarrow CH_3CHO$; $+ H_2 \rightarrow$ no Ru^0	Observed	227
1,1-Difluoroethylene	Catalytic hydration $\rightarrow CH_3COOH$; $+ H_2 \rightarrow$ no Ru^0	Observed	227
Vinyl chloride	Polymerization in absence of Ru^{II}	—	227, 229
Vinyl bromide	Polymerization in absence of Ru^{II}	—	227, 229
Crotonaldehyde	Catalytic hydration and/or polymerization	Observed	229
Crotonitrile	Catalytic hydration and/or polymerization	Observed	229

[a] At 25°.

79

yielded 2,3-dideuterosuccinic acid, showing that the added hydrogen came from the solvent. Deuteration of fumaric acid gave mainly DL-2,3-dideutero-succinic acid, indicating that the addition was stereospecifically *cis*. The maleic and fumaric acid complexes that were hydrogenated showed no isotopic exchange between D_2 and H_2O, that is, neither HD nor H_2 appeared in the gas phase, whereas the norbornene complex that was not hydrogenated did catalyze this exchange at a rate similar to the hydrogenation rates (219, Table 17).

To account for these data reaction 128 was written in more detail as shown in reaction 129. The π-complex (**13**) reacts with hydrogen to form a hydrido-

$$(129)$$

ruthenium(II) complex (**14**) by heterolytic splitting of hydrogen; **14** re-arranges to a σ-alkyl complex (**16**) by "insertion" of the olefin into the metal-hydride bond; electrophilic attack on the metal-bonded carbon atom of **16** by a proton yields the saturated product and regenerates ruthenium(II). The insertion reaction by a four-center transition state (**15**) involves *cis* addition, and hence *cis* addition of hydrogen in the overall reaction requires that step 3 occurs with retention of configuration at the metal-bonded carbon. This result is considered reasonable since it accompanies the electrophilic displacement of carbon from other metals such as mercury or tin (226, 226a; but see also Section XII-C, eq. 410). The fact that the added hydrogen originates in the solvent implies that **14** exchanges its hydrogen with the solvent before undergoing rearrangement; such exchange is consistent with the behavior exhibited by hydridochlororuthenate(III) complexes (see eq. 112, Section IX-B-1).

The requirement of a carboxyl substituent was rationalized in terms of its electron-withdrawing power favoring the nucleophilic attack of the hydrido ligand on the double bond for the hydrometallation $(14 \rightarrow 16)$. Hydrogenation proceeds when this response successfully competes with the reversal of step 1 (the latter constituting the mechanism of D_2-H_2O exchange). To test this hypothesis further, James and coworkers have extended these studies by using other substituted ethylenes (227–229), although the solubility requirements for the aqueous media restricted the range that could be investigated. The data obtained together with that of the earlier work are summarized in Table 17. Besides substrates with carboxyl substituents, hydrogenation was observed for acrylamide with its electron-withdrawing substituent $-CONH_2$. No clear trend between hydrogenation rate and stability of the olefin complex emerges from the limited data on the reducible substrates. Fluoroethylenes were not hydrogenated owing to a competing catalytic hydration (227), the mechanism of which is thought to be very similar to that shown for hydrogenation in eq. 129 but with the hydride ion replaced by hydroxide. Nucleophilic attack by coordinated hydroxide at the carbon atom attached to the fluorine atom(s) predominates over hydride formation, presumably owing to the high electronegativity of the fluorine atom.

Unlike ethylene and propylene, however, the fluoroethylenes do stabilize the ruthenium(II) sufficiently to prevent reduction to the metal. Other olefins containing the electron-withdrawing chloro, bromo, aldehyde, or nitrile substituents were not hydrogenated; hydration and/or polymerization reactions were observed. Similar reactions were noted with butadiene as substrate in the hydrochloric acid media, although Brennan (229a) has reported that ruthenium(II) chloride catalyzes the reduction of conjugated dienes (including butadiene) to monoenes in 3 M sodium chloride solution at 25° under 5 atm hydrogen.

Rose and Wilkinson (1398) have recently presented evidence that the blue ruthenium(II) chloride solutions in water, 10 M HCl, methanol, and dimethylformamide are likely to contain the cluster anion $Ru_5Cl_{12}^{2-}$, although it is difficult to interpret the magnetic behavior of the systems, which show about one unpaired electron per five metal atoms for the ruthenium(II) d^6 configuration. These workers also note that during complex formation with ethylene in 3 M HCl at 80° (230), $\sim 5\%$ of the ruthenium is oxidized by the water to the trivalent state (1398); this oxidation precludes n.m.r. study of the "ethylene" complex, which could, in fact, possibly be an ethyl complex formed by protonation. This would imply that the rate-determining step in the hydrogenations could involve alkyl species (see eq. 135 in Section B-2b):

$$Ru - alkyl + H_2 \xrightarrow{k} RuH + product \qquad (129a)$$

$$RuH + olefin \xrightarrow{fast} Ru - alkyl \qquad (129b)$$

However, it is difficult to reconcile the data from the deuterium studies with this mechanism; hence, the mechanism outlined in eqs. 127 to 129 is preferred.

Hui and James (230a) have shown that hydrochloric acid solutions containing the tetrachloro(bipyridyl)ruthenate(II) anion, $RuCl_4(bipy)^{2-}$, catalyze the hydrogenation of maleic acid (MA) at 80°, 1 atm pressure. At catalyst concentrations of $\sim 10^{-3}\ M$ with a tenfold excess of olefin, the uptake plot, measured at constant pressure, showed initial autocatalysis leading to a region of linear rate; complete hydrogenation took several hours. The initial red $RuCl_4(bipy)^{2-}$ solution turned yellow in the presence of excess substrate, and spectrophotometric measurements indicated a 1:1 complex; then toward the end point of the hydrogen uptake, metal started to precipitate. These observed visible changes follow closely those of the chlororuthenate(II) system. But the autocatalytic region and the kinetics measured in the linear region, which were first-order in catalyst and hydrogen and between zero- and first-order in maleic acid (decreasing order with increasing concentration), indicated differences from the mechanism outlined in 127 to 129. Scheme 2 was suggested.

$$[RuCl_4(bipy)]^{2-} + MA \xrightleftharpoons{K_1} [RuCl_3(bipy)(MA)]^- + Cl^-$$

$$\quad\ \ \mathbf{17} \qquad\qquad\qquad\qquad\qquad\qquad\quad \mathbf{18}$$

$$+H_2 \downarrow k_1$$

$$[RuCl_3H(bipy)]^{2-} \xrightleftharpoons[K_2]{+MA} [RuCl_2H(bipy)(MA)]^-(+Cl^-)$$

$$\mathbf{19} \quad +HCl$$

$$[RuCl_3H(bipy)]^{2-} \xleftarrow[k_2]{+H_2} [RuCl_2(bipy)(alkyl)]^-$$

$$+\ \text{succinic acid} \qquad\qquad\qquad\qquad \mathbf{20}$$

$$\textbf{Scheme 2}$$

The initially formed 1:1 complex **18** did not react with hydrogen. K_1 was estimated to be about 5×10^3, and the autocatalytic region was thought to form the active hydride intermediate, that is, **19** from **17**. In the absence of olefin, **17** is reduced very slowly by hydrogen to metal, presumably through **19** (eq. 125). Reaction of **19** with the olefin forms an alkyl complex **20**, which reacts with hydrogen in the rate-determining step to give product and to regenerate **19**. After hydride formation, the rate law becomes

$$\frac{-d[H_2]}{dt} = \frac{k_2 K_2 [H_2][\text{olefin}][Ru^{II}]_T}{1 + K_2[\text{olefin}]} \tag{130}$$

At 80° in 3 M HCl, k_2 was estimated to be 1.5 $M^{-1}\ s^{-1}$ and $K_2 \sim 200$.

Extension of these studies to substituted dipyridyl systems should give valuable information about the electronic and steric effects on the various steps involved as well as a more detailed understanding of the mechanism. The mechanism outlined in Scheme 2 corresponds to that for hydrogenations catalyzed by the better-characterized $RuCl_2(PPh_3)_3$ systems, which will be discussed in more detail in the following section, B-2b.

The scheme outlined in reaction 129 involved a ruthenium(II) hydride intermediate, and it is interesting that Chatt and Hayter (231) first reported the isolation of such stable hydrides of the type *trans*-HRuX(diphosphine)$_2$, X = halogen, or H, by reaction of *cis*-RuX$_2$(diphosphine)$_2$ with lithium aluminum hydride, although these *cis* compounds were too unreactive to react directly with hydrogen.

b.   Ruthenium(II) Phosphine Systems.   In contrast to the behavior of the chelated phosphine ruthenium(II) complexes (231), Wilkinson's group (44) reported in 1965 that the ruthenium(II) complexes $RuCl_2(PPh_3)_4$ and $RuCl_2(PPh_3)_3$ did react with hydrogen at room temperature in ethanol-benzene solution to give a hydrido complex, $HRuCl(PPh_3)_3$; this system was extremely efficient for the reduction of olefins and acetylenes at 25° and 1 atm hydrogen pressure and is the most active yet discovered for the hydrogenation of alk-1-enes. The $RuCl_2(PPh_3)_3$ complex was also made independently by Vaska who supplied LaPlaca and Ibers crystals for an X-ray diffraction study (232), which indicated that the sixth octahedral site was effectively blocked by a hydrogen atom of a phenyl ring. Extensive studies on hydrogen activation by dichlorotris(triphenylphosphine)ruthenium(II) systems and related ones have been reported since 1965 by a number of groups (221, 222, 233–245a, 641, 644, 660, 686).

Molecular weight measurements (233) in benzene solution showed that the triphenylphosphine complexes dissociate

$$RuCl_2(PPh_3)_4 \rightleftharpoons RuCl_2(PPh_3)_3S + PPh_3$$
$$(S = \text{solvent}) \qquad\qquad \big\updownarrow \qquad\qquad\qquad (131)$$
$$RuCl_2(PPh_3)_3 \rightleftharpoons RuCl_2(PPh_3)_2S_2 + PPh_3$$

Hydrogenation of hept-1-ene by the $RuCl_2(PPh_3)_3$ complex in benzene solution is very slow but in benzene-ethanol (1:1) becomes very rapid after a short initial induction period (234, 235), owing to formation of the active catalyst tris(triphenylphosphine)hydridochlororuthenium(II); this process produces a change in solution color from brown to violet. The ethanol acts as a base to promote formation of the hydride

$$RuCl_2(PPh_3)_3 + H_2 + \text{base} = HRuCl(PPh_3)_3 + \text{base HCl} \qquad (132)$$

A number of other bases, such as triethylamine, sodium phenoxide, and potassium hydroxide, are effective for forming the hydride *in situ* in pure benzene solution. **21** has been prepared stoichiometrically according to eq. 132 as a benzene solvate by such methods, but is more conveniently prepared from $RuCl_2(PPh_3)_3$ and sodium borohydride in benzene solution containing a little water (235). The crystal structure (236, 237) indicates a highly distorted trigonal bipyramid with the phosphine groups approximately equatorial and with the hydride and chloride in approximately axial positions but with the H-Ru-Cl angle much less than 180°—the distortions arise from the closeness of the hydrogen atoms in the phenyl rings. $RuCl_2(PPh_3)_3$ and $HRuCl(PPh_3)_3$ have very similar coordination about the metal atom (237).

In dilute (10^{-4} to 10^{-6} M) benzene or toluene solutions at 25° and 1 atm H_2, $HRuCl(PPh_3)_3$ selectively reduces $RCH{=}CH_2$ alkenes rapidly, but, compared with the findings of earlier work using $RuCl_2(PPh_3)_3$ in ethanol-benzene solution (44), does not effectively reduce acetylenes (235). Unsaturated aldehydes are not reduced (660). Wilkinson's group was unable to study in detail the kinetics of the hydrogenation reactions in benzene solutions because of the limited solubility of the catalyst, oxygen sensitivity, possible diffusion control in the very rapid reactions and some catalyst poisoning during a reaction. However, initial hydrogen uptake rates were measured for some alkene systems, and the data are summarized in Table 18. There was no accompanying isomerization during alkene reduction. Internal, cyclic, and substituted alk-1-enes were reduced very slowly. *Cis*- and *trans*-hexa-1,4-diene were also reduced slowly under similar conditions to those shown in Table 18, giving about a 30% yield of *cis*- and *trans*-hex-2-ene in one hr and essentially no isomerization of the unreacted diene.

The limited kinetic data presented (235) indicated between zero- and first-order in alkene concentration and inhibition by added triphenylphosphine.

Table 18   Hydrogenation Rates of Alkenes Relative to that of Hept-1-ene (235)[a]

Alkene	Initial Rate	Alkene	Initial Rate
Hept-1-ene	1.0[b]	*cis* + *trans*-Hex-2-ene	$<3 \times 10^{-4}$
Pent-1-ene	1.4	*cis* + *trans*-Oct-2-ene	$<3 \times 10^{-4}$
Hex-1-ene	1.6	Hept-3-ene	$<3 \times 10^{-4}$
Dec-1-ene	1.3	Cyclohexene	$<1 \times 10^{-3}$
		2-Methylpent-1-ene	$<3 \times 10^{-4}$

[a] Using $HRuCl(PPh_3)_3$ as catalyst; 5×10^{-5} mole catalyst, 0.07 mole alkene, 50 cm H_2, 25° in benzene, total volume 60 ml.
[b] Quoted in Ref. 235 as 260 ml min^{-1} (i.e., $\sim 7.5 \times 10^{-3}$ mole min^{-1}).

The mechanism suggested, based strongly on that indicated for a comparable rhodium complex, $HRh(CO)(PPh_3)_3$ (see Section XI-D-1) is shown in eqs. 133 to 135:

$$HRuCl(PPh_3)_3 \underset{21}{\overset{K_1}{\rightleftharpoons}} HRuCl(PPh_3)_2 + PPh_3 \quad\quad (133)$$

$$\underset{21}{} \quad\quad\quad\quad\quad \underset{22}{}$$

$$HRuCl(PPh_3)_2 + \text{alkene} \overset{K_2}{\rightleftharpoons} RuCl(PPh_3)_2(\text{alkyl}) \quad\quad (134)$$

$$\underset{23}{}$$

$$RuCl(PPh_3)_2(\text{alkyl}) + H_2 \overset{k}{\longrightarrow} HRuCl(PPh_3)_2 + \text{alkane} \quad (135)$$

The dissociation product, **22**, has *trans* phosphine ligands and reacts with the olefin to produce the alkyl intermediate **23** that reacts with hydrogen in the rate-determining step to give product and regeneration of the catalyst.

Similar studies using some hydridocarboxylatotris(triphenylphosphine) ruthenium(II) complexes, $HRu(OCOR)(PPh_3)_3$, have also been reported by Wilkinson's group (238); again high selectivity was observed for reduction of alk-1-enes (Table 19). Variation of the carboxylate group had little effect on the hydrogenation rate; acetate, benzoate, and salicylate were about two to three times more active than the propionate and trifluoroacetate. The latter was studied in more detail because of its lower sensitivity to oxidation. This trifluoro complex was less than 10% as active as the chloro complex for the hydrogenation of hex-1-ene in benzene, but kinetic measurements were again difficult for reasons previously outlined for the chloro complex. However,

Table 19 Hydrogenation Rates of Alkenes (238)[a]

Alkene	Initial Rate[b]	Alkene	Initial Rate[b]
Hex-1-ene	21.70	*trans*-Hexa-1,4-diene	1.26
Undec-1-ene	20.90	Cyclohexene	1.24
Hept-1-ene	16.38	Penta-1,3-diene	0.71
Allylbenzene	7.14	Cyclo-octene	0.54
Hexa-1,5-diene	4.45	Allyl alcohol	0.53

1-methylcyclohexene, *cis + trans*-hex-2-enes, *cis + trans*-hex-3-enes, *cis*-hept-2-ene, 2-methyl-pent-1-ene reduced very slowly (<0.1)

[a] Using $HRu(CF_3CO_2)(PPh_3)_3$; 0.625×10^{-3} M catalyst, 1.25 M alkene, 50 cm H_2 at 25° in benzene.
[b] The rates are expressed in ml min^{-1}; no solution volume was quoted but the data indicate ∼50 ml.

initial rates indicated a first-order dependence on both hydrogen and catalyst (up to its solubility limit). These data are consistent with the mechanism of reactions 133 to 135, if the initial dissociation of the complex is complete at the concentrations used. The rate law is the same as that given in eq. 130 but with k replacing k_2.

Rate law 130 is consistent with the olefin dependence noted for the hydridochloride system (235), but an apparent anomaly exists in the paper on the trifluoroacetate system (238) where the rate is said to be linearly dependent on hexene concentration, although a linear plot of reciprocal rate against the reciprocal of the hexene concentration is presented showing a positive intercept on the y-axis in agreement with relationship 130. The ratio of this intercept to the slope gives an approximate value for K_2 of 0.3 M^{-1} for hex-1-ene at 25°. This value indicates that measurable amounts of the alkyl intermediate should be produced according to an equation such as 134 at the substrate concentrations used. This is also true in the chloride system where the olefin dependence indicates that measurable amounts of alkyl should be present (i.e., the K_2[olefin] term in the denominator in 130 is significant). However, in neither system is there any evidence for reactions such as 134 under the hydrogenation conditions, although n.m.r. evidence indicated a reversible reaction with ethylene occurring at high pressure ($\sim$35 atm C_2H_4) in deuterated chloroform (235, 238):

$$(Ph_3P)_nLRuH + C_2H_4 \rightleftharpoons (Ph_3P)_nLRu(C_2H_5) \tag{136}$$

($L = Cl^-$ or $CF_3CO_2^-$).

The difficulties in the kinetic measurements clearly limit the analysis of the data obtained, and a further limitation is lack of data concerning the dissociation reaction 133 in the benzene solution. The molecular weight of the chloride could not be ascertained (235) but a determination for the trifluoroacetate indicated dissociation (238), although the appearance of a quartet hydride resonance in the n.m.r. spectrum in benzene solution indicated little dissociation. A similar discrepancy between molecular weight and n.m.r. data has been found for the rhodium catalyst, RhCl(PPh$_3$)$_3$ (see Sections XI-B-1 and XI-B-2). However, in the presence of triphenylarsine, no hydride resonance was detected for HRu(OCOCF$_3$)(PPh$_3$)$_3$, and this was interpreted in terms of line-broadening by an exchange process which must involve dissociation (238).

A crystal structure of the hydridoacetato complex (239) has shown it to be a highly distorted octahedron with a loosely held bidentate acetate group; the arrangement of the phosphorus atoms and hydridic hydrogen are the same as in the chlorocomplex.

Detailed kinetic and spectroscopic studies by James and Hui (221, 222) on the HRuCl(PPh$_3$)$_3$ system in DMA solution strongly support the mechanism outlined in eqs. 133 to 135. The catalyst is prepared *in situ* by reaction of hydrogen with the RuCl$_2$(PPh$_3$)$_3$ complex; the usual base required for the reaction (eq. 132) is not required in this basic solvent system. Catalyst solubility is not a limitation, and complete hydrogen uptake plots for reduction of maleic acid were measured over a range of temperature (30 to 45°) and catalyst, hydrogen, substrate, and phosphine concentrations. A solvated hydride, possibly HRuCl(PPh$_3$)$_3$(DMA), has been isolated (222a); hydride is not formed *in situ* in DMSO solution (222a).

The rate law for the mechanism is in general of the form shown in eq. 137,

$$\frac{-d[\text{H}_2]}{dt} = \frac{kK_2[\text{Ru}^{\text{II}}]_T[\text{olefin}][\text{H}_2]}{1 + K_2[\text{olefin}] + [\text{PPh}_3]/K_1} \tag{137}$$

where [olefin] and [PPh$_3$] refer to the *free* concentration of these species; in terms of total added concentration of these species this rate law becomes

$$\frac{-d[\text{H}_2]}{dt} = \frac{kK_2[\text{Ru}^{\text{II}}]_T[\text{olefin}]_T[\text{H}_2]}{1 + K_2[\text{olefin}]_T + [\text{PPh}_3]_T/K_1 - 2[\text{Ru}^{\text{II}}]_T/K_1} \tag{138}$$

Although this expression is complex, some limiting forms which exist in some of the experimental conditions used are relatively simple and the data can be analyzed accordingly. Kinetics studied at 35° were in complete agreement with this expression. The rate was first-order in hydrogen, between zero- and first-order in olefin; it showed an inverse dependence on triphenylphosphine. The dependence on ruthenium is a result of the conditions used, since reactions 133 and 134 occur at the hydrogenation temperature to an extent that depends on the ratio of olefin to phosphine concentrations. At high ratios, the initial red hydride color turns yellow, all the ruthenium is present as **23**, and the rate law reduces to a simple form giving a first-order dependence on ruthenium:

$$\text{Rate} = k[\text{H}_2][\text{Ru}^{\text{II}}]_T \tag{139}$$

At lower olefin concentrations, equilibrium 134 is not complete, and the effect of the dissociation equilibrium 133 is observed in the ruthenium dependence which becomes less than one at higher ruthenium concentrations in systems where no excess triphenylphosphine has been added. This is reflected in the [PPh$_3$]/K_1 term in eq. 137 which increases with increasing [Ru^{II}]$_T$. In the presence of a constant excess phosphine, a first-order dependence is again observed. The kinetic data at 35° was analyzed to give $K_1 = 0.03$ M, $K_2 = 80$ M^{-1} and the rate constant $k = 3.6$ $M^{-1}\text{s}^{-1}$ ($\Delta H^{\ddagger} = 15.8$ kcal mole^{-1}, $\Delta S^{\ddagger} = -4.5$ eu). K_1 and K_2 values, in close agreement with those from the kinetic data, were estimated from visible spectrophotometric measurements. The K_1 value shows that in DMA solution the catalyst at 10^{-2} M is about 80% dissociated in the absence of olefin and excess triphenylphosphine.

Ethylene was readily hydrogenated in the dimethylacetamide system (221) and, in contrast to the benzene system, reacted with the catalyst to give brown solutions, probably according to eq. 136, at the hydrogenation conditions (1 atm total pressure, 30°); it is possible that the hydride complex dissociates to a greater extent in DMA than in benzene. Benzene solutions of the violet hydridochloro and the yellow hydridocarboxylato complexes react with alk-1-ynes to produce brown solutions, but these are not effective for catalytic hydrogenation (235, 238).

Coordination of alkene is thought to take place before the hydrogen transfer step (reaction 134), which is considered (221, 235, 238) to proceed through a four-center transition state as suggested for the chlororuthenate(II) catalyzed systems (219, eq. 129). The low or zero rates of hydrogenation for nonterminal alkenes are attributed to difficulty in hydride transfer to the coordinated alkene caused by steric interaction of the triphenylphosphine groups (235, 238). This conclusion is based mainly on comparison with studies on the HRh(CO)(PPh$_3$)$_3$ system (Section XI-D-1). In the case of the nonreducible norbornadiene substrate, the hydrido-alkene complex, HRuCl(C$_7$H$_8$)(PPh$_3$)$_2$, was isolated (235).

Some observed hydrogen atom exchange and isomerization reactions (44, 235, 238) add further support to a mechanism involving an alkyl intermediate. In the absence of hydrogen, hex-1-ene is very slowly isomerized to hex-2-enes in benzene solutions of the hydridoruthenium(II) complexes at rates negligible to those of hydrogenation. Such a reaction occurs through equilibrium 134. The nonreducible alk-2-enes, however, show no isomerization, adding some support to the conclusion that hydride transfer in reaction 134 determines whether hydrogenation will proceed or not. A more rapid exchange of hydrogen/deuterium atoms between ruthenium and alk-1- and -2-enes was also observed and followed by the growth of the high-field proton resonance, when alkenes were added to deuterochloroform solutions of the deuteridochloro complex in an n.m.r. tube; such exchange can only proceed via alkyl formation (see also Section XI-D-1).

Hudson and coworkers (241) have studied the isomerization of pent-1-ene and the redistribution of deuterium in ethylene (*trans*-C$_2$H$_2$D$_2$) catalyzed by benzene/hydrochloric acid solutions of RuCl$_2$(PPh$_3$)$_3$ and benzene solutions of HRuCl(PPh$_3$)$_3$, and concluded that alkyl intermediates are involved; rates were higher when a hydride ligand is initially coordinated.

Of related interest is the observed phosphine ligand-metal hydrogen-transfer in the hydridochloro system first reported by Wilkinson's group

(235) and later studied by Parshall and coworkers (242) and Levison and Robinson (960). Equilibration of $HRuCl(PPh_3)_3$ in solution with 1 atm deuterium for long periods at 25° yields $DRuCl[(2,6-D_2C_6H_3)_3P]$; in the presence of excess triphenylphosphine, the system can be used for the selective catalytic *ortho*-deuteration of the phosphine. Proton n.m.r. studies establish the *ortho*-phenyl C—H bonds as the active ligand sites. The mechanism of this exchange, which has been observed for a number of transition-metal phosphine and phosphite complexes (242, 242a) is outlined in Scheme 3.

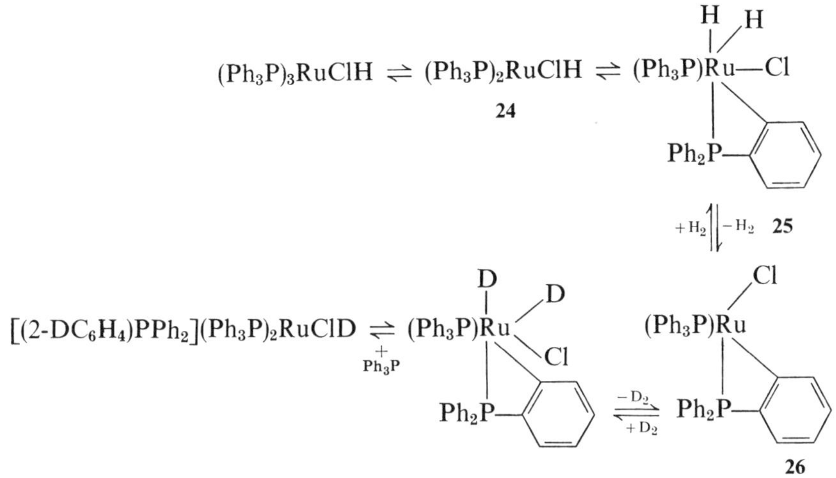

Scheme 3

The exchange involves reversible oxidative addition ($Ru^{II} \rightleftharpoons Ru^{IV}$) of the *ortho*-phenyl C—H bond to the coordination-deficient complex (**24** $\rightleftharpoons$ **25**). Direct evidence for the sequence **24** → **26** was obtained for a triphenyl-phosphite complex (242) where the reversible reaction 140 was observed. The hydrogen exchange with the phosphine ligand is very slow in comparison with

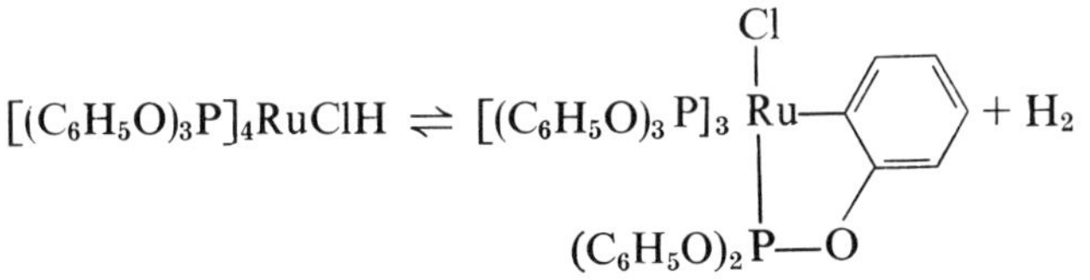

the hydrogenations catalyzed by $HRuCl(PPh_3)_3$ (235). Hydrogen abstraction from phosphine ligands coordinated to ruthenium was first established by

Chatt and Davidson (246) who demonstrated the tautomeric equilibrium shown in 140a for the 1,2-bisdimethylphosphinoethane complex. Related

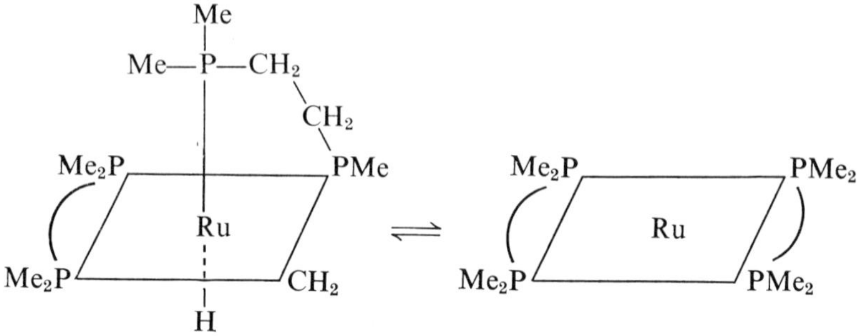

(140a)

is the hydrogen abstraction from a similar complex containing naphthalene (246, 246a):

$$Ru^0(C_{10}H_8)(Me_2PCH_2CH_2PMe_2)_2 \rightleftharpoons$$
$$cis\text{-}HRu^{II}(C_{10}H_7)(Me_2PCH_2CH_2PMe_2)_2 \quad (140b)$$

Such intramolecular aromatic substitution in transition metal complexes has been reviewed by Parshall (242a).

The rate-determining step in the hydrogenation reaction catalyzed by the ruthenium(II) hydrido complexes (reaction 135) is thought (221, 235, 238) to involve the reactions shown in Scheme 4. Either the oxidative addition of

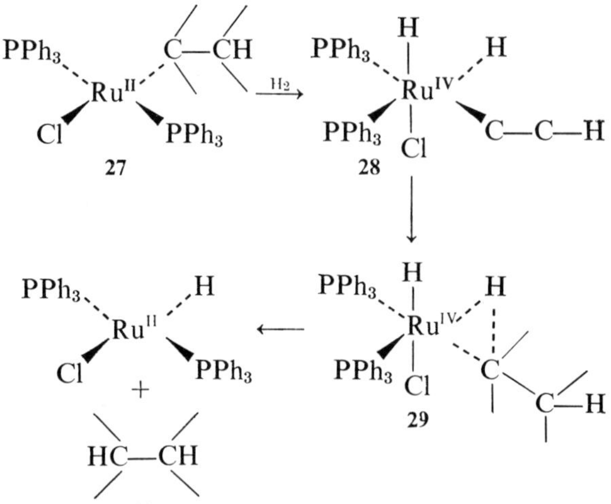

Scheme 4

molecular hydrogen to the square planar alkyl complex to form the octahedral ruthenium(IV) species, **28**, or the reductive elimination of alkane by hydrogen transfer (**28** → **29** → **30**) could be rate-controlling. The former is considered to be more likely; the activation parameters measured for the HRuCl(PPh)$_3$ system in DMA together with the small kinetic isotope effect, $k_H/k_D = 1.05$ (221), are thought to be consistent with such a process, on comparison with data for oxidative addition of hydrogen to square planar iridium(I) complexes (247). Hydride transfer to the coordinated alkyl would involve a three-center transition state, **29**. The possibility of reaction 135 occurring via a direct hydrogenolysis of the ruthenium-carbon bond through a four-center transition state was considered unlikely, although such a possibility cannot be completely ruled out (cf. Scheme 14, Section XIII-C-2). Indeed, one report by Wilkinson's group (234) on the HRuCl(PPh$_3$)$_3$ system suggested that the failure to hydrogenate internal olefins was due to steric hindrance in such a step, since these olefins did undergo hydrogen exchange with the catalyst. This is evidence against hydride transfer to the olefin (reaction 134) determining whether hydrogenation will occur or not.

Jardine and coworkers have reported (240, 240a) that ruthenium trichloride (10^{-2} moles) with triphenylphosphine (6×10^{-2} moles) in methanol solution catalyzes the hydrogenation (presumably at room temperature) of nor-bornadiene to norbornane, cycloalkenes to cycloalkanes, diphenylacetylene to *cis*-stilbene, and 3,5-dimethylhex-1-yn-3-ol to 3,5-dimethylhexan-1-ol. Nishimura and Tsuneda (245a) have selectively reduced 1,4-androstadiene-3,17-dione (and -3-ones) to the 4-monoene stage by using an analogous system in benzene, but at 125 atm pressure. Horner and coworkers (686) have used benzene/ethanol (2:1) solutions of ruthenium(III) chloride/triphenyl-phosphine in the presence of basic amines for the hydrogenation of pent-1-ene. Triphenylphosphine reduces ruthenium halides to the divalent state (233), and these systems undoubtedly involve the HRuCl(PPh$_3$)$_3$ catalyst. The hydrogenation rates reported by Jardine and McQuillin (240) are about 10^{-2} to 10^{-1} ml min^{-1}, and correspond to the slow hydrogenations reported by Wilkinson and coworkers (Tables 18 and 19). The possibility of hydrogen originating from the alcohol in these systems was ruled out, since similar hydrogenation rates were observed for the reduction of norbornadiene to norbornane, and acetylenes (diphenylacetylene, stearolic acid and 2,5-dimethylhex-3-yn-2,5-diol) to the corresponding olefin, using benzene solutions of RuCl$_2$(PPh$_3$)$_3$ (240). However, the extent and manner of the production of the hydrido catalyst in this medium was not considered. The observation of *cis*-stilbene as the reduction product of diphenylacetylene would require that in a reaction such as 134 the production of the vinyl ruthenium intermediate, **31**, involves hydride transfer through a four-center transition state in which both phenyl groups are directed away from the metal center:

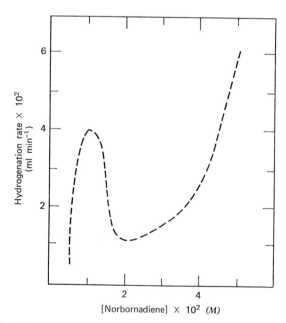

(141)

The dependence of the hydrogenation rate on the norbornadiene concentration as determined by Jardine and coworkers (240a) is shown in Fig. 4. The explanation of this type of behavior remains uncertain, but the maximum and minimum at 1:1 and 2:1 diene to ruthenium ratios, respectively, were thought to be related to the presence of 1 and 2 moles of coordinated diene moieties in reactions such as 134 and 135. However, it seems possible in this system that the olefin could compete with the hydrogen for reaction with $RuCl_2(PPh_3)_3$. At lower olefin concentration, the hydride may be formed, and the hydrogenation could proceed via reactions 134 and 135; at higher concentration, olefin complexing followed by reaction with hydrogen could

Fig. 4   Effect of norbornadiene concentration on rate of hydrogenation by $RuCl_3$ (10^{-2} M) and PPh_3 (6×10^{-2} M) in 50 ml methanol (240a).

predominate (cf. eqs. 127, 128). Two such similar paths have been recognized for the well-known rhodium catalyst RhCl(PPh₃)₃ (see Section XI-B-1). The complex behavior shown in Fig. 4 could well result from competing mechanisms. Similar olefin dependences for some systems with iridium, palladium, and platinum complexes with stannous chloride as cocatalyst were also noted (see Sections XII-C, XIII-B, and XIII-C-3b, respectively).

Benzene-ethanol solutions of $RuCl_2(PPh_3)_3$ have been used at 25° and 1 atm hydrogen by Takahashi and coworkers (641) to reduce 1-methoxyocta-2,7-diene selectively to the 2-ene, and by Augustine and Van Peppen (644) for the hydrogenation of 4-t-butylmethylenecyclohexane. In the latter system a small amount of accompanying isomerization to 4-t-butylmethylcyclohexane was noted, and the overall product distribution was compared to that of a heterogeneous ruthenium catalyst.

Abley and McQuillin have reported (243) that a benzene-methanol (3:2) solution of the chloride-bridged complex $[(Ph_3P)_3RuCl_3Ru(PPh_3)_3]Cl$ hydrogenates oct-1-ene at 20°, 1 atm H₂; a slower isomerization to $trans$-oct-2-ene was also observed under hydrogenation conditions and also under refluxing ethanol in the absence of hydrogen. Little detail was given, but presumably the complex dissociates and forms the hydridochloro catalyst.

A patent by Wilkinson (699) has stated that $RuCl_2(PPh_3)_4$ or $[Ru_2Cl_3(Ph_2PEt)_6]Cl$ is more efficient than $RuCl_3(LPh_3)_2 \cdot MeOH$, L = P, As, or $RuCl_2(diphos)$ for the catalytic reduction of hex-1-ene at ambient conditions.

Workers of Imperial Chemical Industries, Ltd. have also studied hydrogenation (20°, 1 atm H₂) catalyzed by ruthenium(II) complexes containing tertiary phosphines or arsines, with the particular aim of selective hydrogenation of olefin/acetylene mixtures (244, 245). The catalyst system is prepared by reacting a hydride such as lithium aluminum hydride with a complex $RuCl_2L_3$, where L is a tertiary phosphine or arsine; the $RuCl_2(PPh_3)_3$ complex was also used without the hydride treatment in benzene-ethanol solution. Again, terminal alkenes were reduced rapidly and cycloalkenes very slowly, and hydrogenation of a mixture of hex-1-ene and cyclohexene reduced the former selectively. Hex-1-yne was also found to be reduced slowly, but reduction of a mixture of hex-1-yne and oct-1-ene could be stopped at a stage showing the conversion of 99 % of the acetylene to hex-1-ene with the oct-1-ene remaining unchanged. By using the reaction scheme outlined in eqs. 133 to 135, such selectivity and the relative rates of reduction of the single substrates could be explained qualitatively if K_2 (acetylene) $> K_2$ (olefin) and k(acetylene) $< k$(olefin). The observation by Wilkinson's group (235) that a violet benzene solution of $HRuCl(PPh_3)_3$ shows no visible reaction with alk-1-enes but produces yellow to brown solutions with alk-1-ynes gives some support for such a relative magnitude of the K_2 constants.

Dewhurst (244a) has described the reduction of unsaturated moieties such as keto, formyl, and nitrile, as well as nonaromatic carbon-carbon double and triple bonds by using the catalysts $RuCl_2(PPh_3)_2$, $HRuCl(CO)(PPh_3)_3$, $H_2Ru(Ph_2PMe)_4$, and $H_2Ru(CO)(Ph_2PMe)_3$; optimum conditions were 10 to 100 atm hydrogen at 20 to 130° in hydrocarbon, ether, or alcohol solvents. Substrates reduced include hexenes, mesityl oxide, methyl isobutyl ketone, acetophenone, propionaldehyde, and benzonitrile.

Vaska (250, 250a) has reported that the complex $HRuCl(CO)(PPh_3)_3$ undergoes hydrogen-deuterium exchange according to eq. 142, and gives some reduction of both acetylene and ethylene in mixtures with hydrogen at ambient conditions; no details were given but eight-coordinate intermediates were thought likely (see Section IX-C, eq. 149).

$$HRuCl(CO)(PPh_3)_3 + D_2 \rightleftharpoons$$
$$[HRuCl(CO)(PPh_3)_3D_2] \rightleftharpoons DRuCl(CO)(PPh_3)_3 + HD \quad (142)$$

The complex $RuCl_2(CO)(PPh_3)_2(DMF)$, which has been isolated and studied by Rempel (1424), is effective for the catalyzed hydrogenation of alk-1-enes in methanol solution at 30° and 1 atm hydrogen, after initial hydride formation using borohydride.

Coffey (908) has reported that the complexes $HRuBr(CO)(PEt_2Ph)_3$ and $HRuCl(Et_2PC_2H_4PEt_2)$ in refluxing acetic acid catalyze the decomposition of formic acid into carbon dioxide and hydrogen; saturated aldehydes added to these systems are reduced to alcohols but the reductions are not as efficient as when using certain iridium complexes (Section XII-B-1).

Wilkinson and coworkers (187a, 825) have reported that protonation of the mixed valence complex $Ru_2(CO_2Me)_4Cl$ with HBF_4–MeOH, displaces the acetic acid and yields an intense blue solution; after removal of chloride, the solution is dark red. Binuclear cationic species of the type Ru_2^{5+} are believed to be present, and on addition of triphenylphosphine (phosphine: metal = 2:1) the blue or red solutions are efficient catalysts at 25° and 1 atm hydrogen for the reduction of the following compounds: hexa-1,5-diene > hex-1-ene > 3-methylbut-1-yn-3-ol ≫ allyl phenyl ether > hex-2-enes. A related catalytic system was produced via protonation of ruthenium(II) acetate, $Ru_2(CO_2Me)_4$ followed by addition of phosphine. The activity of these cationic catalysts is similar to that of a corresponding Rh_2^{4+}/triphenyl-phosphine catalyst (Section XI-I), although the ruthenium systems appear more selective for reduction of terminal alkenes relative to internal ones. The mechanisms of these ruthenium systems have not been studied.

c.   Other Ruthenium-Catalyzed Hydrogenation Systems.   The first published report by Breslow's group (56) on Ziegler homogeneous hydrogenation

systems (see Section XV-A) included the use of a ruthenium(III) acetylacetonate-triisobutylaluminum mixture in *n*-heptane solution; this effectively reduced oct-1-ene at 40° (3.5 atm H_2). A ruthenium(II) alkyl intermediate is possibly involved since triethylaluminum has been used as reductant in the preparation of ruthenium(II) complexes from ruthenium(III) complexes (252) (Section XV-B).

Rylander and coworkers (248) reported that dimethylformamide solutions containing ruthenium trichloride ($\sim$0.02 M) homogeneously hydrogenate dicyclopentadiene ($\sim$1.5 M) to the fully saturated hydrocarbon at a constant rate ($\sim$1.5 ml min^{-1} for 120 ml of solution at room temperature, 1 atm H_2); the final green solution was thought to contain ruthenium(II) species. The reduction also occurred in dimethylacetamide solutions, but no details were given.

James and Hui (220, 221) have investigated ruthenium chloride systems in dimethylacetamide, since in the corresponding aqueous chlororuthenate(II) system (Section IX-B-2a) it seemed that hydrogenation proceeded only when step 2 competed successfully with the reversal of step 1 (eq. 129). This suggested that less reactive olefins, such as ethylene, could be hydrogenated if the rate of the back-reaction of step 1 could be lowered, for example, by reducing the acidity of the medium. However, since ruthenium(II) is unstable in aqueous solutions of low acidity (249), a nonaqueous polar medium was used. Blue DMA solutions containing ruthenium(II) chloride catalyzed the hydrogenation of ethylene (80°, 1 atm H_2/C_2H_4), but the system was complex because of the accompanying reduction of the metal to the univalent state (see Section IX-B-1, eqs. 116 and 117). The brown ruthenium(I) chloride solutions were also catalytically active both for ethylene and activated olefins. At 80°, 1 atm H_2, fumaric or maleic acid was stoichiometrically reduced to succinic acid at ruthenium concentrations of 0.005 M. The form of the uptake plots measured at constant pressure depended on the substrate concentration used: at low concentration ($<$0.05 M), the reaction was first-order in olefin; at high concentrations (0.2 M), a zero-order dependence was observed. The reaction was first-order in hydrogen, half-order in ruthenium, and independent of acid concentration. The ruthenium(I) solutions gave no observable reactions with either hydrogen or olefin. The following mechanism is consistent with these data:

$$Ru_2^I \underset{}{\overset{K}{\rightleftharpoons}} 2Ru^I \tag{143}$$

$$Ru^I + H_2 \underset{k_{-1}}{\overset{k_1}{\rightleftharpoons}} Ru^{III}H_2 \, (k_{-1} > k_1) \tag{143a}$$

$$Ru^{III}H_2 + olefin \xrightarrow{k_2} Ru^I + saturated\ product \tag{143b}$$

Assuming K is small, the corresponding rate law is

$$\frac{-d[H_2]}{dt} = \frac{k_1 k_2 K'[H_2][\text{olefin}][Ru^I]_T^{0.5}}{k_{-1} + k_2[\text{olefin}]} \qquad (143c)$$

where $K' = (K/2)^{0.5}$. Analysis of the kinetic data at 80° gave k_2/k_{-1} values of 1.8 and 6.2 M^{-1} for fumaric and maleic acid respectively, and a $k_1 K'$ value of about 0.5 s^{-1} for both substrates. The observed half-order dependence on ruthenium even at the lowest concentration used (10^{-3} M) indicated an upper limit of K of about 10^{-5} M. A $\Delta H^{\ddagger}$ value of about 11.0 kcal mole^{-1} was estimated for the k_1 step; a small isotope effect ($k_H/k_D = 1.5$) was observed.

Under hydrogenation conditions, maleic acid was found to isomerize at a rate comparable to that of hydrogenation; catalytic reduction of fumaric acid using deuterium yielded a mixture of DL-sym-1,2-dideuterosuccinic acid and unsymmetrical dideuterosuccinic acid, $HO_2CCH_2CD_2CO_2H$ (221). These data were rationalized in terms of Scheme 5 for the detailed path of reaction 143b. The hydrogen molecule transfer to the coordinated olefin occurs in two consecutive hydrogen transfer steps. The isomerization is readily accounted for by the reversible reaction of an olefin dihydride complex (32) to an alkylhydride species (33). The DL acid (35) results from transfer of the second coordinated deuterium in 33; the unsymmetrical acid (36) results from successive transfer of deuterium and hydrogen in a species (34) formed from 33. The catalyst is regenerated as the monomeric ruthenium(I) species. The overall cis addition of hydrogen (fumaric acid → 35) requires that decomposition of the alkyl hydride 33 involves retention of configuration at the carbon atom attached to the metal. The reaction scheme is analogous to that suggested for some better characterized rhodium(I) systems (Sections XI-B-1 to XI-B-4).

Addition of triphenylphosphine to the brown DMA solutions of ruthenium(I) chloride resulted in a darkening of the coloration; this solution, in contrast to the initial one, slowly absorbed hydrogen at 80° up to a 1:1 ratio of H_2:Ru to produce a hydride (ν(Ru-H) 1900 cm^{-1}). This is consistent with a reaction such as 144, and gives indirect support for the formation of a

$$(PPh_3)_n Ru^I + H_2 \longrightarrow (PPh_3)_n Ru^{III} H_2 \qquad (144)$$

ruthenium(III) dihydride according to 143a. These solutions containing the hydridophosphine species did catalyze the hydrogenation of maleic acid, but they were five times less active than the ruthenium(I) chloride solutions under corresponding conditions (221).

Candlin and coworkers (244, 245) have reported that the compounds $RuCl_3$, $3H_2O$, and $RuCl_3(SEt_2)_3$, on treatment with lithium aluminum hydride in a variety of solvents, yield solutions that have been used for the selective hydrogenation of olefin/acetylene mixtures (Section IX-B-2b).

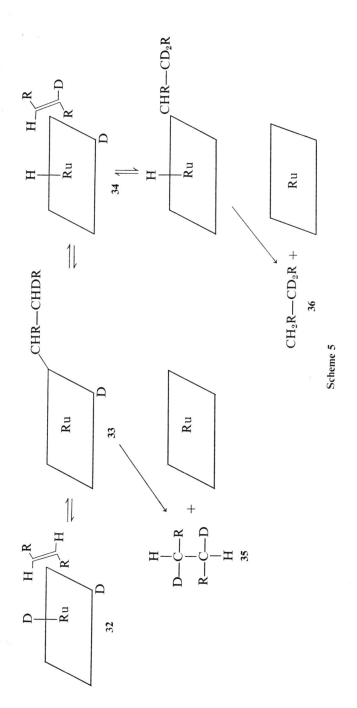

Scheme 5

Hui and Rempel (824) have used DMF solutions of ruthenium(II) acetate, $Ru_2(CO_2Me)_4$ at 80° and 1 atm hydrogen for the reduction of oct-1-ene, dec-1-ene, diethylmaleate, and cycloocta-1,3-diene; the initial rates ranged from $(7.0-18.0) \times 10^{-2}$ ml min^{-1} for 5 ml of a 5×10^{-3} M catalyst solution containing 1 M substrate. The system has been less studied than the corresponding rhodium(II) system (Section XI-I), but unlike the latter system the ruthenium is reduced by hydrogen to the monovalent state; these ruthenium(I) systems are more active in the presence of triphenylphosphine (1424).

A ruthenium catalyst stabilized by quinonoid and aromatic systems has apparently been used by Khidekel and coworkers (188) in DMF solution for hydrogenation of olefins (see Section XI-H for the corresponding rhodium system). Ogata and coworkers (1454) later reported the activity of the $(C_6H_6RuCl_2)_n$ complex at 30° and under 30 atm hydrogen. The complex is partially soluble in DMF, and pent-1-ene is hydrogenated with this system, although the main reaction is isomerization. Pent-2-ene was reduced more slowly. The arylated catalyst was compared to $RuCl_2(PPh_3)_3$, on the basis of the aromatic π-ligand replacing the phosphines, and the mechanism of eqs. 132–135 was invoked, DMF being the base required to form the hydride. A very low activity was observed for a benzene suspension of the complex, although addition of an amine base enhanced the activity, presumably by promoting hydride formation (cf. eq. 132). No activity was observed in DMSO solution but, in the absence of hydrogen or olefin, a complex C_6H_6-$RuCl_2(DMSO)$ was isolated. This complex was water soluble, and it catalyzed the hydrogenation of maleic acid at 25° and 20 atm hydrogen. By hydrogen treatment in the presence of triethylamine, a hydride, believed to be [HRuCl-$(C_6H_6)(DMSO)]_2$, was formed in the aqueous solution. A solvated hydride believed to be $HRuCl(PPh_3)_3(DMA)$ has been mentioned (see Section IX-B-2b).

Porri and coworkers (1319) have reported an interesting reaction of the complex $RuCl_2(C_{10}H_{16})$, which, formed from ruthenium trichloride and isoprene, has a bridged chloro structure and the coordinated bisallylic ligand 2,7-dimethylocta-2,6-diene-1,8-diyl. This complex reacts with hydrogen in dichloromethane at ambient conditions to yield 2,7-dimethyl-2,6-octadiene:

$$\text{CH—(CH}_2)_2\text{—CH}$$
$$\text{Me—C} \quad \text{——Ru——} \quad \text{C—Me} + H_2 \longrightarrow$$
$$\text{CH}_2 \qquad \text{CH}_2$$

$$CH_3C(Me)\!=\!CH\text{—}(CH_2)_2\text{—}CH\!=\!C(Me)CH_3 + \quad Ru \qquad (145)$$

The mechanism of this overall 1,8-addition would be of interest.

James and Hui (220, 221, 1399) have studied the activity of some chloro-carbonyl ruthenium complexes in DMA solution. Solutions containing a monocarbonyl ruthenium(I) species formed by absorption of carbon monoxide by ruthenium(I) solutions (1:1 uptake) catalyzed a slow hydrogenation of maleic acid; solutions containing a dicarbonyl species were completely inactive. Thus catalytic activity toward hydrogen of these complexes falls in the trend $Ru^I > Ru^I(CO) > Ru^I(CO)_2$. A similar trend was found for the corresponding ruthenium(II) complexes $(Ru^{II} > Ru^{II}(CO) > Ru^{II}(CO)_2)$ in dimethylacetamide, and the data in Tables 16 and 17 indicate a similar trend for these complexes in aqueous hydrochloric acid solutions.

Valentine and Collman (260) have observed that although trans-$Ru(CO)_3$-$(PPh_3)_2$ is unreactive toward hydrogen at temperatures up to 150°, it does react when irradiated at 365 nm to form an unstable photoproduct capable of hydrogenating cyclohexene; they suggest that oxidative addition of H_2 to a 4-coordinate species $Ru(CO)_2(PPh_3)_2$ produces the active species. In keeping with this, a relatively stable $H_2Ru(CO)_2(PPh_3)_2$ has been prepared by L'Eplattenier and Calderazzo (260a) from $Ru(CO)_3(PPh_3)_2$ in THF by using 120 atm hydrogen at 130°.

A large range of stable carbonyl, dicarbonyl, and hydridocarbonyl complexes of ruthenium(II) containing neutral ligands such as tertiary phosphines, and halogenoligands have been prepared (e.g., 75, 1311).

Collman's group (258a) reported that the nitrosyl-substituted carbonyl cluster, $Ru_3(CO)_{10}(NO)_2$, catalyzes the hydrogenation of hex-1-ene in benzene at 25° and 4 atm hydrogen, although the major reaction is isomerization. Currently, interest in the chemistry of nitrosyl complexes has renewed, because of their potential as catalysts (Sections XI-G and XII-B-4).

Stiddard and Townsend (259a) have prepared the ruthenium analog of Vaska's compound, $RuCl(NO)(PPh_3)_2$; the complex absorbed hydrogen in benzene solution, but the expected dihydride was not isolated. Other low-valent ruthenium nitrosyl triphenylphosphine complexes with and without coordinated carbonyl, such as $RuX(CO)(NO)(PPh_3)_2$, X = Cl, Br, OH, $RuCl(NO)_2(PPh_3)_2^+$, and $Ru(NO)_2(PPh_3)_2$, have been made (254, 1310, 1357, 1367, 1372, 1410). Laing and Roper (1310) report that $RuCl(CO)(NO)$-$(PPh_3)_2$ undergoes a wide variety of oxidative addition reactions with loss of carbon monoxide, but reactivity toward hydrogen was not mentioned.

L'Eplattenier and coworkers (257a) report that $Ru_3(CO)_{12}$, $Ru(CO)_5$, and $Ru(acac)_3$ promote the homogeneous reduction of nitrobenzene to aniline by a hydrogen-carbon monoxide mixture (~200 atm) at 150°; the postulated mechanism involves a phenylnitrene intermediate stabilized by bonding to ruthenium, this then undergoes hydrogenolysis:

$$C_6H_5NO_2 + 2CO \longrightarrow \left(C_6H_5N \diagup\diagdown \right) + 2CO_2 \qquad (146)$$

$$C_6H_5N \diagup\diagdown + H_2 \longrightarrow C_6H_5NH_2 \qquad (146a)$$

$$C_6H_5N \diagup\diagdown + CO \longrightarrow C_6H_5NCO \qquad (146b)$$

$$C_6H_5NH_2 + C_6H_5NCO \longrightarrow C_6H_5NHCONHC_6H_5 \qquad (146c)$$

The 2,2'-diphenylurea by-product is also thought to be produced via the nitrene.

Reactions 146 to 146c occur under hydroformylation conditions. The use of ruthenium carbonyls for the Oxo reaction has been reported by Pino and coworkers (1325) and in the patent literature, for example, by Smith and Jaeger (256a) and Cox and Whitefield (430). The limited information available suggests that the ruthenium carbonyls are not as efficient as cobalt carbonyls (Section X-E). One of the patents (430) is concerned with formation of the alcohol reduction products from the initial aldehyde product (Section X-E-2), and uses ruthenium(III) stearate in aqueous methanol solutions containing trimethylamine for the formation of n- and i-butanols from propene. A related reaction, which must involve aldehyde reduction, is the report by Anderson and Lindsey (254a) on the production of alkanediols from dienes and formaldehyde by using $RuCl_3$ or $Ru_2(CO)_9$ in diglyme at 140° under 700 atm hydrogen/carbon monoxide (1:2). For example, butadiene and paraformaldehyde produced hexan-1,6-diol with smaller amounts of methanol and methyl formate.

Braca and coworkers (1461) have reported that the hydroformylation of simple olefins using $Ru_3(CO)_{12}$ as catalyst is accompanied by the production of paraffins.

Ruthenium phosphine and arsine complexes with and without carbonyls have been used for the usual olefin-to-aldehyde Oxo reaction, for example, by Wilkinson and coworkers (44, 55, 699) and Slaugh and Mullineaux (524, 525, 527)—the most effective complex observed so far is $Ru(CO)_3(PPh_3)_2$ which is efficient at 100° and 100 atm hydrogen/carbon monoxide (55). James and Markham (255a) have observed slow gas uptake at 80° and 1 atm by solutions containing $HRuCl(CO)_2(PPh_3)_2$ in the presence of alkene substrates, but whether this was hydroformylation or hydrogenation was not established.

Hydroformylation, and related carbonylation and decarbonylation reactions, catalyzed by ruthenium complexes have been reviewed (79e).

d. Hydrogenation during Dimerization (Hydrodimerization). Misono and coworkers (1300–1303) and McClure and coworkers (1304) have used ruthenium(III) chloride in ethanol at 150° under hydrogen pressure (15–40 atm) for the catalytic dimerization of acrylonitrile to *cis*- and *trans*-1,4-dicyanobut-1-ene; the hydrogenated products propionitrile and adiponitrile are also formed (Section IX-A, eqs. 102, 104). Essentially no reaction occurs under a nitrogen atmosphere. The complexes $RuCl_2(CH_2 = CHCN)_n$, $n =$ 3 or 4, and another less well characterized one, containing a coordinated cyanoethyl group, were isolated, and both catalyzed the reaction (1300, 1304). Scheme 6, involving insertion of acrylonitrile into a Ru–H or a Ru-cyanoethyl bond was suggested (1302):

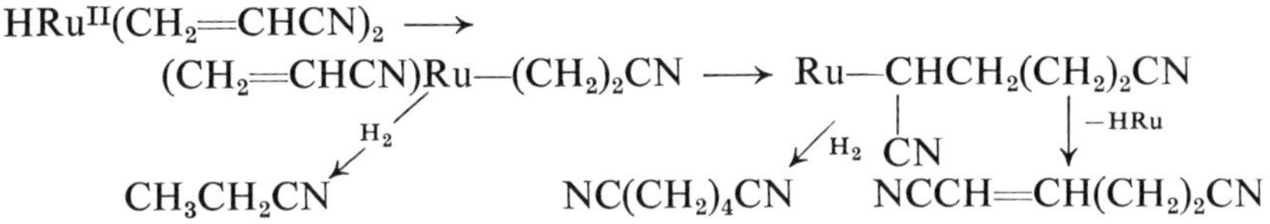

$$HRu^{II}(CH_2{=}CHCN)_2 \longrightarrow$$

$$(CH_2{=}CHCN)Ru{-}(CH_2)_2CN \longrightarrow Ru{-}CHCH_2(CH_2)_2CN$$

$$\swarrow{\scriptstyle H_2} \qquad\qquad \swarrow{\scriptstyle H_2} \ \underset{|}{\overset{|}{CN}} \quad \downarrow{\scriptstyle -HRu}$$

$$CH_3CH_2CN \qquad NC(CH_2)_4CN \quad NCCH{=}CH(CH_2)_2CN$$

Scheme 6

Ru(acac)$_3$ and a complex $RuCl_2(C_{12}H_{18})$, dichloro(dodeca-2,6,10-triene-1, 12-diyl)ruthenium(IV) (1305), exhibited activities similar to RuCl$_3$, 3H$_2$O, but $RuCl_2[P(OPh)_3]_4$, $RuCl_2(PPh_3)_3$, HRuCl(PPh$_3$)$_3$, Ru(O$_2$CCH$_3$)$_2$PPh$_3$, Ru(acac)$_2$(PPh$_3$)$_2$, [RuCl(PPh$_3$)$_2$(CH$_2{=}$CHCN)]$_2$, and $RuCl_2(PPh_3)_2(CH_2{=}$CHCN)$_2$ gave mainly propionitrile, indicating that more than two molecules of coordinated acrylonitrile may be required for dimerization (1303, 1304). Proceeding from RuCl$_2$(PPh$_3$)$_3$ to RuCl$_3$(AsPh$_3$)$_2$ to RuCl$_2$(SbPh$_3$)$_4$ gave increasing yields of dimer, since the arsine and stibine are weaker ligands than the phosphine and may be replaced by acrylonitrile (1304). RuCl$_2$(CO)$_2$-(py)$_2$ with strongly bound ligands was inactive (1303). Using ethanol solutions of ruthenium(II) chloride under hydrogen pressure, only propionitrile is formed (1302), and the system becomes solely a homogeneous hydrogenation (Section B-2a). The reason for this behavior is not too clear, since the intermediates for dimerization are ruthenium(II) complexes, and hydrogen readily reduces ruthenium(III) chloride to the divalent state in ethanol (221).

McClure and coworkers (1304) observed that the catalyzed hydrodimerizations occurred at lower pressures (6 atm) in the presence of *N*-methylpyrrolidine. Billig and coworkers (1306) made a similar finding, using as cocatalyst stannous chloride together with bifunctional amines, such as *N*-methylmorpholine, and/or bifunctional alcohols such as methylcellosolve. From data on deuterium incorporation into products and on acrylonitrile remaining after reaction under deuterium, Billig and coworkers deduced that Scheme 6 is unlikely, and they postulated the following series of reactions:

$$Ru^{II}\underset{CH_2}{\overset{CH(CN)}{\|}} \rightleftharpoons Ru^{IV}\underset{CH_2}{\overset{CH(CN)}{<}} \rightleftharpoons HRu^{IV}—CH{=}CHCN \quad (147)$$

36a   **36b**

36b $+ CH_2{=}CHCN \longrightarrow NCCH_2CH_2—Ru—CH{=}CHCN$ **(36c)** (147a)

36c $+ CH_2{=}CHCN \longrightarrow$ **36a** $+ NCCH{=}CHCH_2CH_2CN$ (147b)

36c $+$ **36b** $\longrightarrow CH_3CH_2CN + NCCH{=}CH—Ru—Ru—CH{=}CHCN$

                                           (36d)   (147c)

             36d $+ H_2 \longrightarrow$ 2. **36b**   (147d)

The oxidative addition reaction 147 involves insertion into a vinylic carbon-hydrogen bond; dimerization occurs through reactions 147 to 147b, and reduction through reactions 147c and 147d. Hydrogen regenerates the catalyst via reaction 147d. Hydrogenolysis of metal-alkyl bonds by a metal hydride (eq. 147c) is well documented (e.g., see Section X-B-2a).

McClure and coworkers (1304) explained the $RuCl_2(PPh_3)_3$ catalyzed reaction in terms of intermediates such as $HRuCl(PPh_3)_2(CH_2{=}CHCN)$, the formation of which was promoted in the presence of the amine base in a reaction such as 132; their scheme involved general insertion and hydrogenolysis reactions of the type shown in Scheme 6, but also included reactions such as 147c between a hydride and cyanoethyl species.

e. Other Ruthenium Complexes of Relevance to the Hydrogenation Catalysts. A number of novel hydridophosphine complexes of ruthenium(II) have been recently synthesized (251–254, 1367, 1368, 1410); although there are few data published on their chemical or possible catalytic properties, their study could be of value for the understanding of ruthenium(II) homogeneous hydrogenation catalysts. Keim and coworkers (251) have reported the preparation of cis-$H_2Ru(Ph_2PMe)_4$; n.m.r. data in chlorobenzene indicate a rapid phosphine ligand exchange. The corresponding triphenylphosphine complex, $H_2 Ru(PPh_3)_4$, was first prepared by Yamamoto and coworkers (252) and later by Knoth (253) and Levision and Robinson (254, 1410); deuterium exchange occurs with the hydride ligands and phenyl hydrogens (1368; cf. Scheme 3 in Section B-2b). Benzene solutions of this complex react with nitrogen at room temperature (252) to produce the molecular nitrogen complex, $(PPh_3)_3Ru(N_2)H_2$, which has also been prepared from the reaction of $HRuCl(PPh_3)_3$ with triethylaluminum under nitrogen in ether (253).

The nitrogen complex readily undergoes reaction with molecular hydrogen to yield $(PPh_3)_3RuH_4$ (252, 253); this reaction involves oxidation addition of hydrogen to produce a ruthenium(IV) complex from ruthenium(II) and provides some evidence for such a step postulated in Scheme 4 (**27** → **28**) for the catalyzed hydrogenations by $HRuCl(PPh_3)_3$. The existence of the seven-coordinate ruthenium(IV) hydride has led to the suggestion by Harrod and coworkers (255), following some studies on the dehydrochlorination of $IrH_2Cl(CO)(PPh_3)_2$, that reaction 148 might involve heterolytic splitting of hydrogen via a dihydro intermediate (Sections XI-E and XIII-C-3a):

$$RuCl_2(PPh_3)_3 + H_2 \rightarrow [H_2RuCl_2(PPh_3)_3] \rightarrow HRuCl(PPh_3)_3 + HCl \quad (148)$$

The complex $(PPh_3)_3Ru(N_2)H_2$, and presumably the tetrahydride, undergo deuterium-hydrogen exchange at the *ortho* positions (253) in the same manner as the $HRuCl(PPh_3)_3$ complex (Scheme 3 in Section B-2b).

Harris and coworkers (1367) have also isolated a complex, $H_2Ru(PPh_3)_3$ from reaction 148 carried out in the presence of triethylamine, although no details were given. This also could possible be formed from the hydridochloride via oxidative addition of hydrogen followed by elimination of hydrogen chloride (cf. eq. 148). The dihydride is coordinatively unsaturated and readily reacts with a variety of small, covalent gaseous molecules, but presumably not hydrogen under which the complex is prepared.

The phosphite complexes, $Ru X_2[P(OR)_3]_4$ (R = aryl), have been prepared (256, 960, 1303), but, except for the acrylonitrile system discussed earlier in Section B-2d, there is no information on the catalytic activity for hydrogenation by these complexes, or the hydrido-chloro complex, $HRuCl[P(OPh)_3]_4$ (242, 257; eq. 140 in Section B-2b).

Wilkinson and Gilbert (258) have prepared nitrile complexes of the type $RuCl_2(RCN)_2(PPh_3)_2$, some amine complexes, $RuCl_2(amine)_2(PPh_3)_2$, and the halogen bridged dimer $[RuCl_2(PPh_3)_2]_2$. The nitriles are reported not to give hydride derivatives. Stephenson (259, 1390) has also isolated anionic ruthenium(III) complexes of the type $(CH_3)_4N[RuCl_4(PPh_3)_2]$.

C. OSMIUM SPECIES

Vaska (261) has reported that toluene solutions containing $HOsCl(CO)(PPh_3)_3$ are active for the hydrogenation of acetylene to ethylene and ethane at 60° and a total pressure of 1 atm. The complex itself did not absorb hydrogen under the hydrogenation conditions but did react with deuterium (250) to give $DOsCl(CO)(PPh_3)_3$; this exchange reaction was thought to go via an eight-coordinate osmium(IV) intermediate:

$$HOsCl(CO)(PPh_3)_3 + D_2 \rightleftharpoons HOsCl(CO)(PPh_3)_3D_2 \rightleftharpoons$$
$$DOsCl(CO)(PPh_3)_3 + HD \quad (149)$$

The hydrogenation of ethylene was written as involving reaction with $H_3OsCl(CO)(PPh_3)_3$ to give ethane and regeneration of the catalyst (261).

Fotis and McCollum (970) have used benzene solutions of osmium(III) hydride complexes, $H_3Os(PPh_3)_3$, $HOsCl_2(PPh_3)_3$, and $HOsCl_2(AsPh_3)_3$, for the reduction of hex-1-ene at $90°$ with 5 atm hydrogen. An accompanying isomerization was also noted—to the extent of 25% of the products for the arsine systems and less than 7% for the phosphine systems.

Dewhirst (244a) has reported that solutions of $[Os_2Cl_3(MePPh_2)_6]Cl$ in m-cresol hydrogenate hex-1-ene between 125 and $175°$ and H_2 pressures of 40 atm. This system was the one example involving osmium quoted in a patent claiming the general use of complexes of the type L_nMX_2 (where M=Os, Ru, X=halogen, L=CO or tertiary phosphine with at least one L being a tertiary phosphine) for the hydrogen reduction of a wide range of unsaturated moieties (Section IX-B-2b). Hydrides of the type H_2ML_n or $HMXL_n$ were suggested as the reactive intermediates.

The systems above appear to be the only ones reported for this metal. Os^0 complexes such as $Os(CO)_3(PPh_3)_2$ show a strong tendency to undergo oxidative addition reactions but require the loss of one ligand to react with a nonpolar molecule such as H_2 (38, 39, 262). The hydrides $H_2Os(CO)_2$-$(PPh_3)_2$, $H_2Os(CO)_3PPh_3$, and $H_2Os(CO)_4$ have been made in this way from $Os(CO)_3(PPh_3)_2$, $Os(CO)_4PPh_3$, and $Os(CO)_5$ respectively, by using high temperatures and pressures (260a, 263). It is not known what the activities are for these hydride complexes as well as for complexes such as $H_2Os(CO)$-$(PPh_3)_3$ (264, 1309), $HOsCl(CO)_2(PPh_3)_2$ (265), $HOs(CO)_3(PPh_3)_2^+$ (265), $H_2Os_2(CO)_6(PPh_3)_2$ (265a), $H_2Os_2(CO)_8$ (265a), $HOs(CO)_5^+$ (1411), and $HOs_3(CO)_{12}^+$ (265b, 1389, 1411). The reactivity of $OsCl(CO)(NO)(PPh_3)_2$ (266, 1310) and $Os(NO)_2(PPh_3)_2$ (1357) toward hydrogen has not been reported. This list of complexes is representative of a range that have been synthesized, most of them quite recently.

Some multihydride complexes of the type $trans$-OsH_4L_3 and OsH_6L_2 (L = tertiary phosphine or arsine) have been prepared (265c, 265d, 265e, 1410) which perhaps gives indirect evidence for reactions such as 149. Douglas and Shaw (265c, 265d,) report that the hydrogen in these complexes undergoes exchange at 0 to $35°$ with the acidic hydrogen of acids and ethanol through protonated species such as $OsH_5L_3^+$; H_2 is subsequently lost, and this an example of generating a vacant coordination site by protonation of a transition metal hydride (39, 39a, 84) which suggests possible catalytic properties for these multihydride species. The dihydride complexes H_2OsL_4 are readily synthesized from the tetrahydride complexes (1309); osmium(II) mono- and dihydrido phosphine complexes without carbonyl ligands have also been made from the halides by reaction with hydrazine (1333) or lithium aluminum hydride (231, 511, 1307).

Some triarylphosphite complexes of osmium have been noted by Robinson and coworkers (959, 960) to abstract hydrogen from the *ortho*-position of the aromatic ring (Scheme 3, Section IX-B-2b) but no details were given.

Chloro complexes of osmium(IV) are essentially inactive toward hydrogen (135) (see Section XVI-B).

Very little information is available concerning the activity of osmium complexes under hydroformylation conditions, suggesting that they are relatively inactive; such studies have been mentioned in the patent literature, for example, by Slaugh and Mullineaux (968).

X

Group VIII Metal Ions and Complexes

COBALT

A. PENTACYANOCOBALTATE(II) COMPLEXES

The absorption of hydrogen by aqueous solutions of cobalt(II) chloride containing potassium cyanide was discovered in 1942 by Iguchi (20); at room temperature the absorption corresponded to one hydrogen atom per cobalt atom. Such solutions have been used for the hydrogenation of various organic compounds, and extensive studies in this area have been reported. The subject has been reviewed fairly extensively to early 1969 (21, 53, 57c, 268c).

The systems involve the pentacyanocobaltate(II) catalyst and are efficient because of the high reactivity of this complex with hydrogen; because of the nature of the cobalt solutions and their somewhat uncertain reaction with hydrogen, however, detailed mechanistic studies are rather difficult.

1. Species in Aqueous Solution. A great deal of investigation has been directed to ascertaining the nature of the species present in an aqueous solution containing cyanide and cobalt(II) (267–279). An initially formed precipitate of $Co(CN)_2$ redissolves at a CN:Co ratio $(R) > 5$ to give a clear solution, which is usually greenish, but may be brown at lower temperatures around 0° (270). Adamson (268) isolated a violet diamagnetic solid of empirical formula $K_3Co(CN)_5$ from the green solutions; Griffith and Wilkinson (271) suggested a dimeric structure, and Chadwick and Sharpe (268b) concluded that it is the hydrate $K_6[Co_2(CN)_{10}]4H_2O$. The green solution has a paramagnetism corresponding to one unpaired electron per cobalt, showing mainly the presence of a monomeric species (268, 272) that undoubtedly is a pentacyanide (267–273). The solutions are difficult to study because of their sensitivity to oxygen (270, 274) and their "decay" or "aging" in an inert atmosphere (272, 273, 277–279). This latter point is a particularly difficult one, and Pratt and Williams (275) have disputed some data indicating the presence of a number of extra species in "fresh" solutions because of the rapid decomposition of the initial complexes (276). However, the recorded visible spectra of the aqueous pentacyanide (271, 275, 279) are in agreement; King and Winfield (279) have reported that the aging process at 25° is very slow at $R = 10$ for $0.002\ M$ cobalt(II). After comparison of this spectrum with that for the isocyanide complexes (280) CoL_5H_2O (L = MeNC, PhNC), Pratt and coworkers (275, 283) concluded that the main complex in freshly prepared solutions is the hexacoordinate $Co(CN)_5H_2O^{3-}$ and not the pentacoordinate $Co(CN)_5^{3-}$; ion pairs such as $[Rb, Co(CN)_5H_2O]^{2-}$ were also formed with alkali metal cations. In very dilute solutions another species was detected which was probably the tetracyanide $Co(CN)_4(H_2O)_2^{2-}$, and in $3\ M$ LiCl a dimeric complex existed (275). These workers (275) concluded that, even under normal conditions, the $Co(CN)_5H_2O^{3-}$ must be in rapid equilibrium with undetectable species such as the tetracyano species, the hexacyano species (which is a required intermediate for some electron transfer reactions (281)), and $Co(CN)_5^{3-}$ which is the intermediate for dimerization.

Birk and Halpern (298a) have concluded from kinetic evidence for some oxidation-reduction reactions that no water is coordinated with the $Co(CN)_5^{3-}$ ion in aqueous solution. E.s.r. studies indicate a square pyramidal ion in the solid state (586) and also in solution (587); any association of water must be very weak.

2. Reaction with Hydrogen in Aqueous Solution. The reaction between hydrogen and cobalt cyanide solutions has been studied by many workers (20, 217, 218, 269, 271–273, 277–279, 282–294, 333, 1384). The results have been sometimes contradictory, and some data (283, 291) have suggested that the systems are complex; a major difficulty is again "the aging" of the solutions.

Winfield and coworkers (269, 282) obtained a second-order dependence on cobalt, and first suggested that the rates of hydrogen absorption that they and Iguchi (20) had measured in the homogeneous solution ($R \geqslant 5$) were due to the small amounts of $Co(CN)_6^{4-}$ present in solution, which was later shown to be incorrect by the same group (279). Incidentally it should be mentioned that solutions containing solid $Co(CN)_2$ ($R < 5$) absorb hydrogen at $0°$ (282), but evolve hydrogen at higher temperatures (20, 273, 290); the present review, however, is concerned only with homogeneous solutions.

Mills and coworkers (272) confirmed Iguchi's finding that freshly prepared solutions ($\leqslant 0.2$ M in cobalt) absorb hydrogen rapidly at 0 to $25°$, 1 atm ($t_{0.5} \sim 1$ min) to an amount corresponding to reduction of Co^{II} to Co^{I}; they also showed that the amount of absorption decreased if the solutions were allowed to "age" *in vacuo*. At higher cobalt concentrations, a slow evolution of H_2 was observed at $25°$ for the *in vacuo* solutions. The fresh solutions lose paramagnetism at a rate paralleling the loss of ability to absorb H_2; this was originally thought (272) to be due to a dimerization process ($2Co^{II} \rightarrow Co_2^{II}$), but later work by King and Winfield (279) showed that aging was partly initiated by a "self reduction" process:

$$2Co(CN)_5{}^{3-} + H_2O \xrightarrow{k} HCo(CN)_5{}^{3-} + Co(CN)_5OH^{3-} \qquad (150)$$
$$37$$

The formulation of the pentacyanocobaltate(II) in eq. 150 is written as it appeared in the literature; it may be $Co(CN)_5H_2O^{3-}$ in aqueous solution (see Sections A-1 and C). Careful spectrophotometric work attributed a characteristic peak at 305 nm for the complex 37 (279). The same species was thought to be formed by reaction of the fresh solutions with H_2 (279):

$$2Co(CN)_5{}^{3-} + H_2 \underset{k_{-1}}{\overset{k_1}{\rightleftharpoons}} 2HCo(CN)_5{}^{3-} \qquad (151)$$
$$\lambda \text{ max 967nm} \qquad\qquad 305nm$$

Indirect evidence for the structure of 37 [a cobalt(III) hydride rather than a cobalt(I) species] was that (*a*) reaction of $Co(CN)_5{}^{3-}$ with H_2 released no protons (20, 277, 282), (*b*) the product did not behave as an acid, (*c*) the stoichiometry of its formation was via eqs. 150 and 151, and (*d*) a high-field n.m.r. hydride signal was detected by Griffith and Wilkinson (271) for a species prepared by reduction of $Co(CN)_5{}^{3-}$ with borohydride; a hydride species was also present to the extent of 3% in a fresh $Co(CN)_5{}^{3-}$ solution, although these workers could detect no signal for the species formed by hydrogenation or by the aging reaction in water. More recently, however, Kemball and coworkers (284) have obtained the expected signal for 37 produced according to reactions 150 and 151. The hydride has been isolated as the salt $Cs_2Na[HCo(CN)_5]$ by Banks and Pratt (283, 285), and as the lithium salt by Pregaglia and coworkers (294, Section X-C).

The reversibility of reaction 151 was demonstrated by removing the H_2 atmosphere after reaction, although complete recovery of $Co(CN)_5{}^{3-}$ is not possible (272, 273, 279, 291); a solution of $HCo(CN)_5{}^{3-}$, formed at $0°$

according to 151, loses H_2 according to the reverse of 151 at 25° (279):

$$2HCo(CN)_5{}^{3-} \xrightarrow{k_{-1}} 2Co(CN)_5{}^{3-} + H_2 \tag{152}$$

Addition of ethanol to the hydride solution also results in the evolution of H_2 with precipitation of the dimer (271):

$$2HCo(CN)_5{}^{3-} = Co_2(CN)_{10}{}^{6-} + H_2 \tag{153}$$

Most of the data reported for the cobalt(II) cyanide solutions and their reaction with H_2 have been discussed in terms of reactions 150 to 152. The very significant effects of added cations and anions on the rates of reactions 150 to 152 (272, 279, 283, 284, 291) are certainly due in part to the formation of ion pairs (275, 279); attempts by Mills and coworkers (272) and King and Winfield (279) to measure the kinetics of the hydrogen reaction were complicated by such salt effects. By using solutions of high ionic strength (~ 1 M KCN), DeVries (277) was able to study the kinetics and thermodynamics of reaction 151. Measurement of the rate of H_2 uptake in a constant-pressure apparatus yielded a rate law reported as

$$-d[Co(CN)_5{}^{3-}]/dt = k_1[H_2][Co(CN)_5{}^{3-}]^2 - k_{-1}[HCo(CN)_5{}^{3-}]^2 \tag{153a}$$

(More precisely, the right-hand side gives $-d[H_2]/dt$, which equals $-0.5d$-$[Co(CN)_5{}^{3-}]/dt$.) The initial uptake rate was given by the k_1 term; at longer reaction times the reverse reaction becomes significant. The equilibrium constant $K(k_1/k_{-1})$ was also determined from separate spectrophotometric and uptake measurements to be 1.6×10^5 M^{-1} at 25°, and temperature dependence data gave the parameters $\Delta H° = -11.2$ kcal mole^{-1}, and $\Delta S° = -14$ eu. The enthalpy change indicates a bond energy of about 58 kcal mole^{-1} for the H—Co bond in $HCo(CN_5)^{3-}$. The k_1 was equal to 0.8×10^3 M^{-2} s^{-1} between 0 and 35° ($k_{-1} = 5.0 \times 10^{-3}$ M^{-1} s^{-1} at 25°). If reaction 151 were termolecular involving homolytic fission of H_2, it would involve a zero activation energy. DeVries (277) considered reaction through small amounts of a dimer more feasible:

$$2Co(CN)_5{}^{3-} \underset{k_{-2}}{\overset{k_2}{\rightleftharpoons}} Co_2(CN)_{10}{}^{6-} \tag{154}$$

$$Co_2(CN)_{10}{}^{6-} + H_2 \underset{k_{-3}}{\overset{k_3}{\rightleftharpoons}} 2HCo(CN)_5{}^{3-} \tag{155}$$

Such a scheme is consistent with the observed kinetics, if $k_{-2} \gg k_2 \gg k_3 > k_{-3}$ and k_1 becomes k_2k_3/k_{-2}, although King and Winfield (279) had found no correlation between concentration of the dimer and rate of H_2 uptake in methanol solution (see also Section X-C).

Simandi and Nagy (292) and Mizuta and Kwan (284a) later measured k_1 and K by the H_2 uptake experiments, and obtained values close to those of DeVries (277); the heat of adsorption of about 12 kcal mole^{-1} for reaction 151 was also confirmed (284a).

The polarograms of a pentacyanocobaltate(II) solution were essentially the same under either a N_2 or H_2 atmosphere, with a single wave at the same half-wave potential and limiting current, while aquopentacyanocobaltate(III) exhibited a two-electron reduction (267, 288, 292–293). From this evidence, Vlcek (293) and Simandi and Nagy (288, 292, 292a) concluded that the oxidation state of the Co in $HCo(CN)_5^{3-}$ was $2+$ and they preferred to think of the hydrogen being present as a stabilized atom, that is, $[Co^{II}(CN)_5(\cdot H)]^{3-}$ rather than $[Co^{III}(CN)_5(:H)]^{3-}$. Simandi and Nagy (297) suggested that reaction 151 occurs by way of the following steps:

$$Co(CN)_5^{3-} + H_2 \overset{K_1}{\rightleftharpoons} H_2Co(CN)_5^{3-} \tag{156}$$

$$H_2Co(CN)_5^{3-} + Co(CN)_5^{3-} \overset{K_2}{\rightleftharpoons} (CN)_5Co..H..H..Co(CN)_5^{6-} \tag{157}$$

$$(CN)_5Co..H..H..Co(CN)_5^{6-} \underset{k_{-6}}{\overset{k_6}{\rightleftharpoons}} 2HCo(CN)_5^{3-} \tag{158}$$

The scheme is consistent with the kinetics, and k_1 becomes $K_1K_2k_6$. Strohmeier and Iglauer (333) reported that reaction 151 was between first- and second-order in cobalt from 0 to 25°, at lower ionic strengths than used by DeVries.

DeVries (277) also followed the slower disproportionation reaction 150 in $\sim 1\ M$ KCN spectrophotometrically at 970 nm and obtained the rate law

$$-d[Co(CN)_5^{3-}]/dt = k[Co(CN)_5^{3-}]^2$$

where $k = 13 \exp[-4000/RT]\ M^{-1}\ s^{-1}$. Again decomposition through the dimer was suggested, that is, reaction 154 was followed by reaction 159:

$$Co_2(CN)_{10}^{6-} + H_2O \overset{k_4}{\longrightarrow} HCo(CN)_5^{3-} + Co(CN)_5OH^{3-} \tag{159}$$

Piringer and Farcas (286) later studied the decomposition of the hydrogenation product in the absence of H_2 (see reaction 152) by purging with N_2 and estimating the liberated H_2 by gas chromatography. They observed a first-order decomposition ($k_5 = 1.2 \times 10^{-4}\ s^{-1}$ at 25°, $\Delta H^{\ddagger} = 12$ kcal mole^{-1}), and concluded that the hydrogenated complex was a dimer $[(CN)_5CoH_2Co(CN)_5]^{6-}$:

$$(CN)_5CoH_2Co(CN)_5^{6-} \overset{k_5}{\longrightarrow} 2Co(CN)_5^{3-} + H_2 \tag{160}$$

The same workers also suggested (286) that the water decomposition of $Co(CN)_5^{3-}$ (see reactions 150 and 152) could occur via a dimeric ion pair, since they also found a second-order dependence on cobalt for this reaction, with an observed rate constant of $1.9 \times 10^{-3} M^{-1} s^{-1}$ at 25° [about 5 times lower than the k value determined by DeVries (277)]:

$$2K^+, Co(CN)_5H_2O^{3-} \xrightleftharpoons{K} K^+, (NC)_5Co^{II}(H_2O)(H_2O)Co^{II}(CN)_5^{6-}, K^+$$
$$\mathbf{38} \qquad (161)$$

$$\mathbf{38} \longrightarrow 2K^+, Co^{III}(CN)_5OH^{3-} + H_2 \qquad (162)$$

Reaction 162 involves homolytic splitting of each H_2O molecule into a H atom and OH radical. Although Piringer and Farcas (286) did not refer to the earlier work of DeVries (277), it is interesting to note that the activation energy of 12 kcal mole^{-1}, which they measured for the decomposition of the hydrogenated species, is consistent with the $\Delta H°$ value of -11.2 kcal mole^{-1} and the apparent zero activation energy for k_1 measured by DeVries.

Kemball and coworkers (284) have since concluded that reactions 150 to 152 are involved in the $Co(CN)_5^{3-}$ system and that the data of DeVries at 25° are essentially correct. They measured $k_1 = 0.6 \times 10^3 M^{-2} s^{-1}$, $k_{-1} = 4.5 \times 10^{-3} M^{-1} s^{-1}$, and an independent K value (k_1/k_{-1}) of $1.5 \times 10^5 M^{-1}$ at an ionic strength of 0.5. The form of the equilibrium constant was inconsistent with the hydride being a dimer. A K value of $2.71 \times 10^5 M^{-1}$ was measured for the higher ionic strength conditions used by DeVries, and Kemball and coworkers applied a number of corrections to DeVries' value of 1.6×10^5 M^{-1} to obtain a value identical to their own. The data of Piringer and Farcas were thought incorrect (284) due to large changes in ionic strength occurring during the measurements. Kemball's group found that k for reaction 150 varied from $(1.6 - 6.7) \times 10^{-3} M^{-1} s^{-1}$ at 25°. DeVries' data give $k = 1.3 \times 10^{-2} M^{-1} s^{-1}$ at this temperature. The spectrophotometric study of this aging reaction appears extremely difficult, although **37** is undoubtedly formed, but not necessarily according to the stoichiometry shown in reaction 150. Some complications noted (284) are: (*a*) $Co(CN)_5OH^{3-}$ may be substantially protonated, since the pK value of $Co(CN)_5H_2O^{2-}$ is about 10 at 40° (298) and the KCN present buffers the solutions at pH 10 to 11; (*b*) $Co(CN)_5OH^{3-}$ and $Co(CN)_5H_2O^{2-}$ can produce $Co(CN)_6^{3-}$; the substitution reaction of the aquo complex is also catalyzed by $Co(CN)_5^{3-}$ (298a) (*c*) transitory formation of solid cobalt(II) cyanide can catalyze reaction 150.

A number of workers have studied the pentacyanocobaltate(II)-catalyzed isotopic exchange of D_2 with H_2O (217, 218, 272, 284, 284a, 286, 287, 290, 316). Mills and coworkers (272) first investigated the exchange by measuring the loss of D_2 in the gas phase (HD and H_2 produced); the exchange rate was much slower than the hydrogenation process and seemed little affected by the "age" of the cobalt solutions. Schindewolf made similar studies using both D_2 (217) and H_2 containing 0.2% HD (218) in the gas phase, and found that the exchange rate was approximately second-order in cobalt and increased a little with increasing H_2 pressure. The mechanism was written

$$2Co + D_2 \overset{K}{\rightleftharpoons} 2CoD \tag{163}$$

$$CoD + H_2O \longrightarrow CoH + HDO \tag{164}$$

$$CoD + CoH \longrightarrow 2Co + HD \tag{165}$$

$$2CoH \overset{k_{-1}}{\longrightarrow} 2Co + H_2 \tag{166}$$

with reactions 165 and 166 being rate controlling. Kemball and coworkers (284) obtained a proton resonance peak for a deuterated $Co(CN)_5{}^{3-}$ solution, which confirms that reaction 164 is fast. The rate law for the exchange (217) becomes

$$\text{rate} = \frac{k_{-1}[Co]_T{}^2(K[H_2])^{0.5}}{1 + (K[H_2])^{0.5}} \tag{166a}$$

The data gave k_{-1} and K values close to those of DeVries (277) and the later work of Kemball's group (284). The activation energy for reaction 166 was determined to be $\sim$7 kcal mole^{-1}, and a low frequency factor of 10^2 M^{-1} s^{-1} was attributed to the high charges of the $HCo(CN)_5{}^{3-}$ ions (217, 218). The K was also measured independently by Schindewolf (217).

The exchange rate increased significantly with added electrolytes, such as KCN, KCl, and NaOH (217, 272), which was considered to be consistent with the known acceleration of a reaction such as 166 between ions of like charges; the ratio of H_2/HD production also increased with increasing NaOH due to enhancement of the exchange reaction 164 (272).

Cobalt cyanide solutions also catalyzed the ortho-para hydrogen conversion at a rate about 1.5 times faster than D_2 exchange (217). Mabrouk and coworkers (316) and Mizuta and Kwan (284a) later made very similar exchange studies with very similar results, and the latter workers found close agreement between the rates of ortho-para hydrogen conversion and D_2 exchange (284a). Piringer (287) has also studied the exchange reaction at high pressure (up to 100 atm), and again he reports an activation energy of about 7 kcal mole^{-1}. The use of high pressure gives more stable solutions of the hydride complex, such as 60% retained after 16 hr at 80°, compared to 30% after 2 hr at 1 atm (287).

Interestingly, the exchange studies (217, 218, 287) give an activation energy of about 7 kcal mole^{-1} for reaction 152($\equiv$166), while the H_2 uptake and evolution studies (277, 284a, 286) give a value of about 12 kcal mole^{-1}. The possibility that other complexes in these solutions catalyze the exchange cannot be completely ruled out, particularly since the "aged" solutions are also equally effective for the exchange.

Pratt and coworkers (283, 291) have also concluded that other complexes, possibly $Co(CN)_4H_2^{3-}$ and $Co^I(CN)_6^{5-}$, are also formed during hydrogenation of cobalt(II) cyanide solutions at low cobalt and cyanide concentrations ($\sim 10^{-2}\,M$) and at high cyanide concentrations ($6\,M$), respectively. This group has also postulated (291) an oxidative addition equilibrium:

$$HCo(CN)_5^{3-} \rightleftharpoons Co^I(CN)_4^{3-} + HCN \qquad (167)$$

because solutions containing a very low excess cyanide were reported to take up H_2 autocatalytically, which was thought to be due to production of $Co(CN)_4H_2^{3-}$ from the $Co(CN)_4^{3-}$ species produced in 167 (see also Section X-B-2e).

The complications and uncertainties in these $Co(CN)_5^{3-}/H_2$ systems were stressed by Williams, Halpern, Simandi, and others at a Faraday Society Meeting (80). Halpern and Pribanic (1384) have since studied the kinetics of hydrogenation of $Co(CN)_5^{3-}$ solutions at higher hydrogen pressures (up to 27 atm) and low ionic strength conditions at which the competing aging reaction is less significant. The kinetic and equilibrium data for reaction 151 agreed well with that of DeVries, Simandi and Nagy, and Kemball's group, and it was concluded that no other kinetically significant species were present. Based on steric grounds, a termolecular reaction was favored over one taking place via a preformed dimer (eqs. 154 and 155).

The catalytic hydrogenation of substrates has generally been discussed in terms of the hydride, $HCo(CN)_5^{3-}$, which is certainly the form of most of the "reduced" cobalt.

The hydridopentacyanocobaltate(III) anion may be prepared from cobaltous cyanide solutions by reaction with hydrogen, borohydride, or the aging reaction. Hydrazine has also been used as the hydride source (Section X-B-2f). Jackman and coworkers (323) have shown, in fact, that the borohydride reduction involves molecular hydrogen produced via hydrolysis of the borohydride (see also Section X-B-5).

B. CATALYTIC REDUCTION (HYDROGENATION AND HYDRO-GENOLYSIS) BY AQUEOUS PENTACYANOCOBALTATE(II) SOLUTIONS

Iguchi (20) observed in 1942 that increased quantities of H_2 were absorbed by the cyanocobaltate(II) solutions in the presence of cinnamate or isatin, but the reaction products were not identified. Twenty years later, Kwiatek and coworkers (295) demonstrated the versatility of this system for the catalytic reduction of a wide range of organic and inorganic substrates at room temperature and 1 atm H_2. Meanwhile, a patent by Spencer and Dowden (296) was issued in 1959 on the reduction of butadiene and cyclopentadiene to monoenes, and DeVries (278) reported in 1960 the selective reduction of sorbic acid to 2-hexenoic acid when using the hydride, prepared under H_2 (eq. 151) or under N_2 from the homolytic cleavage of water (eq. 150).

1. Reduction of Inorganic Substrates. Kwiatek and coworkers (268c, 295, 297, 303) have shown that $Co(CN)_5^{3-}$ solutions under H_2 can reduce hydrogen peroxide, oxygen, sulfur, halogens, ferricyanide, permanganate, persulfate, dichromate, nitrite, hydroxylamine, and pentacyanocobaltate(III) complexes; the reactions were followed by H_2 uptake measurements (297).

Other workers found that H_2O_2 reacted rapidly with $Co(CN)_5^{3-}$ or $HCo(CN)_5^{3-}$ to form $Co(CN)_5OH^{3-}$, or $Co(CN)_5H_2O^{2-}$ depending on the pH (270, 298), identical to the product formed with water (see eq. 150):

$$2Co(CN)_5^{3-} + H_2O_2 \longrightarrow 2Co(CN)_5OH^{3-} \tag{168}$$

$$HCo(CN)_5^{3-} + H_2O_2 \longrightarrow Co(CN)_5OH^{3-} + H_2O \tag{169}$$

Hydrogenation of the H_2O_2 occurred only when the quantity of peroxide added was less than the amount of hydride present. On adding greater quantities, poisoning of the catalyst occurred and there was no H_2 uptake; the poisoned system, however, could be reactivated by adding a further slight excess of $HCo(CN)_5^{3-}$. The results were rationalized by Kwiatek and coworkers (297) in terms of reaction 169 followed by 170:

$$Co(CN)_5OH^{3-} + HCo(CN)_5^{3-} \rightleftharpoons 2Co(CN)_5^{3-} + H_2O \tag{170}$$

Absorption of H_2 by $Co(CN)_5^{3-}$ according to eq. 151 gives the net reaction

$$H_2O_2 + H_2 \longrightarrow 2H_2O \tag{171}$$

Reaction 170 indicates the reversibility of the aging reaction. The autocatalytic H_2 reduction of $Co(CN)_5OH^{3-}$ (or the aquo complex) according to reactions 170 and 150 has been reported by Kwiatek and Seyler (303), although the $Co(CN)_6^{3-}$ complex was not so reduced (297). Jackman and coworkers (323), however, were unable to reduce $Co(CN)_5OH^{3-}$ with the hydride. Also, a later study by Halpern's group (323a) on the reaction of $Co(CN)_5^{3-}$ with hydrogen peroxide indicates that the following free-radical mechanism is operative:

$$Co(CN)_5^{3-} + H_2O_2 \xrightarrow{k} Co(CN)_5OH^{3-} + \cdot OH \tag{171a}$$

$$Co(CN)_5^{3-} + \cdot OH \longrightarrow Co(CN)_5OH^{3-} \tag{171b}$$

Thus the water production could occur also via reaction of the hydride with the $\cdot OH$ radical (see also Section X-B-3):

$$HCo(CN)_5^{3-} + \cdot OH \longrightarrow Co(CN)_5^{3-} + H_2O \tag{171c}$$

In view of the description by Simandi and Nagy (292) of the hydride as containing a stabilized hydrogen atom, it is interesting to note that Halpern and Rabani (299) have reduced $Co(CN)_6^{3-}$ and a range of cobalt(III) pent-ammines to the cobalt(II) state by using hydrogen atoms. These reactions are thought to involve attack of the hydrogen atom on an anionic ligand.

The catalytic reductions of persulfate, nitrite and hydroxylamine were thought to involve the initial formation of $Co(CN)_5OH^{3-}$ according to reactions such as 168 (268c, 303). The hydroxylamine reaction is a free-radical process (323a), involving as well production of $\cdot NH_2$. All these reductions could involve mechanisms similar to that shown above for H_2O_2.

Oxygen is reduced to water by H_2 in the presence of $Co(CN)_5^{3-}$ (297, 300); the hydride undergoes autoxidation to a hydroperoxide complex (300, 300a):

$$HCo(CN)_5^{3-} + O_2 \longrightarrow Co(CN)_5OOH^{3-} \tag{172}$$

According to Bayston and Winfield (300), further reaction yielded the cobalt-(III) complex (reaction 173), which was thought to be reduced according to eq. 170:

$$Co(CN)_5OOH^{3-} + HCo(CN)_5^{3-} \longrightarrow 2Co(CN)_5OH^{3-} \tag{173}$$

Again the system is poisoned on using excess O_2 due to consumption of all the hydride and, consistent with the reaction scheme, when the O_2 supply is cut off, most of the cobalt returns to the hydride form (300). The net reaction is

$$O_2 + 2H_2 \longrightarrow 2H_2O \tag{174}$$

although only about 70% of the O_2 is reduced (303).

Sulfur is catalytically reduced quantitatively by H_2, presumably through cobalt-sulfur bonded intermediates (303). Iodine and bromine are also cata-lytically reduced to the anions probably via reactions 175 to 177 and 151 (303):

$$2Co(CN)_5^{3-} + X_2 \longrightarrow 2Co(CN)_5X^{3-} \tag{175}$$

$$Co(CN)_5X^{3-} + H_2O \longrightarrow Co(CN)_5H_2O^{2-} + X^- \tag{176}$$

$$Co(CN)_5H_2O^{2-} + HCo(CN)_5^{3-} + OH^- \rightleftharpoons 2Co(CN)_5^{3-} + 2H_2O \tag{177}$$

Reactions 175 and 176 are well substantiated (274, 301, 302), while reaction 177 is the reaction corresponding to 170 for the aquocobalt complex.

Catalytic reduction of ferricyanide was thought to proceed via reactions 178 and 179, followed by reactions 170 and 151 (297):

$$Co(CN)_5^{3-} + Fe(CN)_6^{3-} \longrightarrow (NC)_5Co(NC)Fe(CN)_5^{6-} \tag{178}$$
$$\mathbf{38a}$$

$$\mathbf{38a} + OH^- \longrightarrow Fe(CN)_6^{4-} + Co(CN)_5OH^{3-} \tag{179}$$

Complex **38a** has been isolated (270) and may be used as the initial reactant for the reduction process (303). Excess sodium hydroxide is required for the hydrolysis step 179; with no added alkali, H_2 is evolved possibly due to displacement of the equilibrium shown in 151 (297). Permanganate and dichromate are catalytically reduced to manganous hydroxide and chromic hydroxide, respectively (303).

The mechanisms thought to be operative for the reduction of inorganic substrates hinge to a large extent on reaction 170, which has been recently questioned by Jackman and coworkers (323); the reductions could be free-radical processes.

2. Catalyzed Hydrogenation of Organic Substrates. A large number of reports have appeared on the catalytic H_2 reduction of organic substrates using aqueous pentacyanocobaltate(II) systems. The complex $HCo(CN)_5^{3-}$ transfers hydrogen to substrates, and two basic mechanisms, which seem to depend on the substrate, have been postulated. One involves organocobalt intermediates, and the other free-radical species; the former is exemplified by reduction of conjugated dienes, and the latter by reduction of certain carboxylic acids. These mechanisms are considered first in some detail, and then the other reductions effected are documented under the organic functional types. Tables 20 to 29 list the hydrogenations of organic substrates that have been catalyzed by $Co(CN)_5^{3-}$. A wide variety of reaction conditions have been used: reductions have been carried out at 0 to 125°, under hydrogen pressures of 1 to 220 atm, in the absence of hydrogen (usually under nitrogen), using different salts with and without added hydroxide, and with CN/Co ratios (R) of 5 to 100. Aqueous alcohol solutions have sometimes been used to give homogeneous solutions of substrate and catalyst (see also Section X-C).

Carbon-carbon double and triple bonds are only reduced when part of a conjugated system (297). A few studies have been reported on the reduction of unsaturated fatty acids (Section XIV-A).

a. Conjugated Dienes and Polyenes (Tables 20 and 21). Spencer and Dowden (296) first reported the selective reduction of butadiene at 20° and 1 atm pressure to a product rich in but-1-ene at CN/Co > 5 and rich in *trans*-but-2-ene at CN/Co ⩽ 5. Their data were reported for aqueous alcohol solutions, but later studies by Kwiatek and coworkers (297), Suzuki and coworkers (304, 1438), Farcas and coworkers (373), and Burnett and coworkers (305) have confirmed this general finding for purely aqueous solutions, although selectivity also depends on the concentration of the hydride complex (305, 373): at higher concentrations, but-1-ene has been observed as the major product at $R = 5$ (373).

Kwiatek's group (297) showed that butadiene formed a 1:1 adduct with $HCo(CN)_5^{3-}$, and they postulated the following equilibrium:

$$HCo(CN)_5^{3-} + C_4H_6 \underset{k_{-1}}{\overset{k_1}{\rightleftharpoons}} \underset{\textbf{39}}{Co(CN)_5(C_4H_7)^{3-}} \qquad (180)$$

They also showed that **39** reacted with the hydride irreversibly and stoichiometrically:

$$Co(CN)_5(C_4H_7)^{3-} + HCo(CN)_5^{3-} \overset{k_2}{\longrightarrow} 2Co(CN)_5^{3-} + C_4H_8 \quad (181)$$

and k_1 was shown qualitatively to be $>k_2$. Complex **39** did not react with H_2, hence the catalytic reduction was accounted for by reactions 151, 180, and 181. Use of a mixture of deuterium and butadiene gave butenes containing di-, mono- and nondeuterated species. The dideuterated product is consistent with reaction via 180 and 181, although some unreacted butadiene contained small quantities of mono- and dideutero species showing the reversibility of 180, which could give rise to the dideuterated butenes. Reduction also occurred in the absence of hydrogen due to production of the hydride by the aging reaction 150, and it was suggested that hydrolysis of **39** might lead to butene production, although this is not essential:

$$Co(CN)_5(C_4H_7)^{3-} + H_2O \longrightarrow Co(CN)_5OH^{3-} + C_4H_8 \qquad (182)$$

Some hydride formation via the aging reaction also readily accounted for the mono- and nondeuterated butenes.

Complex **39** was isolated by Kwiatek and Seyler (306, 307) and was found to be identical to that formed by reaction of crotyl bromide with $Co(CN)_5^{3-}$ (306–308):

$$2Co(CN)_5^{3-} + CH_3CH{=}CHCH_2Br \longrightarrow$$
$$\underset{\textbf{39}}{Co(CN)_5CH_2CH{=}CHCH_3} + Co(CN)_5Br^{3-} \quad (183)$$

The presence of the σ-2-butenyl group was indicated (306, 308) by its n.m.r. spectrum in D_2O, by but-1-ene evolution on protonation (reaction 184), and by but-1-ene or *trans*-but-2-ene evolution on treatment with hydride complex at either high or low cyanide concentration. Moreover, decomposition in water under nitrogen gave an equimolar mixture of butenes and butadiene, in agreement with the reverse of reaction 180 followed by 181.

$$Co(CN)_5CH_2CH{=}CHCH_3^{3-} \overset{H^+}{\longrightarrow} Co(CN)_5CH_2\overset{+}{C}HCH_2CH_3^{2-} \longrightarrow$$
$$Co(CN)_5^{2-} + CH_2{=}CHCH_2CH_3 \quad (184)$$

The protonation reaction 184 assumes attack at the γ-carbon of the crotyl group to give the carbonium ion which then undergoes 1,2 elimination (308);

such attack on allylic iron complexes and crotylmercuric halides had been demonstrated earlier (309, 310).

The stereoselectivity control by cyanide has been explained in terms of equilibria between σ-complexes and a π-allylic intermediate (21, 297, 305, 306):

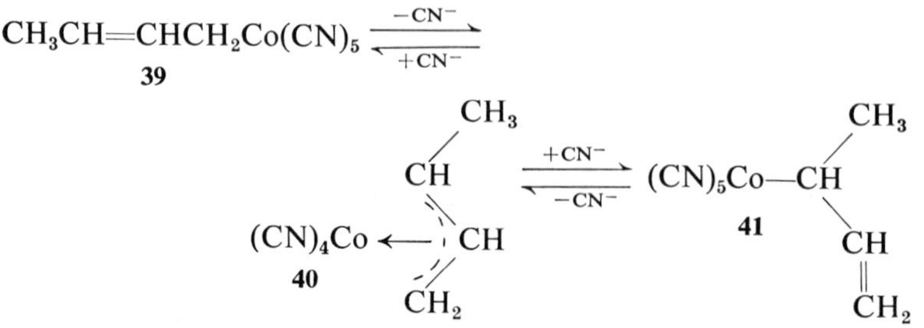

$$CH_3CH{=}CHCH_2Co(CN)_5 \underset{+CN^-}{\overset{-CN^-}{\rightleftarrows}}$$

39

With excess cyanide only σ-complexes **39** and **41** are assumed to be present, and these may also equilibrate directly via internal ion pair complexes. At low cyanide concentration, π allyls will also be involved, but only the *syn* complex **40** is expected to be significant (21). With these assumptions Kwiatek and coworkers (21) explained the observed products on the basis that reaction 181 probably involved γ-attack by the hydride complex on the allylic group; for example:

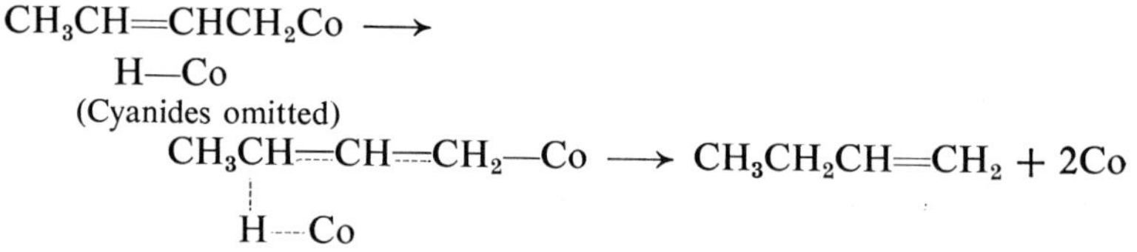

$$CH_3CH{=}CHCH_2Co \longrightarrow$$
$$\quad H{-}Co$$

(Cyanides omitted)

$$CH_3CH{-}\!-\!-CH{-}\!-\!-CH_2{-}Co \longrightarrow CH_3CH_2CH{=}CH_2 + 2Co \quad (186)$$
$$\quad\vdots$$
$$\quad H{-}\!-\!-Co$$

Burnett and coworkers (305), however, consider that in reaction 181 the carbon-hydrogen bond about to be formed acts as a bridge between the two cobalt atoms, and thus the σ-complex **41** is required to produce but-1-ene,

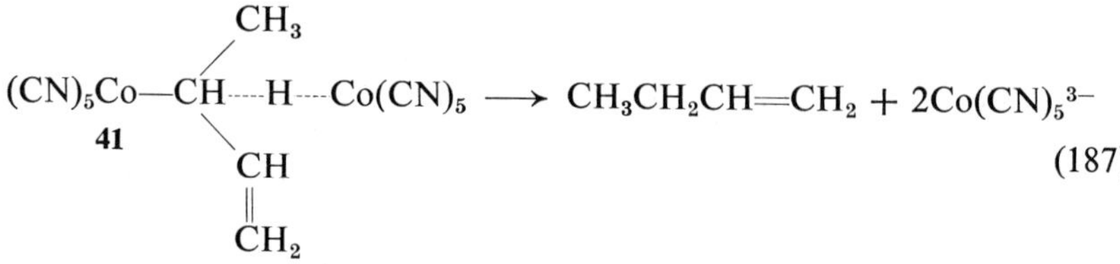

This group concluded that reaction 180 produces **41** since this would more readily isomerize to **40** than would **39** at low cyanide concentration. It should be noted that **41** is an expected product of an insertion reaction of a π-coordinated butadiene into the Co—H bond. The fact that the butene product ratios varied with the concentration of the hydride complex at high cyanide was interpreted in terms of direct isomerization of **41** to **39**.

A detailed kinetic study by Burnett and coworkers (305) showed that reaction 180 was first-order in both hydride and butadiene, and that, at 25°, $k_1 = 7.6\,M^{-1}\,s^{-1}$ and $k_{-1} \sim 2 \times 10^{-5}\,s^{-1}$; an equilibrium constant of $\sim 3.6 \times 10^5\,M^{-1}$ was also measured independently (305). Halpern and Wong (328) have determined k_1 as $1.6\,M^{-1}\,s^{-1}$ at 25° in 50 vol % aqueous methanol. The rate of formation of but-1-ene was shown by Burnett and coworkers (305) to be first-order in hydride and σ-butenyl complex (eq. 181).

The relative and absolute rates of formation of but-1-ene and *trans-* and *cis*-but-2-ene were analyzed in terms of the following reactions (305), (CoH = HCo(CN)$_5$$^{3-}$):

$$\text{CoH} + \text{C}_4\text{H}_6 \underset{k_{-1}}{\overset{k_1}{\rightleftharpoons}} \textbf{41}, \qquad k_1 = 7.6\,M^{-1}\,s^{-1}, \qquad k_{-1} \sim 2 \times 10^{-5}\,s^{-1} \tag{188}$$

$$\textbf{41} \underset{k_{-2}}{\overset{k_2}{\rightleftharpoons}} \textbf{40} + \text{CN}^-, \qquad k_2 \sim 1.5 \times 10^{-3}\,s^{-1},$$
$$k_{-2} \sim 1.5\,M^{-1}\,s^{-1} \tag{189}$$

$$\textbf{41} \underset{k_{-3}}{\overset{k_3}{\rightleftharpoons}} \textbf{39}, \qquad k_3 \sim 4 \times 10^{-5}\,s^{-1}, \qquad k_{-3} \sim 3 \times 10^{-4}\,s^{-1} \tag{190}$$

$$\textbf{41} + \text{CoH} \overset{k_4}{\longrightarrow} 1\text{-C}_4\text{H}_8 + 2\text{Co}, \qquad k_4 \sim 5 \times 10^{-2}\,M^{-1}\,s^{-1} \tag{191}$$

$$\textbf{39} + \text{CoH} \overset{k_5}{\longrightarrow} t\text{-2-C}_4\text{H}_8 + 2\text{Co}, \qquad k_5 \sim 1.7 \times 10^{-1}\,M^{-1}\,s^{-1} \tag{192}$$

$$\textbf{39} + \text{CoH} \overset{k_6}{\longrightarrow} c\text{-2-C}_4\text{H}_8 + 2\text{Co}, \qquad k_6 \sim 1.7 \times 10^{-1}\,M^{-1}\,s^{-1} \tag{193}$$

$$\textbf{40} + \text{CoH} \overset{k_7}{\longrightarrow} 1\text{-C}_4\text{H}_8 + 2\text{Co}, \qquad k_7 \sim 1.7 \times 10^{-1}\,M^{-1}\,s^{-1} \tag{194}$$

$$\textbf{40} + \text{CoH} \overset{k_8}{\longrightarrow} t\text{-2-C}_4\text{H}_8 + 2\text{Co}, \qquad k_8 \sim 1\,M^{-1}\,s^{-1} \tag{195}$$

$$\textbf{40} + \text{CoH} \overset{k_9}{\longrightarrow} c\text{-2-C}_4\text{H}_8 + 2\text{Co}, \qquad k_9 \ll k_8 \tag{196}$$

The data (eq. 190) indicate that only quite small amounts of **39** are present under any conditions. The validity of the application of the steady-state treatment for the intermediates **39**, **40**, and **41** was verified.

A σ to π isomerization, such as reaction 189, was observed for the allyl-pentacyanocobaltate(III) ion in n.m.r. spectra in D$_2$O (307), but not for the butenyl complex because of its decomposition in water.

In the presence of hydroxide, somewhat smaller quantities of cyanide will effect predominant formation of but-1-ene (303, 304, 1438), but only when the complex is hydrogenated before addition of butadiene (304). A catalyst solution prepared by adding initially hydroxides (or ammonia) and potassium cyanide gave mainly but-1-ene, while the converse order of addition favored formation of *trans*-but-2-ene; neutral salts such as sulfate had little effect on the hydrogenation selectivity (311). The hydroxide effect has been attributed to possible changes in ligand coordination on the cobalt (311), and the suppression of cyanide ion hydrolysis (268c).

Reduction of butadiene has also been effected by using $Co(CN)_5^{3-}$ solutions in the absence of hydrogen (21, 304, 312) (see eq. 150); the relation between the selectivity and the CN/Co ratio was the same as in the presence of hydrogen. The addition of potassium hydroxide did not alter this relation (304), although in the absence of inorganic or organic compounds containing hydroxy groups (also bisulfate, bicarbonate, dihydrogenphosphate, borate, phenol, citric acid, and phthalic acid) the hydrogenation rate was very low (312).

The stereoselectivity control by cyanide for reductions under hydrogen or nitrogen has been observed for substituted butadienes. Reduction of isoprene yields 78 to 90% 2-methylbut-2-ene when the CN/Co ratio is <5.1 (21, 295 313, 1439). At higher ratios the major product has been reported by one group as 2-methylbut-1-ene (21) and as 3-methylbut-1-ene by another (313). 1,3-Pentadiene is reduced almost entirely to *trans*-pent-2-ene at CN/Co = 5.0; while at a ratio of 7.0, some (20%) pent-1-ene is produced (21). 2,3-Dimethylbutadiene gives a greater proportion of terminal olefin in the presence of excess cyanide; less stereoselectivity in this reduction at low CN/Co was attributed to a sterically unfavorable methyl group in the π-allylic form (303). Reduction of the homoconjugated norbornadiene at either CN/Co ratios of 5.1 or 7.0 yielded the same (1:13) mixture of nor-bornene and nortricyclene, indicating that a π-allylic complex is not involved in this case (21). Unsuccessful reduction of 2,5-dimethyl-2,4-hexadiene, in which a "cisoid" arrangement is sterically hindered, and successful reduction of norbornadiene with a fixed "cisoid" conformation, had been interpreted originally in terms of a "cisoid" geometry being necessary for reduction (306), but later reduction of 3-methylenecyclohexene, with fixed *trans* conformation, indicates that other steric factors are involved (21, 303). 2,4-Hexadiene, 1,2-dimethylenecyclohexane, and 1,3,7-octatriene have also been reduced (303, Table 20).

Strohmeier and Iglauer (313a) have studied the catalyzed reduction of a series of cyclic polyolefins to monoolefins by hydrogen uptake at 25° and 1 atm (Table 21). The reactions were first-order in substrate, and the listed k values are pseudo-first-order constants. The systems were discussed basically in terms of the mechanism of eqs. 180 and 181. An inverse correlation between rate and the angle $\beta(\hat{CC}_1C)$, which depends on the degree of hybridiz-ation at C_1, was suggested; other geometrical factors (ring size, bond lengths, etc.) seemed unimportant. The reduction of cyclopentadiene (294, 296, 297), 1,3-cyclohexadiene (295, 297), and cyclooctatetrene (303), and the

Table 20   Hydrogenation of Conjugated Dienes Catalyzed by $Co(CN)_5^{3-}$

Substrate	Product	Reference
Butadiene[a]	But-1-ene, *trans*-but-2-ene,	21, 268c, 296, 297, 303–308, 311, 312, 328, 1438
Isoprene[b]	2-Methylbut-1- and -2-enes	21, 295, 313
	3-Methylbut-1-ene	1439
1,3-Pentadiene[b]	Pent-1-ene, *trans*-pent-2-ene	21
2,3-Dimethylbutadiene[c]	2,3-Dimethylbut-1- and 2-enes	303
2,4-Hexadiene[c]	Hex-2-enes	303
2,5-Dimethyl-2,4-hexadiene[c]	No reduction	306
1,3,7-Octatriene[c]	Non-conjugated octadienes	303
Norbornadiene[b]	Nortricyclene, norbornene	21, 306
1,2-Dimethylenecyclohexane[c]	Dimethylcyclohexenes, methyl methylenecyclo- hexane	303
Cyclopentadiene[d]	Cyclopentene	294, 296, 297, 313a
1,3-Cyclohexadiene[b]	Cyclohexene	295, 297, 313a
3-Methylenecyclohexene[c]	1- and 3-Methylcyclohexenes	303
Cyclooctatetrene[b]	A bicyclooctadiene	303, 313a

[a] With 1 atm H_2, 20 to 25°.
[b] 1 atm H_2, 25°.
[c] Conditions not reported.
[d] 1 to 65 atm H_2, 20 to 25°.

nonreduction of 1,3-cyclooctadiene (268c) had been noted earlier. Kwiatek and coworkers (297) have reported that the reduction rate of 1,3-cyclohexadiene was limited by the rate of hydrogen absorption by $Co(CN)_5^{3-}$. Considering this and the nonhomogeneity of the solutions used by Strohmeier and Iglauer (the polyolefins were insoluble in the aqueous media), their interpretation may not be entirely valid.

A mechanism corresponding to reactions 180 and 181 has been considered for the reduction of sorbic acid, $CH_3CH{=}CHCH{=}CHCO_2H$, and sorbate (278, 295, 314–319) (see also Section B-2b). DeVries reported (278) the selective hydrogenation of sorbic acid to 2-hexenoic acid, using $Co(CN)_5^{3-}$ in aqueous solution with and without hydrogen (1 atm); however, the reduction had been observed earlier by De Jonge (314). At 25° the reaction rates, measured in a constant-pressure gas uptake apparatus, were first-order in substrate and cobalt hydride (produced according to either eq. 150 or eq. 151); k was $\sim 4 \times 10^{-3}$ M^{-1} s^{-1} for solutions 0.5 M in KCN and 0.2 M in

Table 21   Hydrogenation Rates of Cyclic Polyolefins (313a)[a]

Substrate		$k \times 10^2$ (s^{-1})	$\beta^0(C\hat{C}_1C)$
C_6H_8,	C_1	3.1	
C_5H_6,	C_1	1.8	109
C_7H_8,	C_1	1.3	121.8
C_7H_{10},	C_1	0.16	
C_8H_{10},	C_1	0.12	
C_8H_8,	C_1	0.05	126.5
C_8H_{12}	C_1	0.0	

[a] Using 0.1 M Co(CN)$_5$$^{3-}$, 0.25 M olefin 1 atm H$_2$, 25°.

KCl at a CN/Co ratio of 5.7 (278), while k was ∼7 × 10^{-2} M^{-1} s^{-1} for solutions with 1 M KCN and added NaOH (278, 315). Mabrouk and coworkers (315) showed that the distribution of products between 2-, 3-, and 4-hexenoates (all *trans*) was 82, 17, and 1 % respectively; the selectivity for 2-hexenoate was increased to 96 % when the hydrogenation was carried out in methanol. The same workers later studied (316) D$_2$—H$_2$O, H$_2$—D$_2$O, and D$_2$-methanol exchange during the reduction of the sorbic acid, and they found that the added hydrogen came from the solvent. Reduction in D$_2$O yielded hexenoate-d_2 and -d_3; mass spectrometric and n.m.r. studies showed that the deuterium in the 2-hexenoate was located on carbon atoms 4 (1 deuterium atom) and 5 (1.7 deuterium atom), although their earlier paper (316) had indicated exchange on C-5 only. These exchange results (316, 320) were not discussed mechanistically, but Frankel and Dutton (81) later presented the following to account for the d_2 product (assuming rapid production of $DCo(CN)_5{}^{3-}$ in D_2O according to equations such as 163 and 164):

$$CH_3CH{=}CH{-}CH{=}CHCO_2{}^- \xrightarrow{DCo(CN)_5{}^{3-}} \underset{\underset{D}{|}\quad\underset{Co(CN)_5}{|}}{CH_3CH{-}CH{-}CH{=}CHCO_2{}^-} \quad \mathbf{42}$$

$$\text{(197)}$$

$$\mathbf{42} + DCo(CN)_5{}^{3-} \rightarrow CH_3CHD{-}CHD{-}CH{=}CHCO_2{}^- + 2Co(CN)_5{}^{3-}$$

$$\text{(198)}$$

Reaction 198 corresponds to the final step postulated by Burnett and co-workers (reaction 187). To account for the major hexenoate-d_3 product (73–87%), Frankel and Dutton (81) suggest exchange of H on C-5 during the addition reaction 197, or directly due to activation by the presence of the conjugated triene system: $C{-}C{=}C{-}C{=}C{-}C{=}O$; as they suggest, analysis of the sorbate during reduction in the D_2O would give evidence for this direct exchange.

Murakami and coworkers (317, 318) have used more severe conditions (70°, ~50 atm H_2) with aqueous cobalt cyanide solutions, and reduced sorbic acid to 60% 4-hexenoic acid.

Possible catalytic isomerization of a reaction product in these sorbic acid reductions does not seem to have been considered. These earlier reported kinetic studies (278, 315) were very limited, and the mechanism via the organocobalt adduct (reaction 197) was by no means well established. A radical mechanism has been suggested from some more recent kinetic studies (see Section B-2b).

b. α,β-Unsaturated Acids, Esters, and Amides (Table 22).   The catalyzed hydrogen reduction of cinnamic acid to β-phenylpropionic acid has been reported by Kwiatek's group (295, 297) and Murakani and coworkers (317, 318). The system has been subjected to a detailed kinetic study by Simandi and Nagy (292a, 321, 322); because hydrogen consumed was replaced electrolytically, electrolysis times were used to construct uptake plots (292, 322). Iguchi and coworkers (322a) have also presented some kinetic data.

The observed kinetics (322) are consistent with the following mechanism involving a radical intermediate (S = cinnamate, $H\dot{S} = PhCH_2\dot{C}HCO_2{}^-$, $H_2S = PhCH_2CH_2CO_2{}^-$, $Co = Co(CN)_5{}^{3-}$, and $CoSH = PhCH_2CHCO_2{}^-$):

$$\underset{\underset{Co}{|}}{\phantom{CoSH = PhCH_2CHCO_2{}^-}}$$

$$2Co + H_2 \underset{}{\overset{K_1}{\rightleftharpoons}} 2CoH \qquad\qquad\qquad \text{(199)}$$

$$CoH + S \xrightarrow{k_1} Co + H\dot{S} \qquad\qquad \text{(200)}$$

$$H\dot{S} + CoH \xrightarrow{k_2} Co + H_2S \qquad\qquad \text{(201)}$$

$$H\dot{S} + Co \underset{}{\overset{K_2}{\rightleftharpoons}} CoSH \qquad\qquad \text{(202)}$$

Table 22 Hydrogenation of α,β-Unsaturated Acids, Esters, and Amides Catalyzed by $Co(CN)_5^{3-}$

Substrate	Condition	Product	Reference
Cinnamic acid	1–50 atm H_2, 20–70°	β-Phenylpropionic acid	292a, 295, 297, 303, 317, 318, 321, 322, 322c
Sorbic acid	1–50 atm H_2, 20–70° or N_2 (25°)	2-, 3-, and 4-Hexenoic acids	278, 295, 314–319, 329–331
Muconic acid	1 atm H_2, 25°	2-Hexendioic acid	331a, 331b
Acrylic acid	1 atm H_2 or CoH^a, 25°	No reduction	295, 297, 323
	1 atm H_2, 125°	Propionic acid, α-methylglutaric acid	295, 297
	30 atm H_2, 70°	Propionic acid	325
	N_2, 125°	3-Methylglutaconic acid	297
Methacrylic acid	CoH, 25°	Reduced	323
	1 atm H_2, 25°	Isobutyric acid	295, 297, 327, 1443, 1444
	N_2, 125°	α-Methylene,γ,γ-dimethylglutaric acid	297
Atropic acid	1 atm H_2, 25°	α-Phenylpropionic acid	297
Itaconic acid	1 atm H_2, 25°	α-Methylsuccinic acid	297, 327
	CoH, 25°	Reduced	323
Tiglic acid	1 atm H_2, 25°	Not reduced	297
	CoH, 25°	Reduced	323
Citraconic acid	CoH, 25°	Reduced	323
Mesaconic acid	CoH, 25°	Reduced	323
Crotonic acid	1 atm H_2, 25°	No reduction	297
	CoH, 25°	Slight reduction	323
Angelic acid	CoH, 25°	Isomerization to tiglic acid	323
2-Ethylpropenic acid	CoH, 25°	Isomerization to tiglic acid	323
Maleic acid	50 atm H_2, 70°	Succinic acid	318, 326
	CoH, 25°	Slight reduction	323
Fumaric acid	CoH, 25°	Slight reduction	323
Acetylenedicarboxylic acid	50 atm H_2, 20°	Fumaric acid	318, 326
	50 atm H_2, 70°	Succinic acid	318, 326
Methylmethacrylate	1 atm H_2, 25°	Methyl isobutyrate, isobutyric acid	295, 327
Ethyl acrylate	30 atm H_2, 70°	Adipic acid	325
	N_2, 70°	Adipic and β-ethoxypropionic acids	325
Methacrylamide	1–200 atm H_2, 20–125°	Isobutyramide, isobutyric acid	327

^{a} Here and elsewhere in this column, CoH = $HCo(CN)_5^{3-}$.

124

Assuming a steady-state concentration of $H\dot{S}$ radicals, such a mechanism yielded the rate law

$$\frac{-d[H_2]}{dt} = k_1[S][Co]_T \cdot \frac{\sqrt{K_1 H_2}}{1 + \{k_1/k_2\}K_2[S]\} + \sqrt{K_1 H_2}} \tag{203}$$

Such an expression was readily analyzed by inverse plots to yield the following values at $20°$ ($CN/Co = 10$, ionic strength $1.0\ M$ with added KCl): $k_1 = 1.95 \times 10^{-4}\ M^{-1}\ s^{-1}$ ($\Delta H^{\ddagger} = 21.7$ kcal mole^{-1}, $\Delta S^{\ddagger} \approx -1$ eu), $K_2/k_2 = 1.61 \times 10^5$ s, and $K_1 = 1.55 \times 10^5\ M^{-1}$. The K_1 value is in good agreement with those determined directly by DeVries and Burnett and coworkers (Section X-A-2). Although the $H\dot{S}$ radical anion forms an organocobalt complex (reaction 202), this plays no essential role in the hydrogenation, and the kinetic data are inconsistent with a mechanism of the type outlined in reactions 180 and 181. Reaction 200 was discussed in terms of hydrogen atom transfer from the hydride $[Co^{II}(CN)_5(\cdot H)]^{3-}$, **43** (Section X-A-2). The original paper (322) indicated the free electron in $H\dot{S}$ to be on the β-carbon atom, and the Co to be bonded to the β-carbon atom in CoSH, but hydrogen atom attack at the β-carbon would be anticipated as shown in reaction 204 (see also Ref. 323 and Section X-B-4).

$$\underset{\underset{\overset{|}{H-Co}}{\overset{\beta}{Ph}}}{Ph-\overset{\beta}{C}H} \overset{\alpha}{\underset{}{\rightleftharpoons}} \overset{\alpha}{C}H - C \overset{\nearrow O}{\underset{\searrow O}{}} \xrightarrow{k_1} Ph-CH_2-\dot{C}HCO_2^- + Co \tag{204}$$

Further support cited (322) for the hydride formulation as **43** were an observed radical polymerization initiated by the hydride (324) (see eq. 200) and the production of dimers during some catalytic hydrogenations; for example, acrylic acid produced some α-methylglutaric acid as well as pro-pionic acid (297) (but see below).

Reaction 200 was likened to the well-known Rideal mechanism for heterogeneous catalysis, the hydrogen atom being "adsorbed" on the active $Co(CN)_5^{3-}$ site (322).

Kwiatek and Seyler (303) have shown by deuteration studies that formation of the radical (eq. 200) is reversible although no details were given. If the reverse reaction rate constant is k_{-1}, rate law 203 is modified and includes an additional term $(+k_{-1}/k_2\sqrt{K_1 H_2})$ in the denominator. The ratio $k_{-1}:k_2\sqrt{K_1 H_2}$, represents the competition between the reverse of reaction 200 and the forward reaction 201. The good analysis of the kinetic data of Simandi and Nagy (322) according to eq. 203 indicates a quite small value of this extra term compared to unity.

Iguchi and coworkers (322a) studied the reaction for CN/Co ratios of 2 to 6 and found optimum activity when the ratio was 6. The hydrogenation rate was given as

$$-d[H_2]/dt = k[Co]_T^{1.2}[\text{cinnamate}]^{0.6} \qquad (205)$$

An overall activation energy of 15.9 kcal mole^{-1} was determined, but k is undoubtedly a composite constant (see eq. 203).

More detailed kinetic studies by Simandi and coworkers (329–331) on the $Co(CN)_5{}^{3-}$-catalyzed hydrogenation of sorbic acid to 2-hexenoic acid yielded a rate law of the type shown in eq. 203. These workers invoked the mechanism outlined in eqs. 199 to 202 and ruled out the previously postulated mechanism of eqs. 197 to 198 involving the organometallic compound (see Section X-B-2a). Their kinetic analysis yielded the following values at 20° for a $CN:Co$ ratio of 10 (see eqs. 199 to 202): $k_1 = 3.1 \times 10^{-2}\ M^{-1}\ s^{-1}$ ($\Delta H^{\ddagger} = 10.3$ kcal mole^{-1}, $\Delta S^{\ddagger} = -31$ eu), $K_2/k_2 = 2.74 \times 10^3$ s, and again $K_1 = 1.55 \times 10^5\ M^{-1}$. Such an entropy of activation value is consistent for a reaction such as eq. 200 between species of charge -3 and -1 (329), although it should be noted that the $\Delta S^{\ddagger}$ value for the cinnamate system with the same charge species was -1 eu (322). The reduction of muconic acid (2,4-hexadienedioic acid) has been analyzed in terms of the same radical mechanism (331a, 331b).

Besides cinnamic acid and sorbic acid, a number of other α,β-unsaturated acids, esters, and amides have been reduced (295, 297, 318, 323, 325–328) (Table 22). The acids have generally been reduced as anions in basic solution. There are some inconsistencies in the reported data (e.g., for tiglic acid). α-Substituted mono- and dibasic acids (methacrylic, atropic, itaconic, tiglic, citraconic, mesaconic) are readily reduced. Acids lacking an α-substituent (acrylic, crotonic, maleic, fumaric, acetylenedicarboxylic) seem to require more stringent conditions. Acrylic acid gives propionic and α-methylglutaric acid, the latter being thought to come from 3-methylglutaconic acid, which is formed in the absence of hydrogen (297). Similarly, methacrylic acid yielded α-methylene-γ,γ-dimethylglutaric acid when heated with $Co(CN)_5{}^{3-}$ under nitrogen (297). The reduction of diphenylmaleic anhydride to *dl*-diphenylsuccinic acid (318, 326) indicates an overall *trans* addition of hydrogen.

Zakhariev and coworkers (1443, 1444) have studied the kinetics of the methacrylate reduction under hydrogen; a first-order dependence on substrate and a 1.5 order in catalyst were interpreted in terms of the radical mechanism. A maximum rate was observed at 60°, since above this temperature side reactions were said to deactivate the catalyst.

Jackman and coworkers (323) and Halpern and Wong (328) have reported data on the kinetics of the addition of the cobalt hydride and deuteride to

α,β-unsaturated acids, and these workers as well as Kwiatek and Seyler (21, 268c, 303) have discussed the nature of the resulting adduct in solution, using particularly n.m.r. data. The carboxylate anions and esters with no α-substituent (acrylate, crotonate, maleate, fumarate, and dimethyl esters of maleic and fumaric acids) form quite stable σ-complexes with bonding between Co and the α-carbon atom, for example, **44** (268c, 303, 323). High stability of such σ-complexes will tie up the cobalt and on both reaction

$$RCH_2\overset{\displaystyle |}{\underset{\displaystyle Co(CN)_5^{3-}}{C}}HCO_2^-$$

44

mechanism schemes considered (eqs. 180, 181, or 200 to 202) will inhibit reduction (268c), in general agreement with the experimental findings. The readily reducible methyl methacrylate forms an unstable adduct (21, 268c, 303), which disproportionates in the same manner as the butadiene adduct (see Section X-B-2a, reverse reaction of 180, followed by reaction 181) to give methyl methacrylate and methyl isobutyrate.

The reaction rates for the adduct formation were first-order in hydride and unsaturated acid, and were unaffected by large changes in cyanide concentration, indicating that π-allyl complexes were not involved (323, 328). The rate for acrylate system showed a pH dependence due to variation in the concentrations of acrylate and acrylic acid (328). The kinetic data and their mechanistic significance will be considered more generally later (Section X-B-4).

Maleate and fumarate yielded the same σ-complex (**44**, R = CO_2^-), and n.m.r. data indicate that one conformer about the central single C—C bond is preferred in which the —$Co(CN)_5^{3-}$ and vicinal —CO_2^- groups are anti-periplanar. The adducts from crotonate and fumarate with deuteride, the latter giving the *threo* isomer **45**, show that σ-complex formation involves a stereospecific *cis* addition (323). However, deuteride reduction of mesaconate and citraconate gave a mixture of the *erythro* and *threo* isomers of $^-O_2CCHDCD(CO_2^-)CH_nD_{3-n}$ ($n = 0.8$), which shows that reduction is nonstereospecific (323).

$$
\begin{array}{c}
CO_2^- \\
| \\
D-\!\!\!-\!\!\!-\!\!\!-\!\!\!-H \\
| \\
H-\!\!\!-\!\!\!-\!\!\!-\!\!\!-Co \\
| \\
CO_2^-
\end{array}
$$

45

c. α,β-Unsaturated Aldehydes (Table 23). The reduction of α,β-unsaturated aldehydes (295, 297, 303, 327, 332) is not very efficient. Methacrolein is reduced (303), possibly slowly, in contradiction to an earlier report

Table 23   Hydrogenation of α,β-Unsaturated Aldehydes Catalyzed by $Co(CN)_5^{3-}$

Substrate	Condition	Product	Reference
Acrolein	1 atm H_2, 25°	No reduction	297
Methacrolein	1 atm H_2, 25°	No reduction	297
Methacrolein	not reported	Isobutyraldehyde	303
Crotonaldehyde	1 atm H_2, 25°	n-Butyraldehyde	297
Tiglaldehyde	1–200 atm H_2, 20–125°	α-Methylbutyraldehyde	295, 297, 327
α-Methylpentenal	1–200 atm H_2, 20–125°	α-Methylvaleraldehyde	297, 327
Propionaldol	1 atm H_2, 25°	α-Methylvaleraldehyde	297
Propionaldehyde	1 atm H_2, 25°	α-Methylvaleraldehyde	295, 297
Butyraldehyde	1 atm H_2, 125°	Reduced condensation product	268c
Benzaldehyde	1–70 atm H_2, 0–125°	Benzyl alcohol	297, 332
Cinnamaldehyde	1 atm H_2, 25°	2-Benzyl-5-phenyl-pent-2-enal	295, 297
Benzaldehyde + acetaldehyde	1 atm H_2, 25°	α-Benzylcinnamaldehyde	297

(297). The saturated propionaldehyde absorbs hydrogen to yield α-methyl-valeraldehyde, **45a**, evidently via a base-catalyzed aldol condensation, since both propionaldol and its dehydration product, 2-methylpentenal, are themselves reduced to the α-methylvaleradehyde (297):

$$2C_2H_5CHO \longrightarrow C_2H_5CH(OH)CH(CH_3)CHO \xrightarrow{-H_2O}$$

Propionaldol

$$C_2H_5CH{=}C(CH_3)CHO \xrightarrow{H_2} \textbf{45a} \qquad (206)$$

2-Methylpentenal

Mixed aldehydes yielded similar condensation products. Benzaldehyde and acetaldehyde form α-benzylcinnamaldehyde presumably via cinnamaldehyde and phenylpropionaldehyde; cinnamaldehyde itself yields 2-benzyl-5-phenyl-pent-2-enal, the aldol condensation product of β-phenylpropionaldehyde.

The benzaldehyde reduction to benzyl alcohol requires the presence of added hydroxide which is consumed during the reaction (297, 332). Adduct formation of aldehydes with the cobalt hydride has not been investigated (297), but the catalytic reduction was rationalized in terms of reactions 207 and 207a followed by 170 and 151:

$$HCo(CN)_5^{3-} + C_6H_5CHO \longrightarrow [Co(CN)_5C_6H_5CH_2O]^{3-} \qquad (207)$$

$$[Co(CN)_5C_6H_5CH_2O]^{3-} + H_2O \xrightarrow{OH^-}$$

$$Co(CN)_5OH^{3-} + C_6H_5CH_2OH \quad (207a)$$

Tiglaldehyde reduction requires no excess base (295); the conditions for the other aldehyde reductions have not been reported.

d. Styrenes (Table 24). Some structural specificities have been noted (21): α-methyl and methoxy styrenes were readily reduced, but a number of β-substituted styrenes (e.g., propenylbenzene, stilbene, and indene) were not reduced. However, cinnamyl alcohol was reduced to 3-phenylpropanol (297), and cinnamyl acetate underwent hydrogenolysis to propenylbenzene (303). HCo(CN)$_5^{3-}$ forms unstable adducts with *p*-cyanostyrene and 2-vinylpyridine; but no complex was isolated from styrene itself, indicating that electronegative groups stabilize the organocobalt complexes (21, 303, 328, see also Section X-B-4).

Some kinetic data for the hydrogenation of styrene have been reported by Strohmeier and Iglauer (333); styrene dissolved in benzene was added to the aqueous catalyst solution and the rate of hydrogen consumption measured. First-order plots were obtained for the loss of styrene, and the radical mechanism of eqs. 199 to 201 was invoked (rate law 203 may give pseudo first-order in substrate under limiting conditions). Half-lives ranging from 18 hr at 0° to 3 hr at 50° were given for 10^{-3} *M* cobalt solutions.

e. Ketones (Table 25). The carbonyl group itself is reduced only when activated by conjugation; thus diacetyl gives 3-hydroxy-2-butanone, benzil gives benzoin, isatin gives dioxindole. Benzophenone and acetophenone are not reduced (268c). Olefinic bonds activated by carbonyl groups are reduced. Thus isopropenyl ketones were readily reduced, and no complexes of the type Co(CN)$_5$C(CH$_3$)$_2$COR^{3-} were formed from the hydride or from a reaction such as 183 using the 1-bromoisopropyl ketones (303). Phenyl vinyl

Table 24 Hydrogenation of Styrenes Catalyzed by Co(CN)$_5^{3-}$

Substrate	Condition	Product	Reference
Styrene	1 atm H$_2$, RT	Ethylbenzene	295, 297, 333
Styrene	50 atm H$_2$, 70°	No reduction	317
α-Methylstyrene	1 atm H$_2$, RT	Isopropylbenzene	297
α-Methoxystyrene	not reported	1-Phenyl-1-methoxy-ethane	303
1,1-Diphenylethylene	1 atm H$_2$, RT	1,1-Diphenylethane	303
Cinnamyl alcohol	1 atm H$_2$, RT	3-Phenylpropanol	297
Cinnamyl acetate	Not reported	Propenylbenzene	303
Propenylbenzene	1 atm H$_2$, RT	No reduction	295, 297
Stilbene	1 atm H$_2$, RT	No reduction	297
Indene	1 atm H$_2$, RT	No reduction	295
p-Cyanostyrene	Not reported	*p*-Cyanoethylbenzene	303
2-Vinylpyridine	Not reported	2-Ethylpyridine	303

Table 25   Hydrogenation of Ketones Catalyzed by $Co(CN)_5^{3-}$

Substrate	Condition	Product	Reference
Methyl isopropenyl ketone	1–200 atm H_2, 20–125°	Methyl isopropyl ketone	303, 327
Phenyl isopropenyl ketone	Not reported	Phenyl isopropyl ketone	303
Phenyl vinyl ketone	Not reported	Phenyl ethyl ketone	307
Benzylidene acetone	1–70 atm H_2, 0–125°	1-Phenylbutan-3-one	334
Dibenzoylethylene	1–70 atm H_2, 0–125°	Dibenzoylethane	334
p-Benzoquinone	1–70 atm H_2, 0–125°	p-Hydroquinone, quinhydrone	295, 297, 332, 335, 336
9,10-Anthraquinone	1 atm H_2, 25°	9,10-Dihydroxyanthracene	297
Indigo	Not reported	Indigo white	303
Benzil	1 atm H_2, 25°	Benzoin	295
Diacetyl	Not reported	3-Hydroxy-2-butanone, 2,5-dimethylhydro-quinone	303
Isatin	1 atm H_2, 25°	Dioxindole	20, 303

ketone was inefficiently reduced and yielded a stable adduct $Co(CN)_5CH(CH_3)COC_6H_5^{3-}$ (307), which could be protonated (308) similarly to the allylic complexes (eq. 184) to yield phenyl ethyl ketone (see also Section X-B-3):

$$Co(CN)_5CH(CH_3)COC_6H_5{}^{3-} \xrightarrow{H^+} [Co(CN)_5CH(CH_3)\overset{OH}{\overset{|}{C}}{}^+C_6H_5] \longrightarrow$$
$$Co(CN)_5{}^{2-} + C_2H_5COC_6H_5 \quad (208)$$

Norbornenone is apparently not reduced but forms adducts with $HCo(CN)_5^{3-}$ (268c); both *endo* and *exo* isomers were thought to be formed; on protonation the *endo* form yields 1-hydroxynortricyclene:

$$(208a)$$

Benzylidene acetone, dibenzoylethylene, benzoquinone, and benzil were catalytically reduced only in the presence of added hydroxide, which was thought (297) to be necessary to hydrolyze intermediates to $Co(CN)_5OH^{3-}$ with displacement of the reduced substrate.

The qualitative data for the benzoquinone system were rationalized in terms of eqs. 209 to 211 (297):

$$HCo(CN)_5^{3-} + C_6H_4O_2 + OH^- \longrightarrow Co(CN)_5(C_6H_4O_2)^{4-} + H_2O \quad (209)$$

$$Co(CN)_5(C_6H_4O_2)^{4-} + OH^- \longrightarrow Co(CN)_5OH^{3-} + C_6H_4O_2^{2-} \quad (210)$$

$$2Co(CN)_5^{3-} + C_6H_4O_2 \longrightarrow \underset{\textbf{46}}{(CN)_5Co(C_6H_4O_2)Co(CN)_5^{6-}} \quad (211)$$

Equation 211 was included to explain inhibition of initial hydride formation in the presence of excess benzoquinone. However, the system is probably much more complex. Vlcek and Hanzlik (335, 336) have since investigated the kinetics and mechanisms of the reaction of benzoquinone with both $Co(CN)_5^{3-}$ and $HCo(CN)_5^{3-}$. The anion **46** is produced rapidly according to eq. 211 via a redox addition mechanism in which the cobalt atoms coordinate a sixth ligand with simultaneous transfer of an electron to the new ligand which in **46** becomes isoelectronic with the dianion of hydroquinone. Species **46** underwent a four-electron reduction at the dropping mercury electrode, which was written (335) as

$$\textbf{46} \xrightarrow{4e} 2Co(CN)_5^{4-} + \text{hydroquinone} \quad (212)$$

$$Co(CN)_5^{4-} + H_2O \rightleftharpoons HCo(CN)_5^{3-} + OH^- \quad (213)$$

The same workers have given polarographic data (337) showing the reversible oxidation to $Co(CN)_5^{2-}$ of the cobalt(I) species, $Co(CN)_5^{4-}$, produced in highly alkaline solutions of the hydride (eq. 213). The pK for the hydride acting as an acid was estimated to be about 17 to 19, which shows that the Co(I) species would be present in kinetically significant concentration even at pH 10 (336, 337). Complex **46** decomposes in solution first to $Co(CN)_5$-H_2O^{2-} and $(CN)_5CoOC_6H_4OH^{3-}$ (**47**), in which the ligand resembles hydroquinone, and then **47** decomposes to $Co(CN)_5H_2O^{2-}$ and hydroquinone. Benzoquinone reacts with the $HCo(CN)_5^{3-}$ complex at pH > 9 to give initially **47**:

$$HCo(CN)_5^{3-} + C_6H_4O_2 \xrightarrow{k} \underset{\textbf{47}}{(CN)_5CoOC_6H_4OH^{3-}} \quad (214)$$

The rate was pH dependent and was attributed to reaction with both the hydride and the $Co(CN)_5^{4-}$ species in equilibrium according to eq. 213 (336). The rate was given as

$$\text{rate} = k_1[\text{quinone}][HCo(CN)_5^{3-}] + k_2[\text{quinone}][Co(CN)_5^{4-}] \quad (215)$$

At 0°, $k_1 = 13.5 \ M^{-1} s^{-1}$ and $k_2 \sim 10^9 \ M^{-1} s^{-1}$. The k_2 reaction is thought to occur via the redox addition mechanism previously mentioned; since **47** was the product at pH 9.2, where the k_1 term predominates, the possibility of direct hydride transfer via a bridge mechanism was excluded, and an insertion involving a 7-coordinate intermediate was suggested. Hanzlik and Vlcek (336) concluded that hydrogenation reactions of this type using solutions of $HCo(CN)_5^{3-}$ were likely to proceed via redox additions (involving Co(I)) followed by protonations of the primary products (see also Section X-B-5).

Halpern and coworkers (1332) have also detected the pentacyanocobaltate(I) species by reaction of $Co(CN)_5^{3-}$ with hydrated electrons in aqueous alkaline solution; the species decays rapidly at pH 13 to $HCo(CN)_5^{3-}$ according to reaction 213.

f.   Nitro and Nitroso Compounds (Table 26).   Nitro compounds have in some cases been reduced to the corresponding amino derivative, but generally bimolecular products are formed. The reactions have not been studied in any detail mechanistically, but it seems that the nitro compound must be added slowly to the catalyst solution to ensure the presence of excess catalyst throughout the reaction; the nitrobenzene system is probably similar to the benzoquinone system but, unlike the latter, partial reduction of nitrobenzene occurred in the absence of alkali (297). For the reduction of nitrobenzene under nitrogen, hydrazine was added to produce the cobalt hydride (338).

N-Nitroso compounds have been reduced to amines. Azoxybenzene and azobenzene have also been reduced.

Corresponding to the redox addition reaction 211, Swanwick and Waters (588) have recently reported that nitroso compounds and nitrobenzene react with $Co(CN)_5^{3-}$ itself to give stable radical-anions, $[(NC)_5Co—N-(Ar)—O\cdot]^{3-}$; for example:

$$(NC)_5Co\cdot^{3-} + Ph—N{=}O \longrightarrow [(NC)_5Co—N(Ph)—O\cdot]^{3-} \quad (215a)$$

g.   α,-β Unsaturated Nitriles (Table 27).   The nitrile group itself is not hydrogenated. α-Phenyl- or α-ethoxyacrylonitrile is reduced to the corresponding propionitrile when added gradually to $HCo(CN)_5^{3-}$ under hydrogen at room temperature; when the acrylonitriles are added in excess to the hydride, *N*-amidocobalt complexes are formed (21). Both the reduction and complex formation are thought to involve radical intermediates (see eqs. 200 and 201).

$$CH_2{=}CRCN + CoH \longrightarrow CH_3\dot{C}RCN + Co \quad (216)$$

$$CH_3\dot{C}RCN + CoH \longrightarrow CH_3CHRCN + Co \quad (217)$$

$$CH_3\dot{C}RCN + Co \longrightarrow CH_3CR{=}C{=}N—Co \xrightarrow{H_2O}$$
$$CH_3CHRCONH—Co \quad (218)$$

Acrylonitrile and methacrylonitrile underwent neither of these reactions under mild conditions; instead they formed $CH_3CR(CN)Co(CN)_5^{3-}$ adducts,

Table 26 Hydrogenation of Nitro and Nitroso Compounds Catalyzed by Co(CN)$_5^{3-}$

Substrate	Condition	Product	Reference
Nitroethane	30 atm H$_2$, 70–90°	Ethylamine, ethanol, acetaldehyde	325
2-Nitro-2-methylpropane	30 atm H$_2$, 70–90°	tert-Butylamine	325
Nitrobenzene	1 atm H$_2$, 25°	Azobenzene, hydrazobenzene	295, 297
	30 atm H$_2$, 70–90°	Hydrazobenzene, azobenzene, aniline	325
	<70 atm H$_2$, 0–125°	Phenylhydroxylamine, azobenzene, hydrazobenzene, azoxybenzene	332
o-Nitrotoluene	N$_2$, 70°	Aniline, azoxybenzene	338
p-Nitrotoluene	1 atm H$_2$, 25°	2,2′-Dimethylazobenzene and 2,2′-dimethylazoxybenzene	297
	1 atm H$_2$, 25°	4,4′-Dimethylazoxybenzene and p-tolylhydroxylamine	297, 332
	<70 atm H$_2$, 0–125°		
o-Chloronitrobenzene	<70 atm H$_2$, 0–125°	2,2′-Dichloroazoxybenzene, 2,2′-dichloroazobenzene, 2,2′-dichlorohydrazobenzene	332
o-Nitrophenol	30 atm H$_2$, 70–90°	o-Aminophenol	325
p-Nitrophenol	30 atm H$_2$, 70–90°	p-Aminophenol	325
m-Nitrophenol	30 atm H$_2$, 70–90°	No reduction	325
o-Nitroanisole	1 atm H$_2$, 25°	2,2′-Dimethoxyhydrazobenzene	297
Azoxybenzene	1 atm H$_2$, 25°	Azobenzene	295
N-Nitrosodimethylamine	<70 atm H$_2$, 0–125°	1,1-Dimethylhydrazine, dimethylamine	332
N-Nitrosodiphenylenamine	Not reported	Diphenylamine	303
Azobenzene	1 atm H$_2$, 25°	Hydrazobenzene	297

Table 27 Hydrogenation of Nitriles, Oximes, and Hydrazones Catalyzed by Co(CN)$_5^{3-}$

Substrate	Condition	Product	Reference
Nitriles			
α-Phenylacrylonitrile	1 atm H$_2$, 25°	α-Phenylpropionitrile	21
α-Ethoxyacrylonitrile	1 atm H$_2$, 25°	α-Ethoxypropionitrile	21
Acrylonitrile	1 atm H$_2$, 0–25°	No reduction	21, 307
	30 atm H$_2$, (or N$_2$), 70°	Adipic acid, β-ethoxypropionic acid	325
Methacrylonitrile	1 atm H$_2$, 25°	No reduction	21
Phenylacetonitrile	30 atm H$_2$, 70°	Phenylacetic acid	325
Benzoyl cyanide	30 atm H$_2$, 70°	Benzoic acid, benzamide	325
Cinnamonitrile	30 atm H$_2$, 70°	β-Phenylpropionic acid and amide	325
	N$_2$ (70°)	Cinnamic acid	325
1-Cyano-allene	Not reported	Crotonitrile	303
Oximes and Hydrazones			
Acetoxime	50 atm H$_2$, 70°	Isopropylamine	317
Acetophenone oxime	50 atm H$_2$, 70°	α-Phenylethylamine	317, 343
Phenylpyruvic acid oxime	50–100 atm H$_2$, 30–100°	Phenylalanine	317, 347
Pyruvic acid oxime	50 atm H$_2$, 40–70°	Alanine	317, 343
α-Ketoglutaric acid oxime	50 atm H$_2$, 70°	Glutamic acid	317, 343
Benzoin oxime	~50 atm H$_2$, 40°	*erythro*-1,2-Diphenylethanolamine	343
Benzoyl cyanide oxime	30 atm H$_2$, 70°	α-Aminophenylacetic and benzoic acids	325
Phenylacetyl cyanide phenyl hydrazone	30 atm H$_2$, 70°	2-(N′-Phenylhydrazino)-3-phenylpropionic acid	325
Dimethylfurazone	Not reported	Tetramethylpiperazine	303

indicating that intermediate radicals were not formed with these substrates (21, 307). Some later kinetic studies, however, do not support this conclusion (Section X-B-4). 1-Cyanocyclohexene behaves similarly (303). At higher temperatures and hydrogen pressures the nitrile group is hydrolyzed, apparently catalytically by the pentacyanocobaltate(II) (325). The stable carbon-cobalt adducts are similarly formed by crotonitrile and cinnamonitrile (303); the latter substrate is reduced and hydrolyzed under more severe conditions (325).

h. *Oximes and Hydrazones (Table 27).* The $>$C$=$NOH linkage in oximes is reduced to $>$CH—NH$_2$ under more severe conditions (317, 325, 343). Reduction of phenylcyanomethylenequinone oxime gave an *N*-amidocobalt complex according to reactions such as 216 to 218, after the oxime group had first been reduced (21):

$$\text{HON}=\!\!\!\left\langle\ \right\rangle\!\!\!=\text{C(C}_6\text{H}_5)\text{CN} \longrightarrow$$

$$\text{H}_2\text{N}\!\!\!-\!\!\!\left\langle\ \right\rangle\!\!\!-\text{CH(C}_6\text{H}_5)\text{CONHCo(CN)}_5{}^{3-} \qquad (218a)$$

A phenylhydrazone group $>$C$=$N–NH— has been reduced to $>$CH—NH—NH— (325). See also Section 2i.

i. *Reductive Amination Reactions (Table 28).* A general useful method for synthesis of amines is reductive amination in which ketones are hydrogenated in the presence of excess ammonia; imines are likely intermediates (340):

$$>\!\text{CO} \underset{\text{H}_2\text{O}}{\overset{\text{NH}_3}{\rightleftarrows}} >\!\text{C}=\!\text{NH} \xrightarrow[\text{catalyst}]{\text{H}_2} >\!\text{CH}-\text{NH}_2 \qquad (219)$$

α-Aminoacids have been obtained by reductive amination of α-ketoacids or esters using the HCo(CN)$_5{}^{3-}$ system in the presence of ammonia (317, 339). The chloropentammine complex [Co(NH$_3$)$_5$Cl]Cl$_2$ also has been used instead of the pentacyano complex, but the process becomes much more efficient in the presence of added cyanide, CN/Co = 5 (317, 341, 342). *N*-Substituted phenylalanines were formed from phenylpyruvic acid by using various amines, RNH$_2$. The yield decreased in the order NMe $>$ NH $>$ NC$_6$H$_{11}$ $>$ NPh $\approx$ N^{t}Bu, and was explained by the decreasing basicity of the amines and steric effects of the substituents (340a). The mechanism was written in terms of a cobalt imino complex, **48**, although the nature of the hydrogen

Table 28 Reductive Amination Reactions[a] Catalyzed by Co(CN)$_5^{3-}$

Substrate	Condition	Product	Reference
Pyruvic acid	50 atm H$_2$, 40–70°	Alanine	317, 339, 341, 342
Phenylpyruvic acid	50 atm H$_2$, 30–70°	Phenylalanine[b]	317, 339, 340a, 341, 342
α-Ketoglutaric acid	50 atm H$_2$, 40°	Glutamic acid	317, 339, 341
Ethylpyruvate	50 atm H$_2$, 40°	Alanine	317
Ethyl phenylpyruvate	50 atm H$_2$, 40°	Phenylalanine	317
Ethyl acetoacetate	50 atm H$_2$, 40°	No reaction	317
Acetophenone	50 atm H$_2$, 40°	α-Phenylethylamine	339
Benzoin	50 atm H$_2$, 40°	erythro-1,2-Diphenylethanolamine	326, 339
Benzil	50 atm H$_2$, 40°	Benzoin, erythro-1,2-diphenylethanol-amine, erythro-1,2-diphenylethylene-diamine	326
Benzalacetone	50 atm H$_2$, 70°	3-Amino-1-phenylbutane	340a
Benzalacetophenone	50 atm H$_2$, 70°	1-Amino-1,3-diphenylpropane	340a
Methylcyclohexanones	50 atm H$_2$, 70°	Methylcyclohexylamines	340a
2-Ethoxycarbonylcyclopentanone	50 atm H$_2$, 70°	cis-2-Aminocyclopentanol (trace)	326
[Co(py)$_2$(O$_2$CCOCH$_2$CH$_2$Ph)$_2$]Cl	90 atm H$_2$, 115°	Phenylalanine	340b
[Co(py)$_2$(O$_2$CCOCH$_2$CH$_2$CO$_2$)OH	90 atm H$_2$, 115°	Glutamic acid	340b
[Co(py)$_2$(O$_2$CCOCH$_3$]O$_2$CCOCH$_3$	90 atm H$_2$, 115°	Alanine	340b

[a] Excess ammonia or amine added.
[b] Use of RNH$_2$ gave N-substituted phenylalanines (340a).

transfer step was not presented. Oximes were thought to be reduced similarly via a nitroso complex:

$$
\begin{array}{c}
\text{RCCOOH (or R$'$)} \\
\parallel \\
\text{NH (or NOH)}
\end{array}
+ \text{Co complex} \longrightarrow
\begin{array}{c}
\text{RC—CO} \\
\parallel \quad | \\
\text{(or HON) HN} \quad \text{O} \\
\diagdown \diagup \\
\overset{\text{Co}}{\underset{\text{complex}}{}} \\
\mathbf{48}
\end{array}
\xrightarrow{\text{H}_2}
\begin{array}{c}
\text{RCH—CO} \\
| \quad | \\
\text{H}_2\text{N} \quad \text{O} \\
\diagdown \diagup \\
\overset{\text{Co}}{\underset{\text{complex}}{}}
\end{array}
$$

$$(220)$$

Similarly some cobalt complexes of α-ketoacids, also containing pyridine ligands, have been reductively aminated to the corresponding amino acids (340b) (see Table 28).

Keto groups in acetophenones, benzil, benzoin, and cyclic ketones have also been reductively aminated (326, 339, 340a). The production of the *erythro* amino alcohol from benzoin, and *cis*-2-aminocyclopentanol from 2-ethoxycarbonylcyclopentanone, indicates that hydrogen addition occurs at the less sterically hindered side of the imine intermediate (326).

The α,β-unsaturated ketones, benzalacetone, and benzalacetophenone, $(C_6H_5CH{=}CHCOR)$, gave the saturated amines (340a); presumably the olefinic bond was reduced first (see Section X-B-2e).

j. Hydrogenation of Some Cobalt Complexes to Amino Acids (Table 29). Closely related to the reductions described in Sections 2h and 2i is the homogeneous hydrogenation of cobalt complexes containing ligands, such as α-oximinocarboxylic acids, salicylaldehyde oxime, and unsaturated amino acid derivatives, to amino acids, as described by Murakami, Kang, and coworkers (317, 344–348). The product yields are poor unless cyanide is added, and the optimum ratio of cyanide ion to complex is 5 to 6. The data suggest that the catalyst is a species in which only part of the ligands is replaced by cyanide, and the hydrogenation occurs easily when cyanide and the unsaturated compound are coordinated to cobalt simultaneously (317, 345, 347). The cobalt intermediates were thought (317) to be chelates of the type shown in eq. 220.

The complex phenylpyruvic acid-oximato-bis(propylenediamine)cobalt(III) carbonate, formulated as $[\text{Co(pn)}_2\text{OCOC}(:\text{NO})\text{CH}_2\text{Ph}]_2\text{CO}_3$ (349), and presumably **48a**, complexes cyanide likely by dissociation of the carboxylate or oximato group(s). Other added ions such as chloride, bicarbonate,

Table 29 Hydrogenation of Cobalt Complexes Containing Unsaturated Compounds[a]

Complex	Medium	Product	Reference
$[Co(pn)_2OCOC(:NO)R]_2CO_3$	MeOH, H_2O	$R = CH_3$, alanine	348
		$R = PhCH_2$, phenylalanine	344, 346, 347
$[Co(en)_2OCOC(NHCOCH_3):CHPh]Cl_2$	MeOH	Phenylalanine	344, 348
$Co(o\text{-}OC_6H_4CH:NOH)_2$	95% acetic acid	2'-Hydroxybenzylamine	344
$Co + PhCH_2C(:NOH)CO_2H$[b]	$EtOH/H_2O$	Phenylalanine	344
$Co + PhCH:C(NHCOCH_3)CO_2H$[b]	$EtOH/H_2O$	Phenylalanine	344, 348
$Co[OCOC(:NOH)R]_2 2H_2O$	H_2O	$R = CH_3$, alanine	345, 345a
		$R = PhCH_2$, phenylalanine	345, 345a
$[Co(pn)_2OCOC(NHCOCH_3):CHPh]Cl_2$	H_2O	Phenylalanine	347
$[Co(pn)_2OCOC(NH_2)CH:CHPh]Cl_2$	H_2O	Benzylalanine	347

[a] 100 to 150 atm H_2, 50–140°.
[b] $Co = Co(OCOCH_3)_2(H_2O)_4$.

138

thiocyanate, nitrate, and cyanate were much less effective, and the last three precipitated metallic cobalt (346).

48a

k. Epoxides. Cyclohexene and styrene oxides are readily reduced to cyclohexanol and β-phenylethanol, respectively, by the $HCo(CN)_5^{3-}$ system at 1 atm H_2 and room temperature (295).

l. Ethylenes and Acetylenes. As mentioned earlier, nonconjugated monoenes are not reduced. Ethylene does not react with either $Co(CN)_5^{3-}$ or $HCo(CN)_5^{3-}$, but perfluoroethylene forms the bridged binuclear complex $(NC)_5CoCF_2CF_2Co(CN)_5^{6-}$ and $(NC)_5CoC_2F_4H^{3-}$, respectively (350–353); the latter has not been reported to yield $C_2F_4H_2$ in any way.

Acetylene dicarboxylic acid has been reduced to fumaric and succinic acids at high hydrogen pressures (318, 326, Table 22). With alkaline solutions of $Co(CN)_5^{3-}$ under H_2, 3-chloropropyne gives some propylene, and 1,4-dichloro-but-2-yne gives a C_4H_8 product (303); these products indicate that catalyzed hydrogenation as well as hydrogenolysis (Section X-B-3), has occurred. The reduction of other acetylenes has not been reported. Phenylacetylene and phenylpropiolate form adducts of the type $(NC)_5Co$—C-(C_6H_5)=CHR (R = H or CO_2^-) with $HCo(CN)_5^{3-}$ (303), but no reduction occurs. Acetylene does not form the expected vinyl complex with the hydride, but does form the bridged binuclear complex $[(NC)_5CoCH=CHCo(CN)_5]^{6-}$ with $Co(CN)_5^{3-}$ (350).

m. Dyes. King and Winfield (354) have noted the catalyzed hydrogenation of a number of dyes, using aqueous pentacyanocobaltate(II) solutions buffered at pH 7.8 with 1 atm hydrogen. A competing and inhibitory reaction, however, is the oxidation of the cobalt(II → III) by the dye, the inhibition decreasing in the order methylene blue > benzyl viologen > methyl violet. These data were pertinent to studies on soluble hydrogenase catalytic systems (see Section XVI-I). Indigo has been hydrogenated (Table 25; 303).

3. Catalyzed Hydrogenolysis of Organic Halides (Table 30). Kwiatek and coworkers (21, 268c, 295, 303, 306, 307) found that a considerable number of organic halides could be catalytically reduced by alkaline solutions of $HCo(CN)_5^{3-}$ under hydrogen. The reductions are thought (21) to involve

Table 30 Catalyzed Hydrogenolysis of Organic Halides by Alkaline Aqueous or Aqueous-Methanol $Co(CN)_5^{3-}$ Solutions (358)

Halide (RX)	Organic Product	Reference
CH_3I	CH_4,[a] none[b]	21, 307, 358
C_2H_5I[a,b]	C_2H_6, C_2H_4	21, 358
$n\text{-}C_3H_7I$[a,b]	C_3H_8, C_3H_6	21, 358
$iso\text{-}C_3H_7I$[a,b]	C_3H_8, C_3H_6	21, 307, 358
$tert\text{-}C_4H_9I$[a,b]	$iso\text{-}C_4H_{10}$, $iso\text{-}C_4H_8$	21, 307, 358
$neo\text{-}C_5H_{11}I$[a]	$neo\text{-}C_5H_{12}$	303
CH_2Br_2[a]	CH_4	303
$I(CH_2)_2I$[b] $Br(CH_2)_2Br$[b]	C_2H_4	307, 358
$Cl(CH_2)_2I$[a]	C_2H_5Cl, C_2H_4	303
$I(CH_2)_3I$	C_3H_8,[a] C_3H_6,[b] cyclo-C_3H_6[b]	303, 358
$I(CH_2)_3Br$[b]	cyclo-C_3H_6	358
$I(CH_2)_4I$[b]	but-1-ene, cis- and trans-but-2-ene	358
$I(CH_2)_5I$[b]	pent-1-ene, cis-pent-2-ene	358
$C_6H_5CH_2Br$[a]	$C_6H_5CH_3$	307
$C_6H_5CH_2I$[b]	None	358
$C_6H_5CH(CH_3)Br$[a]	$C_6H_5C_2H_5$	303
$C_6H_5CH(CH_2OH)Cl$[a]	$C_6H_5CH_2CH_2OH$	303
$(C_6H_5)_3CCl$[a]	$(C_6H_5)_3CH$	303
$p\text{-}BrCH_2C_6H_4CH_2Br$[a]	$p\text{-}CH_3C_6H_4CH_3$	303
$CH_2{=}CHCH_2I(Br, Cl)$[a]	C_3H_6	21, 295, 306, 307
$CH_3CH{=}CHCH_2Br(Cl)$[a,b]	but-1-ene, trans-but-2-ene, butadiene	21, 306
$CH_2{=}CHCH(CH_3)Cl$[a,b]	but-1-ene, trans-but-2-ene, butadiene	21, 306
$CH_2{=}C(CH_3)CH_2Br$[a]	$iso\text{-}C_4H_8$	303
$C_6H_5CH{=}CHCH_2Br(Cl)$[a]	$C_6H_5CH{=}CHCH_3$	303
$CH_2{=}C(C_6H_5)CH_2Br$[a]	$C_6H_5CH(CH_3)_2$	303
$ClCH_2CH{=}CHCH_2Cl$[a]	C_4H_8	303
$HC{\equiv}CCH_2Cl$[a]	$CH_3{\equiv}CH$, $CH_3CH{=}CH_2$	303
$ClCH_2C{\equiv}CCH_2Cl$[a]	C_4H_8	303
$C_6H_5COCH_2Br$[a]	$C_6H_5COCH_3$	303
$C_6H_5COCH(CH_3)Br$[a]	$C_6H_5COC_2H_5$	303, 307
C_6H_5, C_6H_5 cyclopropane with CO_2^- and Br (a)	C_6H_5, C_6H_5 cyclopropane with CO_2^- and H	21
$NCCH_2Cl$[a,b]	CH_3CN[a]	21, 307
$NCCH(CH_3)Br$[a,b]	C_2H_5CN[a]	21
$NCC(CH_3)_2I(Cl)$[a,b]	$(CH_3)_2CHCN$[a]	21
$NCCH(C_6H_5)Br$[a,b]	$C_6H_5CH_2CN$[a]	21
$NC(C_6H_5)_2Cl$[a,b]	$(C_6H_5)_2CHCN$;[a] $[(C_6H_5)_2C(CN)]_2$[b]	21

[a] 1 atm H_2, 25°.
[b] No H_2, 25°.

organic radicals produced from $Co(CN)_5^{3-}$, which is in equilibrium with the hydride; for example,

$$Co(CN)_5^{3-} + RX \longrightarrow Co(CN)_5X^{3-} + R\cdot \qquad (221)$$

$$HCo(CN)_5^{3-} + R\cdot \longrightarrow Co(CN)_5^{3-} + RH \qquad (222)$$

Organic monohalides may react with $Co(CN)_5^{3-}$ to give organo- and halopentacyanocobaltate(III) species according to eq. 223 (21, 268c, 303, 306–308, 356–358); this provided a further method of forming and studying the $Co(CN)_5R^{3-}$ complexes which were also formed during the addition of $HCo(CN)_5^{3-}$ to activated olefins (see Section X-B-2).

$$2Co(CN)_5^{3-} + RX \longrightarrow Co(CN)_5R^{3-} + Co(CN)_5X^{3-} \qquad (223)$$

This reaction, first-order in $Co(CN)_5^{3-}$ and halide (357, 358), proceeds via the rate-determining step shown in eq. 221, followed by a faster reaction 224 (21, 268c, 306, 307, 357, 358):

$$R\cdot + Co(CN)_5^{3-} \longrightarrow Co(CN)_5R^{3-} \qquad (224)$$

Related mechanisms have been suggested for reactions of some organic monohalides which also give hydrocarbons directly in the absence of hydrogen. For example, isopropyl iodide yields equimolar quantities of propylene and propane by reaction with $Co(CN)_5^{3-}$ at ratios of halide to cobalt $\geqslant 1.0$ (21, 358):

$$2Co(CN)_5^{3-} + 2(CH_3)_2CHI \longrightarrow$$
$$2Co(CN)_5I^{3-} + CH_3CH{=}CH_2 + (CH_3)_2CH_2 \quad (225)$$

Propylene is thought to be produced via eq. 221 followed by hydrogen atom abstraction from the radical (eq. 226), while propane is produced according

$$Co(CN)_5^{3-} + (CH_3)_2CH\cdot \longrightarrow HCo(CN)_5^{3-} + CH_3CH{=}CH_2 \quad (226)$$

to eqs. 221 and 222; the overall stoichiometry is given by eq. 225. At lower halide to cobalt ratios, the proportion of propylene increases because reaction 226 predominates (21).

Generally, the competition between the radical combination reaction 224 and the disproportionation reaction 226 determines the composition of the product, in terms of the organocobalt complex and the hydrocarbons. In the presence of hydrogen, the $HCo(CN)_5^{3-}$ formed promotes reduction to saturated product according to eq. 222. The production of only mono-deuterated alkane when a hydrogenolysis is carried out in D_2O is consistent with the hydride being an intermediate (303). Direct evidence for the free radical intermediates $(R\cdot)$ has been obtained by trapping them with acrylonitrile and methylmethacrylate and, of course, with $HCo(CN)_5^{3-}$ (eq. 222).

These competition studies have been discussed in some detail (21, 268c, 358), and will not be considered further here.

The saturated alkyl and benzyl cobalt complexes themselves are not intermediates in the respective halide catalytic reductions, since they are generally stable towards $HCo(CN)_5^{3-}$ (21, 307) (in contrast, e.g., to the butenyl cobalt complex, eq. 181). The dehydrogenation step shown in eq. 226 could, however, involve an unstable isopropylocobalt complex, since such a reaction has been observed for the isobutyl complex (307):

$$Co(CN)_5CH_2CH(CH_3)_2^{3-} \longrightarrow HCo(CN)_5^{3-} + CH_2{=}C(CH_3)_2 \quad (227)$$

Allylcobalt complexes, however, which can be formed from the halides according to eq. 223, are probably involved in the reduction of these halides, since the complexes do react with the hydride or acid to yield monoolefins (21, 306–308), just as the butenyl complex formed from the hydride and butadiene does (eqs. 181, 184). Thus, for example, the reduction of crotyl bromide is intimately related to the hydrogenation of butadiene, since the same intermediate butenyl complex is involved. The hydrogenolysis of other allylic derivates, including acetates, alcohols, esters, and amines, has also been reported (295, 306), but no mechanism has been presented.

The mechanism of the reduction of α-bromopropiophenone, which according to eq. 223 (307), forms the same organocobalt intermediate, $Co(CN)_5CH$-$(CH_3)COC_6H_5^{3-}$ (**49**), as the $HCo(CN)_5^{3-}$ adduct of phenyl vinyl ketone (see Section X-B-2e), is thought not to involve **49**, since reaction of **49** with the hydride gives only low yields of phenyl ethyl ketone (307); presumably the radical mechanism (eqs. 221, 222) operates.

The reduction product of optically active 1-bromo-2,2-diphenylcyclopropane carboxylic acid was optically inactive, and this was consistent with a cyclopropyl radical intermediate (21). Halides of resonance-stabilized radicals that cannot disproportionate according to eq. 226 (di- and triphenylmethyl, γ-phenylallyl, and cycloheptatrienyl) form dimers on reaction with $Co(CN)_5^{3-}$ (268c, 307):

$$2Co(CN)_5^{3-} + 2RX \longrightarrow 2Co(CN)_5X^{3-} + RR \quad (228)$$

Triphenylmethyl chloride is reduced, however, if hydrogen is present (303).

The α-halonitriles undergo halogen abstraction according to eq. 221 to produce a resonance-stabilized radical ($RR'\dot{C}CN \leftrightarrow RR'C{=}C{=}\dot{N}$) which may be reduced according to eq. 222 (21). Hydrogenation of α,β-unsaturated nitriles proceeds via the same radical intermediate (Section X-B-2g). The reduction product of only α-iodoisobutyronitrile was reported, but the others are reduced to varying degrees, depending on the extent of other reactions such as formation of a organocobalt complex (eq. 224), an N-amidocobalt complex (eq. 218), or dimer formation by combination of radicals (21, 307).

The reactions of $Co(CN)_5^{3-}$ with dihalides also follow second-order kinetics; the rate-determining step is thought to be the abstraction of one halogen atom according to eq. 221 (358). Taking $I(CH_2)_3I$ as an example, subsequent reactions, shown in eqs. 229 to 232, accounted for the cyclopropane and propylene products (358):

$$Co(CN)_5^{3-} + ICH_2CH_2\dot{C}H_2 \longrightarrow Co(CN)_5CH_2CH_2CH_2I^{3-} \quad (229)$$

$$Co(CN)_5CH_2CH_2CH_2I^{3-} \longrightarrow Co(CN)_5I^{3-} + \text{cyclo-}C_3H_6 \quad (230)$$

$$Co(CN)_5CH_2CH_2CH_2I^{3-} + Co(CN)_5^{3-} \longrightarrow$$
$$Co(CN)_5CH_2CH_2\dot{C}H_2^{3-} + Co(CN)_5I^{3-} \quad (231)$$

$$Co(CN)_5CH_2CH_2\dot{C}H_2^{3-} \longrightarrow Co(CN)_5^{3-} + CH_3CH{=}CH_2 \quad (232)$$

Under H_2, the propane is presumably produced according to

$$HCo(CN)_5^{3-} + ICH_2CH_2\dot{C}H_2 \longrightarrow Co(CN)_5^{3-} + I(CH_2)_2CH_3 \quad (233)$$

followed by further reaction of the monohalide according to eqs. 221 and 222.

The acetylene mono- and dihalide substrates yielded some olefinic products, indicating some reduction of the triple bond as well as hydrogenolysis. Similarly the unsaturated halide, $CH_2{=}C(C_6H_5)CH_2Br$, gave isopropylbenzene (303); the hydrogenolysis product, α-methylstyrene, is known to be readily hydrogenated (see Section X-B-2d).

Besides the organic halides listed in Table 30, a large number of others have been shown to react with $Co(CN)_5^{3-}$ in the absence of H_2 to form organocobalt complexes according to eq. 223, although as discussed above smaller amounts of hydrocarbons can also be formed. Such halides are likely to undergo more extensive hydrogenolysis when using the alkaline solutions of $Co(CN)_5^{3-}$ under H_2 (268c); such reductions, however, have not been reported. The other halides will not be listed here, but they are given in a review by Kwiatek (268c) and papers by Halpern's group (357, 358).

Vinyl halides appear somewhat exceptional in that they do not undergo hydrogenolysis with alkaline $HCo(CN)_5^{3-}$ solutions; they form vinyl complexes (307), possibly as follows:

$$HCo(CN)_5^{3-} + CH_2{=}CHX \longrightarrow Co(CN)_5CHXCH_3^{3-} \quad (234)$$

$$Co(CN)_5CHXCH_3^{3-} \longrightarrow Co(CN)_5CH{=}CH_2^{3-} + HX \quad (235)$$

Some deuterium studies, however, do not support this mechanism (303).

The rates of the hydrogenolysis reactions will be governed by the rate-determining step shown in eq. 221. The kinetic studies (357, 358) have yielded rate constants at 25° ranging from 2.5×10^{-4} to $9 \times 10^4 \ M^{-1} s^{-1}$. The general trend shows an inverse dependence on the carbon-halogen bond strength, for example, $k_{RI} > k_{RBr} > k_{RCl}$, and the rate decreases in the sequence $C_6H_5CH_2X > (CH_3)_3CX > (CH_3)_2CHX > CH_3CH_2X > CH_3X$.

Reactivity also increases with increasing number of halides (e.g., $RCCl_3 >$ $RCHCl_2 > RCH_2Cl$). For the series of diiodides $I(CH_2)_nI$, the rate showed an inverse dependence on n, and this was rationalized in terms of stabilization of the iodoalkyl radical possibly through cyclic bridging (50)

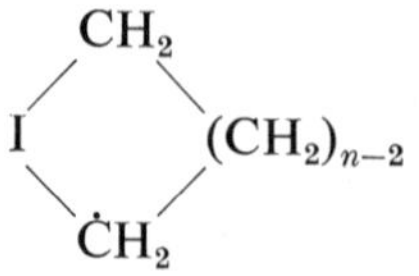

$$
I\underset{\dot{C}H_2}{\overset{CH_2}{\diamond}}(CH_2)_{n-2} \quad \textbf{50}
$$

4. Mechanism of Catalytic Hydrogenation by Pentacyanocobaltate(II) under Hydrogen.

The two basic mechanism postulated for reduction by HCo $(CN)_5{}^{3-}$ involve either organocobalt complexes (eqs. 180 and 181) or radical species (eqs. 200 and 201) as intermediates. A general scheme incorporating both is depicted below (268c, 323, 328):

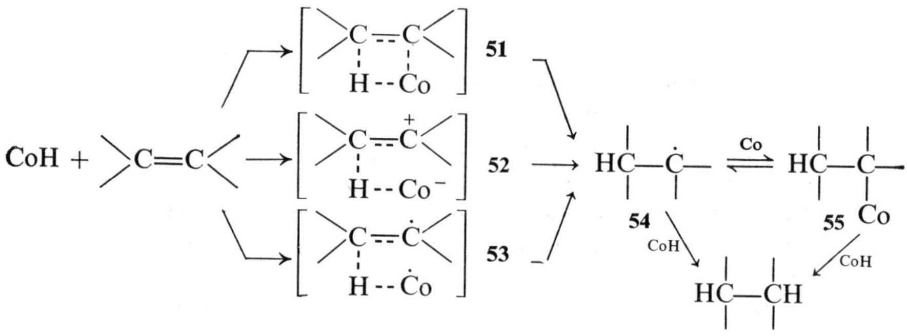

$$
CoH + \;\;>\!\!C\!=\!\!C\!\!<\;\;\longrightarrow\;\;
\begin{bmatrix} >\!C\!-\!\!-\!C\!< \\ \;\vdots\quad\vdots \\ H\!-\!-\!Co \end{bmatrix} \textbf{51}
$$

$$
\begin{bmatrix} >\!C\!=\!\!\overset{+}{C}\!< \\ \;\vdots \\ H\!-\!-\!Co^- \end{bmatrix} \textbf{52}
$$

$$
\begin{bmatrix} >\!C\!-\!\!-\!\dot{C}\!< \\ \;\vdots\quad\; \\ H\!-\!-\!\dot{Co} \end{bmatrix} \textbf{53}
$$

Scheme 7

In this scheme the α-carbon atom is defined as the one containing the activating group, commonly CO_2H, CN, monoene, pyridine, phenyl, $CO_2{}^-$.

Additions of $DCo(CN)_5{}^{3-}$ to butadiene (Section X-B-2a), sorbate (Section X-B-2a), cinnamate (Section X-B-2b), other α,β-unsaturated acids (323), and styrene, α-methylstyrene, α-vinylpyridine, methyl methacrylate (303) was reversible, as indicated by the deuterium content in the hydrogenation products and recovered olefins. Product incorporation of >1 deuterium $β$, and <1 deuterium α to the activating group shows the deuterium adds initially to the β-carbon. This direction of addition was also found with olefins giving stable organocobalt complexes (268c, 307). Systematic kinetic studies on the reaction of the hydride with olefins have been reported by Jackman and coworkers (323) and Halpern and Wong (328), and their data for the second-order rate constants are given in Table 31.

The addition of the hydride could involve a four-center transition state (51), initial proton transfer from CoH (52), or initial hydrogen atom transfer

Table 31 Kinetic Data for Reactions of $HCo(CN)_5^{3-}$ with Olefins

Substrate	$k^{a,b}$	$k^{a,c}$
CH_2=CHCO$_2^-$, acrylate	1.2×10^{-3}	$(1.5 \pm 0.5) \times 10^{-3}$
CH_2=CHCO$_2$H		$(2.0 \pm 0.5) \times 10^{-2}$
CH_2=CMeCO$_2^-$, methacrylate	1.8×10^{-2}	
trans-MeCH=CHCO$_2^-$, crotonate	$<5.3 \times 10^{-6}$	
cis-MeCH=CHCO$_2^-$, angelate	1.5×10^{-4}	
trans-MeCH=CMeCO$_2^-$, tiglate	9.2×10^{-5}	
CH_2=CEtCO$_2^-$, 2-ethylpropenate	3.7×10^{-3}	
cis-O$_2$CCH=CHCO$_2^-$, maleate	1.03×10^{-5}	
trans-O$_2$CCH=CHCO$_2^-$, fumarate	7.1×10^{-5}	
cis-O$_2$CCH=CMeCO$_2^-$, citraconate	7.4×10^{-6}	
trans-O$_2$CCH=CMeCO$_2^-$, mesaconate	1.8×10^{-5}	
CH_2=C(CO$_2^-$)CH$_2$CO$_2^-$, itaconate	1.4×10^{-3}	
CH_2=CHCN, acrylonitrile		$(1.8 \pm 0.2) \times 10^{-1}$
CH_2-2-C$_5$H$_4$N, 2-vinylpyridine		1.0 ± 0.1
CH_2=CHC$_6$H$_5$, styrene		1.1 ± 0.5
CH_2=CHCH=CH$_2$, butadiene		1.6 ± 0.2; 7.6^d
CH_2=CMeCH=CH$_2$, isoprene		2.5 ± 0.2
CH_2=CMeCN, methacrylonitrile		4.0 ± 1.0
CH_2=CH-2-C$_5$H$_4$NH$^+$, 2-vinylpyridinium		$(4.7 \pm 0.5) \times 10^2$
CH$_3$CH=CHCH=CHCO$_2^-$, sorbate	$^e4 \times 10^{-3}$, $^f7 \times 10^{-2}$, $^g3.1 \times 10^{-2}$	
C$_6$H$_5$CH=CHCO$_2^-$, cinnamate	$^h1.95 \times 10^{-4}$	

[a] The k values are second-order rate constants in $M^{-1}\,s^{-1}$.

[b] Reference 323. Calculated from reported half-lives in H$_2$O; temperature not stated, but reactions followed by n.m.r., presumably at ambient temperatures.

[c] Reference 328. At 25° in 50 vol % H$_2$O—MeOH, $\mu = 0.5$ M (KCl).

[d] Reference 305 gives $k = 7.6\ M^{-1}\,s^{-1}$ at 25° in H$_2$O; $\mu = 0.5$ M (KCN or KCN/KCl).

[e] Reference 278. At 25° in H$_2$O, 0.5 M KCN, 0.2 M KCl.

[f] Reference 278. At 25° in H$_2$O, 1 M KCN, 2 M NaOH.

[g] Reference 329, 331. At 20° in H$_2$O, $\mu = 1.0$ M (KCl).

[h] Reference 322. At 20° in H$_2$O, $\mu = 1.0$ M (KCl).

(53). Besides the electrostatic effects between the anionic $HCo(CN)_5^{3-}$ species and substrates, which increase rates in the expected direction (acrylate < acrylic acid < 2-vinylpyridinium), other appreciable effects are apparent (323, 328). The rate decreases with increasing electron-withdrawing power of X in $CH_2 = CHX$ (compare butadiene or styrene with 2-vinylpyridine, acrylonitrile, and acrylic acid). On the other hand, the rate increases on α-methylation of neutral substrates (acrylonitrile, butadiene) and (except for maleate) of carboxylate anions (acrylate, crotonate, fumarate). These trends suggest an electrophilic attack on the double bond, and this tends to argue against a mechanism involving **51** (328). The very low basicity of $HCo(CN)_5^{3-}$ (see Section X-B-2e) makes formulation **52** unlikely, hence **53** involving the hydrogen atom transfer seems the most plausible (328). Using the bond dissociation energy of 58 kcal mole^{-1} for the H—Co bond in the hydride (Section X-A-2), Halpern and Wong (328) concluded that production of **54** (with generation of $Co(CN)_5^{3-}$) via this mechanism would be close to thermoneutral. α-Methylation is expected to stabilize the radical **54** (323). The retardation in rate with β-substitution (by methyl or carboxylate) on unsaturated carboxylates such as acrylate, methacrylate was attributed to steric hindrance for the approach of the hydride (323).

The observed *cis*-addition of $HCo(CN)_5^{3-}$ to carboxylate during σ-complex formation (323, Section X-B-2b) indicates that combination of $Co(CN)_5^{3-}$ with the radical **54** occurs within a "cage" similar to **53** at a rate greater than that of rotation about the α,β-carbon-carbon single bond (323).

The kinetic trends in Table 31 do not follow the stabilities of the organo-cobalt adducts. Those of acrylonitrile and 2-vinylpyridine were the most stable of the series studied by Halpern and Wong (328). Similarly there was no correlation in the carboxylates studied by Jackman and coworkers (323), since the most stable adducts were formed with those substrates lacking an α-substituent (acrylate, crotonate, maleate, and fumarate; Section X-B-2b); the last three gave, in fact, three of the slowest rates. These findings further indicate the unimportance of the cobalt-carbon bonding in the transition state of these reactions and suggest that the adducts are formed via inter-mediate radical species; Kwiatek and Seyler, however, (21, 307) have suggested the contrary for acrylonitrile and methacrylonitrile (see Section X-B-2g).

The rates shown in Table 31 were unaffected by changing the CN:Co ratio from 6 to 500 (323, 328), showing that the mechanisms do not involve prequilibria such as

$$HCo(CN)_5^{3-} \rightleftharpoons HCo(CN)_4^{2-} + CN^- \qquad (236)$$

$$HCo(CN)_5^{3-} + \,\underset{/}{\overset{\backslash}{C}}{=}\underset{\backslash}{\overset{/}{C}}\, \rightleftharpoons HCo(CN)_4\left(\underset{/}{\overset{\backslash}{C}}{=}\underset{\backslash}{\overset{/}{C}}\right)^{2-}+ CN^- \qquad (236a)$$

Thus production of the adduct **55** from the hydride and olefin (Scheme 7), a net "insertion" reaction, apparently requires no prior coordination of the unsaturated moiety to the metal (328). Such coordination is, in most cases, probably a requirement (359).

It is clear from the discussions in sections B-2a to B-2m and B-3 that the effectiveness of hydrogenation will depend on the stability of the organo-cobalt complex **55**, even if radical intermediates (**54**) are directly involved. Kwiatek and Seyler (21, 307, 308) have studied the complexes $LCo(CN)_5^{3-}$ and find that the order of stabilities is $L=XCH_2 > XCHCH_3 > XC(CH_3)_2$ (Table 32). The dependence of stability on X is in the order $CN > COOR >$

Table 32   Stabilities of Organopentacyanocobaltates, $LCo(CN)_5^{3-}$ (21)

Type	Primary	Secondary	Tertiary
Benzyl	$C_6H_5CH_2{}^a$	$C_6H_5CHCH_3{}^b$	$C_6H_5C(CH_3)_2{}^b$
Alkyl	$CH_3CH_2{}^a$	$CH_3CHCH_3{}^c$	$(CH_3)_3C^b$
Allyl	$CH_2=CHCH_2{}^a$	$CH_2=CHCHCH_3{}^c$	$CH_2=CHC(CH_3)_2{}^b$
Ketone	$RCOCH_2{}^a$	$RCOCHCH_3{}^a$	$RCOC(CH_3)_2{}^b$
Ester	$ROOCCH_2{}^a$	$ROOCCHCH_3{}^a$	$ROOCC(CH_3)_2{}^c$
Nitrile	$NCCH_2{}^a$	$NCCHCH_3{}^a$	$NCC(CH_3)_2{}^a$

[a] Stable complexes isolated.
[b] Complexes not isolated.
[c] Unstable complexes isolated.

$COR > CH=CH_2 \sim R > C_6H_5$. Thus electropositive α-substituents de-stabilize the complexes, while electronegative groups stabilize them. The studies of Halpern and Wong (328) and Jackman and coworkers (323) discussed above, and the studies of adduct formation with substituted styrenes (Section X-B-2d) and substituted acrylonitriles (Section X-B-2g) support these findings.

The studies of Vlcek and Hanzlik (335, 336) on the reduction of benzo-quinone (Section X-B-2e) show that cobalt(I) intermediates are involved for this substrate. The possible role of such intermediates in the other hydro-genations has not been ascertained (see also Section X-B-5).

5.   Borohydride Reductions Using Pentacyanocobaltate(II) and Other Cobalt Complexes.   Kasahara and Hongu (360, 361) have reported that, at room temperature under nitrogen, aqueous borohydride solutions containing

$Co(CN)_5^{3-}$ readily reduce α,β-unsaturated acids (acrylic, 2-phenylpropenic β-methylcinnamic, cinnamic, crotonic, fumaric and phenylpropiolic, and maleic anhydride) to the fully saturated acids. Similarly, azoxybenzene and azobenzene are reduced to hydrazobenzene (361), nitrobenzene to a mixture of aniline and hydrazobenzene (361), and N-nitrosodimethylamine to 1,1-dimethylhydrazine (362). Dyes have also been reduced in this manner (363). $HCo(CN)_5^{3-}$ was considered to be the reducing agent for the unsaturated acids (360), since this species has been reported to be produced rapidly by the borohydride reduction of $Co(CN)_5^{3-}$ (271, 279); Jackman and coworkers (323) have stated without details that this reaction involves molecular hydrogen formed by slow hydrolysis of the borohydride. Also since crotonic, fumaric and phenylpropiolic acids are not reduced by $HCo(CN)_5^{3-}$ under the mild conditions reported for the borohydride reductions other intermediates may well be involved. Cobalt(I) intermediates present an attractive possibility, particularly in view of the studies of Vlcek and Hanzlik (335, 336) on the benzoquinone reduction (Section X-B-2e). Vlcek and coworkers (364, 365) have also used $Co(dipy)_3^{2+}$ with borohydride as an effective catalytic reducing system; the univalent cobalt complex $Co(dipy)_3^+$ is formed, and this reduces substrates by an electron transfer reaction followed by protonation of the primary products.

A similar redox process could be involved in the reduction of butadiene and isoprene reported by Mizuta and Kwan (355) who used the dimethyl-glyoxime complex of cobalt(III), $CoCl(DMGH)_2py$, in the presence of borohydride at room temperature. Cobalt(III) can be reduced to the univalent state with borohydride (364). The intermediate cobalt(III) complexes (**55a**) were isolated with these dienes, and the reductions were thought to occur via the hydride $HCo(DMGH)_2py$ in the same manner as catalytic reduction via $HCo(CN)_5^{3-}$ (eqs. 180 and 181).

$$R—Co(DMGH)_2py, \textbf{55a};$$

$$R = —CH_2CH{=}CHCH_3 \text{ and } —CH_2CH{=}C(CH_3)$$

Isoprene gave amounts of butenes in the order: 3-methylbut-1-ene > 2-methylbut-1-ene > 2-methylbut-2-ene. No hydrogenation occurred if hydrogen gas replaced the borohydride.

These dimethylglyoximate cobalt complexes (cobaloximes), used as models for reactions of vitamin B_{12}, are considered further in Section X-H. The discussions there suggest that complexes such as **55a** can be formed via cobalt(I) species, and that they can undergo hydrogenolysis by reaction with borohydride or by reduction with further cobalt(I) species followed by protonation.

C. HOMOGENEOUS HYDROGENATION BY PENTACYANO-COBALTATE(II) IN NONAQUEOUS SOLVENTS

Most of the work on the $Co(CN)_5^{3-}$ complex has been carried out in aqueous media due to limited solubility in other solvents. However, some of the studies described in Section X-B have involved aqueous methanol or ethanol media for more convenient preparations of reduction products (315, 316, 325, 326, 340a, 344, 348) or for some systematic kinetic studies (328, 357, 358).

King and Winfield (279) reported that the $HCo(CN)_5^{3-}$ species was formed more rapidly in methanol than water, even after allowing for the tenfold greater solubility of the H_2 in methanol (but see below); the initial rates in methanol were about second-order in cobalt. The u.v. and visible spectra of the pentacyanocobaltate(II) in methanol and glycerol were similar to that in water (279). A dimethylsulfoxide solution, which gave bands displaced considerably to higher wavelengths, was said to age rapidly and react weakly with H_2 (279). Kwiatek and Seyler (303) also mention the use of methanol-glycerine solvent, but no details are given.

Tarama and Funakibi (366) have reported the catalyzed reduction of butadiene in methanol-glycerine at 20° under H_2 or N_2 at 1 atm. As in water, at CN:Co ratios of <5.2, a product rich in *trans*-but-2-ene is formed. At ratios >5.2, the product is about a 50/50 mixture of *cis*-but-2-ene and but-1-ene, whereas in water but-1-ene is formed to a large extent (Section X-B-2a). Addition of water to the mixed solvent greatly suppressed the *cis*-but-2-ene formation, but addition of ethylenediamine increased the formation of the *cis* isomer. From the reaction scheme proposed by Burnett and coworkers (Section X-B-2a, eqs. 188 to 196), it would appear that formation of the σ-2-butenyl complex (**39**) could be favored in these nonaqueous solvents at high cyanide concentrations. Pregaglia and coworkers (294) have reported that butadiene is slowly hydrogenated to butenes in ethanol solution under H_2 when added to the cobalt system in less than a 1:1 stoichiometry. In contrast to the aqueous system, reduction did not occur when excess butadiene was added, since a stable complex, thought to be a π-allyl compound, was formed.

Pregaglia and coworkers (294) were able to study the pentacyanocobaltate-(II)-H_2 system in ethanol by using the more soluble lithium salt, $Li_3[Co(CN)_5]$. The hydride was also isolated as a lithium salt, formulated $Li_3[Co(CN)_5H]$-$(CH_3COCH_3)_{1.5}(C_2H_5OH)_{1.5}(LiCl)_{0.5}$. This salt decomposed with evolution of H_2 to $Co_2(CN)_{10}^{6-}$ in the presence of traces of water, but in excess water more stable solutions of the hydride resulted; this is consistent with second-order decomposition (Section X-A-2, eqs. 152 and 153). The rate of hydrogen absorption by $Li_3[Co(CN)_5]$ in ethanol increased on adding water, and was given by the expression $(k_1 + k_2[H_2O]^2)[Co]^2[H_2]$. The following scheme was presented to account for the kinetics:

$$2Co(CN)_5ROH^{3-} + 2H_2O \overset{K}{\rightleftharpoons} 2Co(CN)_5H_2O^{3-} + 2ROH$$

$$+H_2 \Big| k_2K^{-1}$$

$$\xrightarrow{+H_2,\, k_1} \quad 2HCo(CN)_5^{3-} + 2ROH(H_2O) \quad (237)$$

At $25°$, k_1 was $0.2\ M^{-2}\,s^{-1}$ and k_2 was $0.35\ M^{-4}\,s^{-1}$. Extrapolating the data for a purely aqueous media gave an overall rate constant $(k_1 + k_2[H_2O]^2)$ of $10^3\ M^{-2}\,s^{-1}$, which is very similar to that measured in pure water by DeVries (277) and others (Section X-A-2). The overall activation energy was about 20 kcal mole^{-1} in ethanol and about 10 kcal mole^{-1} in a 6% water solution, compared to the value of zero in aqueous solutions (277). The role of the water in the scheme may be involved with ion pairs; hydrogenation of the lithium salt was slower in methanol than water (294), although the contrary had been observed by others with the potassium salt (279).

Cyclopentadiene is reduced to cyclopentene by ethanolic solutions of $HCo(CN)_5^{3-}$ (lithium cations) at rates similar to those measured in water (potassium cations). The mechanism is likely to be that shown for butadiene (Section X-B-2a); however, the ethanol solutions maintained their activity for months while the activity of the aqueous solutions decreased with time (294), presumably due to an aging process (Section X-A-2). A homogeneous ethanol solution, resulting at a CN:Co ratio = 4 in the presence of cyclopentadiene, absorbed hydrogen about three times faster than the $Co(CN)_5^{3-}/$ cyclopentadiene system, indicating the presence of a different catalytic complex (294). A π-allyl compound might be anticipated at the lower cyanide concentration (eq. 185, Section X-B-2a).

The $HCo(CN)_5^{3-}$ anion can be produced in acetic acid, and in a 4:1 ethanol-acetic acid mixture the ion loses only 30% of its hydrogen at room temperature (294).

D. HOMOGENEOUS HYDROGENATION BY OTHER CYANO-COBALT(II) COMPLEXES

A carbon dioxide complex, $Co_2(CN)_{10}CO_2^{6-}$, prepared by Kwan and Suzuki (367), has been used by these workers as a source of nascent hydrogen to reduce butadiene (368). The complex is readily formed at $10°$ according to eq. 238, and decomposes more slowly in water according to eq. 238a:

$$2Co(CN)_5^{3-} + CO_2\ (1\ atm) \xrightarrow{H_2O} Co_2(CN)_{10}CO_2^{6-} \quad (238)$$

$$Co_2(CN)_{10}CO_2^{6-} + H_2O \longrightarrow Co_2(CN)_{10}CO_3^{6-} + H_2 \quad (238a)$$

Butenes are produced when reaction 238a occurs in the presence of butadiene. The reaction is not catalytic; a maximum of 64% of the available hydrogen was consumed with production of the inactive carbonate complex (368). In basic solution with a CN:Co ratio >6, but-1-ene was formed; in acid solution with a ratio <6, but-2-enes were formed. The mechanism involved is not known, but the ratio effects are the same as those found for the HCo-$(CN)_5^{3-}$ system in water (Section X-B-2a).

The effect of added hydroxide on hydrogenations catalyzed by $HCo(CN)_5^{3-}$ (e.g., see Section X-B-2a, 304, 311, 312) could be due to changes in ligand coordination; hydrogenation of cobalt complexes containing some cyanide ligands (<5) has been discussed in Sections B-2i and B-2j.

Schindewolf (218) first suggested that the D_2-H_2O exchange catalyzed by $Co(CN)_5^{3-}$, which was governed by the slow decomposition of the hydride, would be more rapid by lowering the charge of the catalyst ion, since two reacting ions were involved (see Section X-A-2, eqs. 163 to 166). A less stable hydride would also be expected to be a more active catalyst.

Piringer, Farcas, and coworkers (369–374) have studied the use of cyano-amine cobaltate (II) species, particularly for the catalytic exchange in aqueous solution. The precipitates formed at CN:Co ratios <5 dissolve in ethylene-diamine, diethylenetriamine, and triethylenetetramine (as well as propylene-diamine and 2,2'-dipyridyl) to give the ions $Co_2(CN)_8en^{4-}$, $Co(CN)_2dien$ and $Co(CN)trien^+$ at maximum concentrations when the CN:Co ratio = 4, 2, and 1 respectively. At ratios $\geqslant 5$, the pentacyanocobaltate(II) is always formed (369, 371). The cyanoamine complexes in solution are much less stable than $Co(CN)_5^{3-}$, and are oxidized in inert atmospheres quite rapidly at 30° with liberation of hydrogen (369); for example,

$$(CN)_4Co(en)(CN)_4^{4-} + 2H_2O \xrightarrow{\text{en}} 2Co(CN)_4(en)^- + 2OH^- + H_2 \quad (239)$$
$$\mathbf{56}$$

This "aging" reaction was first-order in cobalt (372), with $k \sim 10^{-3}$ s^{-1} at 25°, in contrast to the $Co(CN)_5^{3-}$ system which was second-order and probably involved small amounts of a dimeric intermediate (Section X-A-2, eqs. 154, 159, 161, and 162). At lower temperatures (0°) some hydrogen absorption was measured for the ethylenediamine system (372), but it is clear that a hydride, whether formed directly or as an intermediate in reaction 239, is much less stable than the $HCo(CN)_5^{3-}$ species in an equilibrium:

$$2CoH \text{ (or } Co_2H_2) \underset{k_1}{\overset{k_{-1}}{\rightleftharpoons}} 2Co \text{ (or } Co_2) + H_2 \quad (240)$$

H_2-HDO exchange studies (370–374), carried out at high pressure (10–90 atm H_2) to stabilize the hydrides, indeed show that the exchange rate is now determined by the formation of the hydride (k_1), rather than its decomposition (k_{-1}) as was found for the pentacyano system (Section X-A-2, eqs. 163 to 166). The exchange rate, measured by HD production in the gas phase from 20 to 40°, was always a maximum at a CN:Co ratio = 4 for all the amine complexes, and generally decreased in the presence of added salts, although KCl had no effect; moreover, the $Co(CN)_5^{3-}$ system was subject to a strong positive salt effect (Section X-2a). The exchange rate also increased with increasing H_2 pressure (<first-order) in contrast to the pressure-independent rate for $Co(CN)_5^{3-}$. These data indicated that formation of tetracyano-amine-cobaltate(II) complexes such as 57 with two active centers was rate-determining (k_1):

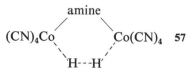

$$\text{(CN)}_4\text{Co} \underset{\text{H}--\text{H}}{\overset{\text{amine}}{\diamond}} \text{Co(CN)}_4 \quad 57$$

Because of the rapidity of the reaction, the rate was subject to some diffusion control; later papers of these workers (373, 374) discussed the influence of the velocity of hydrogen transport from the gas phase to the solution on the overall reaction velocity. The overall exchange rate (R_0) showed less than first-order in total cobalt; to account for this, the dimeric amine complex such as 56 was assumed to be in equilibrium with monomeric $Co(CN)_4$-amine^{2-} species. Thus the exchange reaction was formulated as

$$2Co(CN)_4\text{amine}^{2-} \underset{\mathbf{I}}{\overset{K}{\rightleftharpoons}} (CN)_4Co(\text{amine})\,Co(CN)_4^{4-} + H_2 \underset{k_{-1}}{\overset{k_1}{\rightleftharpoons}} 57 \quad (241)$$

which gives rise to the following rate law:

$$\text{exchange rate } (R_e) = \frac{k_1 p(Co_T - [\mathbf{I}])}{2(1 + p \cdot K_1)} \quad (242)$$

where $p = H_2$ pressure, $K_1 = k_1/k_{-1}$, Co_T = total cobalt as monomer, and $[\mathbf{I}] = (\sqrt{1 + 8K(1 + p \cdot K_1)Co_T} - 1)/4K(1 + p \cdot K_1)$. The diffusion rate ($R_D$) was given by $k_D p$ where k_D is the diffusion constant. Since $(R_0)^{-1} = (R_e)^{-1} + (R_D)^{-1}$, the complete rate law was given by

$$\frac{1}{R_0} = \frac{2(1 + p \cdot K_1)}{k_1 p(Co_T - [\mathbf{I}])} + \frac{1}{k_D p} \quad (243)$$

This accounts for observed dependences on H_2 and cobalt. The rate law simplified at higher Co_T since $[\mathbf{I}]$ could be neglected. The results obtained are summarized in Table 33. Much higher exchange rates are evident for the amine complexes than for the $Co(CN)_5^{3-}$ species. The data give k_{-1} values

Table 33   Kinetic Data for H_2-HDO Exchange Catalyzed by Cyano-Amine-Cobaltate(II) (374)

Catalyst[a]	$R_e \times 10^4$ (M s^{-1})	k_1 (atm^{-1} s^{-1})	K (M^{-1})	$K_1 \cdot 10^2$ (atm^{-1})
$Co_2(CN)_8en^{4-}$	3.9	7.9	55	2.7
$Co_2(CN)_8pn^{4-}$	3.2	8.0	8	2.5
$Co_2(CN)_8dien^{4-}$	12	190	1.2	19
$Co_2(CN)_8trien^{4-}$	14	20	30	1.3
$Co(CN)_5^{3-}$	0.013			

[a] $[Co]_T = 0.05$ M at $30°$ and 90 atm H_2.

>300 s^{-1}, in contrast to the second-order exchange rate constant $k_{-1} \approx$ 5×10^{-3} M^{-1} s^{-1} at $25°$ for the $Co(CN)_5^{3-}$ system, and show clearly the instability of the cyanoamine hydrides and why their formation now determines the exchange rate.

Using solubility data for hydrogen (277), k_1 at $30°$ for $Co_2(CN)_8en^{4-}$ can be calculated to be about 16×10^3 M^{-1} s^{-1}, and an activation energy of 14 kcal mole^{-1} was measured (374). For $Co(CN)_5^{3-}$, the termolecular rate constant k_1 was 0.8×10^3 M^{-2} s^{-1} with an apparent zero activation energy, although this could have been a composite constant resulting from an initial preequilibrium to a dimer (eqs. 154 and 155, Section X-A-2). The true activation energy would then equal the heat evolved from the dimerization reaction 154 (277). These studies by Piringer, Farcas, and coworkers on the cyanoamines perhaps give some indirect evidence for a dimer intermediate in the formation of $HCo(CN)_5^{3-}$ from the reaction of hydrogen and $Co(CN)_5^{3-}$.

Wymore (1326) and Schwab and Mandre (375, 376) have studied in particular the cobalt(II) cyanide systems in the presence of 2,2'-dipyridyl and o-phenanthroline, formed in situ by using Co/dip/CN of 1:2-3 :R. Spectroscopic studies (1326) indicate that a mixture of species such as $Co(dip)_2(CN)_2$, $Co(dip)(CN)_3^-$, and $Co(CN)_5^{3-}$ exists, the pentacyano complex predominating at $R > 5$. Hydrogen absorption increases with increasing number of cyanides from nil for the neutral species, about 0.1 mole per cobalt for the tricyanide, and close to 1:1 stoichiometry for $Co(CN)_5^{3-}$. The most active system for olefin reduction (see below), as in the exchange systems, is with $R = 4$, when the major species present is thought to be $Co(dip)(CN)_3^-$. The hydride, presumably $HCo(dip)(CN)_3^-$, is formed, but is less stable and more reactive than $HCo(CN)_5^{3-}$. Slight absorption of hydrogen by $Co(phen)_2(CN)_2$ solutions was observed at elevated temperatures (375).

Farcas and coworkers (373), Schwab and Mandre (375, 376), and Wymore (1326) have studied the hydrogenation of activated olefins by using aqueous or aqueous ethanolic solutions of the cyanoamine systems at 25 to 30° with 1 atm hydrogen. With $R \leqslant 4$, the saturated amine catalysts produced mainly *trans*-but-2-ene from butadiene hydrogenation, while at $R \geqslant 4.5$ but-1-ene was mainly formed. The dipyridyl catalyst, however, gave mainly but-1-ene at $R = 2$ (373). When using the dipyridyl system (Co:CN:dip = 1:2:3), cyclohexa-1,3-diene, sorbate, and cinnamalacetate (PhCH = CHCH = $CHCO_2^-$) were rapidly reduced to monoenes; the rate with the cyclohexadiene was some three times faster than when using $Co(CN)_5^{3-}$ under corresponding conditions. Cinnamyl alcohol, cinnamate, and acrylate were reduced, the last two very slowly; some hydrogenation rates, first-order in catalyst and independent of substrate, were ascribed to slow formation of a dihydride followed by rapid reaction with substrate (375, 376). Methacrylate was not reduced in contrast to the $Co(CN)_5^{3-}$ system where α-substituents on unsaturated acids promote hydrogenation via a radical mechanism (Section X-B-4). *p*-Benzoquinone, nitrobenzene, and to some extent sorbate, were said to poison the catalyst by oxidation to the trivalent state (376). The hydrogenation of styrene (1326) using a Co/2 dip/2 CN system was between zero- and first-order in styrene; the order decreased with increasing concentration. At zero-order, the reaction is second-order in cobalt and first-order in hydrogen; at first-order, the reaction is first-order in cobalt and about half-order in hydrogen. The mechanism outlined by eqs. 151, 180, and 181 for the $Co(CN)_5^{3-}$-butadiene system, was invoked here rather than the radical mechanism of eqs. 199 to 201, which has been considered for the $Co(CN)_5^{3-}$-styrene system (Section B-2d). At high styrene concentration, monohydride formation could be rate-determining (cf. eq. 151), while at low concentration, alkyl formation (cf. eq. 180) could be rate-determining. Crotonaldehyde, acrylate, and butadiene were also studied with this system.

Use of 4,4′-dipyridyl instead of the 2,2′-amine gave a less active catalyst (375). Wymore (1326) noted that glyoxime could also be used as the nitrogen donor; such a system, and indeed even the amine systems, are closely related to cobaloxime systems, which are considered in Section X-H.

Turco and coworkers (1342, 1343) have reported that the five-coordinate complexes $CoL_3(CN)_2$, where L is a tertiary phosphine, are unreactive toward hydrogen in 1,2-dichloroethane solution, although they react with organic halides and dihalides in a manner analogous to $Co(CN)_5^{3-}$; thus ethyl iodide yields ethylene and ethane in a reaction such as eq. 225.

E. ACTIVATION OF HYDROGEN BY DICOBALT OCTACARBONYL

Dicarbonyl octacarbonyl, $Co_2(CO)_8$, has been used extensively as a catalyst for the hydroformylation (Oxo) reaction which is represented basically thus:

$$RCH{=}CH_2 + CO + H_2 \xrightarrow[\substack{100-160° \\ 100-300 \text{ atm}}]{Co_2(CO)_8}$$

$$RCH_2CH_2CHO + RCH(CHO)CH_3 \quad (244)$$

The Oxo reaction has been extensively reviewed elsewhere (51–55), and will not be discussed in detail here. However, hydrogenation of aldehyde products to alcohols, and of the starting olefin, were found to be competing reactions which could become significant, and such studies led to the use of the carbonyl as a hydrogenation catalyst.

Most features of the Oxo reaction can be accounted for in terms of a mechanism summarized in eqs. 245 to 250 for formation of the straight chain aldehyde (53, 54); overall rate expressions have been presented (e.g., Ref. 962):

$$Co_2(CO)_8 + H_2 \underset{k_{-1}}{\overset{k_1}{\rightleftharpoons}} 2HCo(CO)_4 \quad (245)$$

$$HCo(CO)_4 \rightleftharpoons HCo(CO)_3 + CO \quad (246)$$

$$HCo(CO)_3 + RCH{=}CH_2 \longrightarrow \begin{bmatrix} HCo(CO)_3 \\ | \\ RCH{=}CH_2 \end{bmatrix} \longrightarrow$$

$$RCH_2CH_2Co(CO)_3, \textbf{58} \quad (247)$$

$$RCH_2CH_2Co(CO)_3 + CO \longrightarrow RCH_2CH_2Co(CO)_4, \textbf{59} \quad (248)$$

$$RCH_2CH_2Co(CO)_4 \rightleftharpoons RCH_2CH_2COCo(CO)_3 \quad (249)$$

$$RCH_2CH_2COCo(CO)_3 + H_2 \longrightarrow RCH_2CH_2CHO + HCo(CO)_3 \quad (250)$$

Reaction 245 involves hydrogen activation, and reaction 247 involves reaction of a hydride with an unsaturated moiety; such reactions are intimately related to homogeneous hydrogenation systems. The overall reaction products will depend on the fate of alkyl intermediates such as **58** and **59**. Reactions 245 to 250 have been discussed at length in reviews of the Oxo reaction (51–55). This section will summarize briefly the studies relevant to the hydrogenation systems.

1. A Hydrocarbonyl as the Active Catalyst. Orchin and coworkers (377, 377a) first demonstrated the presence of the hydrocarbonyl, $HCo(CO)_4$, under conditions of the Oxo reaction, although previously its presence had often been postulated (378, 378a, 383) since the discovery of the reaction by Roelen in 1943 (379). The hydrocarbonyl decomposes rapidly above its melting point ($-26°$), but is relatively stable in the gas phase and decomposes in a second-order process for which k_{-1} is about 3×10^{-3} M^{-1} s^{-1} at $25°$ and 1 atm total pressure, with either CO/H_2 or He as diluent gas (380, 381). The formation of $HCo(CO)_4$ according to eq. 245 can occur at room temperature, but it is inhibited by carbon monoxide, suggesting an unsaturated intermediate (382, 383, 389, 390); this could involve an equilibrium with a more reactive lower carbonyl (383). Also, carbon monoxide can react with Co_2-$(CO)_8$ to form $Co_2(CO)_9$ (384), and $Co_2(CO)_8$ can coexist with $Co_4(CO)_{12}$ (378a, 382). The $HCo(CO)_4$ formation rate is reported to increase with increasing hydrogen pressure up to 100 atm (389, 390), but is retarded at pressures above 200 atm (383). The conversion to hydrocarbonyl has also been reported to be heterogeneous involving activation by cobalt metal (385–387), which is known to be present under hydroformylation conditions (388). The equilibrium constant for reaction 245 (k_1/k_{-1}, with concentrations of $Co_2(CO)_8$ and $HCo(CO)_4$ expressed as M and p_{H_2} in psi) is of the order of 10^{-4} at 110 to 160° with 10 to 100 atm of both CO and H_2 in hexane and di-n-hexyl ether (377, 387). Qualitative data have been reported (381) on the decomposition rate of the hydrocarbonyl at room temperature in the absence of oxygen in hexane ($t_{1/2} \sim 5$ hr for a 0.02 M solution), and in aqueous solution in which it is a strong acid (20% decomposition in 14 days for a 0.02 M solution).

That $HCo(CO)_4$ is the active catalyst for the reactions occurring under hydroformylation conditions, including hydrogenation, is established by its reactions at room temperature with olefins, aldehydes and unsaturated aldehydes, ketones, ethers, esters, and nitriles (377a, 383, 391–401, 403). Orchin and coworkers (377a) and Pino and coworkers (383) showed that $HCo(CO)_4$ in the absence of hydrogen and carbon monoxide reacts with substrates at room temperature and 1 atm to give products obtained from the same substrates under hydroformylation conditions. Besides obtaining the expected aldehyde products from olefins, Orchin and coworkers (377a) reported that α-methylstyrene was reduced to isopropylbenzene, benzyl alcohol gave toluene, benzhydrol gave diphenylmethane, and triphenylcarbinol gave triphenylmethane (Table 34); decomposition of $HCo(CO)_4$ furnishes some carbon monoxide under these conditions. Orchin and coworkers (391–393, 521) further studied the conversion of olefins to aldehydes using the hydrocarbonyl under 1 atm carbon monoxide as well as inert atmospheres; Breslow and Heck (394–398) and Marko and coworkers (399) showed that such reactions occurred via alkyl- and acyl-cobalt carbonyls (eqs. 247 to 250). Goetz and Orchin (400, 401) have reported stoichiometric reactions of the hydrocarbonyl at 25° under 1 atm carbon monoxide with saturated aldehydes to give the corresponding alcohols (400), with unsaturated aldehydes and ketones to give saturated aldehydes and ketones (400, 401), and with acrylonitrile to give propionitrile (400). The reductions also proceed under an atmosphere of hydrogen or nitrogen, but the carbon monoxide retards decomposition of the hydrocarbonyl. Takegami and

Table 34 Stoichiometric Hydrogenations and Hydrogenolyses Using $HCo(CO)_4$ under Mild Conditions in Hydrocarbon Solution[a]

Substrate	Product	$10^2 k^b (s^{-1})$	Reference
Hept-1-ene	n-Heptane[c]		522, 523
Butadiene	Butenes		522, 523
Isoprene	Methylbutenes		522, 523
Cyclohexa-1,3-diene	Cyclohexene		522, 523
Cyclopentadiene	Cyclopentene		522, 523
Styrene	Ethylbenzene[c]		402
α-Methylstryene	Isopropylbenzene		377a
Benzyl alcohol	Toluene		377a, 400
Benzhydrol	Diphenylmethane		377a
Triphenylcarbinol	Triphenylmethane		377a
n-Butyraldehyde	1-Butanol		400
n-Hexaldehyde	1-Hexanol		400
Benzaldehyde	Benzyl alcohol, toluene		400
Acrolein	Propionaldehyde	0.02	400, 401
Crotonaldehyde	n-Butyraldehyde	0.02	400, 401
2-Methyl-2-butenal	2-Methylbutanal	0.06	401
Cinnamaldehyde	3-Phenylpropanal	0.43	401
Methyl vinyl ketone	Ethyl methyl ketone	>2.5	401
4-Methylpent-3-ene-2-one	Isobutyl methyl ketone	0.06 (0.2[d])	401
4-Phenylbut-3-ene-2-one	4-Phenyl-2-butanone	0.37	401
Acrylonitrile	Propionitrile		400
n-Butyl vinyl ether	n-Butyl ethyl ether[c]	0.57	401, 402
Allyl ethyl ether	Ethyl propyl ether[c]	0.63	401
1,1-Dichloroethylene	1,1-Dichloroethane		470
Epoxyacrolein	A hydroxypropionaldehyde		460

[a] $<25°$, 1 atm CO (or N_2 or H_2).
[b] Pseudo-first-order rate constant for loss of $HCo(CO)_4$ in presence of fivefold excess of substrate, under 1 atm CO.
[c] The hydroformylation product was also observed.
[d] Under argon.

coworkers (402) and Ungvary and Marko (522, 523) have also studied reactions of $HCo(CO)_4$ with various olefins to give hydroformylation and reduction products; Gut and coworkers (470) have made similar studies on chlorinated ethylenes.

Table 34 lists the stoichiometric reductions with some rate data that have been reported under mild conditions (Section E-3).

The evidence for reaction 246, showing that a tricarbonyl species is involved, is reasonably strong. Karapinka and Orchin (393) found that reaction 251 was inhibited by carbon monoxide at 0° and 1 atm.

$$2HCo(CO)_4 + CO + olefin \longrightarrow Co_2(CO)_8 + aldehyde \qquad (251)$$

This was confirmed by Heck and Breslow (394, 395), who showed that the inhibition also retarded the formation of the alkyl and acylcobalt carbonyl intermediates. The inhibition could be due to formation of an olefin intermediate via an associative mechanism (393),

$$HCo(CO)_4 + \quad \underset{/}{\overset{\backslash}{C}} = \underset{\backslash}{\overset{/}{C}} \quad \rightleftharpoons \quad HCo(CO)_3 + CO \qquad (252)$$
$$\underset{/ \quad \backslash}{\overset{|}{C} = C}$$

or via a dissociative mechanism (395) involving formation of $HCo(CO)_3$ (eqs. 246 and 247). These workers and Takegami and coworkers(402) also found that an observed isomerization of an olefin, pent-1-ene, under similar conditions, was inhibited by carbon monoxide; the isomerization and inhibition are readily accounted for in terms of eqs. 246 and 247 (written as equilibria) involving the tricarbonyls. Some data of Takegami and coworkers (402) include autocatalytic uptake plots of carbon monoxide by olefins to give acylcobalt carbonyls; Chalk and Harrod (54) have suggested that this could be due to buildup of $HCo(CO)_3$.

Wender and coworkers (403) presented evidence that hydrogenation of n-butyraldehyde under hydroformylation conditions (185°, 140 atm H_2) was homogeneous involving the octacarbonyl or the hydrocarbonyl. With no added carbon monoxide, heterogeneous reduction with cobalt metal occurred. At 20 atm, carbon monoxide poisoned the cobalt metal catalyst and was insufficient to produce the homogeneous catalyst, and very little reduction occurred; at 70 atm, the homogeneous reduction occurred. Marko (404) with similar studies later confirmed that the reduction of aldehydes and olefins during hydroformylation was not heterogeneous. The observation that the addition of metallic bismuth, gold, lead, mercury, and zinc, or their compounds, inhibits aldehyde reduction (385, 405–408) had been interpreted as implying a heterogeneous reaction on cobalt (385), because these metals were known to poison solid catalysts. But apparently, the inhibition results from the formation of compounds such as $M[Co(CO)_4]_2$ (M=Zn, Hg) which are inactive catalysts (407–409).

$Co_2(CO)_8$ has sometimes been studied as a homogeneous hydrogenation catalyst under purely hydrogen atmospheres, although it generally decomposes to metal in these conditions. Such studies have been carried out for comparison with systems involving organophosphine-cobalt carbonyls and will be considered in Sections X-F-1 and X-F-2.

2. Reduction of Aldehydes to Alcohols. The temperatures normally employed in the hydroformylation reaction are about 100 to 120°; at higher temperatures reduction of the saturated aldehyde product to a primary alcohol is commonly observed. Table 35 lists such reported reductions.

Table 35 Hydrogenation of Saturated Aldehydes and Unsaturated Esters, Acids, and Aldehydes, under Hydroformylation Conditions[a]

Substrate	Product	Reference
2-Ethylhexenal	2-Ethylhexanol	387
Propionaldehyde	1-Propanol	49, 416
n-Butyraldehyde	1-Butanol	415
n-Heptaldehyde	1-Heptanol	415, 417
2-Phenylpropionaldehyde	2-Phenyl-1-propanol	415, 418
Benzaldehyde	Dibenzyl ether, benzyl alcohol, toluene	415–417
Crotonaldehyde	1-Butanol	415, 416
2-Thiophenealdehyde	2-Thenyl alcohol, 2-methylthiophene	415
Cyclohexylaldehyde	Cyclohexylcarbinol and formate	417
Diphenylacetaldehyde	2,2-Diphenylethanol	418
Methyl acrylate	γ-Butyrolactone	411, 412
Ethyl acrylate	γ-Butyrolactone, ethyl propionate	410–412
Methyl methacrylate	α-Methyl-γ-butyrolactone	411, 412
Methyl crotonate	δ-Valerolactone, β-methyl-γ-butyrolactone	410–412
Ethyl crotonate	δ-Valerolactone, β-methyl-γ-butyrolactone	410, 412
Ethyl vinylacetate	δ-Valerolactone, β-methyl-γ-butyrolactone	412
Ethyl tiglate	α-Methyl-δ-valerolactone, α-ethyl-γ-butyrolactone, α,β-dimethyl-γ-butyrolactone	412
Ethyl β,β-dimethylacrylate	β-Methyl-δ-valerolactone, β,β-dimethyl-γ-butyrolactone	410–412
Ethyl α,β,β-trimethyl-acrylate	α,β-Dimethyl-δ-valerolactone, α-isopropyl-γ-butyrolactone	410, 412
Ethyl α,α-dimethylvinyl-acetate	α,α-Dimethyl-δ-valerolactone, α,α,β-trimethyl-γ-butyrolactone	410–412
Diethylmaleate	β-Carbethoxy-γ-butyrolactone, diethylsuccinate	410–412
Diethylfumarate	β-Carbethoxy-γ-butyrolactone, diethylsuccinate	410–412
Ethyl cinnamate	β-Phenyl-γ-butyrolactone, ethyl β-phenyl-propionate	411, 412, 419
Ethyl cyclohex-1-ene carboxylate	2-Hydroxymethylcyclohexanecarboxylic acid γ-lactone	412
Cyclohept-1-eneylisobutyric acid (or ethyl ester)	2-Hydroxymethylcycloheptylisobutyric acid lactone	410–412
Ethyl sorbate	γ-Ethyl-δ-valerolactone, β-propyl-γ-butyrolactone	410, 412
Methyl 6-methyl-5,6-dihydro-4H-pyran-2-carboxylate	6-Methyl-3-hydroxymethyltetrahydropyran-2-carboxylic acid γ-lactone, methyl 6-methyl-tetrahydropyran-2-carboxylate	420
1-Cyclohexen-4-al	1,3 and 1,4-bis(hydroxymethyl)cyclohexane, hexahydrobenzyl alcohol	762

[a] At 160 to 350°, up to 300 atm using $Co_2(CO)_8$ or $HCo(CO)_4$, hydrocarbon or ether solvents. The unsaturated ester systems involve saturated aldehyde intermediates (eqs. 252a and 252b).

The hydroformylation of unsaturated esters and acids at higher temperature gives good yields of lactones, whose formation also involves reduction of the aldehyde group (410–412, 420, Table 35). Equation 252a shows the production of γ-butyrolactone from methyl acrylate:

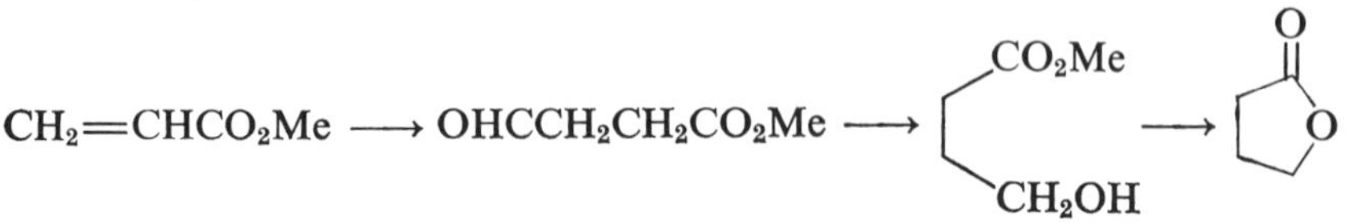

$$CH_2{=}CHCO_2Me \longrightarrow OHCCH_2CH_2CO_2Me \longrightarrow$$

(252a)

An $\alpha\beta$-unsaturated ester with a β-substituted alkyl group gives mainly the δ-lactone, because of migration of the hydrocarbon double bond to the $\beta\gamma$-position. Thus methyl crotonate gives 72% δ-valerolactone and 20% of the β-methyl-γ-butyrolactone:

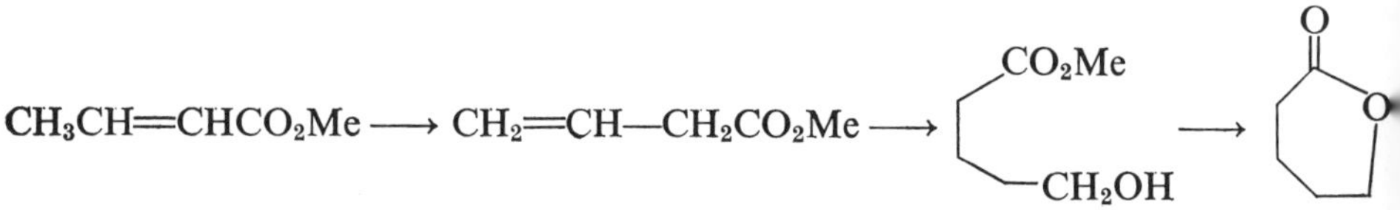

$$CH_3CH{=}CHCO_2Me \longrightarrow CH_2{=}CH{-}CH_2CO_2Me \longrightarrow$$

(252b)

Sorbate gives a mixture of γ- and δ-lactones, one of the double bonds having been reduced. Accompanying reduction of the olefinic bond (see Section E-3) has also been noted for ethyl acrylate, cinnamate, maleate, fumarate, and the 5,6-dihydro-4H-pyran ester system (Table 35). With cinnamate, the lactone yield is increased if a stoichiometric amount of $Co_2(CO)_8$ is used; presumably this is due to less hydrogenation of the olefinic bond.

The production of alcohols from olefins under hydroformylation conditions also proceeds via reduction of the aldehyde group. These reactions are listed in Table 36. Alcohols have been formed from simple monoenes, cyclic mono- and polyolefins, furans, 5,6-dihydro-4H-pyrans, terpenes, steroids, carbohydrates, and lignins. If the substrate contains further carbon-carbon double bonds to the one being hydroformylated, these may be reduced. For example, 3-vinylcyclohexene gives ethylcyclohexylcarbinols, 2,3-dimethylbutadiene gives 3,4-dimethylpentanol, cycloheptatriene gives cycloheptylcarbinol, cis, trans, trans,-cyclododeca-1,5,9-triene gives cyclododecylcarbinol, furans give tetrahydrofurfuryl alcohols, and terpenes give some saturated alcohols. The catalytic reduction of olefins under hydroformylation conditions will be considered in Section E-3.

The studies by Rosenthal's group on the production of hexitols by hydroformylation of 3,4-di-O-acetyl-D-xylal, **59a** (453), and heptitols from 3,4,6-tri-O-acetyl-D-glucal **59b** (455) show that *cis* addition occurs across the double bond; this was established by using deuterium, and by analyzing the products by n.m.r. (eqs. 253 and 254):

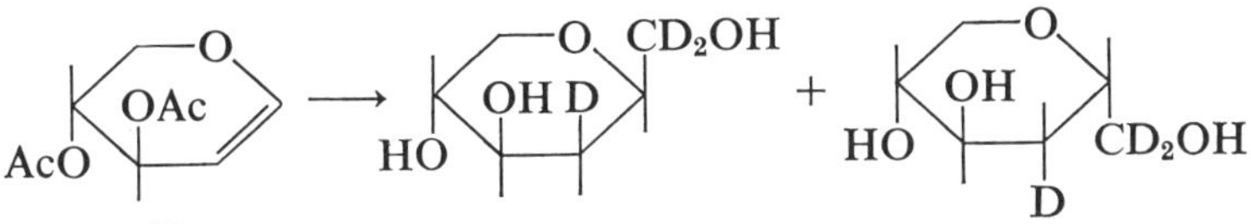

(253)

(254)

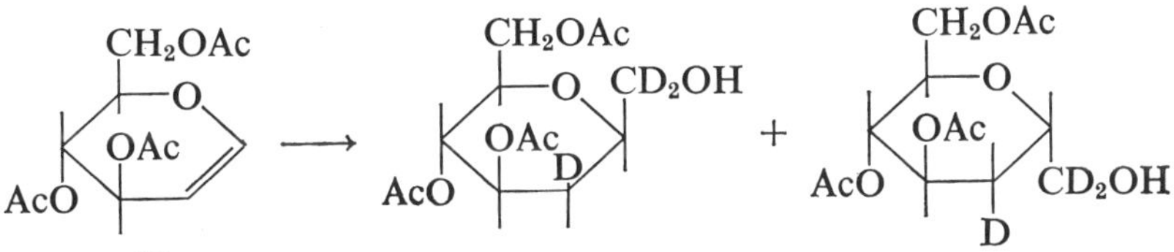

The hydroformylation products containing hydroxmethyl groups obtained from Δ^5-unsaturated steroids (446–448) are also consistent with *cis*-addition from the α-side (**59c** → **59d**)

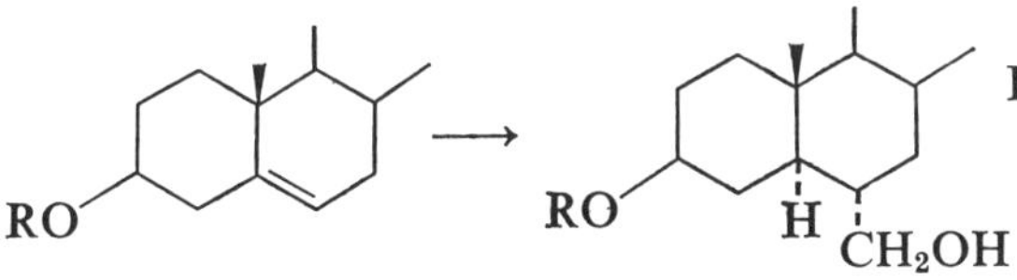

$$R = H, CH_3CO \quad (255)$$

In the hydroformylation of olefins, the two stages of reduction to aldehyde (2 moles of gas consumed per mole of substrate) and alcohol (a further mole) are readily distinguishable. A study by Rosenthal and coworkers (459) has shown that this is not so with glycals such as **59a** and **59b**; the aldehyde products (aldoses) are readily reduced to the alditol, and they suggest that the oxygen atom in the ring possibly activates the aldose. The same may be true in the hydroformylation of 5,6-dihydro-4H-pyrans reported by Falbe and Korte (420, 452). The rate of reduction of aldose increased with increasing hydrogen pressure (459).

The methods for preparing the ether product from α-pinene (441) and the dioxabicyclo [3, 2, 1] octane from 2,6-bishydroxymethyl-5,6-dihydro-4H-pyran (420) have not been elucidated; they could involve hydroxymethyl

Table 36 Production of Alcohols from Olefins under Hydroformylation Conditions

Substrate	Product	Reference
Propylene	n- and iso-butanol	423–429, 489, 1457
But-1-ene	1-Pentanol, 2-methylbutanol	423, 425
But-2-ene	1-Pentanol, 2-methylbutanol	423, 431
Isobutylene	3-Methylbutanol, neopentyl alcohol, isobutane	423, 427, 432, 433, 489
Pent-1-ene	1-Hexanol, 2-methylpentanol, 2-ethylbutanol	423, 538
Pent-2-ene	As above	429, 538
2-Methylbut-1-ene	4-Methylpentanol, 3-methylpentanol, 2,3-dimethylbutanol	423
3-Methylbut-1-ene	As above	423
2-Methylbut-2-ene	As above	423
Hex-1-ene	1-Heptanol, 2-methylhexanol	415, 423
4-Methylpent-1-ene	5-Methylhexanol, 3-methylhexanol, 2,4-dimethylpentanol	423
2,3-Dimethylbut-1-ene	3,4-Dimethylpentanol	423
2,3-Dimethylbut-2-ene	3,4-Dimethylpentanol	423
3,3-Dimethylbut-1-ene	4,4-Dimethylpentanol	423
Hept-1-ene	Octanols	425
2-Methylhexene and 2-ethylpentene	3-Methylheptanol and 3-ethylhexanol	435
Oct-1-ene	Isomeric nonyl alcohols	415
2,4,4-Trimethylpent-1-ene (diisobutylene)	3,5,5-Trimethylhexanol, 2,4,4-trimethylpentane	423, 427, 436, 489
2-Ethylhexene	3-Ethylheptanol	435
Tripropylene	Alcohols and saturated hydrocarbons	489
C₇–C₉ olefins	C₈-C₁₀ alcohols	1462
Cyclopentene	Cyclopentylcarbinol	423, 438
Cyclohexene	Cyclohexylcarbinol	423, 427, 434, 438, 465, 489
3-Vinylcyclohexene	C₁₀ glycols, 3- and 4-ethylcyclohexylcarbinols	439
α-Methylstyrene	Isopropylbenzene, alcohols	449, 450

Starting material	Product	Page
Dipentene (limonene)	Some saturated alcohols and aldehydes	440, 441
α-Terpinene	Some saturated alcohols and aldehydes	441
2,3-Dimethylbutadiene	3,4-Dimethylpentanol	479
Cycloheptatriene	Cycloheptylcarbinol	442
Cyclooctene	Cyclooctylcarbinol	438, 443
Cyclododecene	Cyclododecylcarbinol	438
cis, trans-Cyclododeca-1,5,9-triene	Cyclododecylcarbinol	444
α-Pinene	Methyl 3-(4′methylcyclohexyl)butyl ether, some alcohols	441
Dihydrodicyclopentadiene	C_{11} alcohol and C_{12} diol	439
	Tricyclo[5,2,1,0^{2,6}] decane-4-methylol	445
3β,20β-Dihydroxypregn-5-ene	6α-Hydroxymethylallopregane-3β,20β-diol	446
3β-Acetoxypregn-5-en-20-one	3β-Acetoxy-6α-hydroxymethyl-5-α-pregnan-20-one	447, 448
11-Oxoprogesterone 3,20 bis(ethyleneketal)	6-Hydroxymethylpregnan-3,11,20-trione 3, 20 bis(ethyleneketal)	447
Androst-5-ene-3β,17β-diol diacetate	6-Hydroxymethyl-5α-androstane-3β,17β-diol diacetate	447
Sitosterol	6-Hydroxymethylsitostanol	447
Eugenol (3-methoxy-4-hydroxyallylbenzene)	3-Methoxy-4-hydroxy-1-propylbenzene, 4-(3-methoxy-4-hydroxyphenyl)butanol	451
Furan	2-Tetrahydrofurfuryl alcohol	415, 431
2,5-Dimethylfuran	2,5-Dimethyl-3-tetrahydrofurfuryl alcohol	415
5,6-Dihydro-4H-pyran	2-Hydroxymethyl- and 3-hydroxymethyl tetrahydropyran	420
6-Hydromethyl-5,6-dihydro-4H-pyran	2,6-Bishydroxymethyltetrahydropyran	420, 452
2,6-Bishydroxymethyl-5,6-dihydro-4H-pyran	1-Hydroxymethyl-2,8-dioxabicyclo [3,2,1] octane	420
2,6-Dimethyl-5,6-dihydro-4H-pyran	2,6-Dimethyl-3-hydroxymethyltetrahydropyran	420
3,4-Di-O-acetyl-D-xylal	1,5-Anhydro-4-deoxy-D-arabino-hexitol, 1,5-anhydro-4-deoxy-L-xylo-hexitol	453, 454
3,4,6-Tri-O-acetyl-D-glucal	4,5,7-Tri-O-acetyl-2,6-anhydro-3-deoxy-D-manno-heptitol and -D-gluco-heptitol	455, 456
3,4,6-Tri-O-acetyl-D-galactal	2,6-Anhydro-3-deoxy-D-galacto-heptitol and -D-talo-heptitol	456–458
Propylene oxide	Some butanols	462, 464
Hexafluoropropylene	Alcohols, hexafluoropropane	495
3-Butene-1,2-diol	Tetrahydrofurfuryl alcohol	479, 504

intermediates. The production of tetrahydrofurfuryl alcohol from 3-butene-1,2-diol occurs via reaction 255a (479, 504):

$$\text{HOH}_2\text{C—CH—CH=CH}_2 \longrightarrow \underset{\text{OH} \quad \text{OH}}{\text{HOH}_2\text{C—}\overset{\text{CH}_2\text{—CH}_2}{\text{CH} \quad \text{CH}_2}} \longrightarrow$$

(with OH on first carbon)

$$\underset{\text{HOH}_2\text{C}}{\overset{\text{CH}_2\text{—CH}_2}{\text{CH} \quad \text{CH}_2}}\text{O} + \text{H}_2\text{O} \quad (255a)$$

Epoxides can undergo hydroformylation in the expected manner to give hydroxyaldehyde products (54, 55):

$$\text{RCH——CH}_2 \xrightarrow{\text{HCo(CO)}_n} \text{RCH(OH)CH}_2\text{Co(CO)}_n \xrightarrow{\text{CO/H}_2}$$

(epoxide O)

$$\text{RCH(OH)CH}_2\text{CHO} + \text{HCo(CO)}_n \quad (256)$$

At higher temperatures some saturated alcohols are formed (462, 464); thus propylene oxide gives some n-butanol presumably by dehydration and then hydrogenation of the α,β-unsaturated aldehyde at both the olefinic bond and the aldehyde group.

$$\text{CH}_3\text{CH(OH)CH}_2\text{CHO} \longrightarrow \text{CH}_3\text{CH=CHCHO} \longrightarrow$$

$$\text{CH}_3\text{CH}_2\text{CH}_2\text{CH}_2\text{OH} \quad (257)$$

The unsaturated aldehyde, 1-cyclohexen-4-al, can give rise to a saturated alcohol and saturated diols (Table 35). α,β-Unsaturated aldehyde intermediates are also possible in the hydroformylation of acetylenes to alcohols, a reaction studied by Wender and coworkers (466). The acetylene reactions proceed sluggishly; pent-1-yne gives only small amounts of n-hexanol and 2-methylpentanol:

$$\text{RC≡CH} \longrightarrow \text{RCH=CHCHO} \longrightarrow \text{RCH}_2\text{CH}_2\text{CH}_2\text{OH} \quad (258)$$

Alternatively the acetylene could first be reduced to pent-1-ene, followed by hydroformylation and then hydrogenation to alcohol. Diphenylacetylene undergoes hydrogenation only, under similar conditions (466, see Section E-3).

The production of alcohols from propylene and pent-2-ene is promoted by the addition of tertiary phosphines, arsines, and stibines such as Bu_3P, Et_2PhP, Et_3As, and $(\text{Ph}_2\text{PCH}_2)_2$ (e.g., 429), whereas the addition of triphenylphosphite is said to decrease alcohol production from cyclohexene (465). The use of tertiary organophosphine-cobalt carbonyl catalysts will be considered more generally in Section X-F.

The stoichiometric reduction of saturated aldehydes according to eq. 259 has been mentioned in Section E-1 (Table 34):

$$RCHO + 2HCo(CO)_4 \longrightarrow RCH_2OH + Co_2(CO)_8 \qquad (259)$$

Goetz and Orchin (400) showed that the reaction was first-order in $HCo(CO)_4$, the disappearance being followed by titration with base, and they proposed the following steps:

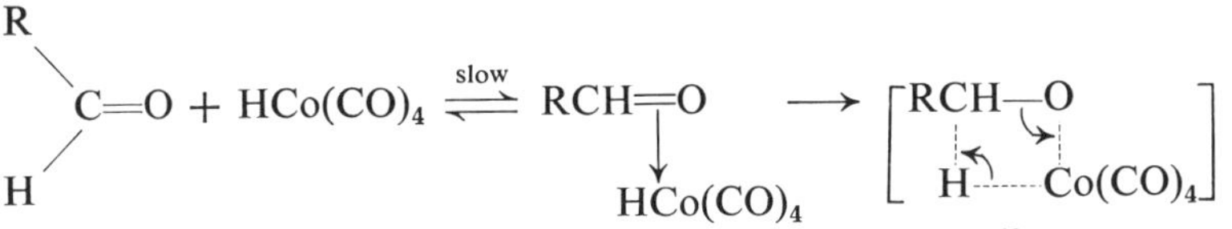

$$
\underset{H}{\overset{R}{\diagdown}}C{=}O + HCo(CO)_4 \underset{slow}{\rightleftharpoons} \underset{\underset{HCo(CO)_4}{\downarrow}}{RCH{=}O} \longrightarrow \left[\begin{array}{c} RCH{-}O \\ \overset{\curvearrowleft}{H\cdots\cdots Co(CO)_4} \end{array} \right] \qquad (260)
$$
$$\mathbf{60}$$

$$\mathbf{60} \longrightarrow RCH_2OCo(CO)_4 \xrightarrow[fast]{HCo(CO)_4} RCH_2OH + Co_2(CO)_8 \qquad (260a)$$

Marko and coworkers (49, 421, 422) have studied alcohol formation under hydroformylation conditions, in particular the hydrogenation of propion-aldehyde to propyl alcohol by using $Co_2(CO)_8$ in toluene at 150° and high pressures. Marko (49) found that a maximum rate was observed at a low carbon monoxide partial pressure (~20 atm at 150°); maximum rates for hydroformylation are found also (53, 54) at low carbon monoxide pressure (~10 atm at 110°). From 32 to 200 atm carbon monoxide, the hydrogenation rate (49) was given by

$$\frac{d[R'OH]}{dt} = k[RCHO][Co]p_{H_2}(p_{CO})^{-2} \qquad (261)$$

The CO dependence was established at a hydrogen pressure of 95 atm. At $p_{CO} < 32$ atm, carbonyl formation is affected and the rate falls off as metal precipitates (49). Analogous to the hydroformylation mechanism (eqs. 246 to 250), Marko suggested the following scheme:

$$OHCCo(CO)_4 \underset{-CO}{\overset{CO}{\rightleftharpoons}} HCo(CO)_4 \rightleftharpoons HCo(CO)_3 + CO \qquad (262)$$

$$RCHO + HCo(CO)_3 \rightleftharpoons \underset{\underset{HCo(CO)_3}{\downarrow}}{RCH{=}O} \rightleftharpoons RCH_2OCo(CO)_3 \qquad (263)$$
$$\mathbf{61}$$

$$\mathbf{61} + H_2 \xrightarrow{slow} [RCH_2OCoH_2(CO)_3] \longrightarrow RCH_2OH + HCo(CO)_3 \qquad (264)$$

$$\mathbf{61} + CO \rightleftharpoons RCH_2OCo(CO)_4 \rightleftharpoons$$
$$\mathbf{62}$$

$$RCH_2OCOCo(CO)_3 \underset{-CO}{\overset{CO}{\rightleftharpoons}} RCH_2OCOCo(CO)_4 \qquad (265)$$
$$\mathbf{63} \qquad\qquad\qquad\qquad \mathbf{64}$$

$$RCH_2OCOCo(CO)_3 + H_2 \longrightarrow RCH_2OOCH + HCo(CO)_3 \qquad (266)$$

Here H_2 is activated by coordinatively unsaturated complexes such as **61** and **63**; this was thought to involve coordination of hydrogen, followed by insertion with resulting cobalt-carbon bond fission (eqs. 264 to 266). The coordination may be considered to be an oxidative addition to a cobalt(I) species. To account for the inverse second-order CO dependence, the cobalt has to be present principally as $OHCCo(CO)_4$ and **64**, which contain two more molecules of carbon monoxide than **61**, the precursor to alcohol formation. Marko (49) also pointed out that in hydroformylation the active species $RCH_2CH_2COCo(CO)_3$ has one molecule of carbon monoxide less than its inactive counterpart $RCH_2CH_2COCo(CO)_4$, which was consistent with the observed inverse first-order CO dependence for this reaction. Reaction 266 explained the observed formation of formates from aldehydes, carbon monoxide, and hydrogen (413, 414).

Aldridge and Jonassen (387) have also studied aldehyde reduction (2-ethylhexanal to 2-ethylhexanol) under hydroformylation conditions at 160° in di-n-hexyl-ether. They demonstrated that the reaction was not a free-radical chain process, which Wender and coworkers (415) had suggested in earlier days as a possibility; for example,

$$Co_2(CO)_8 \rightleftharpoons 2\dot{C}o(CO)_4 \tag{267}$$

$$\cdot Co(CO)_4 + H_2 \rightleftharpoons H\cdot + HCo(CO)_4 \tag{268}$$

$$H\cdot + RCHO \rightleftharpoons R\dot{C}HOH \tag{269}$$

$$R\dot{C}HOH + HCo(CO)_4 \rightleftharpoons RCH_2OH + \cdot Co(CO)_4 \tag{270}$$

Aldridge and Jonassen (387), in contrast to Marko (49), found a direct inverse first-order dependence on carbon monoxide in the p_{CO} range 20 to 70 atm. The rate was half-order in hydrogen up to ~ 7 atm H_2, with a decreasing dependence at higher pressures; the order in $Co_2(CO)_8$ was about 0.8. This reaction scheme involved eq. 245 followed by these reactions:

$$RCHO + HCo(CO)_4 \overset{K}{\rightleftharpoons} \underset{\underset{\underset{\textbf{65}}{HCo(CO)_3}}{\downarrow}}{RCH{=}O} + CO \tag{271}$$

$$\textbf{65} \overset{k_2}{\longrightarrow} RCH(OH)Co(CO)_3, \textbf{66} \tag{272}$$

$$\textbf{66} + HCo(CO)_4 \overset{fast}{\longrightarrow} RCH_2OH + Co_2(CO)_7 \tag{273}$$

$$Co_2(CO)_7 + CO \overset{fast}{\rightleftharpoons} Co_2(CO)_8 \tag{274}$$

This gives the rate law

$$\frac{d[RCH_2OH]}{dt} = \frac{k_2 K[RCHO][HCo(CO)_4]}{[CO]} \tag{274a}$$

The $[HCo(CO)_4]$ in solution could be expressed in terms of the octacarbonyl concentration (Co_i) after experimental determination of both the partition coefficient (K_x) for the hydrocarbonyl between the liquid and gas phases and the equilibrium constant (K_1) for reaction 245:

$$[HCo(CO)_4] = K_1^{0.5}[H_2](Co_i + aK_1[H_2])^{0.5} - bK_1[H_2]$$

$$(274b)$$

(where a and b are constants determined by K_x). Substitution of 274b into 274a gave a rate law consistent with the experimental findings. A rate law of the same form results if carbon monoxide loss (giving $HCo(CO)_3$) precedes addition to aldehyde.

Equation 272 differs from eqs. 260 and 263 in the direction of addition of the hydrocarbonyl to C=O. Some evidence in favor of eq. 272 is the fact that epoxides add the hydrocarbonyl in the same way (460–463):

$$HCo(CO)_4 + RCH\!\!-\!\!-\!\!CH_2 \longrightarrow RCH(OH)CH_2Co(CO)_4 \quad (275)$$

$$O$$

The cobalt-oxygen bonded intermediate **61** is not essential to produce the formate ester (eq. 266) which could equally result from the reaction

$$OHCCo(CO)_4 + RCH_2OH \longrightarrow RCH_2OOCH + HCo(CO)_4 \quad (276)$$

Chalk and Harrod (54) conclude that both modes of addition of the hydrocarbonyl to C=O seem feasible.

Marko also preferred hydrogenolysis of the metal-oxygen bonded intermediate **61**, rather than reaction with $HCo(CO)_4$ as favored by Aldridge and Jonassen for the metal-carbon bonded intermediate **66**. The hydrogenolysis reaction is favored in hydroformylation (eq. 250) because the concentration of hydrocarbonyl is very low (54).

In their study on the reduction of 2-ethylhexanal, Aldridge and Jonassen (387) concluded, by examining poisoning effects of thiophene and heavy metals, that eq. 245 was heterogeneously catalyzed by cobalt metal.

3. Reduction of Olefinic Bonds. Hydrogenation of an olefinic substrate is a competing reaction under hydroformylation conditions and seems significant when the olefin is branched, conjugated, or has electronegative substituents.

The concurrent reduction of conjugated double bonds during alcohol formation from 3-vinylcyclohexene, some dienes, trienes, furans, terpenes and some epoxide and acetylene systems, and some direct reduction of α,β-unsaturated esters during lactone formation were mentioned in Section E-2 (Tables 35 and 36). Direct reduction products were also observed during production of alcohols from certain olefins listed in Table 36: isobutylene, diisobutylene, α-methylstyrene, hexafluoropropylene, and eugenol. Other systems where appreciable reduction of olefinic bonds has been observed are listed in Table 37. A selective reduction product, cyclododecene, is produced from 1,5,9-cyclododecatriene (548, Section X-F-2). Frankel and coworkers (1132) and Ogata and Misono (1133) have hydrogenated unsaturated fats using $Co_2(CO)_8$ as catalyst (see Section XIV-C).

For simple monoolefins, hydrogenation to paraffins seems negligible during hydroformylation to the aldehyde stage, although hydrogenation of a double bond is more favored thermodynamically than its hydroformylation (490, 491). Marko and coworkers (427, 489), however, reported that while propylene, cyclohexene, and Fischer-Tropsch olefins give <10% saturated hydrocarbons, either isobutylene or diisobutylene give as much as 65% paraffin

Table 37   Reduction of Olefinic and Acetylenic Bonds under Hydroformylation Conditions

Substrate	Product	Reference
Butadiene	n-Valeraldehyde, α-methylbutyraldehyde	467, 468, 500
1,2-Dimethylbutadiene	C_7 aldehydes	467
2,3-Dimethylbutadiene	3,4-Dimethylpentanal	467
Myrcene (7-methyl-3 methylene-1,6-octadiene)	4,8-Dimethylnonanal	441, 469
Alloocimene (3,7-dimethyl-1,3,6-ocatatriene)	4,8-Dimethylnonanal	469
Cyclopentadiene	Cyclopentanealdehyde	467
1;5,9-Cyclododecatriene	Cyclododecene	548
Styrene	Ethylbenzene	402, 499
1-Phenylbutadiene	n-Butylbenzene (trace)	419, 467
Isoeugenol (2-methoxy-4-propenylphenol)	3-Methoxy-4-hydroxy-1-propylbenzene	451
1,1-Dichloroethylene	1,1-Dichloroethane	470
1-Vinylnaphthalene	1-Ethylnaphthalene	490
Dimethyl-6-methyl-5,6-dihydro-4H-pyran-2,3-dicarboxylate	Dimethyl-6-methyltetrahydropyran-2,3-dicarboxylate	420
Acrolein	Propionaldehyde	419
Crotonaldehyde	Butyraldehyde	419
Methyl vinyl ketone	Ethyl methyl ketone	419
Mesityl oxide	Isobutyl methyl ketone	419
Diphenylacetylene	1,2-Diphenylethane	466
cis-Stilbene	1,2-Diphenylethane	471
Ethyl β-(2-furan)-acrylate	Ethyl β-(2-furan)propionate	419
Acrylonitrile	2-Methylglutaronitrile, propionitrile	482, 494, 508

products during hydroformylation to alcohols. They concluded that the ratio of hydrogenation to hydroformylation products increases with increasing branching of the olefin. This was consistent with earlier findings by Keulmans and coworkers (423) and Wender and coworkers (433), who showed that hydroformylation of a tertiary carbon atom occurs to a very small extent. Marko's group (427, 489) proposed a mechanism to explain the effect of branching on the rates of transformation of the intermediate RCo(CO)$_4$ formed from an olefin and the hydrocarbonyl (eqs. 247, 248). Their mechanism was similar to that suggested by Marko for aldehyde reduction (eqs. 262 to 264):

$$RCo(CO)_4 \overset{k_3}{\rightleftharpoons} RCOCo(CO)_3 \overset{H_2}{\rightleftharpoons} RCOCoH_2(CO)_3 \longrightarrow$$
$$RCHO + HCo(CO)_3 \quad (277)$$

$$RCo(CO)_4 \overset{k_4}{\rightleftharpoons} CO + RCo(CO)_3 \overset{H_2}{\longrightarrow} RH + HCo(CO)_3 \quad (278)$$

Here k_4/k_3 must increase with increasing branching of R.

The decarbonylation of acylcobalt tetracarbonyls, which is similar to reaction 278, has been studied by Heck (492), who showed that reaction with triphenylphosphine proceeds via a predissociation to the tricarbonyl;

$$RCOCo(CO)_4 \overset{k_5}{\rightleftharpoons} RCOCo(CO)_3 + CO \quad (279)$$

A steric acceleration in the rate by a factor of about 90 was observed as R went from CH$_3$ to (CH$_3$)$_3$C. This is consistent with the mechanism of eqs. 277, 278, but, as pointed out by Chalk and Harrod (54), is not strong evidence for the mechanism since such a relation between R and decarbonylation rate also exists; for example, for the decarbonylation of acyl radicals (493).

Heck (492) also observed a decrease in k_5 with electronegative substituents on R (halogen and methoxy groups). But this did not rationalize the significant hydrogenation of olefins with electronegative substituents, for example, 1, 1-dichloroethylene, hexafluoropropylene, acrylonitrile, and styrenes (Tables 36 and 37). Increasing electronegativity in R, however, would be expected to stabilize both alkyl-metal and acyl-metal bonds, particularly the former. Hence the equilibrium

$$RM(CO)_n + CO \overset{k_6}{\rightleftharpoons} RCOM(CO)_n \quad (280)$$

should be displaced to the left with a more electronegative R, and this could explain the hydrogenation of the mentioned olefins (54, 492). Calderazzo and Cotton (496, 497) have measured the equilibrium constant, K, and the rate constant, k_6, of reaction 280 for the formation of acylmanganese penta-carbonyls. K decreased in the order R=C$_2$H$_5$ > CH$_3$ > C$_6$H$_5$ and k_6 varied similarly, R=C$_2$H$_5$ > C$_6$H$_5$ $\approx$ CH$_3$ $\gg$ C$_6$H$_5$CH$_2$, CF$_3$. Thus the tendency for fluoroolefins and styrenes to be hydrogenated could result from an unfavorable thermodynamic or kinetic control of carbonylation of the alkyl cobalt carbonyls (54). Heck and Breslow (498) have reacted alkyl and acyl halides with sodium cobalt tetracarbonylate under carbon monoxide; acylcobalt carbonyls were formed except for the case of chloroacetonitrile which gave a cyanomethylcobalt carbonyl, indicating the electronegative nitrile group can stabilize the alkylcobalt carbonyl. This similarly accounts for the tendency of acrylonitrile to be hydrogenated, although the major product is still the hydroformylation product 2-cyanopropanal (53, 54, 508). The proportion of propionitrile formed decreased with use of a higher initial concentration of acrylonitrile, a smaller H_2/CO pressure ratio, a lower temperature, and a more polar solvent (benzene, cyclohexane, and isopropanol). The catalyst concentration and a variety of additives (pyridine, thiosemicarbazide, sulfur, carbon disulfide, and iron pentacarbonyl) had no effect (508). A stoichiometric reaction of α,β-unsaturated nitriles with $Co_2(CO)_8$ in acidic methanol under nitrogen at room temperature to give saturated aldehydes with the original carbon skeletons has been reported by Wakamatsu and Sakamaki (580). Misono and coworkers (212) have used $Co_2(CO)_8$ under hydrogen in the absence of carbon monoxide for the catalytic hydrodimerization of acrylonitrile (Section IX-A; eq. 102).

Sato and coworkers (1450) have reported that a butenolactonedicobalt heptacarbonyl complex in benzene reacts with molecular hydrogen under pressure at 120° to give some butyrolactone. This reaction involves hydrogenation of the coordinated unsaturated moiety, and it gives evidence for attack of a cobalt-carbon bond by molecular hydrogen.

Rudkovskii and Imyanitov (434, 450) have also found that the ratio of hydrogenation to hydroformylation products for α-methylstyrene was independent of the catalyst concentration. Both reactions were first-order in catalyst and greater than first-order in substrate. The yield of isopropylbenzene varied with the solvent used: methanol (93%) > acetone (84%) > pentane (77%), diethyl ether (77%) > toluene (66%).

In the stoichiometric reductions of olefinic bonds using $HCo(CO)_4$ (Table 34, styrenes, vinyl ethers, vinyl ketones, heptene, and conjugated dienes), the hydrogenolysis of the alkylcobalt carbonyl intermediate must involve $HCo(CO)_4$, and not hydrogen as in eq. 278 (522, 523). Ungvary and Marko (522, 523) report that the highest yield of heptane from hept-1-ene is observed at a low olefin to $HCo(CO)_4$ ratio, which is consistent with a large hydrocarbonyl concentration favoring hydrogenolysis of the alkylcobalt carbon before carbon monoxide insertion. Takegami and coworkers (402) showed that in such a reduction of styrene the yield of ethylbenzene was independent of the carbon monoxide pressure, and Harrod and Chalk (54) have thus suggested that for this system the hydrogenolysis step (eq. 278) involves the alkylcobalt tetracarbonyl and that the k_3 step is probably S_N1 in character.

The stoichiometric reduction of conjugated dienes gave monoenes to a high degree of selectivity (Table 34) but small amounts of π-allyl cobalt tricarbonyls were also formed and the following reaction scheme was presented (522, 523):

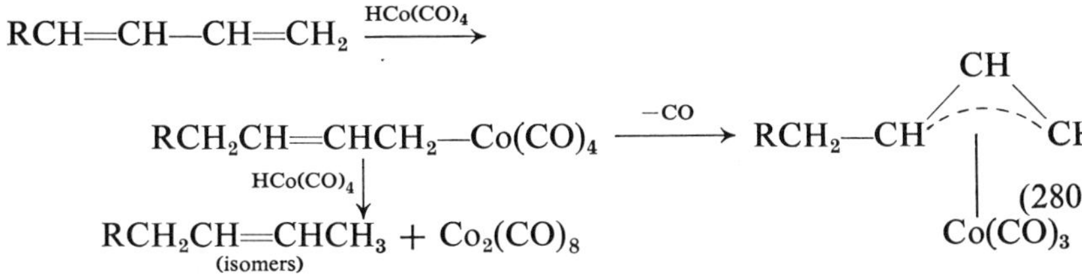

$$RCH{=}CH{-}CH{=}CH_2 \xrightarrow{HCo(CO)_4}$$

$$RCH_2CH{=}CHCH_2{-}Co(CO)_4 \xrightarrow{-CO} RCH_2{-}CH{\overset{CH}{\underset{Co(CO)_3}{\diagup\;\diagdown}}}CH_2$$

$$\downarrow HCo(CO)_4$$

$$RCH_2CH{=}CHCH_3 + Co_2(CO)_8 \qquad (280a)$$
$$\text{(isomers)}$$

Again a relatively large $HCo(CO)_4$ concentration favored the hydrogenolysis of the σ-allyl intermediate.

Hydroformylation of conjugated dienes (including furans) usually gives saturated aldehydes (Table 37) or, at higher temperatures, saturated alcohols (Section E-3, Tables 35 and 36). The saturated aldehydes could result from hydrogenation followed by hydroformylation:

$$\text{Butadiene} \xrightarrow{H_2} CH_3{-}CH{=}CH{-}CH_3 + CH_3{-}CH_2{-}CH_2{=}CH_2 \quad (281)$$
$$\qquad\qquad\qquad \downarrow CO/H_2 \qquad\qquad\qquad\qquad CO/H_2 \downarrow$$
$$\qquad\qquad \alpha\text{-methylbutyraldehyde} \qquad\qquad n\text{-valeraldehyde}$$

Reaction of $HCo(CO)_4$ with butadiene may produce π-methylallyl cobalt tricarbonyl complexes (501, 502, 522, 523, eq. 280a); this behavior is similar to that observed for $HCo(CN)_5^{3-}$ (e.g., eq. 185, Section B-2a) but, unlike in the latter system, there are no reports of further reaction of these complexes to give butenes, although this seems quite feasible. The scheme of eq. 281 is perhaps less likely than initial hydroformylation to α,β-unsaturated aldehydes, which are known to be readily reduced (Tables 34 and 37).

An olefinic bond conjugated to a ketonic carbonyl is also readily hydrogenated (Table 37). Aliphatic α,β-unsaturated esters usually undergo hydroformylation (eq. 252a) although ethyl β-(2-furan)-acrylate undergoes reduction to the corresponding propionate (419), and maleate and fumarate give some succinate (410–412). If the olefinic bond is conjugated with an aromatic ring, reduction again seems favored. Thus, cinnamate gives β-phenylpropionate (Table 35) and styrenes are reduced; the latter were also considered above as olefin systems with electronegative substituents. Diphenylacetylene is reduced to 1,2-diphenylethane, presumably via cis-stilbene since this is also readily reduced (471). Orchin (490) has also suggested that hydroformylation of 1-vinylnaphthalene (419) probably gives appreciable quantities of 1-ethylnaphthalene.

Goetz and Orchin (400, 401) have studied the reduction of unsaturated aldehydes and ketones and acrylonitrile by cobalt hydrocarbonyl (Table 34). Unsaturated ethers could also be reduced, and this seems reasonable considering the electron-withdrawing nature of alkoxy groups, although the major products were those of hydroformylation. Goetz and Orchin (401) suggested that the hydrogenation of α,β-unsaturated aldehydes and ketones proceeded via a π-oxapropenyl (or pseudo-π-allyl) complex, 67:

$$\underset{/}{\overset{\backslash}{C}}{=}\overset{|}{C}{-}\overset{|}{C}{=}O + HCo(CO)_4 \underset{}{\overset{k}{\rightleftharpoons}} \underset{/}{\overset{\backslash}{C}}{=}\overset{|}{C}{-}\overset{|}{C}{-}O \longrightarrow$$
$$\downarrow$$
$$H\overset{\cdot}{C}o(CO)_4$$

$$\underset{/ \ 67}{\overset{\backslash}{CH}}{-}C \underset{}{\overset{\nearrow C}{\overset{\backslash}{\underset{|}{\underset{Co(CO)_3}{O}}}}} + CO \quad (282)$$

$$HCo(CO)_4 + 67 \longrightarrow \underset{/}{\overset{\backslash}{CH}}{-}\overset{|}{C}{=}\overset{|}{C}(OH) + Co_2(CO)_7 \quad (283)$$

$$\downarrow \qquad\qquad \downarrow CO$$

$$\underset{/}{\overset{\backslash}{CH}}{-}\overset{|}{CH}{-}\overset{|}{C}{=}O \qquad Co_2(CO)_8$$

They suggested that 67 was stable to carbonylation but was reducible by $HCo(CO)_4$, by proton addition at the oxygen atom to give the enol form of the carbonyl product. The rate data for k (Table 34) were rationalized in terms of electron donors stabilizing 67. For example, k was much greater for methyl vinyl ketone than for crotonaldehyde; for the former case 67 would contain two methyl substituents while for the latter there would be only one ethyl group. The faster rate under argon compared with under carbon monoxide was consistent with the absence of the latter as a competing nucleophile for the cobalt center. The more stable π-oxapropenyl complexes were thought to decrease the rate of reaction 283 and give lower product yields. These reactions could alternatively involve chelate intermediates, which have been substantiated for hydrogenation of unsaturated ketones by using some iridium complexes (Section XII-C).

The direction of addition of $HCo(CO)_n$ ($n = 4$ or 3) to olefinic bonds (eq. 247) seems to be determined by the substrate and temperature (375). Even though most simple olefins are not hydrogenated, their hydroformylation reaction involves alkylcobalt carbonyls and, for example, the reversibility of the addition of $HCo(CO)_n$ to ethylene has been investigated (395):

$$C_2H_5Co(CO)_4 \rightleftharpoons C_2H_4 + HCo(CO)_4 \quad (284)$$

By a consideration of hydroformylation products at 120°, and products isolated by reaction with $HCo(CO)_4$ at 0°, Heck and Breslow (395) have concluded that a straight chain 1-ene is approximately neutral toward hydrocarbonyl addition at both temperatures:

$$CH_3(CH_2)_2CH{=}CH_2$$
$$CH_3(CH_2)_4Co(CO)_4 + CH_3(CH_2)_2CH(CH_3)Co(CO)_4 \quad (285)$$

At 0° the high electron density of the double bond of isobutylene results in an acid type addition (eq. 286), while the low electron density in methyl acrylate leads mainly to a hydridic addition (eq. 287):

$$(CH_3)_2C{=}\overset{\delta-}{C}H_2 + HCo(CO)_4 {\longrightarrow} [(CH_3)_3C\,Co(CO)_4],\mathbf{68} \quad (286)$$

$$\overset{\delta+}{C}H_2{=}CHCO_2CH_3 + HCo(CO)_4 \longrightarrow \left[\begin{array}{c} CH_3{-}CHCO_2CH_3 \\ | \\ Co(CO)_4 \end{array} \right], \mathbf{69} \quad (287)$$

68 and **69** were not isolated as such, but as monophosphine adducts of the corresponding acyl compounds (see eq. 249). At 120° the directions of addition in eqs. 286 and 287 were reversed, and this probably reflects the relative stability of the adducts.

The addition to an activated olefin is shown in eq. 282 to involve hydrogen attack at the β-carbon (the α one is defined as having the activating group); such attack has been well demonstrated for $HCo(CN)_5^{3-}$ systems (see Section A-4).

4. Reduction of Aromatic Hydrocarbons and Heterocyclic Compounds.

Unlike most homogeneous hydrogenation catalysts (including $Co(CN)_5^{3-}$) $Co_2(CO)_8$ can reduce various aromatic systems, including naphthalene, anthracene, phenanthrene, pyrene, naphthacene, chrysene, perylene, thiophene, pyridine, and the pyrrole ring in indoles (Table 38). The hydrocarbon reductions were discovered during attempts by Friedman and others (742) to hydroformylate those containing "reactive double bonds." Phenanthrene is only slightly reduced at 200°, while anthracene is quantitatively reduced to the 9,10-dihydro derivative at 135°. Isolated benzene rings and phenanthrenoid systems are very resistant to reduction, and highly condensed systems tend to form phenanthrene derivatives, e.g., chrysene, a benzphenanthrene, is reduced slowly to the 5,6-dihydro compound (eq. 288), while perylene and pyrene are reduced to phenanthrene derivatives (eqs. 289 and 290). Hydroformylation of benzanthrone (Table 38) gives reduction of an aromatic ring as well as the ketonic group (Section E-6).

Of interest is that both hydrogen and carbon monoxide can be added to coal under hydroformylation conditions at the relatively low temperature of 200° (503).

Table 38 Reduction of Aromatic Systems, Ketones, Nitriles, Nitro and Azo Compounds, Imines, and Epoxides under Hydroformylation Conditions

Substrate	Product	Reference
Naphthalene	Tetralin	472
1- or 2-methylnaphthalene	Methyltetralins	418, 472
Acenaphthene	2a,3,4,5-Tetrahydroacenaphthene	472
Fluorene	No reaction	472
Anthracene	9,10-Dihydroanthracene	472
Phenanthrene	Some di- and tetrahydrophenanthrenes	415, 472
Fluoranthene	1,2,3,10b-Tetrahydrofluoranthene	472
Pyrene	4,5-Dihydropyrene	472
Triphenylene	No reaction	472
Naphthacene	5,12-Dihydronaphthacene	472
Chrysene	5,6-Dihydrochrysene	472
1,1-Dinaphthyl	No reaction	472
Perylene	1,2,3,10,11,12-Hexahydroperylene	472
Coronene	No reaction	472
2-Methylindole	2,3-Dihydro-2-methylindole	487
2-Phenylindole	2,3-Dihydro-2-phenylindole	487
Pyridine	N-Formyl- and N-methylpiperidine	488
Thiophene	Thiolane	50, 415
2-Methylthiophene	2-Methylthiolane	50
2-Ethylthiophene	2-Ethylthiolane	50
2,5-Dimethylthiophene	2,5-Dimethylthiolane	50
2-Acetylthiophene	2-Ethylthiophene, 2-ethylthiolane	50, 418
Acetone	Isopropanol	415
Diethylketone	Limited reduction	417

Benzyl methyl ketone	Limited reduction	417
Acetophenone	Ethylbenzene	417, 418
p-Methoxyacetophenone	p-Ethylanisole, 2-(p-methoxyphenyl)-propanol	418
Pinacolone	Pinacolyl alcohol	479
Cyclohexanone	Limited reduction	417
2-Acetylnaphthalene	2-Ethylnaphthalene	417
Benzophenone	Diphenylmethane	417, 418
Fluorenone	Fluorene, 9-fluorenol	418
Benzanthrone	1,10-Trimethylenephenanthrene	418
Nitrobenzene	Aniline, diphenylurea	474, 481, 1437
Azobenzene	Aniline, diphenylurea	481
Hydrazobenzene	Aniline, diphenylurea	481
N-Cyclohexylidenecyclohexylamine	Dicyclohexylamine	483
N-Hexahydrobenzylidenecyclohexylamine	N-Hexahydrobenzylcyclohexylamine	483
N-Cyclohexylideneaniline	N-Cyclohexylaniline	483
N-Benzylidenecyclohexylamine	N-Benzylcyclohexylamine	483
N-Benzylideneaniline	N-Benzylaniline	417, 481, 483–485
N-Benzylidene-p-toluidine	N-Benzyl-p-toluidine	481, 485, 486
N-Benzylidene-2,6-diethylaniline	N-Benzyl-2,6-diethylaniline	484
N-Benzylidene-p-chloroaniline	N-Benzyl-p-chloroaniline	481, 485, 486
N-Benzylidene-p-nitroaniline	N-Benzyl-p-nitroaniline	481, 485, 486
N-Benzylidene-p-anisidine	N-Benzyl-p-anisidine	481, 485, 486
Benzonitrile	Limited reduction to amines	481
Benzylcyanide	Limited reduction to amines	481
Ethylene oxide	Some ethanol	507
Trimethylene oxide	Some 1-propanol	507
1-4-Dioxane	No reaction	507
4-Phenyl-1,3-dioxane	Some phenylpropanol and higher alcohols	507

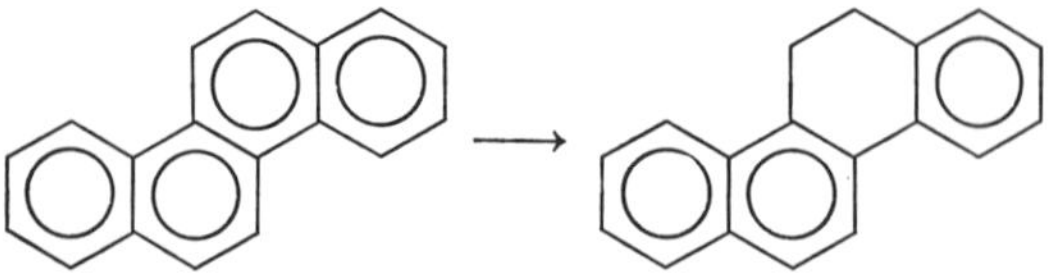

(288)

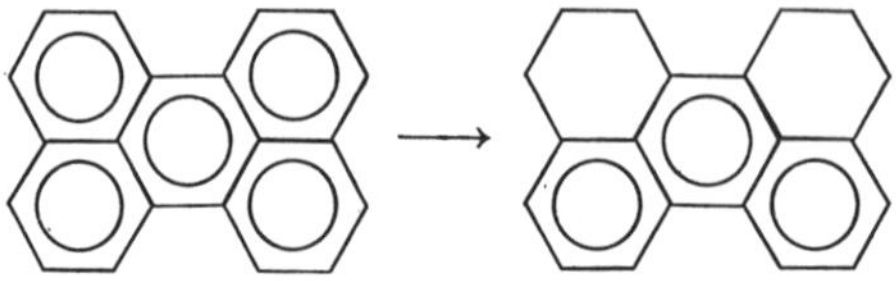

(289)

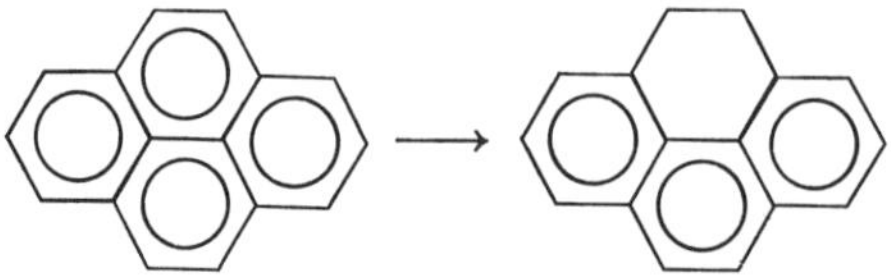

(290)

Shaw and Ryson (487) showed that indoles were readily hydrogenated at 180°. Pino and Ercoli (488) converted pyridine to N-formyl and N-methyl piperidine. No piperidine was detected in the products, and it is likely that hydroformylation is first followed by reduction of the aldehyde group to hydroxymethyl (Section E-2), then reduction to methyl (Section E-5), and finally complete reduction of the ring. Greenfield and coworkers (50) have shown that the thiophene ring system is also reduced completely. The successive reduction of 2-thiophenealdehyde to 2-thenyl alcohol (415, Table 35) to 2-methylthiophene (50, Table 39, Section E-5) to 2-methylthiolane (50, Table 38) is shown in eq. 291:

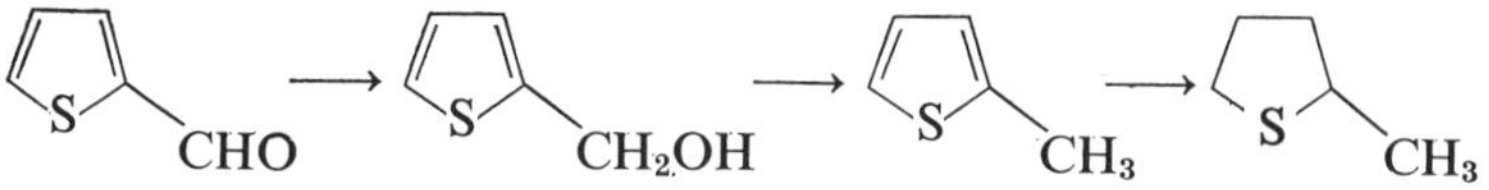

(291)

5. Homologation and Hydrogenolysis of Alcohols.   Wender and coworkers (473) first reported on the hydroformylation of alcohols, which were found to give the next higher alcohol (the homologation reaction). However, some hydrogenolysis of alcohols occurred, particularly for aromatic compounds (Table 39).

For the homologation reaction, an olefin intermediate generally seems likely, (433, 473, 477–479, 490) and then this is hydroformylated to aldehyde followed by reduction to alcohol (Section E-2); for example,

$$(CH_3)_3OH \xrightarrow{-H_2O} CH_3-C(CH_3)=CH_2 \xrightarrow[H_2]{CO}$$

$$CH_3-CH(CH_3)CH_2CHO \xrightarrow{H_2} (CH_3)_2CHCH_2CH_2OH$$

(292)

Table 39 Hydroformylation of Alcohols: Homologation and Hydrogenolysis

Substrate	Product	Reference
Methanol	Ethanol, methylacetate, methane, etc.	475–478
n-Propanol	Butyl and amyl alcohols	473, 477
Isopropanol	Butyl alcohols	473
t-Butyl alcohol	Isoamyl and neopentyl alcohols, isobutane, isobutene	433, 473, 477
Cyclohexanol	Cyclohexylcarbinol	477
Pinacol	3,4-Dimethylpentanol, pinacolone, pinacolyl alcohol, 2,2,3-tri-methyltetrahydrofuran	479
Benzyl alcohol	2-Phenylethanol, toluene	417, 418, 473, 474, 477
p-Methylbenzyl alcohol	2-(*p*-Methylphenyl) ethanol, *p*-xylene	474
m-Methylbenzyl alcohol	2-(*m*-Methylphenyl) ethanol, *m*-xylene	474
p-t-Butylbenzyl alcohol	2-(*p-t*-Butylphenyl) ethanol, *p-t*-butyl toluene	474
2,4,6-Trimethylbenzyl alcohol	2-(2,4,6-Trimethylphenyl) ethanol, 1,2,3,5-tetramethylbenzene	474
p-Hydroxymethylbenzyl alcohol	2-(*p*-Methylphenyl) ethanol, *p*-phenylene-β,β'-diethanol, *p*-xylene	474
p-Methoxybenzyl alcohol	2-(*p*-Methoxyphenyl) ethanol, *p*-methoxytoluene	474
m-Methoxybenzyl alcohol	2-(*m*-Methoxyphenyl) ethanol, *m*-methoxytoluene	474
p-Chlorobenzyl alcohol	2-(*p*-Chlorophenyl) ethanol, *p*-chlorotoluene	474
m-Trifluoromethylbenzyl alcohol	*m*-Methylbenzotrifluoride	474
p-Carbethoxybenzyl alcohol	*p*-Carbethoxytoluene	474
p-Nitrobenzyl alcohol	Polymer of *p*-aminobenzyl alcohol	474
1-Phenylethanol	Ethylbenzene	417, 418
1-Naphthalenemethanol	1-Methylnaphthalene, methyltetralins	418
Benzhydrol	Diphenylmethane	418, 505
Triphenylcarbinol	Triphenylmethane	418
Benzopinacol	Diphenylmethane	418
2-Thenyl alcohol	2-Methylthiophene, 2-methylthiolane	50

177

Wender and coworkers (473, 479) have suggested that the dehydration step involves production of a carbonium-ion intermediate catalyzed by the acid $HCo(CO)_4$. Thus the products derived from pinacol were accounted for via the following scheme (479):

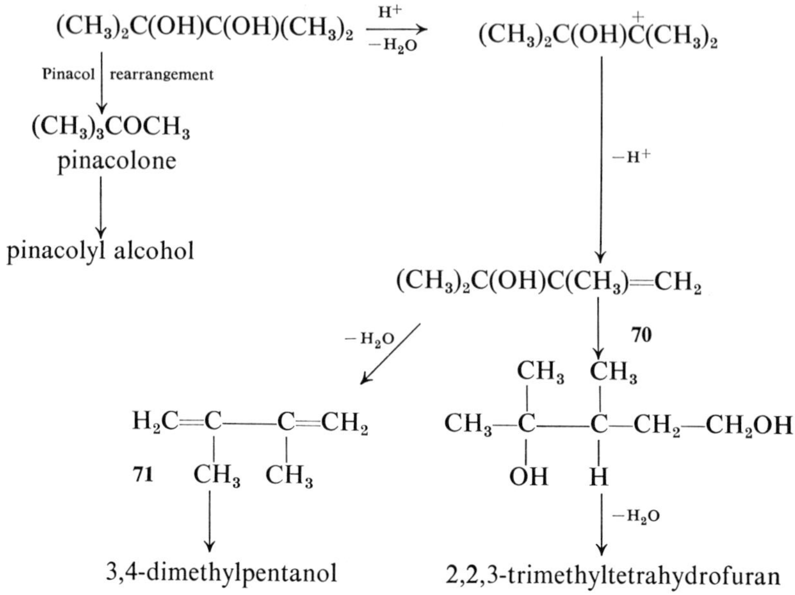

$$(CH_3)_2C(OH)C(OH)(CH_3)_2 \xrightarrow[-H_2O]{H^+} (CH_3)_2C(OH)\overset{+}{C}(CH_3)_2$$

Pinacol | rearrangement

$(CH_3)_3COCH_3$
pinacolone

$-H^+$

pinacolyl alcohol

$(CH_3)_2C(OH)C(CH_3){=}CH_2$

$-H_2O$

70

$$H_2C{=}C\underset{CH_3}{\overset{|}{}}\!\!\!-\!\!\!C\underset{CH_3}{\overset{|}{}}\!\!\!=CH_2$$

71 CH_3 CH_3

$$CH_3{-}\underset{OH}{\overset{CH_3}{\overset{|}{\underset{|}{C}}}}\!\!\!-\!\!\!\underset{H}{\overset{CH_3}{\overset{|}{\underset{|}{C}}}}\!\!\!-CH_2{-}CH_2OH$$

$-H_2O$

3,4-dimethylpentanol   2,2,3-trimethyltetrahydrofuran

Scheme 8

The production of the trimethyltetrahydrofuran from **70** is similar to reaction 255a; the formation of the dimethylpentanol from **71** has already been discussed (Sections E-2 and E-3). The reduction of ketones (pinacolone) is considered in Section E-6. Scheme 8 is of interest in that it shows the wide variety of reactions that may occur under hydroformylation conditions.

Methanol and benzyl alcohol cannot dehydrate to an olefin, although they undergo homologation reaction as well as reduction; based on the alcohol consumption, 8% methane and 60% toluene were produced respectively (Table 39). Wender and coworkers (418, 474) in particular have studied the hydrogenolysis of aromatic alcohols. Substitution into the —CH_2OH group of benzyl alcohol gives more reduced product; thus 1-phenylethanol gives 70% ethylbenzene, while benzhydrol and triphenylcarbinol give exclusively hydrocarbon. The relative reaction rates of a series of nuclear-substituted benzyl alcohols (measured by pressure decrease for the overall reduction and homologation reactions) were in the following order (474):

$p\text{-}OCH_3 \gg p\text{-}CH_3 > m\text{-}CH_3 > p\text{-}t\text{-butyl} > H > p\text{-}Cl > m\text{-}OCH_3$

$\gg m\text{-}CF_3$

Thus the rates are increased by *para* or *meta* electron donors. The ratio of homologation to reduction products also decreased in the same order. A *p*-nitro group, expected to retard the reaction, was itself reduced to an amino group with subsequent polymerization products of the aminobenzyl alcohol. The rate trend above is that expected for a carbonium ion intermediate. For example a *p*-methoxy group would stabilize the benzyl carbonium ion (produced by the acid $HCo(CO)_4$ as in Scheme 8).

$$CH_3O\text{—}\langle\!\!\!\bigcirc\!\!\!\rangle\text{—}\overset{+}{C}H_2 \longleftrightarrow CH_3\overset{+}{O}\text{=}\langle\!\!\!\bigcirc\!\!\!\rangle\text{=}CH_2$$

72

The ensuing steps to hydroformylation and reduction products have not been established. The former could involve nucleophilic attack on **72** by carbon monoxide followed by hydrogenation (474), while the latter could involve reduction by $Co(CO)_4^-$ to a radical followed by hydrogen atom abstraction from $HCo(CO)_4$ (418, 474, 490). Orchin (490) had suggested that an intermediate such as $R\overset{+}{C}_6H_4CH_2Co(CO)_4^-$ might be formed; the present reviewer suggests that possibly the neutral σ-alkyl complex is formed (eq. 293),

$$RC_6H_4CH_2OH + HCo(CO)_4 \xrightarrow{-H_2O} RC_6H_4CH_2Co(CO)_4 \qquad (293)$$

and that this undergoes carbon monoxide insertion (eq. 249) or homolytic cleavage by hydrogen (eq. 278) or hydrocarbonyl.

The rates of reduction of benzhydrol to diphenylmethane have been measured in benzene solution (505); the rate increases with $Co_2(CO)_8$ concentration; it is first-order in substrate and probably half-order in hydrogen An activation energy of 16 kcal mole^{-1} was given for some determined pseudo-first-order rate constants at 125 to 145° and 250 atm (H_2:CO=1). In different solvents the rate decreased in the order: ethanol > benzene > cyclohexane, which is consistent with charged intermediates. Pyridine solvent completely inhibited the reaction, and this could be consistent with an acid-catalyzed process.

The production of diphenylmethane from benzopinacol, $Ph_2C(OH)C(OH)$-Ph_2, is explained by the fact that the pinacol decomposes on heating to benzhydrol and benzophenone, both of which are readily reduced (Tables 38 and 39).

The possibility that the hydroformylation products of an aromatic alcohol are formed via an intermediate aldehyde according to eq. 294 seems unlikely (418, 490)

$$\begin{array}{l} \quad\;\; R \\ \quad\;\; | \\ Ph\text{—}CHOH \longrightarrow PhCH_2CHO \begin{array}{l} \nearrow_{H_2} Ph\text{—}CH_2CH_2OH \\ \\ \searrow Ph\text{—}CH_3 + CO \end{array} \qquad (294) \\ (R = H,\, Ph) \end{array}$$

The product composition from hydroformylation of relevant aldehydes (e.g., phenylacetaldehyde and diphenylacetaldehyde) showed that little decarbonylation had occurred. Also, phenylacetaldehyde polymerized and gave no alcohol production, while diphenylacetaldehyde produced 2,2-diphenylethanol which was not formed from benzhydrol itself.

Hydrocarbon products are sometimes observed in hydroformylation of aldehydes together with the alcohol reduction product (Tables 34 and 35).

6.   Reduction of Ketones, Nitriles, Nitro and Azo Compounds, Imines, and Epoxides.   The reduction of ketones under hydroformylation conditions has been studied principally by Orchin and coworkers (418, 490) and Dawydoff (417) (Table 38). Acetone, diethylketone, pinacolone, and cyclohexanone are reduced with difficulty, but aromatic ketones with a carbonyl adjacent to an aromatic ring are readily reduced to hydrocarbons via the carbinols (Section E-5). In the reduction of fluorenone, fluorenol was in fact isolated. *p*-Methoxyacetophenone gave small amounts of the next higher alcohol 2-(*p*-methoxyphenyl)propanol, presumably produced by the homologation reaction of the intermediate 1-(*p*-methoxyphenyl)ethanol. Benzanthrone gives a high yield of 1,10-trimethylenephenanthrene, one aromatic ring having been reduced:

(295)

The reduction of nitro and azo compounds, nitriles, and imines has been studied principally by Murahashi and coworkers (481, 484–486), Takesada and Wakamatsu (1437), and Nakamura and Hagihara (483); however, Wender and coworkers (474) first reported reduction of nitro compounds (Table 38). Nitro-, azo-, and hydrazobenzene give mainly aniline along with diphenylurea which is thought to result from insertion of carbon monoxide into hydrazobenzene (506). The imine group in Schiff bases is readily reduced. Nitriles give small quantities of primary and secondary amines. Benzaldehyde phenylhydrazone is reported to give no definite products (481).

Epoxides can be hydroformylated to hydroxyaldehydes in the expected manner following eq. 275 and then eqs. 249, 250 (53, 54). Hydrogenation products of epoxides have been observed, however, under hydroformylation conditions (Table 38), and on reaction with HCo(CO)$_4$ (Table 34). These presumably result from hydrogenolysis by hydrogen or hydrocarbonyl of hydroxyalkyl cobalt carbonyls formed in reaction 275.

F. TERTIARY ORGANOPHOSPHINE-COBALT CARBONYL CATALYSTS AND THEIR ANALOGS

Cobalt carbonyl complexes containing a tertiary phosphine, arsine, or phosphite ligand have been studied as possible hydroformylation and hydrogenation catalysts.

1. Hydroformylation Catalysts. Studies using organophosphine-cobalt carbonyl complexes for hydroformylation were first reported mostly in the patent literature by Slaugh and coworkers (429, 524–528). The catalysts were prepared *in situ* by adding the required ligand to $Co_2(CO)_8$ These systems differ from those of $Co_2(CO)_8$ itself in that they are active at lower pressures (7–20 atm), and, except for the phosphite systems, generally have a greater hydrogenation activity and so produce alcohols rather then aldehydes; they also show a high tendency to react at the terminal carbon position (528). Under the hydroformylation conditions, part of the olefin substrate is hydrogenated (528, 529), the degree of hydrogenation increasing with temperature and with branching on the olefin (528), as observed for the $Co_2(CO)_8$ systems (Section E-3). Work on the organophosphine carbonyl systems by various other investigators has led to the same general conclusions (428, 465, 465a, 521, 529–536, 539, 540, 549, 1431).

Complexes such as $[Co(CO)_3PBu_3]_2$ (**73**) and $[Co(CO)_3PBu_3)_2]^+[Co(CO)_4]^-$ (**74**) have been isolated by treatment of $Co_2(CO)_8$ with the phosphine under hydroformylation conditions (e.g., 528, 529). Compound **74** readily converts to **73**; the active catalyst is thought to be the hydride $HCo(CO)_3$-PBu_3, which is formed by H_2 cleavage of **73** (528, 529, 535). The hydroformylation mechanism is thought to be the same as that for $Co_2(CO)_8$ (eqs. 245 to 250); the product isomer distribution could reflect the direction of addition of the hydride to the double bond (Section X-E-3). Replacing a CO group of $HCo(CO)_4$ by a tertiary phosphine greatly reduces the acidity of the metal hydride; for example, with Ph_3P, the acidity is reduced by 7 pK_a units (84). Thus $HCo(CO)_3PBu_3$ acts as a hydridic reagent and, unlike $HCo(CO)_4$ (Section X-E-3), probably adds, say, to pent-1-ene, to give predominantly $CH_3(CH_2)_4Co(CO)_3PBu_3$, with resulting production of mainly nonbranched *n*-hexanol and *n*-hexanal. (528, 529). The reduction in branched isomers could also result from a decrease in isomerization of a terminal olefin before hydroformylation. Tucci (529, 532, 539) has suggested that the enhanced electron density on the cobalt atom induced by PR_3 reduces double-bond migration and favors terminal olefin formation. However, Fell and coworkers (536, 540) found no inhibition of oct-1-ene isomerization when tricyclohexylphosphine was added to modify the $HCo(CO)_4$ catalyst.

Kniese and coworkers (534) and Hershman and Craddock (533) made rate and product studies on the hydroformylation of oct-1 and -2-enes and hex-1 and -2-enes respectively in the presence and absence of tri-n-butylphosphine; they concluded that the bulky PBu_3 ligand in the catalyst favors reaction with the terminal olefin due to steric interaction and that isomerization is not significantly inhibited. According to another report both change in the hydride addition mechanism and degree of isomerization may operate generally for olefins higher than propene, but a decrease in formation of branched isomers with propene must result from an alteration in the addition mechanism (529).

Triarylphosphine derivatives, for example, $HCo(CO)_3PPh_3$ whose $pK_a = 2.7$, are more acidic (i.e., less hydridic) than trialkyl derivatives, such as $HCo(CO)_3PR_3$ (R = n-butyl, cyclohexyl) whose $pK_a = 8$ to 10; such triaryl-phosphine derivatives are less effective for the formation of straight chain products, and they give lower alcohol yields (529). The increase of hydro-genating activity with increasing hydridic character is consistent with aldehyde reduction through the mechanisms suggested by Marko, and Goetz and Orchin (Section X-E-2). Tucci (529) has presented a similar scheme for the organophosphine derivatives; he preferred a final reaction of alkoxide intermediate, $RCH_2OCo(CO)_3PR_3$, with $HCo(CO)_3PR_3$ rather than hydro-genolysis to alcohol (cf. eqs. 260a, 264).

Tucci (530) has reported that the hydroformylation of propene using $HCo(CO)_3PBu_3$ gives increased formation of 1-butanol with increasing temperature, catalyst concentration and H_2 to CO ratios.

The addition of triphenylphosphite is claimed to decrease alcohol pro-duction during hydroformylation of cyclohexene (465, 465a).

Hydroformylation rates can be varied by a factor of about 15 by using different added phosphine ligands (ligand: cobalt mole ratio of 2:1). PBu_3, $PEtPh_2$, and PBu_2Ph were the most active, and Ph_3P and diphos were the least active (521, 528). Systems with tertiary arsines are more active than the corresponding tertiary phosphine ones; a system with the tertiary phosphite $(EtO)_3P$ was some ten times less active than the triethylphosphine system (528). Rate studies on pent-1-ene (528) and oct-1 and -2-enes (534) show that $Co_2(CO)_8$ itself has greater initial activity than the derivatives at the lower pressures, but, unlike the derivatives, it decomposes to metal under these conditions and gives mainly aldehyde products (528).

The hydride $HCo(CO)_3PBu_3$ adds to olefins much more slowly than $HCo(CO)_4$ (535). This could be reflected in either a slower CO dis-sociation in a reaction corresponding to eq. 246 or a slower addition reaction (eq. 247), and would explain the greater activity of the $Co_2(CO)_8$ systems (528). Imyanitov and Rudkovskii (576) consider that the catalytic activity of metal carbonyls used for hydroformylation generally increases with the acidic nature of the hydrocarbonyl due to electrophilic attack by the hydrogen on the initial addition to olefin. However this may not necessarily be the mode of addition for all olefins (see Section E-3, eqs. 285 to 287, and Section F-2).

The hydrogenation of aldehydes by using the complexes HCo(CO)$_2$-(PBu$_3$)$_2$ and HCo(CO)(PBu$_3$)$_3$ (550) under hydroformylation conditions has been announced by Andreetta and coworkers (549) but no details were given; the complexes might be expected, however, to be more active than HCo(CO)$_3$PBu$_3$ (see Section F-2).

Pruett and Smith (552) have reported that hydroformylation of olefins using ionic cobaltate complexes [Co(CO)$_3$PR$_3$]$^-$ (R = alkyl) produces mainly aldehydes, which again would be consistent with the acidic properties of any hydride intermediate.

Tucci and coworkers (531, 542) and Palm (541) have also studied the effect of adding various classes of organophosphorus ligands: tetraalkyl-(aryl)diphosphines, P$_2$R$_4$, P$_2$Ar$_4$; dialkyl(aryl)chlorophosphines, R$_2$PCl, Ar$_2$PCl; cyclic secondary organophosphines, PH[CH(R)O]$_2$CHR; bicyclic tertiary organophosphines, P(CH$_2$O)$_3$CR; and bicyclic organophosphites, P(OCH$_2$)$_3$CR. The carbonyl derivatives were generally also selective in forming linear products from propene; but complexes with P$_2$Ph$_4$, Ph$_2$PCl, R$_2$PCl, and P(OCH$_2$)$_3$CR were such poor hydrogenation catalysts that aldehydes were the dominant products (90%). Clearly, the dual character exhibited by these modifiers cannot be rationalized totally in terms of basicity.

Van Winkle (1431) has reported the conversion of dodec-1-ene into n-tridecanol and dodecane by using cobalt salts in the presence of secondary phosphines and hydroxide under hydroformylation conditions.

2. **Hydrogenation Catalysts.** Misono and Ogata (543–546, 1446) have studied the hydrogenation of polyunsaturated compounds to monoenes by using a variety of organophosphine-cobalt carbonyl catalysts (Table 40). These workers have studied in particular the selective reduction of 1,5,9-cyclododecatriene (CDT) to cyclododecene in benzene solutions at 110 to 180° and 20 to 30 atm hydrogen, using the complexes [Co(CO)$_3$(PR$_3$)$_2$][Co-(CO)$_4$] or [Co(CO)$_3$PR$_3$]$_2$. For R = n-C$_4$H$_9$ and cyclo-C$_6$H$_{11}$, complete and highly selective reduction to cis and $trans$-monoene (~1:2) occurred, and the catalyst could be recovered after reaction. For R = PPh$_3$, a weaker σ-donor phosphine, the complex decomposed to metal and hydrogenation was incomplete and nonselective; addition of about 5 atm carbon monoxide prevented catalyst decomposition, and complete selective reduction (at least for 1,4-cyclohexadiene to cyclohexene) was observed at the higher temperatures. Limited data on these three tertiary phosphine systems, and the observed inactivity of one involving the much more weakly basic triphenylphosphite ligand, suggested greater reactivity with increasing σ-donor strength. However, the complex with triphenylarsine, of similar σ-donor

Table 40 Hydrogenations Using Organophosphine-Cobalt Complexes

Catalyst[a]	Substrate	Product	Reference
$HCo(CO)_3PR_3$ (R$=CH_3$, n-C_4H_9, C_6H_{11})	1,5,9-Cyclododecatriene	Cis- and $trans$-cyclododecene	543–546
$HCo(CO)_3PPh_3$	1,5,9-Cyclododecatriene	C_{12}-monoenes and dienes	543, 545
$HCo(CO)_3PPh_3$[b]	1,4-Cyclohexadiene	Cyclohexene	545
$HCo(CO)_3PBu_3$	1,3-Cyclooctadiene	Cyclooctene, cyclooctane	543–545
$HCo(CO)_3PBu_3$	1,5-Cyclooctadiene	Cyclooctene(87%), cyclooctane	543–545
$HCo(CO)_3PBu_3$	1,5-Hexadiene	Hexenes (84%), hexane	543–545
$HCo(CO)_3PBu_3$	Cyclohexene	Cyclohexane	545
$HCo(CO)_3PBu_3$	Oct-1-yne	No reaction	551
σ-$C_3H_5Co(CO)_3PBu_3$	1,5,9-Cyclododecatriene	Cyclododecenes	546
π-$C_3H_5Co(CO)_2PBu_3$	1,5,9-Cyclododecatriene	Cyclododecane, cyclododecenes	546
π-$C_3H_5Co(CO)_2PBu_3$[c]	1,5,9-Cyclododecatriene	Cyclododecenes	546
π-$C_4H_7Co(CO)_2PBu_3$	Pentenes	Pentanes	584
$[Co(CO)_2PBu_3]_3$	Pentenes	Pentanes	584, 1350
$[Co(CO)_2PBu_3]_3$	Propene	Propane	584, 1350
$[Co(CO)_2PBu_3]_3$	Cyclohexene	Cyclohexane	584, 1350
$HCoCO(PBu_3)_3$	Butadiene	Butenes, butane	584, 1350
$HCoCO(PPh_3)_3$	Oct-1-yne	Oct-1-ene, octane	551
$H_3Co(PPh_3)_3$[d]	Cyclohexene	Cyclohexane	553
$H_3Co(PPh_3)_3$	Ethylene	Ethane	553
$H_3Co(PPh_3)_3$	n-Heptaldehyde	Heptyl alcohol, n-hexane	554
$H_3Co(PPh_3)_3$	Cyclohexene	Cyclohexane	553, 555
$H_2Co(PPh_3)_3$	1,5,9-Cyclododecatriene	Cyclododecane, dienes, monoenes	553
H_3CoL_3[e]	Hept-1-ene	n-Heptane	556
$(Bu_3P)_2CoX_2 + BH_4^-$	Cyclooctadiene	Not reported	567
$(Bu_3P)_2CoX_2 + BH_4^-$	Oct-1-ene	Octane	575
$[(OEt)_3P]_{3.4}CoX$ (X = halide)	Hex-1-ene	Hexane	575
	Acetylenes, monoolefins	—	57d, 594, 595
$HCo(diphos)_2$[f,g]	Butadiene or isoprene	Monoenes	581
$CoCl_2(diphos)_2$-$LiAlH_4$[f]	Butadiene	Butenes	581

[a] $HCo(CO)_3PR_3$ catalyst added as $[Co(CO)_3(PR_3)_2][Co(CO)_4]$ or $[Co(CO)_3PR_3]_2$.
[b] More effective in presence of added CO.
[c] With added CO.
[d] Initially formulated in Ref. 554 as $H_2Co(PPh_3)_3$.
[e] Reported to be in equilibrium with $HCoL_{-3}$ (eq. 306); L$=PPh_3$, $PMePh_2$, $PEtPh_2$, $PBuPh_2$, PEt_2Ph, PBu_2Ph, PBu_3, $PPh(OPh)_2$, $PPh_2(OPh)$, $P(OPh)_3$, $AsPh_3$.
[f] In chlorophenol.
[g] With Et_2AlCl as cocatalyst in toluene.

strength to triphenylphosphine (547), was inactive, suggesting the importance of π-acceptor properties to catalytic activity.

The hydride species $HCo(CO)_3PR_3$ were proposed as the true catalysts (Section F-1) since excess $HCo(CO)_3PBu_3$ reacted with 1,3-cyclooctadiene at 0° to give a solution containing $R'Co(CO)_3PBu_3$, which at higher temperatures yielded cyclooctene presumably via reaction 296 (545):

$$R'Co(CO)_3PBu_3 + HCo(CO)_3PBu_3 \longrightarrow R'H + [Co(CO)_3PBu_3]_2 \quad (296)$$

Here R' is the appropriate allylic grouping which could be σ- or π-bonded to the cobalt. Cyclooctene similarly reacted with the hydride solution to give the σ-alkyl complex, but heating this complex regenerated cyclooctene and not cyclooctane. These facts accounted for the reasonably selective reduction to the monoene stage. Under hydrogen atmospheres the final hydrogenolysis step could involve molecular hydrogen:

$$R'Co(CO)_3PBu_3 + H_2 \longrightarrow R'H + HCo(CO)_3PBu_3 \quad (297)$$

Ogata and Misono (545) presented the general mechanism shown in eqs. 298 to 300 where R' was considered initially to be a π-bonded allylic group:

$$-CH{=}CH{-}CH{=}CH{-} + HML_n \longrightarrow$$

$$-CH_2{-}\underset{\overset{|}{ML_n}}{CH}{-}CH{=}CH{-} \underset{+L}{\overset{-L}{\rightleftharpoons}} -CH_2{-}\underset{\overset{\downarrow}{ML_{n-1}}}{C} \quad (298)$$

$$-\underset{\overset{\downarrow}{ML_{n-1}}}{C} \xrightarrow[L]{\overset{\delta-\delta+}{H\,ML_n}} -CH_2{-}CH{=}CH{-} + [ML_n]_2 \quad (299)$$

$$-\underset{\overset{\downarrow}{ML_{n-1}}}{C} \xrightarrow{H_2}$$

$$\underset{\overset{|}{M(H_2)L_{n-1}}}{\overset{CH{=}CH{-}}{-CH}} \longrightarrow -CH_2{-}CH{=}CH{-} + HML_{n-1} \quad (300)$$

$$\mathbf{75}$$

Complexing with the olefin (eq. 298) protects the catalyst from decomposition; eq. 299 involves nucleophilic attack by the anionic metal hydride, and eq. 300 involves coordinative activation of molecular hydrogen. The effect of stronger σ-donation of a phosphine was rationalized in terms of enhanced nucleophilic attack by hydride in eq. 299, and assistance in labilizing the olefin to form the dihydride intermediate **75** in eq. 300.

Cyclooctadienes and 1,5-hexadiene were not reduced as selectively as was CDT (Table 40); it was suggested that with cyclododecene a sufficiently stable alkyl complex may not be formed for steric reasons. In a mixed solution of 1,5-hexadiene and CDT, the former was hydrogenated much faster than the latter when using the tri(n-butyl)phosphine complex; again, steric factors were thought to be involved (545). In the presence of excess phosphine, reduction of CDT was nonselective and considerable amounts of cyclododecane as well as monoenes were formed. Other complexes such as $HCo(CO)_2(PR_3)_n$ were thought to be involved (545), and later work by Ogata (546) confirmed this (see below).

The use of other solvents (hexane, ether, and THF) had little effect on the hydrogenation of CDT when using the tri(n-butyl)phosphine complex but a much slower rate was observed in ethanol, where dissociation of the hydride catalyst to a proton and cobalt-containing anion was suggested; the catalysts decomposed in carbon tetrachloride with formation of chloroform (525).

CDT appears to react with $Co_2(CO)_8$ itself under hydrogenation conditions at $\sim 140°$ to give a catalytically inactive complex, possibly **76**.

76

At temperatures above 170° this decomposes to form a nonselective hydrogenation catalyst (525). In the presence of added carbon monoxide, a selective reduction to the monoene along with Oxo and polymer products has been claimed (548, Section X-E-3, Table 37).

Ogata (546, 1446) extended these studies and used the σ- and π-allylcobalt-carbonyl-tributylphosphines, σ-$C_3H_5Co(CO)_3PBu_3$, **77**, and π-C_3H_5Co-$(CO)_2PBu_3$, **78**, as catalysts for the hydrogenation of CDT in order to establish more definitely the mechanism of selective reduction (eqs. 298 to 300). Complex **77** showed essentially the same selective activity at 130° and 30 atm H_2 as the $HCo(CO)_3PBu_3$ system; **78** at 60° gave nonselective reduction, but at about 100° in the presence of added carbon monoxide (1 atm), 98% cyclododecene production was observed. Moreover, **77** reacted with $HCo(CO)_3PBu_3$ or acetic acid under nitrogen, or with 1 atm hydrogen at 60°, to give only propylene, while **78** on reaction with hydrogen gave a 4:1

mixture of propylene and propane. Finally, the propyl complex n-C_3H_7Co-$(CO)_3PBu_3$ under hydrogen regenerated only propylene. Ogata concluded reasonably from these data that (a) selectivity results from reaction of a σ-allylic tricarbonyl intermediate (**79**) and he favored final hydrogenolysis by $HCo(CO)_3PBu_3$ (eq. 301) and (b) nonselectivity involves the coordinatively unsaturated hydride $HCo(CO)_2PBu_3$ (**80**) directing the reaction via dihydride formation (eqs. 302 and 303):

$$HCo(CO)_3L + C{=}C{-}C{=}C \longrightarrow C{=}C{-}\overset{\overset{\displaystyle CH}{|}}{C}{-}Co(CO)_3L$$

$$\textbf{79}$$

$$+ \text{HCo(CO)}_3\text{L} \qquad (301)$$

$$\Big|\; H_2$$

$$\text{---}[Co(CO)_3L]_2 + C{=}C{-}CH{-}CH \longleftarrow$$

$$\pi\text{-}C_3H_5Co(CO)_2L \xrightarrow{\;H_2\;}$$

$$[H_2C{=}CH{-}CH_2CoH_2(CO)_2L] \xrightarrow{-C_3H_6} HCo(CO)_2L \quad (302)$$

$$\textbf{80}$$

$$\textbf{80} + \overset{|}{\underset{|}{-C}}{=}\overset{|}{\underset{|}{C-}} \longrightarrow \overset{|}{\underset{|}{CH}}{-}\overset{|}{\underset{|}{C}}{-}Co(CO)_2L \xrightarrow{\;H_2\;}$$

$$\overset{|}{\underset{|}{CH}}{-}\overset{|}{\underset{|}{C}}{-}CoH_2(CO)_2L \longrightarrow \textbf{80} + \overset{|}{\underset{|}{CH}}{-}\overset{|}{\underset{|}{CH}} \quad (303)$$

$$(L = PBu_3)$$

Pregaglia and coworkers (584, 1350) have isolated the trimeric cobalt(o) carbonylphosphine cluster complexes $[Co(CO)_2PR_3]_3$ ($R_3 = Bu_3$, Ph_2Bu, Ph_3), which are active hydrogenation catalysts for olefins in n-heptane solutions at $66°$, 15 atm hydrogen (Table 40); saturated aldehydes were slowly reduced under more severe conditions. The hydrogenation is not selective between terminal and internal olefins, but butadiene is reasonably selectively reduced to butenes before further reduction to butanes. The suggested catalyst was a solvated monomer $Co(CO)_2PR_3$ formed by dissocia-tion of the cluster. These workers also reported hydrogenations catalyzed by the π-methylallyl complex, π-$C_4H_7Co(CO)_2PBu_3$ (Table 40); they suggested the same solvated monomer catalyst formed by hydrogenation of the π-allyl compound:

$$\pi\text{-}C_4H_7Co(CO)_2PBu_3 + \tfrac{1}{2}H_2 \longrightarrow C_4H_8 + Co(CO)_2PBu_3 \quad (303a)$$

$$\Big\Updownarrow$$

$$\text{Cluster}$$

This representation differs somewhat in detail from that proposed by Ogata (eq. 302), which is probably better from a mechanistic point of view.

Andreetta and coworkers (549, 551) have used $HCo(CO)_2(PBu_3)_2$ and $HCo(CO)(PR_3)_3$ (where R = Bu, Et, and hexyl) in heptane solution under 20 atm hydrogen as catalysts for the reduction of alkynes and olefins (Table 40). Using $HCo(CO)(PBu_3)_3$ at 50°, selective reduction of oct-1-yne to oct-1-ene occurred; at 100° there was complete reduction to octane (551). Conditions for the selective hydrogenation of straight chain 1-olefins in a mixture of unsaturated hydrocarbons have been identified (549). The [Co-$(CO)_3PBu_3]_2$—$HCo(CO)_3PBu_3$ system is ineffective for octyne reduction at 75° (551).

Misono and coworkers (553) report that the complex $HCo(CO)(PPh_3)_3$, unlike the rhodium analog (Section XI-D) is not a very efficient hydrogenation catalyst. At 150° and 50 atm hydrogen in benzene solution, slow hydrogenation of cyclohexene occurs; the rate is enhanced by addition of triethylaluminum, which possibly removes a phosphine to give an active, coordinatively unsaturated species.

It was noted in Section F-1 that the triphenylphosphine derivative HCo-$(CO)_3PPh_3$ is an inefficient hydrogenation catalyst (at least for aldehydes) because of the relatively acidic hydrogen atom. A wide range of tertiary organophosphine hydridocarbonyls have been made, and the order of acidity for the triphenylphosphine complexes has been established (565) as HCo-$(CO)_4 > HCo(CO)_3PPh_3 > HCo(CO)_2(PPh_3)_2 > HCo(CO)(PPh_3)_3$. Thus the hydrogenation activity for aldehyde reduction should be in the reverse order. The general order of efficiency, for olefin reduction, however, may well depend on the nature of the substrate and the mechanism of hydride addition as discussed earlier for $HCo(CO)_4$ (Section E-3, eqs. 285 to 287). Other factors will also be involved. For example, the hydrogen atom in $HCo(diphos)_2$ is extremely basic (566), but the complex is unlikely to be an effective hydrogenation catalyst for any substrate, because of the presence of the chelated phosphine and the difficulty of establishing a vacant coordination site for substrate coordination (but see Section G).

Marko and coworkers (1451) have patented the use of $Co_2(CO)_8$ with trialkylphosphines under hydrogen only (50 atm, 200°), for the reduction of propionaldehyde to butanol in toluene solution.

G.   TERTIARY ORGANOPHOSPHINE-COBALT HYDRIDE CATALYSTS WITH NO COORDINATED CARBONYLS

A number of cobalt hydride complexes containing tertiary phosphines have recently been prepared and some studies on their hydrogenating activity have been reported (553–556, 567, Table 40). $H_3Co(PPh_3)_3$ is reported by Misono and coworkers (553, 554) to catalyze the reduction of ethylene and cyclohexene in benzene at room temperature and 1 atm hydrogen, the rate increasing with hydrogen pressure and temperature. A nonselective reduction of 1,5,9-cyclododecatriene is reported at 80° and 50 atm hydrogen (553). Aldehydes (*n*-heptaldehyde) are reduced noncatalytically at room temperature under argon, with some hydrogen to prevent decompositon of the catalyst, to give alcohol (heptyl alcohol), a cobalt carbonyl product and a decarbonylation product (*n*-hexane) (553, 555); the irreversible decarbonylation reaction prevents a catalytic reduction of aldehydes.

$H_3Co(PPh_3)_3$ is of particular interest in view of its conversion under nitrogen to give reversibly $HCo(N_2)(PPh_3)_3$, a molecular nitrogen complex (554–559, 573):

$$H_3Co(PPh_3)_3 + N_2 \rightleftharpoons HCo(N_2)(PPh_3)_3 + H_2 \qquad (304)$$

There has been considerable discussion and confusion over the formulation of the nitrogen complex, first reported by Yamamoto and coworkers (557, 560) and Misono and coworkers (554, 563), regarding whether or not hydrogen was present in the molecule and whether or not this complex under hydrogen gave the trihydride complex $H_3Co(PPh_3)_3$ or the dihydride complex $H_2Co-(PPh_3)_3$ (555, 556, 561, 562). The relevant reports are summarized and discussed in a paper by Speier and Marko (556), and it appears that all the reported work had involved the $H_3Co(PPh_3)_3$ and $HCo(N_2)(PPh_3)_3$ complexes and that the formulations $H_2Co(PPh_3)_3$ and $N_2Co(PPh_3)_3$ reported by Yamamoto's group and Misono's group were incorrect. Nevertheless, Speier and Marko have shown that the complexes $H_2Co(PPh_3)_3$ and $N_2Co-(PPh_3)_3$, and the equilibrium corresponding to eq. 304, do exist:

$$H_2Co(PPh_3)_3 + N_2 \rightleftharpoons N_2Co(PPh_3)_3 + H_2 \qquad (305)$$

The dihydride rapidly reduces hept-1-ene in a 1:1 stoichiometry at room temperature under argon in benzene (556); one expects that the system would become catalytic under hydrogen.

Both the di- and trihydride complexes have a strong irreversible affinity for carbon monoxide (553–556, 564), which could inhibit the efficiency of any catalytic hydrogenation of substrates containing carbonyl groups and even possibly oxygen atoms. The reaction of $H_2Co(PPh_3)_3$ with carbon monoxide gives the complex $Co_2(CO)_2(PPh_3)_6$ (556), which also can be prepared by the reaction of carbon monoxide with a solution obtained from cobalt stearate, alkyl Grignard reagents, and triphenylphosphine (569). Speier and Marko (556) suggested that the dihydride is present in the Grignard solutions, which is consistent with the suggestion by Marko and coworkers (570) that an intermediate L_xCoH_2 is formed in solutions of stearate and alkyl Grignard reagent alone. Moreover, the intermediate also catalyzes olefin hydrogenation as a typical Ziegler system (570–572, Section XV-A).

Rossi and Sacco (567) have described other H_3CoL_3 complexes, where L comprises a range of tertiary phosphines, triphenylarsine, and triphenylphosphite (Table 40). They also report the existence of equilibria 306 and 307, which are strongly dependent on the π-acceptor and σ-donor properties of the ligand. H_3CoL_3 and $HCo(N_2)L_3$ are said to be less stable with the more π-acid ligands.

$$H_3CoL_3 \rightleftharpoons HCoL_3 + H_2 \qquad (306)$$

$$HCoL_3 + N_2 \rightleftharpoons HCo(N_2)L_3 \qquad (307)$$

These workers have also prepared the 4- and 5-coordinate complexes $HCoL_3$ and $HCoL_4$ (567, 578). The complex $HCo(diphos)_2$ has been prepared by the action of hydrogen on the cobalt(o) complex, $Co(diphos)_2$, at room temperature in benzene (578). No details were given for this reaction; presumably it involves 1 mole of H_2 per 2 moles of complex, and it seems worthy of a kinetic study for comparison with that of the corresponding pentacyanocobaltate(II) reaction. Sacco and coworkers (579) have also reported that the 4-coordinate complexes $[Co(diphos)_2]X$ ($X = ClO_4^-$, BPh_4^-) oxidatively add hydrogen to give $[H_2Co(diphos)_2]X$. These hydrides with chelated phosphines as such are unlikely to be active hydrogenation catalysts, although addition of Lewis acids (e.g., aluminum alkyls and phenols) can create activity by attacking a basic coordinated phosphorus and produce coordinative unsaturation for olefin attachment. $HCo(diphos)_2$ has been used in this manner by Iwamoto (581) to reduce dienes to monoenes (see Table 40) at 80 to 120° and 30 to 50 atm hydrogen; the $CoCl_2(diphos)_2$-$LiAlH_4$ system was used similarly in chlorophenol solution.

The combination of eqs. 306 and 307 represents the equilibrium shown in eq. 304. The complex $HCo(N_2)(PPh_3)_3$ also undergoes a reversible reaction with other covalent gas molecules including ethylene, presumably (557, 564) as follows:

$$HCo(N_2)(PPh_3)_3 + C_2H_4 \longrightarrow HCo(C_2H_4)(PPh_3)_3 + N_2 \qquad (308)$$

Such hydridoolefin or alkyl intermediates must be involved in the hydrogenation processes. A similar equilibrium has been given for the 4-coordinate hydrides with cyclooctadiene (567):

$$HCoL_3 + C_8H_{12} \rightleftharpoons Co(C_8H_{13})L_3 \qquad (308a)$$

Equilibria 306 and 308a must be key steps in hydrogenations catalyzed by H_3CoL_3, although no mechanistic studies have so far been reported. Rossi and Sacco (567) did not report specifically on the hydrogenating activity of the various complexes, but did mention that under hydrogen atmospheres cyclooctadiene was hydrogenated. Similarly the ethylene in $[(C_2H_4)Co-(PPh_3)_3]_2$ has been reduced stoichiometrically with hydrogen (567a). Tyrlik and Stepowska (1346) have studied stoichiometric hydrogenation of conjugated dienes and α,β-unsaturated aldehydes and ketones, using the $HCo-(N_2)(PPh_3)_3$ complex.

Clearly the molecular nitrogen complexes $HCo(N_2)L_3$ and N_2CoL_3 will act as hydrogenation catalysts under hydrogen atmospheres because of their ready conversion to H_3CoL_3 and H_2CoL_3 (eqs. 304 to 306). The ready reaction of the $HCoL_3$ system with hydrogen, nitrogen, carbon monoxide, and ethylene has been mentioned above; reactions with ammonia (557), carbon dioxide (555, 564), and organonitriles (585) have been reported. Reactions of "CoL_3" (stabilized as N_2CoL_3) with hydrogen, carbon monoxide, oxygen, carbon dioxide, and carbon disulfide (556, 568) have also been reported. An interesting catalytic reduction of nitrous oxide to nitrogen using the $HCo(N_2)(PPh_3)_3$ or $H_3Co(PPh_3)_3$ complexes has been reported by Ikeda and coworkers (583). A suggested mechanism is shown below:

$$\left.\begin{array}{l} HCo(N_2)(PPh_3)_3 + N_2O \\ H_3Co(PPh_3)_3 \quad + N_2O \end{array}\right\} \longrightarrow$$

$$[HCo(N_2O)(PPh_3)_3] \longrightarrow HCo(PPh_3)_2 + N_2 + OPPh_3 \quad (309)$$
$$\underset{N_2O,\ PPh_3}{\underline{\qquad\qquad\qquad\qquad\qquad\qquad\qquad}}$$

Excess phosphine is necessary, since it is simultaneously catalytically oxidized to triphenylphosphine oxide. Continued activity in this general area should benefit the field of homogeneous catalysis.

Heck (Table 40; 575) has disclosed that the homogeneous solutions resulting from borohydride reduction of $(Bu_3P)_2CoX_2$ (X = halide) in heptane are effective for hydrogenation of oct-1-ene and hex-1-ene at 50° and 2 atm hydrogen. A patent (582) claims that tetrakis(trialkylphosphite) complexes Co-$[P(OR)_3]_4$ (R = Me, Et, β-chloroethyl, 2-ethylhexyl, p-tolyl) are useful hydrogenation catalysts, and that under hydrogen pressure they give rise to hydride derivatives $HCo[P(OR)_3]_4$ (582), which have been prepared by other methods (784, 785, 991, 1345). On the other hand, Volpin and Kolomnikov (57d, 595) have reported that $Co[P(OEt)_3]_4$, which is probably dimeric (592, 593, 595), does not catalyze hydrogenation even at 150° and 120 atm hydrogen. Few details on the nature of the solvent, substrate, or concentrations are available for these phosphite systems. The Russian workers, however, report that the univalent cobalt complexes $CoXL_3$ and $CoXL_4$ (L = $P(OEt)_3$ and X = halide) absorb about 1 mole hydrogen under mild conditions (592, 593), and that they act as catalysts in alcohol under more severe conditions

($>75°$, high pressure) for hydrogenation of alkynes and alkenes with activated double bonds, such as vinyl ethyl ether and vinyl acetate (Table 40; 57d, 594, 595). The following reactions summarize their proposed reaction scheme for an acetylene (57d, 595):

$$L_4CoCl + H_2 \longrightarrow L_4CoH + HCl \tag{309a}$$

$$L_4CoH + RC{\equiv}CH \longrightarrow L_4Co{-}CH{=}CHR \tag{309b}$$

$$L_4Co{-}CH{=}CHR + H_2 \longrightarrow L_4CoH + H_2C{=}CHR \tag{309c}$$

Only the inactive $Co_2[P(OEt)_3]_8$ complex, formed by decomposition of the hydride, could be isolated from reaction 309a (no substrate). The possibility of product formation by alcoholysis

$$L_4Co{-}CH{=}CHR + R'OH \longrightarrow L_4CoOR' + H_2C{=}CHR \tag{309d}$$

$$L_4CoOR' + H_2 \longrightarrow L_4CoH + R'OH \tag{309e}$$

was ruled out, because using C_2H_5OD as solvent gave a product containing no deuterium. It also seems unlikely that a further mole of hydride would be used for the final hydrogenation step, since this would generate the inactive dimer. L_3CoCl was said to be less effective than L_4CoCl. Substrate hydrogenation rates decreased thus: terminal acetylenes $>$ disubstituted acetylenes $>$ monoolefins; however, addition of amines accelerated the reaction (595). Tolman (1370) has noted that $HCo[P(OEt)_3]_4$, unlike the isoelectronic $HNi[P(OEt)_3]_4^+$ complex (Section XIII-A), does not readily react with butadiene; this was attributed to the lower lability of the phosphite ligand.

Martino (1349) has reported that a cobalt(II) hydride, $HCo(NCS)(PPh_3)_2$, does not react with hydrogen; on the other hand, reaction with ethylene gives ethane production, the complex disproportionating to metal and $Co(NCS)_2$-$(PPh_3)_2$.

Tyrlik and Stepowska (574) have reported that $HCo(N_2)(PPh_3)_3$ reacts with acid chlorides, RCOCl, in solution to produce RH products, (R = CH_3, C_6H_5, $C_6H_5CH_2$, $C_6H_5OCH_2$, $(CH_3)_2CH$):

$$2HCo(N_2)(PPh_3)_3 + 2RCOCl \longrightarrow$$
$$2RH + 2N_2 + (Ph_3P)_2CoCl_2 + (Ph_3P)_2Co(CO)_2 + 2PPh_3 \tag{310}$$

For a 1:1 molar reaction, the RH yield was 80% (based on RCOCl); small amounts of coupling products R—R also formed.

Sacco and Rossi (559) have reported the phosphine ligand-metal hydrogen-transfer in the $HCo(N_2)(PPh_3)_3$ complex (and therefore under hydrogen, the $H_3Co(PPh_3)_3$ complex). The mechanism, corresponding to that considered for the $HRuCl(PPh_3)_3$ system (Section IX-B-2b, Scheme 3), involves reversible oxidative addition ($Co^I \rightleftharpoons Co^{III}$) of the *ortho*-phenyl C—H bond to the Co atom within the $HCo(PPh_3)_3$ complex, the labile N_2 ligand being likely to dissociate (242). An analogous hydrogen exchange has been reported by Parshall and coworkers (242) for the hydridotetrakis(triphenylphosphite) complex $HCo[P(OC_6H_5)_3]_4$ (see eq. 140, Section IX-B-2b). Added triphenylphosphite suppresses the exchange by suppressing the dissociation into the 4-coordinate species necessary for the subsequent oxidative addition reaction involving the C—H bond:

$$HCoL_4 \rightleftharpoons HCoL_3 + L \quad (L = (C_6H_5O)_3P) \qquad (310a)$$

The $HCo(diphos)_2$ chelate complex undergoes no such exchange reaction (559), presumably because of the difficulty in forming the required 4-coordinate intermediate.

In the thermal decomposition of $HCo(N_2)(PPh_3)_3$, benzene is produced together with an intermediate which on reaction with hydrochloric acid yields some diphenylphosphine (559):

$$HCo(N_2)(PPh_3)_3 \xrightarrow{\Delta} N_2 + PhH + [CoPPh_2(PPh_3)_2] \xrightarrow{HCl}$$
$$HPPh_2 + CoCl_2 + PPh_3 \quad (310b)$$

It seems likely that the hydrogen transfer in the thermal reaction also involves metal-assisted interaction at the *ortho* position of a phenyl group.

An outstanding problem in homogeneous catalysis is the activation of saturated hydrocarbons such as methane. Gol'dshleger and coworkers (577) have attempted to demonstrate exchange between CH_4 and D_2 in the presence of benzene solutions of $H_3Co(PPh_3)_3$ at room temperature, since a plausible mechanism would exist if CH_4 can form a metal alkyl hydride species with the $DCo(PPh_3)_3$ species:

$$H_3CoL_3 \xrightarrow{D_2} D_3CoL_3 \rightleftharpoons DCoL_3 + D_2 \qquad (311)$$

$$DCoL_3 + CH_4 \rightleftharpoons H_3C-\overset{\overset{\displaystyle H}{|}}{\underset{\underset{\displaystyle D}{|}}{Co}}L_3 \rightleftharpoons HCoL_3 + CH_3D \qquad (312)$$

Small amounts of CH_3D (2% after 6 days) were observed, but definite conclusions could not be drawn because of the gradual decomposition of the initial complex over this period. More definite results indicating exchange of CH_4 with deuterons were observed with a chloroplatinate system (Section XIII-C-1).

H. COBALOXIMES

The reactions of vitamin B_{12} are closely paralleled by the reactions of bis-dimethylglyoximatocobalt compounds (cobaloximes); particularly important is the reduced state, vitamin B_{12_s}, which contains a highly reactive cobalt(I) center (eq. 312a). Such studies have been reviewed recently by Schrauzer (830, 831, 1068) and Hill and coworkers (834).

$$\overset{\displaystyle OH}{\underset{\displaystyle \text{Vitamin } B_{12_a}}{\overset{\displaystyle |}{Co^{III}}}} \overset{e}{\longrightarrow} \underset{\displaystyle \text{vitamin } B_{12_r}}{Co^{II}} \overset{e}{\longrightarrow} \underset{\displaystyle \text{vitamin } B_{12_s}}{Co^{I}} \qquad (312a)$$

Schrauzer and Windgassen (832) reported that although vitamin B_{12_r} is not reduced by molecular hydrogen, cobaloximes(II), of formulation $Co(DMGH)_2 \, 2B$ or the dimeric form $B(DMGH)_2—Co—Co(DMGH)_2B$ (where B is a Lewis base, H_2O, pyridine, CN, tertiary phosphines and arsines) can activate molecular hydrogen. The dimers are particularly susceptible to reaction with hydrogen in mildly alkaline solution when the cobaloxime(I) products $Co(DMGH)_2B^-$ are formed, presumably via a cobaloxime(III) hydride:

$$[Co(DMGH)_2B]_2 + H_2 \rightleftharpoons 2HCo(DMGH)_2B \xrightarrow[]{2OH^-}$$

$$2Co(DMGH)_2B^- + 2H_2O \quad (312b)$$

Schrauzer (831) suggests a scheme in which the initial products are cobalt(I) and a cobalt(III) hydride:

$$HO^{-}\cdots H\!\!\frown\!\!H \cdot\overset{\frown}{\cdots} Co^{II}\overset{\frown}{\frown}Co^{II} \longrightarrow H_2O + H—Co^{III} + Co^I \qquad (312c)$$

$$\underset{\displaystyle \xrightarrow{OH^-} Co^I + H_2O}{\bigg|}$$

Cobaloximes(III) do not react with hydrogen in alkaline solution, but addition of a trace of cobaloxime(II) gives an autocatalytic conversion to the univalent state, due to reaction 312b and the existing equilibrium shown in 312d.

$$Co^I + Co^{III} \rightleftharpoons 2Co^{II} \qquad (312d)$$

The reverse of this equilibrium is evident in some disproportionation of the cobaloxime(II) species (832):

$$2Co(DMGH)_2B \xrightarrow[-2B]{OH^-} Co(DMGH)_2B^- + HOCo(DMGH)_2B \quad (312e)$$

Schrauzer and coworkers (833) have demonstrated the reversibility of reaction 312b by decomposing cobaloximes(I) in water to the hydride which can further decompose to the cobaloxime(II) with liberation of hydrogen (834). Hydrides with tertiary phosphines, such as $HCo(DMGH)_2PPh_3$, were isolated, however.

The equilibria outlined in eq. 312b encompass the corresponding equilibria reported for the pentacyanocobaltate(II) system (Section X-A-2, eqs. 151 to

153, and Section X-B-2e, eq. 213). The pK for $HCo(CN)_5^{3-}$ acting as an acid (eq. 213) was $\sim$17 to 19; hence the predominant species even in alkaline solution is the cobalt(III) hydride. The pK of $HCo(DMGH)_2\ P(n\text{-}C_4H_9)_3$ is 10.5; hence alkaline solutions contain cobaloxime(I) (833).

Studies by Schrauzer and coworkers (831, 833, 835–838) suggest that the use of cobaloximes as hydrogenation catalysts is limited. Cobaloximes(I), for example, via reaction 312b, do react with acetylenes and activated olefins to give cobalt(III)-vinyl or cobalt(III)-alkyl species, respectively, via a carbanion intermediate; for example:

$$Co(DMGH)_2B^- + CH_2{=}CHX \longrightarrow \left[B{-}Co{-}CH_2{-}\overset{-}{C}HX \right] \xrightarrow[-OH^-]{+H_2O}$$

$$B{-}Co{-}CH_2{-}CH_2X \quad (312f)$$

Alkyl halides react similarly with liberation of halide. The cobalt-carbon bond in these complexes is remarkably stable, and does not readily undergo hydrogenolysis with hydrogen. However, this hydrogenolysis may be accomplished (836) by using borohydride (Section X-B-5), a heterogeneous hydrogenation catalyst, or a cobaloxime(I) (eq. 312g). This suggests that

$$R{-}Co(DMGH)_2B + Co(DMGH)_2B^- \xrightarrow{-H^+}$$

$$RH + B(DMGH)_2Co{-}Co(DMGH)_2B \quad (312g)$$

reduction of acetylenes, certain olefins and alkyl halides may occur, if excess cobaloxime(II) is present on adding the substrate under hydrogen. Reductions of monoenes, dienes, unsaturated esters and nitriles, alkyl halides and 1,2-diketones by using cobaloximes (831) have been reported. Little detail is given, but the reductions probably involve final hydrogenolysis with borohydride or heterogeneous catalysts. One reduction that is catalytic when using cobaloxime(II) and molecular hydrogen is the conversion of disulfides to thiols (837). This results because the cobalt-sulfur bond is readily cleaved by hydrogen:

$$RS{-}Co(DMGH)_2py + H_2 \longrightarrow RSH + \text{cobaloxime(I or II)} \quad (312h)$$

The nature of cobaloxime product will depend generally on the pH, and is governed by the equilibria shown in 312b via the hydride. In neutral or alkaline solution, the mercaptocobaloxime is presumably formed according to a reaction such as 312i (see eq. 312f), then a catalytic reduction of the disulfide results.

$$Co(DMGH)_2py^- + RSSR \xrightarrow[-OH^-]{+H_2O} RS{-}Co(DMGH)_2py + RSH \quad (312i)$$

Aoki and Mayake (1445) have also used the dimeric cobaloxime(II) complex $[Co(DMGH)_2py]_2$ in benzene at 160° under 100 atm hydrogen for the catalytic reduction of isoprene. In this medium, formation of the cobaloxime(III) hydride (eq. 312b) and then hydrogenation via a mechanism exemplified by eqs. 180 and 181, for $HCo(CN)_5^{3-}$ (Section X-B-2a) seems likely. The major product is 2-methylbut-1-ene. This contrasts with the borohydride reduction using $CoCl(DMGH)_2py$, which can form $[Co-(DMGH)_2py]_2$ (1445), where the major product is 3-methylbut-1-ene (Section X-B-5); cobalt(I) intermediates may be involved here.

Khidekel and coworkers (839, 840, 1472) have reported that addition of cobaloximes, such as $HCo(DMGH)_2PPh_3$, promotes hydrogen reduction of nitrobenzene to azobenzene, azoxybenzene, and aniline in alkaline alcoholic solution at 1 atm hydrogen, and in the presence of a heterogeneous platinum catalyst or an unspecified homogeneous rhodium catalyst. The cobaloxime is referred to as a hydrogen carrier in a catalytic chain, but from the limited information available, the extent of the role of the other catalyst, which must activate the hydrogen, is not clear (Section XVI-I). One would expect that the cobaloxime system itself will catalyze the reduction under these conditions (see above).

Simandi and Budo-Zahonyi (1327) have reported on the kinetics of the reaction between hydrogen and $Co(DMGH)_2$ in aqueous methanol at 20° and 1 atm; the initial rates are second-order in cobalt and first-order in hydrogen. The same kinetics occur in the presence of pyridine, but the rate increases to a limiting value with increasing pyridine concentration due to production of a more active $Co(DMGH)_2py$ complex. The total hydrogen uptake in both cases indicates possible hydrogenation of the dimethylglyoxime ligand. Furthermore, the variation of rate with hydroxide shows a maximum at about 1 mole excess per cobalt. This was attributed to ionization of a ligand proton to give another active species $Co(DMGH)(DMG)^-$ with or without pyridine; such ionization equilibria have been observed with cobalt(III)-dimethylglyoxime complexes (1328). Hydrogen activation was written in terms of both homolytic and heterolytic reactions:

$$Co(DMGH)_2 + H_2 \rightleftharpoons$$

$$H_2Co(DMGH)_2 \xrightarrow[\text{slow}]{Co(DMGH)_2} 2HCo(DMGH)_2 \quad (312j)$$

$$Co(DMGH)(DMG)^- + H_2 \rightleftharpoons$$

$$HCo(DMGH)_2^- \xrightarrow[\text{slow}]{Co(DMGH)_2} HCo(DMGH)_2 + Co^I(DMGH)_2^- \quad (312k)$$

$$Co^I(DMGH)_2^- + H^+ \rightleftharpoons HCo(DMGH)_2 \quad (312l)$$

Reaction 312j corresponds to that proposed by the same group for hydrogen activation by $Co(CN)_5^{3-}$ (Section X-A-2). In more basic solution, disproportionation to cobalt(I) occurs (see eq. 312b).

I. OTHER COBALT-CATALYZED HYDROGENATION SYSTEMS

Marko (589) had studied the effect of adding the thioethers Me_2S and Bu_2S to $Co_2(CO)_8$ on their activity for hydroformylation of olefins and subsequent aldehyde hydrogenation. The observed retarding effects, which can be compensated for by raising the carbon monoxide pressure, are thought to be due to the production of less active catalysts, $Co_2(CO)_7SR_2$.

A wide variety of carboxylation and carbonylation reactions catalyzed by metal carbonyls, particularly $Co_2(CO)_8$, under carbon monoxide or carbon monoxide/hydrogen atmospheres, have been reported (51, 52, 590, 591). Hydrogenation as a parallel reaction is not uncommon; this has been discussed in detail under the subject of hydroformylation reactions (Section X-E). Such hydrogenations, involving hydridocarbonyls as catalysts, will not be presented in further detail here.

Soluble Ziegler catalysts containing cobalt(Section XV) have been used extensively for the reduction of unsaturated carbon-carbon bonds (e.g., 56, 147, 163, 167, 168, 175, 177, 195, 196, 570–572), including the hydrogenation of fats (206, 207). Cobaltic acetylacetonate (144, 145) has also been used to hydrogenate fats (Section XIV-E).

The cobaltous ion in perchloric acid, and cobaltous heptanoate in heptanoic acid are reported unreactive toward hydrogen at up to $150°$ (25, 92).

XI

Group VIII Metal Ions and Complexes

RHODIUM

A. ACTIVATION OF MOLECULAR HYDROGEN IN AQUEOUS SOLUTION

1. Rhodium Complexes with Halides, Cyanide, and Nitrogen Donor Ligands.
The first report on hydrogenations catalyzed by rhodium(III) complexes
was that of Iguchi (19) in 1939. It is of interest that Ipatieff and Tronev (993)
had earlier reported the hydrogen reduction of rhodium chloride solutions to
the metal. In aqueous acetate solutions at 25° and 1 atm hydrogen, $RhCl_3$,
$[Rh(NH_3)_4Cl_2]Cl$, and $[Rh(NH_3)_5H_2O]Cl_3$ were found to be active catalysts
for the reduction of fumaric acid, quinone, methylene blue, sodium nitrite,
and hydroxylamine; ammonia was the reduction product from nitrite and
hydroxylamine. The more substitution-inert complexes $[Rh(NH_3)_6]Cl_3$ and
$[Rh(en)_3]Cl_3$ were inactive. The kinetics were not extensively studied, nor
were possible mechanisms presented. Hydrogen-catalyzed substitution
reactions (Section A-2) may be involved under these experimental conditions
which quite readily lead to traces of metallic rhodium (596), a powerful
heterogeneous catalyst for olefin reduction (603).

Wilkinson and coworkers (597–599) had detected a rhodium(III) ammino-
hydride species in solution by reacting $[Rh(NH_3)_5Cl]^{2+}$ with borohydride,
and later Powell, Wilkinson, and coworkers (600–602) isolated a variety of
salts of the cationic hydrides $HRh(NH_3)_5^{2+}$ and $HRh(H_2O)(NH_3)_4^{2+}$ by

reaction of the chloropentammine with metallic zinc in ammoniacal solution. The hydridopentammine in aqueous solution readily gives the hydrido-aquotetrammine species, and it effectively catalyzes the hydrogenation of water-soluble unsaturated carboxylic acids, such as maleic, crotonic, and acrylic, at 60° and 1 atm (601, 1433). Little detail was reported except that linear rates were observed, presumably indicating zero-order in olefin; metallic rhodium precipitated only in the absence of substrate. Solutions of the hydrides absorbed unsaturated hydrocarbons (ethylene, propene, but-1-ene, tetrafluoroethylene, hexafluoropropene, octafluorobut-2-ene, acetylene, and hexafluorobut-2-yne) at 1 atm and 25° fairly slowly to give 1:1 complex cations containing σ-bonded alkyl or alkenyl groups; for example, with ethylene, $Rh(C_2H_5)(NH_3)_5^{2+}$ (602, 842, 1434). The reaction rates decreased in this order: $C_2H_4 > C_3H_6 \sim C_4H_8$, and $C_4F_6 > C_2F_4 > C_3F_6 \sim C_4F_8$; the rates were inhibited by ammonia. The 5-coordinated $HRh(NH_3)_4^+$ was thought to be the most likely intermediate for these insertion reactions. Since metal alkyls must be involved in these catalytic hydrogenations, it would be of interest to discover the manner of their conversion to the saturated organic product and regenerated catalyst in the aqueous solutions; for example, attack by hydrogen, proton, or further metal hydride (Section XI-D-I). No reduction of dec-1-ene was detected when using an aqueous alcohol solution of borohydride and chloropentammine (598). If the alkyl were formed, this would be consistent with molecular hydrogen being involved:

$$Rh(NH_3)_5R^{2+} + H_2 \longrightarrow HRh(NH_3)_5^{2+} + RH \qquad (313)$$

Aqueous solutions of $HRh(NH_3)_5^{2+}$ reduce permanganate, cupric to copper, silver(I) to silver(o)—one atom of copper or two atoms of silver per rhodium—and form a hydroperoxo complex, $Rh(NH_3)_4(OH)(O_2H)^+$, with molecular oxygen (600–602, 633). The following dissociation reaction shows the "equivalence" of a rhodium(III) hydride and a rhodium(I) species:

$$Rh^{III}H \rightleftharpoons Rh^I + H^+ \qquad (314)$$

Both species can result from a two-electron reduction of a rhodium(III) species (604, 617, 632, 633, 1413). Johnson and Page (633) have prepared the hydridopentammine complex by electrolytic reduction of $Rh(NH_3)_5OH^{2+}$; this reaction proceeds via a rhodium(I) intermediate, that is, by the reverse of reaction 314. The position of this equilibrium will depend partly on the nature of other coordinated ligands, in particular their ability to stabilize the lower oxidation state, for example, via π-acceptor properties (Section XI-B-9). Gillard and coworkers (1413) similarly isolated the trans-$HRh(en)_2OH^+$ complex by electrochemical reduction of trans-$Rh(en)_2Cl_2^+$ at pH 0 to 12. At pH $\geqslant 13$, equilibrium 314 lies well to the right, and the hydride is not formed. This rhodium(III) hydride can also form a hydro-peroxide complex with molecular oxygen.

Wilkinson and coworkers (598, 599, 609, 610, 1387) have also reported on other rhodium(III) monohydrides, such as cis- and trans-HRh(NH$_3$)(en)$_2{}^{2+}$ cis-HRh(trien)Cl$^+$, cis- and trans-HRh(en)$_2$Cl$^+$, trans-HRhpy$_4$Cl$^+$, trans-HRh(dimethylglyoximate)$_2$Cl$^-$, and HRh(CN)$_4$H$_2$O^{2-}, as well as some dihydrides, such as cis- and trans-H$_2$Rh(en)$_2{}^+$ and cis-H$_2$Rh(trien)$^+$. Krogmann and Binder (1239) have isolated HRh(CN)$_5{}^{3-}$. Borohydride reduction of Rh(bipy)$_2$Cl$_2{}^+$ leads to production of the rhodium(I) species Rh(bipy)$_2{}^+$, the bipyridyl stabilizing the lower valency state in reaction 314 (598, 599, 611). The cyanohydride complexes seem remarkably stable considering the strong π-capacity of the cyanide ligand; aquotetracyanoalkylrhodate(III) complexes have been made (609, 631).

None of the isolated rhodium(III) hydrides have been obtained by using molecular hydrogen, and except for the hydridoammine species, no reports have appeared on their potential as hydrogenation catalysts. The complexes that form the hydrides are efficient catalysts for the reduction of nitrobenzene and quinone by borohydride (598). These reductions could involve hydride transfer or the electron transfer-protonation mechanism discussed for the cobalt(III)–borohydride systems in Section X-B-5. HRh(CN)$_4$H$_2$O^{2-} reacts with oxygen to give a hydroperoxide complex, HO$_2$Rh(CN)$_4$H$_2$O^{2-}, which is decomposed by acid to hydrogen peroxide and Rh(CN)$_4$(H$_2$O)$_2{}^-$ (300a). Unfortunately this system is not catalytic, since the bis-aquo complex is not reconverted to hydride (e.g., by borohydride). Hydrogen can react with rhodium(III) complexes in ethanol and aqueous ethanol media to give a hydride with elimination of hydrogen chloride (Sections XI-G and XI-H).

Harrod and Halpern (605) studied the hydrogen reduction of ferric ions catalyzed by chlororhodate(III) species in aqueous acid solutions at around 80° and 1 atm hydrogen, and this system was further investigated by James and Rempel (606) after data on characterization of the rhodium complexes had been published (607). At constant hydrogen pressure, linear uptake plots were observed to the point of complete reduction to iron(II); at this stage rhodium metal precipitated, and a sharp increase in uptake rate was observed. Spectrophotometric data showed that the rhodium(III) catalyst remained unchanged throughout the linear uptake region, which gave rise to the rate law $-d[\text{H}_2]/dt = k_1[\text{H}_2][\text{Rh}^{\text{III}}]$. The mechanism was written as involving heterolytic splitting of hydrogen followed by a rapid reaction of the intermediate hydride with ferric:

$$\text{RhCl}_6{}^{3-} + \text{H}_2 \xrightarrow{k_1} \text{HRhCl}_5{}^{3-} + \text{H}^+ + \text{Cl}^- \qquad (315)$$

$$\text{HRhCl}_5{}^{3-} + 2\text{Fe}^{3+} \xrightarrow[\text{Cl}^-]{\text{fast}} \text{RhCl}_6{}^{3-} + 2\text{Fe}^{2+} + \text{H}^+ \qquad (316)$$

Only the anionic, labile chlororhodate(III) species were effective catalysts, the activity increasing with increasing number of coordinated chlorides. Presumably increasing strength of the metal-chloride bonding in the lower chloro species makes its replacement by hydride more difficult. $\Delta H^{\ddagger}$ and $\Delta S^{\ddagger}$ were estimated to be 24.5 kcal mole^{-1} and 9 eu, respectively, for the k_1 step in 3 M HCl, when the major species present is $Rh(H_2O)Cl_5^{2-}$. Bromorhodate(III) species in hydrobromic acid are more effective catalysts than the corresponding chloro complexes (613).

Water-soluble unsaturated carboxylic acids were not hydrogenated when using the chlororhodate(III) system(606). Hydrogen reduced the rhodium(III) to the univalent state, this being stabilized by complexing with the olefin:

$$Rh^{III} + H_2 \xrightarrow{k_1} Rh^{I} + 2H^+ \tag{317}$$

$$Rh^{I} + \text{maleic acid} \xrightarrow{\text{fast}} \text{complex} \tag{318}$$

Here k_1 corresponds to that for reduction of the ferric substrate, indicating a common rate-determining step as in reaction 315. The ferric reduction then, and those noted above with the hydridoammine systems, could equally well involve a rhodium(I) intermediate rather than a rhodium(III) hydride species. This will depend on the relative rates of the two fast steps exemplified by reaction 316 and the forward reaction of 314.

Aqueous acid solutions of $Rh(CO)_2Cl_2^-$ are ineffective as hydrogenation catalysts for olefins. As with the chloro complexes (eq. 318), rhodium(I) olefin complexes are formed, but these do not activate hydrogen. Rhodium(I) chlorides are, however, effective in some nonaqueous solvents (Section XI-G).

The hydrogen reduction of rhodium(III) chlorides to metal in the absence, or after consumption, of substrate (605, 606) could involve hydride production (eq. 315), and dissociation to rhodium(I) (eq. 314), followed by further hydrogen reduction or disproportionation:

$$2Rh^{I} \rightleftharpoons Rh^{II} + Rh^{0} \tag{319}$$

The filtrate from such reductions gives an e.s.r. signal which could arise from a paramagnetic, monomeric rhodium(II) species (608). Further evidence for eq. 319 is the observation that the hydride ammines are decomposed in acetic acid to yield metal and rhodium(II) acetate in a 1:1 ratio (602).

Closely related to the catalytic hydrogenation by chlororhodate(III) complexes is the work of Charman (614, 615) on a homogeneous dehydrogenation of isopropanol catalyzed by the same species in the presence of hydrochloric acid. The kinetic data, measured by hydrogen evolution, were analyzed in terms of eqs. 320 to 322.

$$Me_2CHOH + RhCl_6^{3-} \underset{k_{-2}}{\overset{k_2}{\rightleftharpoons}} Me_2CO + HRhCl_5^{3-} + HCl \tag{320}$$

$$\text{HCl} + \text{HRhCl}_5^{3-} \xrightarrow{k_3} \text{H}_2 + \text{RhCl}_6^{3-} \tag{321}$$

$$\text{HRhCl}_5^{3-} \xrightarrow{k_4} \text{Rh}^0 \tag{322}$$

The evolution rate falls as metal is precipitated. The dehydrogenation is thought to involve transfer of hydride from the α-carbon atom to the rhodium to give the intermediate hydrido species which either reacts with a proton to give hydrogen or decomposes to metal. Equation 321 is the reverse of reaction 315; the forward reaction is assumed to be negligible under a low hydrogen concentration. The similar reversibility of the reaction of hydrogen with chlororuthenate(III) complexes has been discussed in Section IX-B-1. Metal precipitation in the rhodium system is avoided by the addition of stannous chloride, which can coordinate as SnCl_3^- or solvated SnCl_2 (616, 681). Whether this stabilizes a rhodium(III) hydride or a rhodium(I) product (eq. 314) was not discussed (615), but a later paper by Charman (1463) suggested the hydride. Even in the presence of stannous chloride the reaction rate decreased because of the buildup of acetone, the k_{-2} step becoming more important (1463). An observed hydrogen exchange between the α-CH and the hydroxide of the isopropanol was analyzed in terms of exchange with the rhodium(III) hydride (eq. 112, Section IX-B-1):

$$\text{HRh} + {}^i\text{Pr—OD} \rightleftharpoons \text{DRh} + {}^i\text{Pr—OH} \tag{322a}$$

2.   Substitution Reactions at Rhodium(III) Catalyzed by Molecular Hydrogen.

Gillard and coworkers (618) reported that substitution reactions of some inert rhodium(III) complexes were catalyzed by hydrogen at room temperature in aqueous solution:

$$1,2,6\text{-Rhpy}_3\text{Cl}_3 + \text{py} \longrightarrow \textit{trans-}[\text{Rhpy}_4\text{Cl}_2]\text{Cl} \tag{323}$$

$$\text{RhCl}_3, 3\text{H}_2\text{O} + \text{H}_2\text{O} \longrightarrow \textit{trans-}[\text{Rh(H}_2\text{O)}_4\text{Cl}_2]^+ + \text{Cl}^- \tag{324}$$

Such catalyzed reactions are clearly relevant to the nature of possible species in solution under homogeneous hydrogenation conditions, particularly if substitution inert complexes are being used as catalysts. This should be kept in mind when considering possible catalysis by other relatively inert complexes, for example, iridium(III).

Gillard and coworkers suggested a hydridic mechanism involving heterolysis of hydrogen:

$$\text{Rhpy}_3\text{Cl}_3 \xrightarrow{\text{H}_2} \text{H}^+ + \text{HRhpy}_3\text{Cl}_3 \xrightarrow{\text{py}} \text{Rhpy}_4\text{Cl}_2^+ + \text{H}^-$$

$$\text{H}^+ + \text{H}^- \longrightarrow \text{H}_2 \tag{325}$$

Intermediate hydrides are almost certainly produced, since ethanolic solutions of $RhCl_3$ or $Rhpy_3Cl_3$ were found to catalyze hydrogenation of hex-1-ene under the same mild conditions (Section XI-H).

Similar syntheses of rhodium(III) complexes that involve nucleophilic replacement of halide ions by nitrogen donor ligands and are catalyzed by other hydridic reducing agents (such as ethanol, borohydride, hydrazinium chloride, and hypophosphorous acid) have been reported by a number of workers (610, 619–624, 629, 630, 634, 1354).

Basolo and coworkers (622, 625) first suggested that these catalyzed substitution reactions might involve rhodium(I) intermediates, and Rund (625a) found that for reaction 326 rhodium(I) compounds were more active catalysts than those containing hydride-producing reagents, and that some rhodium(III) hydrides were inactive.

$$RhCl_6^{3-} + 4py \longrightarrow Rhpy_4Cl_2^+ + 4Cl^- \qquad (326)$$

The following mechanism was proposed. The labile rhodium(I) catalyst coordinates four pyridines; this complex then forms a bridged intermediate through its axial position with the inert rhodium(III) and also coordinates a sixth ligand to give **81**. After a two-electron transfer the bond breaks at the

81

new labile rhodium(I) center to give the product and regenerated catalyst. It is likely that the hydridic reducing agents generate an active rhodium(I) intermediate, depending on equilibrium 314. A 10^{-6} M concentration of rhodium(I) is sufficient for the catalysis; hence there is no observable consumption of the reducing agent. The nonhydridic reducing reagents carbon monoxide and ethylene are also effective catalysts (626–628), and Rh^I . . . Rh^{III} bridged intermediate were again postulated. Some catalysis actually arises from deoxygenation resulting from bubbling the gases through the solutions, since nitrogen, argon, or cyclopropane may also be effective, as is complete evacuation (608, 628, 630). This leads to the interesting conclusion that certain rhodium(III) complexes in water must establish an equilibrium with small amounts of a rhodium(I) species (630).

The possibility that these catalyzed substitution reactions may involve rhodium(II) intermediates cannot be completely ruled out.

B.   TRIS(TRIPHENYLPHOSPHINE)CHLORORHODIUM(I), RhCl(PPh₃)₃, AND ANALOGS

This important catalyst was first reported by Wilkinson's group in 1965 (43, 43a), although its hydrogenating activity was also discovered simultaneously and independently by Coffey (45). The compound, which exists in two crystalline forms, was also made independently by Bennett and Longstaff (635), Vaska and Rhodes (730), and Dewhirst (731). Mason and coworkers (1316) reported that in the solid state the coordination of the rhodium is strongly distorted toward tetrahedral. Extensive studies have since been carried out by many workers on the hydrogenating ability of this complex and related ones, $RhXL_3$, where X = halogen and L = a tertiary phosphine, arsine, stibine, or phosphite.

1.   Mechanism.   Ethanol-benzene solutions of $RhCl(PPh_3)_3$, **82**, at 25° and hydrogen pressures up to 1 atm, were at first reported to give rapid reductions of hex-1-yne and hex-1-ene (43, 43a). Wilkinson and coworkers (612, 637, 638, 660) extended the studies and found generally that isolated olefinic and acetylenic linkages were readily reduced; the mechanism has been considered in some detail. Molecular weight studies in benzene or chloroform (612, 635) had originally suggested that **82** dissociated reversibly to a solvated species, which slowly dimerizes to **83**.

$$RhCl(PPh_3)_3 + solvent(S) \overset{K}{\rightleftharpoons} RhCl(PPh_3)_2(S) + PPh_3 \qquad (327)$$

$$2RhCl(PPh_3)_2(S) \longrightarrow [RhCl(PPh_3)_2]_2 + 2S \qquad (328)$$
$$\textbf{83}$$

A later note by Shriver and coworkers (859) shows, however, that with rigorous exclusion of oxygen, molecular weight and chemical evidence point to no dissociation in benzene at $(2.4–5.8) \times 10^{-3}\ M$, or in dichloroethane at $3 \times 10^{-2}\ M$. In the presence of traces of oxygen, which may well occur during catalytic hydrogenations, extensive dissociation can occur presumably to give the solvated species (Section B-8), although in the presence of further oxygen new oxidized species may be formed (Sections B-4 and B-8).

Complex **82** in these solvents is extremely reactive and, for example, forms reversibly 1:1 isolable complexes with hydrogen, $H_2RhCl(PPh_3)_2$, and with ethylene, $RhCl(PPh_3)_2(C_2H_4)$. Spectroscopic data indicate that the hydride is present in solution as a solvated octahedral cis-dihydride complex **84**; solvated species were obtained from chlorinated hydrocarbons. The mechanisms of formation of the dihydride and ethylene complexes have not been elucidated, but the important conclusion is that a phosphine ligand is dissociated at some stage (Section B-7a). The dimer **83** also reacts with hydrogen to give a dimeric hydride believed to be **85**:

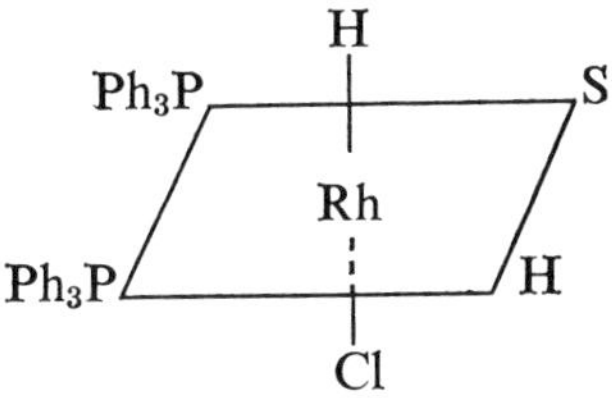

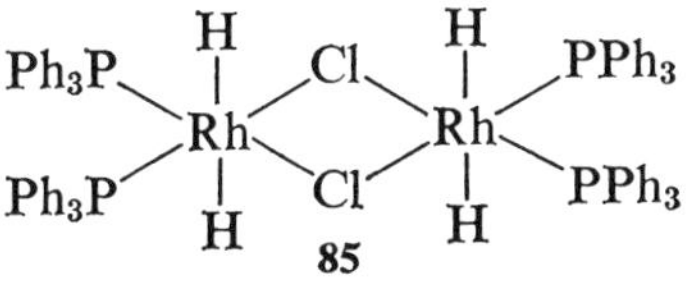

The ethylene complex, with likely *trans* phosphine ligands, has a stability constant of ~100 M^{-1} at 25° (estimated with respect to RhCl(PPh$_3$)$_2$(S), which was thought to be present) but contains a quite labile ethylene molecule (612) as do other rhodium(I) ethylene complexes (537). Other simple olefins formed much weaker complexes, e.g., K for propylene <0.05 M^{-1}. A paraffin resulted from addition of ethylene or hex-1-ene to the dihydride complex **84**, thus indicating that the dihydride was an intermediate in the catalysis (43a, 612); **85** did not transfer hydrogen to olefins.

Kinetic studies have been reported for the hydrogenation of hept-1-ene, cyclohexene, and hex-1-yne in benzene or benzene-hexane; the decrease in hydrogen pressure with time was measured in a constant-volume apparatus (612). The rate was between zero- and first-order in substrate and hydrogen, and was approximately, but somewhat less than, first-order in catalyst up to its solubility limit. The postulated mechanism in terms of the RhCl(PPh$_3$)$_2$(S) catalyst is outlined in Scheme 9 (612).

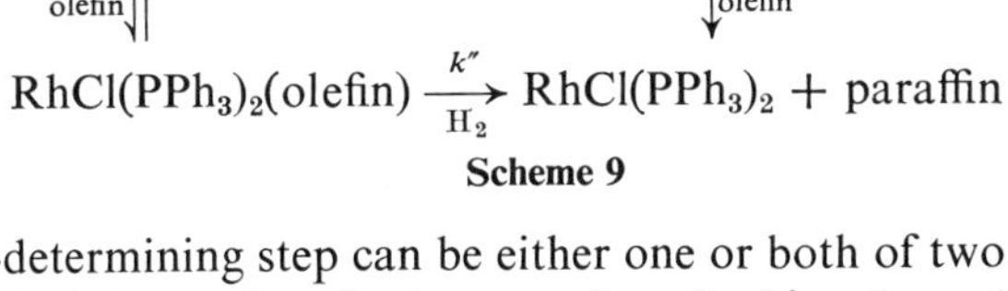

Scheme 9

The rate-determining step can be either one or both of two possible paths: (1) attack of olefin on the dihydro complex, the k' path, or "hydride route" (636); (2) attack of hydrogen on the olefin complex, the k'' path, or "unsaturate route" (636). The rate law for the scheme, assuming that no undissociated RhCl(PPh$_3$)$_3$ species remains, is given in the following equation, where [A] is the substrate concentration:

$$\text{Rate}(R) = \frac{-d[\text{H}_2]}{dt} = \frac{-d[A]}{dt} = \frac{(k'K_1 + k''K_2)[\text{H}_2][A][\text{Rh}]}{1 + K_1[\text{H}_2] + K_2[A]} \quad (329)$$

The kinetic data are essentially consistent with this rate law. At higher catalyst concentrations the presence of small amounts of undissociated RhCl(PPh$_3$)$_3$, the dimer **83**, or H$_2$RhCl(PPh$_3$)$_3$ (Section B-7a) could give rise to a less than first-order dependence on catalyst [see the HRuCl(PPh$_3$)$_3$ system and eq. 137 (Section IX-B-2b) for the dissociation effect]. The catalyst dependence which was not studied in detail initially because of solubility problems (612), will be considered again later (Section B-7a).

It should be noted that if K_1 and K_2 refer to complexing with the undissociated complex, for example eq. 329a, then the rate law of eq. 329 is modified with the unity term in the denominator replaced by [PPh$_3$]; the kinetic dependences would still be satisfied. The hydride H$_2$RhCl(PPh$_3$)$_3$,

$$\text{RhCl(PPh}_3)_3 + \text{H}_2 \overset{K_1}{\rightleftharpoons} \text{H}_2\text{RhCl(PPh}_3)_2 + \text{PPh}_3 \qquad (329a)$$

prepared by Candlin and Oldham (636) by using excess phosphine, is a likely intermediate in reaction 329a.

Wilkinson's group (612) reported that the relatively stable ethylene complex showed no reaction with hydrogen. Ethylene was not catalytically hydrogenated, and it was an effective poison for the hydrogenation of other olefins such as propylene. Osborn (841) later reported that benzene solutions of the allene complex RhCl(PPh$_3$)$_2$(allene) did not react with hydrogen. These workers thus concluded that the "unsaturate route" generally did not operate; putting $k'' = 0$ in eq. 329, they analyzed the data by inverse plots of (rate)$^{-1}$: (substrate)$^{-1}$, which were reasonably linear for hept-1-ene, cyclo-hexene, and hex-1-yne. The hydrogen dependence data were similarly analyzed for the cyclohexene system. For this system at 25,° k' was determined to be 0.15 M^{-1} s^{-1} in benzene-hexane with $\Delta H^{\ddagger} = 22.3$ kcal mole^{-1}, $\Delta S^{\ddagger} = +12.9$ eu (Table 41) and $K_2/K_1 \sim 1.6 \times 10^{-3}$. Independent measurement of K_2 and K_1 would be useful to confirm this analysis. Since the hydride is essentially fully formed at 25° under 50 cm hydrogen (concentration $\sim 2.5 \times 10^{-3}\ M$), K_1 must be $> 4 \times 10^3\ M^{-1}$, which suggests $K_2 \geqslant 6\ M^{-1}$ (again with respect to RhCl(PPh$_3$)$_2$S). If reactions such as 329a are involved, this method of analysis will yield the same k' and K_2/K_1 values; the K_1 and K_2 values obtained above would be modified by a multiplication factor of [PPh$_3$].

Contrary to the data of Wilkinson's group (612), Candlin and Oldham (636) have reported that the ethylene complex in benzene may be used as effectively as RhCl(PPh$_3$)$_3$ itself for the catalytic hydrogenation of cyclo-pentene, hex-1-yne, and ethylene; also the presence of ethylene did not seriously hinder propylene hydrogenation catalyzed by RhCl(PPh$_3$)$_3$. These latter workers also observed stoichiometric reduction of hex-1-yne and oct-1-ene when using the dihydride, hence they concluded that both the k' and k'' paths may operate in Scheme 9 (see also Section B-11).

Table 41 Kinetic Data for Hydrogenation of Olefins Catalyzed by RhCl(PPh$_3$)$_3$[a]

Substrate	k' at 25° $(M^{-1}\,s^{-1})$	$\Delta H^{\ddagger}$ (kcal mole^{-1})	$\Delta S^{\ddagger}$ (eu)	k'_{H_2}/k'_{D_2} (25°)	Reference
Cyclohexene	0.15[b]	22.3	12.9	0.84	612, 637, 638
	0.25[c]	—	—		638
	0.32	18.6	1.3	0.90	637
	0.30[d]	16.6	−5.3	1.03	637
	0.28	—	—		672
Cyclopentene	0.34	16.7	−4.7		637
Cycloheptene	0.22	21.7	11.0		637
Hex-1-ene	0.29	18.6	1.1		637
Hex-2-ene	<0.29	—	—		612
Dodec-1-ene	0.34	16.9	−4.1		637
1-Methylcyclohexene	0.006	12.7	−26.0		637
cis-Pent-2-ene	0.23	12.6	−20.2		637
2-Methylpent-1-ene	0.27	12.7	−18.5		637
Styrene	0.93	11.1	−21.5		637
cis-4-Methylpent-2-ene	0.10	16.1	−9.3	0.78	637
trans-4-Methylpent-2-ene	0.02	11.6	−27.6	0.97	637

[a] In benzene solution unless otherwise stated.
[b] Originally quoted for benzene solutions (612); later stated (637, 638) to refer to a benzene-hexane solvent.
[c] Using [RhCl(C$_8$H$_{14}$)$_2$]$_2$ + 3PPh$_3$.
[d] In ethyl methyl ketone-hexane solvent.

The k' path can be given in more detail as shown in eqs. 330 and 331, either of which could be rate-determining (637):

$$\text{H}_2\text{RhCl(PPh}_3)_2(\text{S}) + \text{olefin} \underset{k_{-3}}{\overset{k_3}{\rightleftharpoons}} [\text{H}_2\text{RhCl(PPh}_3)_2(\text{olefin})] + \text{S} \quad (330)$$
$$\mathbf{84} \qquad\qquad\qquad\qquad\qquad\qquad \mathbf{86}$$

$$\mathbf{86} \longrightarrow \text{RhCl(PPh}_3)_2 + \text{paraffin} \quad (331)$$

A measured isotope effect, k'_{H_2}/k'_{D_2}, of about 0.9 for the cyclohexene system was originally rationalized (612) in terms of hydrogen transfer (eq. 331) being rate-determining with lengthening of a Rh—H bond, occurring synchronously with formation of the new C—H bond, since the zero point energy of a C—H bond was estimated to be greater than that of the Rh—H bond by about 1.4 kcal mole^{-1}. However, further studies (637) on a variety of olefinic substrates (Section B-3) led the same authors to conclude that eq. 330 is probably rate-determining and that the observed isotope effect is secondary (Section XI-D).

2.   Reproducibility of Kinetic Data and Initial Rates; the Dissociation reaction.   It should be noted at this stage that the reproducibility of kinetic data for the hydrogenation of substrates by these rhodium(I) systems is generally inadequate. Candlin and Oldham (636) report a value of $\pm 10\%$; a discussion involving small differences in constants derived from kinetic data should, therefore, be considered guardedly. From reading the literature, this $\pm 10\%$ figure seemingly refers to the same catalyst sample and solvent batch. Variation in rates reported by different workers for similar systems is sometimes larger than this, which will become evident as these systems are discussed.

The lack of reproducibility is due to several factors. The purity of the olefin, particularly its peroxide content, is an important factor. As prepared from rhodium trichloride and triphenylphosphine, $RhCl(PPh_3)_3$ contains a small amount of paramagnetic impurity (612, 636). This may be due to contamination by a rhodium(II) species, since the corresponding reaction with tri(o-tolyl)phosphine can yield $RhCl_2[P(o\text{-tolyl})_3]_2$ (703, 704). Catalytic activity by this species has not been reported, but examples of hydrogenating activity by other rhodium(II) species are known (Section XI-I). Candlin and coworkers (245) report that recrystallization of the triphenylphosphine complex, except from acetonitrile, results in some decomposition. The complex in solution rapidly and irreversibly absorbs oxygen. Although not realized until quite recently, the behavior of $RhCl(PPh_3)_3$ seems markedly different in benzene than in benzene-ethanol (the two media commonly used for hydrogenation studies), particularly concerning sensitivity and effect toward oxygen (see below). The complex is also extremely reactive generally, undergoing a wide variety of oxidative addition reactions (683, 705, 706); there is also the problem of dimerization.

It should also be pointed out that a discussion involving comparison of initial rate data must at best be qualitative for systems (with variations in substrate, halide, tertiary phosphine, etc.) where a rate law of the type shown in eq. 329 applies. Assuming that the hydride route only operates, values of k', K_1, and K_2 (or at least K_2/K_1) are really required. This has been done for a limited number of systems (Tables 41 and 49). Different rate data may also be obtained depending on whether the catalyst is subjected to hydrogen treatment (dihydride formation) before addition of olefin (Section B-5c).

As indicated in Section B-1, the extent of the dissociation reaction 327 has been the subject of some discussion. Eaton and Suart (716) studied the n.m.r. of the phosphorus and hydrogen for the complex in CH_2Cl_2 and $CDCl_3$ respectively, and concluded that there was $<5\%$ dissociation at complex concentrations $>10^{-2}$ M, although their molecular weight measurements in these solvents indicated complete dissociation in agreement with the findings of Wilkinson's group (612). Augustine and Van Peppen (710), by monitoring phosphine *after* pretreatment of RhCl(PPh$_3$)$_3$ with oxygen, have shown that the complex dissociates quite readily in benzene with loss of a phosphine, but not at all in benzene-ethanol if $>4\%$ of ethanol is present. Triphenylphosphine oxide was detected in the ethanolic media, but it did not result from oxidation of dissociated phosphine. Traces of oxygen in these systems indeed have marked effects on the catalyst solutions (Sections B-4 and B-8). Eaton and Suart (716) observed triphenylphosphine oxide in their n.m.r. studies, despite stringent precautions including solvent purification. Taken together with the findings of Shriver (Section B-1), it would seem that the discrepancies are due to oxygen and/or solvent impurities, but it is still difficult to draw any definite conclusions about whether the dihydride is formed according to eq. 329a or via reaction 327. The dissociation equilibrium constant K can be readily included in the rate law which will have a similar form to eq. 329 (see eq. 336, Section B-4). A small K value should also manifest itself in the catalyst dependence which should be less than first-order (Section B-7a). The similar discrepancy between molecular weight and n.m.r. data for the HRu(OCOCF$_3$)(PPh$_3$)$_3$ catalyst has been noted in Section IX-B-2b.

By studying the methyl resonance of tri(p-tolyl)phosphine (effectively a labeled phosphine) in solutions of the triphenylphosphine complex, Eaton and Suart (716) have also shown that *cis-trans* isomerization in the square-planar complex is faster than ligand exchange, and they have suggested that an intermediate, [RhCl(PPh$_3$)$_2$]PPh$_3$, with one more loosely bound phosphine ligand may be involved. The possibility of loosely held moieties in a second coordination sphere is particularly attractive for homogeneous catalytic reactions (716, 768).

3. Variation of Rate with Olefin. The kinetic data obtained by Wilkinson's group (612, 637, 638) for hydrogenation of olefins catalyzed by RhCl(PPh$_3$)$_3$ in benzene, assuming that the hydride route only is operative, are summarized in Table 41. Semiquantitative data in terms of initial rates have also been presented for some olefinic substrates containing other functional groups (Table 42). Candlin and Oldham (636) have also reported the rates for the hydrogenation of substrates relative to that of terminal alkenes, using RhCl(PPh$_3$)$_3$ in benzene at 22° and 1 atm. These data are given in Table 43 for conditions with 1.0 M substrate concentration when the reactions are said to be first-order in substrate, although it should be noted that Wilkinson's data at quite similar conditions give a somewhat less than first-order dependence (see the cyclohexene data in Tables 41 and 42). The data from Wilkinson's group have been rationalized in terms of the hydride route, with eq. 330 being rate-determining. This could involve an associative-type mechanism with olefin displacing the solvent or a dissociative mechanism with olefin attacking a 5-coordinate intermediate.

Table 42 Initial Hydrogenation Rates of Some Unsaturated Substances Using RhCl(PPh₃)₃ (637)[a]

Substrate	Rate (mmoles min^{-1})	Substrate	Rate (mmoles min^{-1})
Cyclohexene	0.80	Penta-1,3-diene	0.06
2,3-Dimethylbut-2-ene	0.01	Hexa-1,5-diene	0.21
3-Ethylpent-2-ene	0.02	Octa-1,7-diene	0.65
trans-Hex-3-ene	0.05	Cyclohexa-1,3-diene	0.13
Acrylamide	0.22	Styrene	2.56
Allyl alcohol	3.02	4-Fluorostyrene	3.70
Allyl cyanide	0.45	4-Methoxystyrene	3.22
		Acenaphthylene	1.76

Acrylic acid, allylamine, allyl chloride, norbornadiene, butadiene (612), cyclohexa-1,4-diene, cycloocta-1,5-diene (612), maleic anhydride, tetraphenylethylene, tetrachloroethylene, and trans-stilbene (612) were reduced very slowly (<0.01).

[a] In 80 ml benzene at 25°; 1.25 × 10^{-3} M catalyst, 1.25 M substrate, 50 cm H₂.

Table 43 Relative Rates of Hydrogenation Using 10^{-2} M RhCl(PPh₃)₃ (636)[a]

Substrate	Relative Rate[b]	Substrate	Relative Rate[b]
Diethyl maleate	1.7	Cycloheptene	0.46
Allyl acetate	1.35	2-Methylpent-1-ene	0.43
Acrylonitrile	1.3	Cyclooctene	0.42–0.27
i-Butyl vinyl ether	~1.05	Methyl methacrylate	0.36
C₆—C₁₂ alk-1-enes	1.0	Diisobutene	0.30
Cyclopentene	1.0	Oct-2-ene	0.25
C₆—C₈ alk-1-ynes	0.85	Phenylacetylene	0.15
2,4,4-Trimethylpent-1-ene	0.70	1,3-Pentadiene	~0.12
Cyclohexene	0.71–0.55	Diethyl fumarate	0.085
Cycloocta-1,3-diene	0.60–0.50	Isoprene	0.045
Octa-1,3-diene	0.60	Cycloocta-1,5-diene ⎫	
Hexa-1,5-diene	~0.55	Cyclopenta-1,3-diene ⎬ very slow	
		Norbornadiene ⎭	

[a] In benzene at 22°, 10^{-2} M catalyst, 1.0 M substrate, 1 atm H₂; $t_{0.5}$(octene) = 10 min.
[b] Calculated from initial rates or half-lives of the substrates.

Tables 41 to 43 show the importance of steric hindrance about the C=C bond—cyclohexene, for example, is reduced about fifty times as fast as 1-methylcyclohexene; this is reflected in the unfavorable $\Delta S^{\ddagger}$ value of the methyl derivative which in fact is favored in terms of $\Delta H^{\ddagger}$. Table 41 suggests that this is generally true, with $\Delta S^{\ddagger}$ becoming more negative and $\Delta H^{\ddagger}$ decreasing with increasing substitution at the C=C bond. The analogy was drawn to stability constant data for some silver(I)-olefin complexes (639), where the greater the steric restraint, the more negative the entropy of formation, but these effects are quite small. Activation energies for the heterogeneous hydrogenation of olefins also decrease with increasing substitution (640). The data for hex-1-ene and dodec-1-ene suggest that the $\Delta H^{\ddagger}$ value for the latter compensates for a slight additional steric hindrance to complex formation. The data in Table 43, however, indicate that terminal olefins are reduced at the same rate, and that there is no additional hindrance on increasing the chain length. According to equations such as 329 with $k'' = 0$, effective hydrogenation requires a large k' and for the olefin complex with RhCl(PPh₃)$_n$, a small K_2. Sterically hindered olefins, which have smaller K_2 values, are not more rapidly hydrogenated, however, and this was attributed to the more dominant changes in k' (637). In any case, there seemed to be no substantial variation in the K_2 values which, relative to hex-1-ene as unity, were estimated as 0.5 M^{-1} for *trans*-4-methylpent-2-ene and 2.3 M^{-1} for cycloheptene (for complexation with RhCl(PPh₃)₂S).

Strongly sterically hindered olefins such as tetraphenylethylene and *trans*-stilbene are reduced, but very slowly.

Terminal olefins, including nonconjugated, nonchelating dienes, are reduced more rapidly than internal olefins. Table 43 suggests that cyclic olefins are generally reduced more slowly than terminal olefins, the rate decreasing with increasing ring size. The k' values in Table 41 do show the same trend, but the small variance for cyclopentene, -hexene, and -heptene is thought to be consistent with eq. 330 being rate-determining, where no release of ring strain energy would be envisaged (639).

Cis-olefins are reduced faster than *trans*-olefins: maleate > fumarate, *cis*-hex-2-ene > *trans*-hex-2-ene (612). The data in Table 41 indicate a greater steric restriction of the *trans*-4-methylpent-2-ene in passing to the transition state compared to the corresponding *cis*-olefin. Bond and Hillyard (643) have also reported relative reduction rates: pent-1-ene > *cis*-pent-2-ene > *trans*-pent-2-ene, and Morandi and Jensen (714) the rates *cis*-decenes > *trans*-decenes.

Strohmeier and Endres (1453) have listed initial rate data for some reductions in toluene at 25° and 1 atm hydrogen when using 2×10^{-3} M catalyst and 0.8 M substrate. The substrates reduced together with the rates in (M min^{-1} × 10^3) were as follows: ethyl acrylate, 8.6; hex-1-ene, 7.8; styrene, 7.8; hept-1-ene, 5.7; cycloheptene, 5.4; dimethyl maleate, 5.3; dimethyl fumarate, 4.0; cis-hept-2-ene, 1.6; β-bromostyrene, 0.14; trans-hept-2-ene, 0.05; trans-hept-3-ene, 0.03; the same general trends are evident: 1-olefins > 2 cis ≫ 2 trans > 3 trans olefin. However, discrepancies with the data of Tables 41 to 43 are readily seen; for example, the relative rates of cycloheptene, maleate, and fumarate in Table 43.

Conjugated dienes are reduced more slowly than terminal olefins; butadiene and cyclohexa-1,3-diene are reduced very slowly at 1 atm hydrogen, but are effectively reduced at ~60 atm (612, 637). Diene reductions yield monoenes and then the fully saturated product. Unsaturated polymers and copolymers of butadiene have also been partially reduced (701).

The data on the styrenes (Tables 41 and 42) suggest that electronic factors override the steric restriction factors which are apparent in the $\Delta S^{\ddagger}$ value for styrene itself. Yet it is surprising that both electron acceptors (p-fluoro) and donors (p-methoxy) accelerate the reduction. Table 43 shows that substrates with the electron-withdrawing carboalkoxy and nitrile groups are hydrogenated rapidly.

Substrates such as allyl chloride, norbornadiene, cycloocta-1,5-diene, allylamine, maleic anhydride, and acrylic acid probably form quite stable complexes with RhCl(PPh$_3$)$_2$(S) and are not readily hydrogenated (635, 637). Allyl alcohol and allyl cyanide are readily reduced, which may be due to relatively weak complexing through the double bond only, that is, K_2 is small (637). Allyl bromide is reduced in a few hours under mild conditions in 1:1 benzene-ethanol (699).

Hussey and Takeuchi (667, 668) have studied reduction of some cyclo-alkenes, and have also discussed their data in terms of eq. 330 as rate-determining. For the hydride route via RhCl(PPh$_3$)$_2$(S), rate law 329 reduces to

$$\text{rate} = \frac{k'K_1[\text{H}_2][A][\text{Rh}]}{1 + K_1[\text{H}_2] + K_2[A]} \tag{332}$$

The initial rate can thus be written k[catalyst] where k is a pseudo-first-order constant. Table 44 gives such constants for benzene and benzene-ethanol (3:1) media, and also the isomer compositions for the mixed solvent. The decreasing rates correlated generally with increasing steric hindrance and lower stability of π-complexes. Horner and coworkers (686) have reported that only cis-products are formed from the reduction of cyclohexene and the methylene-, 4-methyl-, 3-methyl-, and 1-methyl-derivatives in benzene; the relative rates of reduction were 1.0:0.87:0.80:0.63:0.01, respectively.

Table 44   Kinetic Data for Hydrogenation of Cycloalkenes Catalyzed by RhCl(PPh$_3$)$_3$ (667, 668)[a]

Substrate	[Substrate] (M)	k[b] (min^{-1})	k[c] (min^{-1})	% cis[b]
Cyclohexene	1.10	17(22[d])	13.7[e]	—
1-Methylcyclohexene	0.94	0.49	0.13	—
1,4-Dimethylcyclohexene	0.82	0.16	0.09	50
1-Methyl-4-isopropylcyclohexene	0.67	0.25	0.11	30
4-Methylmethylenecyclohexane	0.81	7.2	—	67
1,3-Dimethylcyclohexene	0.82	0.0	—	—
2,3-Dimethylcyclohexene	0.74	0.20	—	50
1,2-Dimethylcyclohexene	>0.60	0.0	—	—
2,4-Dimethylcyclohexene	0.41	0.16	—	48
Norbornene	1.22	—	4.4	—

[a] In benzene-ethanol (3:1) and benzene at 25°; 2.4 × 10^{-3} M catalyst, 1 atm H$_2$. Initial rate = k [catalyst].
[b] Benzene-ethanol (3:1).
[c] Benzene.
[d] Using [RhCl(C$_8$H$_{14}$)$_2$]$_2$ + 3PPh$_3$.
[e] The data of Table 42 give a k[c] value of 8.0 min^{-1} for a similar cyclohexene concentration.

4. Isotope Studies and Hydrogen Transfer. Wilkinson and coworkers (612, 637) and Biellman and Liesenfelt (649) have reported that on exposing benzene or alcohol solutions of RhCl(PPh$_3$)$_3$ to deuterium gas no exchange was observed with any component of the catalyst system, and that solvent hydrogens generally play no part in the hydrogenation process (612, 649). When H$_2$–D$_2$ mixtures are used at lower pressures (15 cm), very little HD is formed in the gas phase, which shows that any exchange is slow compared to substrate hydrogenation rates. Reduction of hex-1-ene by an H$_2$–D$_2$ mixture at 15-cm pressure gave 50% C$_6$H$_{14}$ and 44% C$_6$H$_{12}$D$_2$; at 60 cm, however, some exchange was indicated (612). The production of some monodeuterated cholestane from catalytic deuteration of 2-cholestene in benzene-methanol has been attributed by Voelter and Djerassi (664) to exchange of the dideuteride catalyst with methanol (Section B-5c). Reduction of propene and styrene at 15-cm pressure by using HD in benzene gave the monodeuterated hydrocarbon; addition occurred both ways, and both isomers were formed (637). Catalytic deuteration of maleic and fumaric acids in benzene-ethanol gave mainly *meso*-2,3-dideuterosuccinic acid and *DL*-2,3-dideuterosuccinic acid, respectively, indicating that predominately *cis*-addition had occurred (612). Addition to norbornene has also been shown to be *exo, cis* (668). Stereospecific *cis* addition of deuterium using RhCl(PPh$_3$)$_3$ has been accomplished at a number of olefinic bonds (see Section B-5, Tables 46 to 48). Morandi and Jensen (714) have used the complex to deuterate a number of linear C$_{10}$–C$_{20}$ monoolefins in benzene.

The kinetic isotope effects reported in Table 41, which were solvent dependent and decreased with increasing temperature, were considered secondary, with eq. 330 as rate determining (637).

Using a H$_2$/HT mixture, Simon and Berngruber (679, 680) have measured the isotope effects in benzene for the hydrogenation of dimethyl fumarate, ethyl crotonate, and 4-methylcyclohexene. The k_H/k_T was about unity for the α,β-unsaturated esters; there was no exchange and the tritium was equally distributed at the C-2 and C-3 carbon atoms of the reduced products. For 4-methylcyclohexene, the isotope effect was 0.7, and some exchange with the substrate was observed. The values of the isotope effects were similar to those reported by Wilkinson's group (Table 41). Simon and Berngruber (680) consider that the inverse isotope effects arise by the stronger σ-donors, deuterium and tritium, favoring the olefin complexing in reaction 330.

Wilkinson and coworkers (612, 637, 638) suggested that in reaction 331 both hydrides might be transferred simultaneously. But evidence, particularly concerning accompanying isomerization and exchange, has been presented by other workers indicating that the hydrogens are transferred consecutively and a σ-alkyl hydride intermediate **87** is involved:

$$[H_2RhCl(PPh_3)_2(\text{olefin})] \underset{k_{-4}}{\overset{k_4}{\rightleftharpoons}} \underset{\mathbf{87}}{HRhCl(PPh_3)_2(\text{alkyl})} \overset{k_5}{\underset{S}{\longrightarrow}}$$

$$RhCl(PPh_3)_2(S) + \text{paraffin} \quad (333)$$

Beillman and Jung (642) and later Ruesch and Mabry (1126) noted the exocyclic methylene group in damsin, **88**, migrates into the ring in preference to being hydrogenated by the RhCl(PPh$_3$)$_3$/H$_2$ system. No isomerization occurs in the absence of hydrogen, and in the presence of EtOD no incorporation of deuterium takes place (642). Using D$_2$RhCl(PPh$_3$)$_3$ and damsin in equivalent amounts led to a 58% incorporation of one deuterium atom into the product isodamsin **90**. Using catalytic amounts of the rhodium complex, 70% of **89** contained no deuterium; and 30%, one deuterium atom. Neither system gave dideuterated product. Scheme 10, incorporating the usual reversible "olefin + metal-hydride $\rightleftharpoons$ metal-alkyl" equilibrium for isomerization, was presented (642):

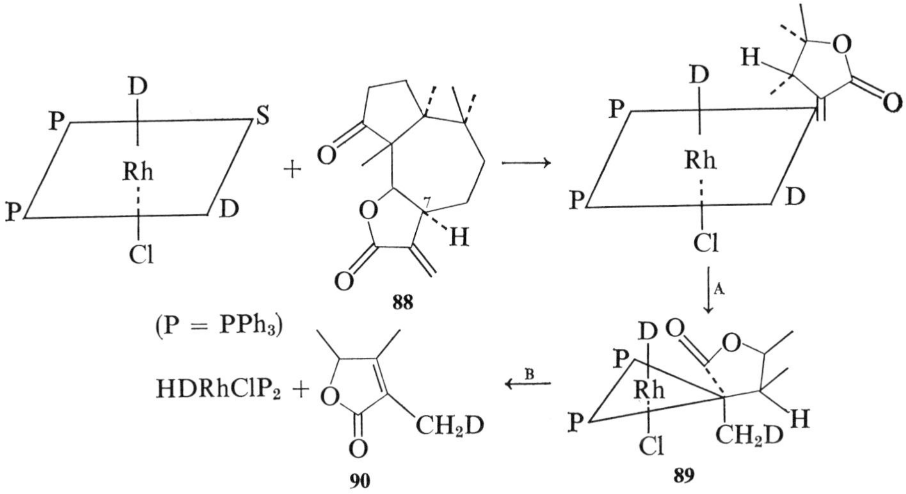

Scheme 10

The hydrogen transferred to the rhodium is considered to come from the C-7 position of damsin. The reaction of HDRhCl(PPh$_3$)$_2$ with a second mole of **88** gives **90** with one or no deuterium atom, and D$_2$RhCl(PPh$_3$)$_2$ is gradually replaced by H$_2$RhCl(PPh$_3$)$_2$. This accounts for 58% labeled isodamsin product. The absence of dideuterated product implies that the initial hydrogen transfer (step A) is irreversible. The σ-alkyl hydride **89** can either transfer a second hydrogen (hydrogenation) or undergo reaction B to give isomerization. Similarly, Augustine and Van Peppen (644) found appreciable isomerization of 4-t-butylmethylenecyclohexane to 4-t-butylmethylcyclohexene in benzene-ethanol or ethanol. Bond and Hillyard (643) noted isomerization of cis-pent-2-ene to trans-pent-2-ene and pent-1-ene in benzene-ethanol; Abley and McQuillin (243) found isomerization of oct-1-ene to trans-oct-2-ene in the later stages of hydrogenation in benzene-methanol. It should be noted that slow isomerization of oct-1-ene has been observed when it was refluxed in the absence of hydrogen with RhCl(PPh$_3$)$_3$ in benzene, ethanol, or chloroform (243). A cocatalyst hydride source will be required for the isomerization process (645). This could be the solvent or possibly the olefin itself via an abstraction in the allylic position (646). The reported isomerization of a number of substituted cyclohexa-1,4-dienes to the corresponding 1,3-dienes (670) by refluxing chloroform or benzene solutions of RhCl(PPh$_3$)$_3$ has been quoted (671) as evidence for stepwise addition. But these systems are not reported to involve molecular hydrogen, and they are unlikely to involve the dihydride complex.

Hussey and Takeuchi (667, 668) studied the deuteration and the deuterium content of both *cis* and *trans* isomer products of the more slowly reduced cycloalkenes listed in Table 44 (benzene-ethanol solution): some d_0, d_1, d_3, and d_4 products were found besides the major d_2 ones, showing that the reductions were so slow that exchange reactions, hence stepwise transfer of H or D, becomes observable. Schemes involving the reversible formation of alkylhydride from the dihydride-olefin complex were presented, and a rationale was advanced for the various proportions of deuterated products in both *cis* and *trans* isomers arising from dideuteride, deuterohydride, and dihydride catalysts. The exchange with the substrate in benzene-ethanol and benzene was also evidenced by production of HD in the gas phase equivalent to the d_3 plus d_4 species. The isomeric products from the quite rapidly reduced 4-methylmethylenecyclohexane were only very slightly exchanged.

Smith and Shuford (671) have studied the distribution of ethylbenzenes obtained from catalytic deuteration of styrene at room temperature. Pure 1,2-dideuteroethylbenzene is formed in benzene or dichloromethane; in chloroform or benzene-ethanol (3:1); however, 22% d_0, 36% d_1, and 39% d_2 products were observed. These workers consider that both simultaneous *cis*-addition and stepwise addition occur, but that two catalysts are involved. They suggest that the monomeric dihydride catalyst gives the specific addition, while the dimeric hydride, **85**, $[H_2RhCl(PPh_3)_2]_2$ might give rise to the exchange. Increasing catalyst concentration, which favors formation of the dimer, did decrease the $d_2:d_0$ product ratio. However, the dimer is essentially an inactive hydrogenation catalyst (612, 671) and the evidence for its participation is not strong. The postulation of competition between two different catalyst systems seems attractive in terms of the recent findings of Augustine and Van Peppen (709–712; see below).

Odell and coworkers (678) have investigated the hydrogenation of cyclohexene in benzene using hydrogen-tritium mixtures, and they have observed accompanying exchange in the form of tritiated cyclohexene. Hydrogenation is more favored at 60° than at 30°, suggesting that the activation energy for exchange is higher than for hydrogenation. In contrast to the observations of Smith and Shuford (671), increasing catalyst concentration at 30° essentially eliminated exchange.

Zeeh and coworkers (715) have deuterated *cis*-4,7,8,9-tetrahydro-indan-1-one in benzene-CH_3OD media to give mainly *cis*-hydrindan-1-one deuterated at the expected 5,6 positions (Table 46), but some mono- and trideuterated products were also observed.

Heathcock and Poulter (669) have provided evidence for stepwise addition of hydrogen in benzene solution by examining the hydrogenation products of some cyclopropyl alkenes (Table 45). The first hydrogen may add alpha to the cyclopropane ring to give an alkyl hydride species which yields the cyclopropylalkane:

$$
\begin{array}{c}
\text{H} \\
| \\
-\text{Rh}- \\
| \\
\text{H}
\end{array}
\quad \longrightarrow \quad
\begin{array}{c}
\text{H} \\
| \\
-\text{Rh}-\text{CH}_2\text{CH}_2-\triangleleft \\
\end{array}
\quad \longrightarrow \quad -\text{Rh} + \text{CH}_3\text{CH}_2-\triangleleft
$$

$$(334)$$

Initial β-addition gives a cyclopropylcarbinyl rhodium species, which is considered to rearrange to a homoallylrhodium compound **91** before final collapse:

$$
\begin{array}{c}
| \\
-\text{Rh}- \\
|
\end{array}
\quad \longrightarrow \quad
\begin{array}{c}
\text{H} \\
| \\
-\text{Rh}-\text{CH} \\
\qquad\quad \text{CH}_3
\end{array}
\quad \longrightarrow
$$

$$
\begin{array}{c}
\text{H} \\
| \\
-\text{Rh}-\text{CH}_2\text{CH}_2\text{CH}{=}\text{CH}-\text{CH}_3 \text{ , } \mathbf{91} \\
\downarrow \\
-\text{Rh} + \text{CH}_3\text{CH}_2\text{CH}{=}\text{CHCH}_3 \quad (335)
\end{array}
$$

The pent-2-ene is subsequently reduced to pentane; the intermediate *trans*-pent-2-ene observed in the vinylcyclopropane reduction rules out a synchronous 1,4-addition of hydrogen with attack at the ring. Similarly hex-3-ene and 4-methylpent-2-ene were detected during the reduction of **93** and **96** respectively (Table 45). With the reasonable assumption that the ease of reaction paths through intermediates decreased in the order primary $>$ secondary $>$ tertiary alkylrhodium species, the relative ratios of products from **92**, **93**, and **94** and the direction of ring cleavage for **95** and **96** are readily accounted for.

Including the stepwise hydrogen transfer shown in eq. 333 together with the mechanism outlined in Scheme 9 and eq. 330, Hussey and Takeuchi (668) have presented the following complete rate law:

$$
\frac{-d[\text{H}_2]}{dt} = \frac{k_3 k_4 k_5 K_1 [\text{H}_2][A][\text{Rh}_T]}{(1 + [\text{PPh}_3]/K + K_2[A] + K_1[\text{H}_2])(k_{-3}k_{-4} + k_{-3}k_5 + k_4 k_5)} \quad (336)
$$

The $[\text{PPh}_3]/K$ term arises from equilibrium 327. Wilkinson's expression (eq. 329) assumes that K is large, k_{-3} is very small, and the hydrogen transfer steps (k_4, k_5) are fast. If reactions such as 329a are involved, the $1 + [\text{PPh}_3]/K$ term should be replaced by $[\text{PPh}_3]$.

Table 45   Catalyzed Hydrogenation of Cyclopropylalkenes Using RhCl(PPh$_3$)$_3$ (669)[a]

Substrate	Products (%)				
92			85		14
93			70		27
94			97		1.5
95			99		1
96			86		11
			93		6.7
			82		17

[a] Using benzene at 25°, 1 atm H$_2$.

Augustine and Van Peppen (709, 1474) have recently stated that previous work reporting isomerization or deuterium exchange has involved benzene-ethanol media, while in benzene or benzene-hydrocarbon media straight-forward hydrogenation or deuteration occurred. From the data presented in this section this is seen to be generally true, although exchange has been observed in benzene alone; for example, by Odell and coworkers (678), Hussey and Takeuchi (667), and Simon and Berngruber (680). Augustine and Van Peppen (709, 1474) have shown that considerable isomerization of hept-1- and 2-enes occurs under mild hydrogenation conditions when using RhCl(PPh$_3$)$_3$ (82) in benzene-ethanol, but that no isomerization was observed if the dihydride 84 was used in the mixed solvent, or if either catalyst was used in benzene alone. The conclusions were that rapid transfer of both hydrogens from the dihydride precluded isomerization, and that the dihydride was readily formed in benzene systems from 82 but not in benzene-ethanol, where some different catalytic intermediate formed with the olefin was thought to be involved. Indeed, the rate of hydrogenation of 1-methyl-4-t-butylcyclohexene to the cis- and trans-cyclohexanes in benzene-ethanol was independent of pressure from 1 to 90 atm when using **84**, but was dependent when using **82**. These workers seem to suggest that the isomerization catalyst in the alcoholic media likely results from the presence of traces of oxygen, because they found that isomerization was further enhanced in benzene-ethanol if the dihydride was treated with 1 mole of oxygen. On the other hand, the oxygen treatment in benzene alone did not yield the isomerization catalyst. The presence of oxygen in benzene-ethanol solutions of RhCl(PPh$_3$)$_3$ gives oxidized species which are effective hydrogenation and isomerization catalysts (711, 712). The finding by Augustine and Van Peppen (710) that the catalyst does not dissociate in benzene-ethanol (Section B-1) is important. The role that the undissociated species might play in the isomerizations observed in ethanol is not clear, but the dissociated phosphine in benzene solution is thought to inhibit isomerization (see Section B-8). Benzene-ethanol solutions of RhCl(PPh$_3$)$_3$ did react with hydrogen to give H$_2$RhCl(PPh$_3$)$_2$, but this product was thought to be formed by the addition of hydrogen *followed* by phosphine dissociation.

5. Reduction of Olefinic Substrates. Carbon-carbon double bonds of a large number of olefinic substrates have been reduced by using the RhCl(PPh$_3$)$_3$ catalyst. Tables 46 to 48 summarize all such reductions that have not already been listed in Tables 41 to 45 (see also Section B-11). Benzene or benzene-ethanol solutions have commonly been used under mild conditions for selective reduction of an olefinic linkage. Some substrates are reduced quite slowly because of steric restrictions, and some compounds reportedly not reduced, particularly involving trisubstituted olefinic bonds, may in fact be hydrogenated but very slowly (Section B-3). The deuterium studies discussed in Section B-4 clearly indicate the possibilities of specific *cis*-deuteration without scrambling, and such additions are also noted in Tables 46 to 48.

A particular advantage of the RhCl(PPh$_3$)$_3$ system is that under mild conditions it does not reduce other functional groups present in the olefinic substance. Groups reported unaffected include keto, hydroxy, cyano, nitro, chloro, azo, ether, ester, carboxylic (acid) (612, 647, 650, 659), although cinnamyl chloride does undergo some hydrogenolysis to propenylbenzene (Table 46), and saturated aldehydes have been reduced to alcohols at higher pressures (Section B-5b).

a. Simple Olefins, Cyclic Monoenes and Dienes, Terpenes, and Exocyclic Methylene Groups. Table 46 lists some reductions of carbon-carbon double bonds of simple olefins, cyclic monoenes and dienes, terpenes, and exocyclic methylene groups. Terminal olefins up to 1-eicosene (C$_{20}$H$_{40}$) have been hydrogenated and deuterated. Birch and Walker (647) have shown that of

Table 46 Hydrogenation (and Deuteration) of Olefinic Bonds Catalyzed by RhCl(PPh$_3$)$_3$[a]

Substrate	Product	Reference
Ethylene	Ethane	636
Propene	Propane	612, 636, 637, 702
Pentenes	Pentane	643
Hex-1-ene	Hexane	699, 1453
Hept-1-ene	Heptane	234, 612, 709, 1453, 1474
Hept-2-ene	Heptane	709, 1453, 1474
Hept-3-ene	Heptane	1453
Oct-1-ene	Octane	243, 650, 657
Dec-1-ene	Decane 1,2-d_2	650, 714
cis- and trans-Dec-n-ene (n = 2, 3, 4, and 5)	Decane n, (n + 1) − d_2	714
C$_{11}$–C$_{20}$ terminal olefins	Paraffins 1,2-d_2	714
Dodec-1-ene	Dodecane	650
Cyclohexene	Cyclohexane 1,2-d_2	647, 667, 668, 672, 678
Cycloheptene	Cycloheptane	1453
1-Methyl-4-t-butylcyclohexene	cis- and trans-1-methyl-4-t-butylcyclohexane	709, 1474
4-Methylcyclohexene	4-Methylcyclohexane	680
Norbornene	[2,2,1]Bicycloheptane 2,3-d_2	668
Butadiene polymer	Partially reduced polymer	701
1-Methoxycyclohexa-1,4-diene	1-Methoxycyclohexene, cyclohexanone	647
1-Methoxy-5-methylcyclohexa-1,4-diene	3-Methylcyclohexanone	647
2,5-Dihydrobenzylamine toluene-p-sulfonate	Cyclohexylmethylamine toluene-p-sulfonate	647
2,3-Dihydropyran	Tetrahydropyran	647
Thebaine (97)	8,14-Dihydrothebaine	647
Substituted norbornadienes		1458
α-Phellandrene	No reduction	647
Isotetralin	9,10-Octalin(80%),1,9-octalin	656
1,4-Dihydrotetralin	9,10-Octalin(80%),1,9-octalin	656

Table 46 (*Continued*)

Substrate	Product	Reference
Methyl 1,4,5,8-tetrahydro-1-naphthoate	9,10-Octalin ester	656
cis-4,7,8,9-Tetrahydro-indan-1-one	*cis*-Hydrindan-1-one 5,6-d_2	715
cis-8-Methyl 4,7,8,9-tetra-hydro-indan-1-one	*cis*-8-Methyl-hydrindan-1-one 5,6-d_2	715
(+)-Carvone (**98**)	Carvotanacetone	647
Geraniol (**103**)	Some reduction at both double bonds	647
Nerol	Some reduction at both double bonds	647
Neryl acetate	3,7-Dimethyloctyl acetate	647
Linalool (**102**)	2,2-Dihydrolinalool	647
Lupenyl acetate	Lupanyl acetate	650
γ-Gurjunene (**99**)	Isopropylidene group reduced	649
(+)-*trans*-2-Methyl-*p*-mentha-2,8-diene (**100**)	(+)-*trans*-2-Methyl-*p*-mentha-2-ene	653
Eremophilone (**101**)	13,14-Dihydroeremophilone	655
4-*t*-Butylmethylenecyclohexane	4-*t*-Butylmethylcyclohexane (*cis* and *trans*)	644
Damsin	Isodamsin, dihydrodamsin	642, 1126
Psilostachyine	Dihydro derivative	642
Confertiflorin	Dihydro derivative	642
Coronopilin	Dihydrocoronopilin	1126
A diene coronopilin	Dihydrocoronopilin	1126
104	Norketone from β-cubebene	652
Styrene	Ethylbenzene (1,2-d_2)	671, 684, 685, 1453
β-Bromostyrene	β-Bromoethylbenzene	1453
Cinnamyl chloride	Phenylpropyl chloride, propenylbenzene, propylbenzene	647
Allyl bromide	Propyl bromide	699
Methyl oleate	Methyl stearate 9,10-d_2	647
Methyl linoleate	Methyl stearate 9,10,12,13-d_4	647
Prostaglandin E_2	Prostaglandin E_1	1460
Tetrafluoroethylene	No reduction	612
Tetracyanoethylene	No reduction	612

[a] In benzene, toluene, benzene-ethanol, or ethanol at about 25° and 1 atm H_2.

some cyclohexa-1,4-dienes, a 2,5-dihydrobenzylamine compound was completely reduced, while 1-methoxy derivatives took up only 1 mole of hydrogen but gave some cyclohexanone products as well as the expected cyclohexene derivative. On the other hand, α-phellandrene (5-isopropyl-2-methylcyclohexa-1,3-diene) was not hydrogenated possibly because of complexing through the diene; however, the more complex conjugated diene thebaine (97) was readily hydrogenated. Kheifits and coworkers (1458) have reported on the hydrogenation of substituted norbornadienes but details are not yet available.

Sims and coworkers (656) have studied the reduction of some 1,4-dihydroaromatic compounds and have demonstrated selective reduction of two disubstituted double bonds in the presence of a tetrasubstituted one; for example, some tetralin derivatives to 9,10-octalin compounds. With isotetralin and 1,4-dihydrotetralin, some isomerization product, 1,9-octalin, is also formed. The use of heterogeneous catalysts for these tetralin reductions gives complex mixtures containing aromatic compounds.

The isopropenyl group in carvone (98), γ-gurjunene (99), *trans*-2-methyl-*p*-mentha-2,8-diene (100) and eremophilone (101), and the vinyl group in linalool (102) were specifically reduced. Carbon monoxide abstraction occurred from geraniol (103) and its *cis*-isomer, nerol, to give the well-known and relatively inactive complex RhCl(CO)(PPh₃)₂ (Section XI-E), but some nonselective reduction occurred. Neryl acetate was reduced completely without loss of carbon monoxide. The isopropenyl group in 104 has been reduced to give a ketone derived from β-cubebene. The exocyclic methylene groups in methylenecyclohexanes, psilostachyine, confertiflorin, and coronopilin are readily reduced; these compounds are structurally similar to damsin **88** (Scheme 10 in Section B-4). A methylene reduction was also involved in the stereoselective total synthesis of seychellene (1113).

A stereoselective hydrogenation of the double bond in **104a** to a single isomer has been carried out (658).

The few miscellaneous reductions listed at the end of Table 46 include the selective deuteration of oleate and linoleate. Koch and Dalenberg (1460) have synthesized tritium-labeled prostaglandin E$_1$ by catalytic tritiation of prostaglandin E$_2$ (this involves reduction of a 5-heptenoic acid moiety). Tetrafluoro- and tetracyanoethylene are not reduced; they form relatively stable complexes, but these ligands are strong π-acids and withdraw electron density from the metal. The so-called promotional energy (61) required for dihydride formation, which involves oxidative addition to a d^8 rhodium(I) system to give a d^6 rhodium(III) system, is increased, so that the species cannot activate molecular hydrogen (612) (see Section XII-A-1 for further discussion on this point).

Rony (702, 721) has recently described the dispersion of hydrocarbon solutions of RhCl(PPh$_3$)$_3$ in porous solids to provide a catalyst bed for hydrogenation, as well as for isomerization, polymerization, hydroformylation, hydration, and oxidation of simple olefins such as propene.

b. *α,β-Unsaturated Carboxylic Acids, Esters, Nitro Compounds, Nitriles, Ketones (Including Quinones), and Aldehydes.* The title compounds generally seem to be readily hydrogenated in the expected manner (Table 47), although more severe conditions are required for internal and highly substituted carbon-carbon double bonds. Steric factors presumably prevent the reduction of 1-menthyl-α-phenylcinnamate, 3,4-diphenyl-3-buten-2-one, and 2,3-diphenylacrylonitrile. Besides the quinones listed, others of higher oxidation potential, such as diphenoquinone, β-naphthoquinone, and 2,6-naphthoquinone, destroy the catalyst and are not reduced (666). Santonin (**105**) is readily reduced selectively to dihydrosantonin A (**106**), and the keto ester **107** is similarly reduced to a substituted octalone.

105   106   107

The reduction of α,β-unsaturated aldehydes has been studied by Jardine and Wilkinson (660), Harmon and coworkers (659), and Birch and Walker (647). The usefulness of the catalyst, however, is limited by the accompanying decarbonylation of the aldehyde to the corresponding saturated hydrocarbon (e.g., Refs. 661–663). This reaction is more pronounced for the more hindered and slowly reduced aldehydes such as but-2-enal and *trans*-2-methylpent-2-enal. Citral (the aldehyde corresponding to **103**) was not reduced because of decarbonylation. Decarbonylation is reported to be minimized when using absolute ethanol as solvent; up to 60% of some reduced cinnamaldehydes have been recovered (659). This incidentally gives some support to the finding that RhCl(PPh$_3$)$_3$ is undissociated in ethanol (Section B-4), since the active decarbonylation catalyst is the bis(triphenylphosphine) species (662). Higher pressures of 55 atm increase the aldehyde hydrogenation rate, but for but-2-enal about 18% conversion to the saturated alcohol was then observed. This additional reduction may be due to activation of hydrogen by the rhodium product of the decarbonylation, RhCl(CO)(PPh$_3$)$_2$ (Section XI-E).

A mixture of 1,4-dicyanobutenes has been reduced to adiponitrile by

Table 47 Hydrogenation of Olefinic Bonds Catalyzed by RhCl(PPh$_3$)$_3$[a]

Substrate	Product	Reference
Maleic acid	Succinic acid (meso-2,3 d_2)	612
Fumaric acid	Succinic acid (*DL*-2,3 d_2)	612, 650
Dimethyl maleate	Dimethyl succinate	1453
Dimethyl fumarate	Dimethyl succinate	680, 1453
Ethyl acrylate	Ethyl propionate	1453
2-Pentenoic acid	*n*-Valeric acid	679
Itaconic acid	α-Methylsuccinic acid	659, 696
Citraconic acid	α-Methylsuccinic acid	659
Cinnamic acid	Hydrocinnamic(β-phenylpropionic) acid	650, 659, 696
p-Methylcinnamic acid	*p*-Methylhydrocinnamic acid	659, 696
α-Methylcinnamic acid	α-Methylhydrocinnamic acid	695
α-Phenylcinnamic acid	2,3-Diphenylpropionic acid	659
α-Acetamidocinnamic acid	*N*-Acetyl-β-phenylalanine	696
Ethyl cinnamate	Ethyl hydrocinnamate	659
Benzyl cinnamate	Benzyl hydrocinnamate	650
1-Menthyl-α-phenylcinnamate	No reduction	659
α-Phenylacrylic acid	2-Phenylpropionic acid	696
β-Nitrostyrene	2-Phenylnitroethane	647
p-Nitro-β-nitrostyrene	2-(*p*-Nitrophenyl)nitroethane	659
3,4-Methylenedioxy-β-nitrostyrene	2-(3,4-Methylenedioxyphenyl)-nitroethane	659, 697
3-Methoxy-4-benzyloxy-β-nitrostyrene	2-(3-Methoxy-4-benzyloxyphenyl)-nitroethane	659
Cinnamonitrile	Hydrocinnamonitrile	659
2,3-Diphenylacrylonitrile	No reduction	659

Table 47 (*Continued*)

Substrate	Product	Reference
1,4-Dicyanobutenes	Adiponitrile	698
Benzal acetone	4-Phenyl-2-butanone	659
3,4-Diphenyl-3-buten-2-one	No reduction	659
1,4-Naphthoquinone	1,2,3,4-Tetrahydro-1,4-dioxonaphthalene	666
2,3-Dimethoxybenzoquinone	2,3-Dimethoxy-1,4-dioxocyclohex-2-ene	666
Benzoquinone	Quinhydrone, quinol	666
Juglone(5-hydroxy-1,4-naphthoquinone)	β-Hydrojuglone(1,4,5-trihydroxynaphthalene)	666
Santonin (**105**)	Dihydrosantonin A (**106**)	656
107	Substituted octalone	654
Propenal	Propanal, propanol ($<5\%$)	660
But-2-enal	Butanal	660
trans-2-Methylpent-2-enal	2-Methylpentanal	660
Cinnamaldehyde	Hydrocinnamaldehyde(60%), ethylbenzene(40%)	659
o-Nitrocinnamaldehyde	*o*-Nitrohydrocinnamaldehyde(60%), *o*-nitroethylbenzene	659
α-Methylcinnamaldehyde	No reduction	659
p-Dimethylaminocinnamaldehyde	No reduction	659
Citral	No reduction	647

^{a} In benzene, toluene, or benzene-ethanol solution; about 25° and 1 atm H$_2$ except Ref. 659 (40–60°, 4–7 atm) and Ref. 698 (see Section B-5b).

using RhCl(PPh$_3$)$_3$ in the presence of added inorganic bases such as sodium carbonate, cyanide, hydroxide, hydrogen phosphate, and benzoate (698); temperatures at 35 to 175° and pressures of 30 to 350 atm were used. Sodium cyanide and carbonate were particularly effective; reduction was slow in the absence of these additives. Cuprous cyanide was less effective than sodium cyanide.

c. **Steroids.** Hydrogenations and deuterations of some steroids have been reported by Birch and Walker (647, 648), Biellman and coworkers (649, 650), Laing and Sykes (514), Wieland and Anner (515), Brodie and coworkers (516), and Djerassi and coworkers (651, 664, 665, 852); these are summarized in Table 48. Unhindered disubstituted olefins such as 1-, 2-, and 3-cholestene and Δ^1-3-keto steroids are readily reduced. 2-Cholestene readily gave 2,3-dideuterocholestane when using D$_2$ in EtOH or EtOD, but not when using H$_2$ in EtOD (649). The more hindered disubstituted olefin, 5β-pregn-11-ene-3,20-dione is less readily reduced. Ergosterol somewhat surprisingly undergoes reduction at the trisubstituted 5,6 position with

Table 48 Hydrogenation (and Deuteration) of Olefinic Bonds in Steroids Catalyzed by RhCl(PPh$_3$)$_3$[a]

Substrate	Product	Reference
2-Cholestene	5α-Cholestane(2,3-d_2)	649–651, 664
1-Cholestene	5α-Cholestane	651
3-Cholestene	5α-Cholestane	651
1-Cholesten-3-one	5α-Cholestane-3-one(1α,2α-d_2)	651
Δ^1-3-keto-5α-steroids	Saturated 3-keto-5α-steroids	651
Ergosterol	5α,6-Dihydro (or d_2) derivative	647, 657
22-Dihydroergosterol acetate	5α,6α-Dihydro (or d_2) derivative	648
5β-Pregn-11-ene-3,20-dione	5β-Pregnane-3,20-dione	651
Androsta-1,4-diene-3,17-dione	Androst-4-ene-3,17-diones (1α,2α-d_2)	651, 516
Androsta-4-6-diene-3,17-dione	Androst-4-ene-3,17-dione	651
17α,21-Dihydroxypregna-1,4-diene-3,20-dione 21-acetate	17α,21-Dihydroxypregn-4-ene-3,20-dione 21-acetate	647
Pregna-5,16-diene-3β-ol-20-one acetate	Pregnenolone acetate	647
Testosterone	17β-Hydroxy-5α-androstan-3-one	647
Δ^4-3-ketones	Very slow reduction	647, 651
4-Androsten-3,17-dione[b]	Androstan-3,17-dione(slow)	664
Δ^5-7-keto 3β-acetates	Very slow reduction	651
4-Cholestene[b]	Very slow reduction	664
4-Cholesten-3-one	No reduction	650
4-Cholesten-3-one[b]	Slow reduction	664
3β,6β-Dihydroxy-4-cholestene	No reduction	650
Cholesterol	No reduction	647, 650
Cholesteryl acetate	No reduction	650
3-Methyl-2-cholestene	No reduction	650
Dimethyl-2-cholestene-3-phosphonate	No reduction	650
4-Androstene	No reduction	651
14-Ergostene	No reduction	651
Δ$^{8(14)}$-Ergostene	No reduction	651
3α,5-Cyclo-5α-cholest-6-ene	3α,5-Cyclo-5α-cholestane	514
1α,5-Cyclo-5α-cholest-2-ene	1α,5-Cyclo-5α-cholestane	514
11α-Acetoxy-20-ethylenedioxy-Δ^4-pregnen-3-one	The pregnane-3-one derivative	515

[a] At room temperature, 1 atm H$_2$, in benzene solution (647, 648) or benzene/ethanol (649–651).
[b] Acetone solution.

neither the 7- nor 22-double bonds being affected; deuterium adds stereo-specifically *cis* to 22-dihydroergosteryl acetate to give the 5α,6α derivative. The trisubstituted 5-double bond in cholesterols and cholesteryl acetate was not reduced, unlike the similar 5-double bond of the 5,7-diene system of ergosterol. Thus, although conjugated dienes are not usually readily reduced because of stronger coordination to the catalyst (Section B-3), it seems that in more complex molecules conjugation may assist the hydrogenation process. This was also noted in the reduction of thebaine (Section B-5a). The trisubstituted 5-double bond in some cholestenes, ergostenes and 4-andro-stenes was reduced very slowly or not at all. Selective reduction at the unhindered bond of diene-keto steroids gives the corresponding monoene-keto derivatives which, however, may be reduced subsequently. For example, the 4-ene-3-one grouping in testosterone is slowly reduced to the 4,5α-dihydro derivative.

The selective hydrogenation of vinylcyclopropanes in 3α,5-cyclo-5α-cholest-6-ene, and 1α,5-cyclo-5-cholest-2-ene was achieved with the cyclo-propane ring remaining intact (514) (cf. Table 45).

Voelter and Djerassi (664) have studied the hydrogenation and deuteration of certain steroids in a range of solvents under mild conditions. The hydro-genation rates for 2-cholestene decreased in the order: acetone > ethanol-benzene (8:2) > methanol-benzene (7:3) > tetrahydrofuran > benzene. No hydrogenation occurred in acetone solution unless the hydride was first preformed in solution; this is again consistent with the mechanism via the hydride route. The isotopic product distribution after deuteration depended on the solvent and catalyst concentration. For 2-cholestene, highly specific deuteration was effected in ethanol-benzene (1:1), benzene, tetrahydrofuran, or acetone. In methanol-benzene (7:3) up to 30% of the product was only monodeuterated; this was thought to result from exchange of the dideuteride catalyst with the methanolic proton via a coordinated methanol (cf. **84**). Slow hydrogenation of 4-androsten-3,17-diene and 4-cholesten-3-one was observed in acetone; deuteration usually gave a mixture of d_0, d_1, d_2, d_3, and d_4 products. At lower catalyst concentrations, mainly equal amounts of mono- and dideuterated cholestane-3-one product could be obtained; subsequent treatment with methanolic KOH resulted in exchange of the 4α deuterium to give a specific 5α-monodeuterated product:

(337)

The same workers (665) have reported that in methanol alone a catalytic acetalization of steroid ketones takes place. Thus hydrogenation of α,β-unsaturated-3-oxo steroids yields the diacetal of the saturated ketone; for example, 4-cholesten-3-one gives the dimethylacetal of cholestane-3-one:

(338)

The catalyst for both the hydrogenation and the acetal formation is thought to be the dihydride **84** (Section B-1) with oxygen-bonded CH_3OH.

Keto groups, including those in cholesterone and 4-cholesten-3-one (650), are not reduced by the $RhCl(PPh_3)_3/H_2$ system, but highly selective and stereospecific reductions of steroid 2- and 3-ketones have been attained by using the rhodium catalyst without hydrogen in isopropanol containing trimethylphosphite at 80° (713). Orr and coworkers (713) reduced 5β-androstan-3,17-dione to the axial 3-alcohol with no reduction of the 17-keto group, and 17β-hydroxyandrostan-2-one to the axial 2β,17β-diol. The systems must involve hydrogen transfer from the alcohol, and are similar to, but even more stereospecific than, those using chloroiridate in the same conditions (Section XII-C). No reduction occurs without trimethylphosphite, which is in sharp contrast to the inactivation of molecular hydrogen by $RhCl[P(OMe)_3]_2$ (Section B-7b).

d. *Sulfur Compounds.* Although the poisoning effects of sulfur compounds have been observed (612, 657; Section B-12), these effects are not very serious, since Hörnfeldt and coworkers (517) have reduced unsaturated thiophene derivatives to the saturated thiophene at 20° and 3 to 4 atm hydrogen in benzene-hexane. Terminal olefinic groups in 5-(2-thienyl)pentene-1, 2-propylidenethiophene, and 4-(2-thenoyl)butene-1 and the conjugated olefinic bond in 2-crotonylthiophene were readily reduced. Birch and Walker (657) have reduced allyl phenyl sulfide to phenyl propyl sulfide, and they mention that the reduction rate of ergosterol (Section B-5c) is unaffected by the addition of phenyl propyl sulfide.

6.   Variation in the Halide Ligand.   The bromide and iodide are more soluble in organic solvents than the chloride. By molecular weight data, they appear to dissociate in solution and behave chemically very similarly to the chloride (612); whether traces of oxygen are necessary for dissociation is not established. The stability of the ethylene complexes decreases in the order Cl > Br > I, presumably because of increase in the trans effect of the halide, with the halide probably *trans* to ethylene (612). Assuming that the hydride route operates, this could be one factor contributing to increased catalytic activity (for cyclohexene in benzene) with $Cl < Br < I$ in a ratio of about $1:2:3$ (612, 668, 686); this is reflected by a decreasing K_2 value in eq. 332. Less predictable effects on k' for the olefin complexing step (eq. 330) must also play a role. For example, the complex $Rh(SnCl_3)(PPh_3)_3$ (681), although containing the strong π-acceptor $SnCl_3^-$ ligand with its accompanying large trans effect, is much less active than $RhCl(PPh_3)_3$ (612). The relative initial rates of hydrogenation of oct-1-ene in benzene have been given as decreasing in the order $Br > I > Cl > SnCl_3$ in the ratios $2.4:2.3:1.0:0.6$ (245). The initial rates for reduction of 1-methylcyclohexene in benzene-ethanol (3:1) when using the chloro or bromo complex are reported to be the same (608). But for pentenes in benzene-ethanol (1:1), the iodo and bromo complexes were both about twice as active as the chloro complex (643). As mentioned in Section B-2, the use of initial rates for comparison is very limited, because of the complex rate law and the number of variable parameters.

The iodo complex will slowly catalyze the hydrogenation of ethylene; this has been rationalized in terms of the weaker $RhI(PPh_3)_2(C_2H_4)$ complex dissociating sufficiently to give solvated $RhI(PPh_3)_2$, which can then generate the active dihydride catalyst (612).

The activation parameters for k' for cyclohexene reduction catalyzed by the bromo complex in benzene-hexane are $\Delta H^{\ddagger} = 19.4$ kcal mole^{-1} and $\Delta S^{\ddagger} = 4.1$ eu (637). Comparison with data for the chloride system (Table 41) indicates a greater activity due to a lower activation energy.

The bromide has been used for the reduction of 1,4-dicyanobutenes to adiponitrile in the presence of added carbonate at about 40 atm hydrogen pressure and 100° (698). The iodo catalyst has been used under mild conditions for the reduction of the steroids 1-, 2-, and 3-cholestene (651). The iodo and bromo complexes are ineffective catalysts for the reduction of unsaturated aldehydes, since their greater hydrogenation rates are more than offset by their much greater rates of carbonyl abstraction (660).

The complexes $Rh(PPh_3)_3L$, where L is hydride, methyl, phenyl, phenoxide, hydroxide, and nitrosyl are considered in Section XI-G.

7. Variation in the Group V Donor Ligand.   a.   *Phosphines. Rate Inhibition by Added Phosphines.*   The effect of the nature of the tertiary phosphine ligand L in the complex $RhClL_3$ has been studied by various groups (245; 638, 682, 687, 693; 684, 685; 667, 668; 686). The use of optically active phosphine ligands to give catalysts for asymmetric hydrogenation is considered in Section XI-C.

Catalyst solutions have generally been made *in situ* by treating a labile rhodium(I) complex, such as $[RhCl(C_2H_4)_2]_2$, $[RhCl(1,5-C_6H_{10})]_2$, or

$[RhCl(C_8H_{14})_2]_2$, with n mole equivalents of L (638, 651, 668, 682, 684, 686). For cyclohexene reduction in benzene and with L $=$ PPh$_3$, $n = 3$, the k' value at 25° was 0.25 M^{-1} s^{-1} compared to 0.32 when using the prepared RhCl(PPh$_3$)$_3$ complex (Table 41); in benzene-ethanol, the corresponding k values were 22 and 17 min^{-1} (Table 44). Essentially identical rates have been measured for reduction of hex-1-ene in benzene by Horner and coworkers (686). Candlin and coworkers (245) found that the displacement method yields more active catalysts.

Table 49 summarizes some constants obtained by analyzing rate data

Table 49   Data for the Hydrogenation of Cyclohexene Catalyzed by RhCl(PR$_3$)$_3$[a] (638)

Catalyst	k' (M^{-1} s^{-1})	$\Delta H^{\ddagger}$ (kcal mole^{-1})	$\Delta S^{\ddagger}$ (eu)	K_2/K_1[b]
RhCl[P(C$_6$H$_4$OMe)$_3$]$_3$	0.34	25.8	+26	1.75×10^{-3}
RhCl(PPh$_3$)$_3$	0.25	18.6	+1.3	2.1×10^{-3}
RhCl[P(C$_6$H$_4$F)$_3$]$_3$	0.02	—	—	2.1×10^{-3}

[a] In 80 ml benzene at 25°; 1.25×10^{-3} M catalyst, 50 cm H$_2$.
[b] Calculated from the data in Ref. 638, and a hydrogen solubility at 50 cm pressure of 2.5×10^{-3} M quoted in Ref. 612.

(Section B-1) according to eq. 332 for systems using triphenylphosphines substituted on the phenyl rings. The electron-releasing p-methoxy group will increase electron density on the rhodium atom in RhCl(PR$_3$)$_n$ and favor formation of the dihydride (Section B-5a). The lower effective oxidation state of the metal will also favor olefin complexing. Thus K_1 and K_2 are both expected to increase; indeed, the data in Table 49 indicate that K_1 increases more so, favoring a higher activity. The somewhat higher k' value suggests that increasing electron density on the rhodium(III) atom of the dihydride favors olefin coordination (eq. 330). The very low activity induced by the electron-withdrawing p-fluorine atom results from a decrease in k'. Augustine and Van Peppen (710) have suggested that the greater activity of the p-methoxy derivative is in fact due to a greater degree of dissociation of the phosphine ligand(s) compared to the triphenylphosphine complex, but these measurements were made under oxygen (Sections B-1 and B-2). Some initial rate data given in Tables 50 to 52 again indicate that electron donors parasubstituted on the aromatic nuclei (methyl, methoxy, and dimethyl-amino) enhance activity, while electron acceptors (halogeno, acetyl, phenyl, and naphthyl) decrease activity. Orthosubstituted methyl groups inhibit

Table 50 Hydrogenation of Hex-1-ene and Cyclohexene Catalyzed by RhClL$_n$

	Hex-1-ene		Cyclohexene
	Initial Rates,[a] (ml min^{-1})		Relative Rates[b]
Phosphine	L/Rh = 3.0,	L/Rh in parenthesis	L/Rh = 3.0
Triphenyl	16		1.0
Tri(p-tolyl)	29	38 (2.4)	
Diphenyl-p-dimethylaminophenol	—	56 (2.7)	
Tri(p-methoxyphenyl)	33	47 (2.5)	1.7
Tri(p-chlorophenyl)		1.7 (2.2)	0.04
Tri(p-fluorophenyl)	—	—	0.21
Diphenyl-p-acetylphenyl		~6 (2.2)	
Phenyl-bis(p-biphenylyl)		~1 (2.2)	
Tri(α-naphthyl)		<0.1 (2.2)	
Diphenylethyl	0.2	10 (2.0)	0.23–0.14
Tri(n-butyl)	0.2	2.5 (2.2)	0.10
Trismorpholino		~1 (2.2)	
Ligand L			
Triphenylarsine			0.03
Triphenylphosphite			0.0

[a] 0.5 M hex-1-ene in 20 ml benzene, 30°, 1 atm H$_2$, [Rh] = 5 × 10^{-3} M (686).
[b] 1.0 M cyclohexene in THF, 22°, 1 atm H$_2$, [Rh] = 10^{-2} M (245).

activity, presumably for steric reasons. A maximum activity is commonly observed for a ligand/rhodium ratio of about 2. Under these conditions the labile solvated coordination site in RhClL$_2$(S) or H$_2$RhClL$_2$(S) cannot be occupied by phosphine (see below).

The trend in selectivity toward alkenes (alk-1-ene > cycloalkene > *cis*-alk-2-ene > *trans*-alk-2-ene) is independent of the phosphine (Table 51) and this seems reasonable if reaction 330 is rate-determining (687).

Initial rate data for some other phosphine systems are also given in Tables 50 to 53. In contrast to the findings above, replacement of phenyl groups by more basic ethyl, butyl, or cyclohexyl groups gives decreasing activity for the RhCl(PR$_3$)$_3$ complexes. Molecular weight data (638) show, however, that while the diphenylethyl complex dissociates completely in solution to the bisphosphine complex, the dihydride formed therefrom reassociates the phosphine ligand to give an inactive, isolable hexacoordinate species:

$$H_2RhCl(PEtPh_2)_2 + PEtPh_2 \xrightleftharpoons{K_3} H_2RhCl(PEtPh_2)_3 \qquad (339)$$

In agreement with this, when only 2 moles of phosphine are used, a marked

Table 51   Hydrogenation of Olefins Catalyzed by RhClL$_n$ (638, 682, 687, 693)^a

Phosphine	Cyclohexene L/Rh = 3,	L/Rh = 2	Hex-1-ene L/Rh = 2	cis-C$_6$H$_{12}$^{b} L/Rh = 2	trans-C$_6$H$_{12}$^{b} L/Rh = 2
Tri-p-methoxyphenyl	19.8	68.0	99.5	34.7	8.23
Tri-p-tolyl		58.8	85.3	29.0	6.85
Triphenyl	12.4	28.1($>$17.9^c)	38.9	14.1	3.12
Diphenyl ethyl	1.8	18.5	17.5		
Tri-p-fluorophenyl		3.93	5.78	1.75	0.35
Phenyl diethyl	0	1.7			
Tri-p-chlorophenyl		1.20	1.58	0.44	0.10
Triethyl	0	0.5			
Tri-2-phenylethyl			1.39		
Tri-o-tolyl			0.11		
Tri-2,3-dimethylphenyl			0.12		
Tri-2,4,6-trimethyl-phenyl			0.09		
Ligand L					
Triphenylarsine	~0		4.63		
Triphenylstibine	~0		2.59		
Triphenylphosphite			0.02		

^{a} In 80 ml benzene at 25°; 0.6 M substrate, 1.25 × 10^{-3} M catalyst, 50 cm H$_2$. Initial rates in ml min^{-1}.
^{b} 4-Methylpent-2-ene.
^{c} Inactive dimer pptd.

Table 52   Hydrogenation of Styrene Catalyzed by RhClL$_n$ (684)^a

Catalyst(L)	Initial Rate (M min^{-1}) L/Rh = 2,	L/Rh = 3
Triphenylphosphine	0.38(0.57)	0.27(0.37)
Diphenylcyclohexylphosphine	0.39	0.12
Phenyldicyclohexylphosphine	0.09	0.16(0.37)
Tricyclohexylphosphine	0.024(0.08)	— (0.07)
Diphenylbornylphosphine	—	0.02(0.06)
Tri-o-tolylphosphine	—	— (0.04)
Phenyldiisobutylphosphine	0.017(0.06)	—
Diphenylbenzoylphosphine	0.014	0.01(0.06)
Trithienylphosphine	0.006	— (0.04)
Tri(cyanoethyl)phosphine	0.002	0.002
Triphenylarsine	0.40	0.34(0.56)
Triphenylstibine	0.017	0.005(0.05)

^{a} [Rh] ≃ 8 × 10^{-3} M, 20 ml styrene, 4 to 6 ml benzene, 40°.
Values in brackets refer to 60°.

increase in activity is observed (Tables 50, 51, and 53). Under these conditions, however, dimerization to the inactive complexes $(R_3P)_2RhCl_2$-$Rh(PR_3)_2$, **83**, may occur, which was sometimes observed for the PPh$_3$ system (Table 51). Wilkinson's group (638) consider that the more basic σ-donor phosphines favor reassociation in the higher oxidation state and lead to lower activity. The inhibition of catalysis by addition of excess PPh$_3$ in the RhCl(PPh$_3$)$_3$ systems (612) was also attributed to the reassociation equilibrium 339 (638). Inclusion of an equilibrium such as 339 into Scheme 9 (Section B-1) would introduce a further term, $K_1K_3[H_2][phosphine]$ into the first denominator term of eq. 336. Similar adjustments can be made to modified rate laws, if the initial complex is undissociated RhCl(PPh$_3$)$_3$.

There remains the question why the alkylphosphines are less efficient than the arylphosphines under conditions where no reassociation can occur $(PR_3/Rh = 2)$. It was suggested that the hydrogen transfer step (eq. 333) rather than olefin complexation (eq. 330) might now become the rate-determining step, with steric factors playing some role (638). The fact that the hydrogen is not readily removed by pumping out solutions containing H$_2$RhCl(PEtPh$_2$)$_3$ gives indirect evidence for this suggestion.

Again there is the problem of drawing definite conclusions from initial rate data. For example, it is possible that higher K_2 values for the alkylphosphine complexes could be the major factor determining the low rates. The inclusion of a K_3 term in the rate law further complicates the issue.

For reduction of styrene, replacement of phenyl groups of triphenylphosphine by *o*-tolyl, isobutyl, cyanoethyl, benzoyl, thienyl, or bornyl

Table 53 Hydrogenation of Cyclohexenes ($\sim$1.0 M) Catalyzed by RhClL$_n$ (667, 668)^a

Catalyst (L)	L/Rh	Initial Rate (10^6 moles min^{-1})	
		Cyclohexene	1-Methylcyclohexene
PPh$_3$	3.0	233	5.3
	2.5	405	8.8
PEtPh$_2$	2.7	94	3.2
	2.1	166	3.5
PPh(Pip)$_2$^{b}	3.0	300	0
PPh$_2$(Pip)	3.0	670	0
AsPh$_3$	3.0	1.2	0
	2.5	2.0	0

^{a} [Rh] $= 2.4 \times 10^{-3}$ M, 4.5 ml benzene-ethanol (3:1), 25°, 1 atm.
^{b} Pip = piperidyl.

groups also markedly decreased catalytic activity. The catalysts generally remained stable at higher temperatures around 60°, where increased activity was observed (Table 52).

Stern and coworkers (684, 685) have measured the hydrogenation rates of styrene in styrene-benzene (~4:1) by using a range of aminophosphines of the type Ph_2PA, $PhPA_2$ or PA_3 where A is a secondary amine. The initial rate data are summarized in Table 54. Tables 50 and 53 also include some data on such systems. Substitution of one and particularly two, but not three, phenyl groups of PPh_3 by basic amines, especially piperidine, pyrollidine, and morpholine, increases activity. The more weakly basic amines (indole, methylaniline, and valerolactam) inhibit activity almost completely (Table 54). The low activity of the diisopropylamino phosphine systems (Table 54) has been attributed to the sterically hindered catalyst; more specifically, the

Table 54   Hydrogenation Rates of Styrene Using $RhClL_n$ (684, 685)[a]

Aminophosphine	Initial Rate (M min^{-1}) L/Rh = 2, L/Rh = 3		Initial Rate (L/Rh) (M min^{-1})
Phenyldipiperidyl	1.12	1.11	1.10(15)
Phenyldipyrollidyl	1.0	0.64	
Phenyldimorpholino	0.72	0.62	
Phenyldihexamethyleneimino	0.70	—	0.60(3.5)
Phenylbisdimethylamino	0.66	0.61	0.58(5)
Phenylbisdiethylamino	0.60	0.58	0.33(14)
Phenyldiisopropylamino-piperidyl	0.03	0.03	0.03(6)
Triphenyl	0.44	0.39	0.05(4)
	0.38	0.27	
Diphenylhexamethyleneimine	0.84	—	
Diphenylpiperidyl	0.59	0.43	0.37(4.5)
Diphenylpyrollidyl	0.67	0.38	0.23(3.5)
Diphenylmorpholino	0.41	0.34	
Diphenyldiisopropylamino	0.04	0.05	0.05(4.5)
Diphenylindolyl	0.02	0.03	
Diphenylmethylaniline	0.015	0.0	
Diphenylvalerolactam	0.0	0.0	
Trisdibutylamino	—	0.03	
Trispiperidyl	0.10	—	
Trisdiethylamino	0.02	0.04	
Diethyldiethylamino	0.45	0.37	0.01(6)

[a] L = aminophosphine; [Rh] = 8.35 or 10 × 10^{-3} M, 20 ml styrene, 4 to 6 ml benzene, 40°, 1 atm.

inactivity of the phenylpiperidyl catalysts for the reduction of 1-methyl-cyclohexene (Table 53) has been attributed to steric interaction with the methyl group of the substrate (668). If A is a nonhindered, sufficiently basic amine and R is an alkyl group, one can establish from the data of Tables 50 to 54 the following general activity trends: PPhA$_2$, PPh$_2$A > PPh$_3$ > PA$_3$ $\geqslant$ PR$_3$; PPh$_3$ > PPh$_2$R $\approx$ PR$_2$A > PPhR$_2$ > PR$_3$. The relative activity of PPhA$_2$ and PPh$_2$A (and presumably other groups) may depend, however, on the substrate and solvent system used. For styrene in benzene, the phenyl-dipiperidyl system is about twice as active as the diphenylpiperidyl one (Table 54); while for cyclohexene in benzene-ethanol (3:1), the converse is true (Table 53).

Several aminophosphine complexes RhClL$_3$ were isolated; the activity of the phenyldipiperidylphosphine complex for hydrogenation of cyclopentene in benzene was identical to that for the system prepared *in situ* for a L/Rh ratio of either 3 or 2:1 (Table 54). Interestingly, in this particular phosphine system, styrene reduction is not inhibited by adding phosphine up to a L/Rh ratio of 15; for cyclopentene reduction only a 10% reduction in rate occurs on adding a twentyfold excess of this phosphine (685). The phenylbisdi-methyl and phenylbisdiethylamine systems behave somewhat similarly, suggesting that these three phosphines coordinate very weakly to rhodium(III) (eq. 339) and possibly to rhodium(I) (eq. 327). Stern and coworkers (685) suggest that such systems should be less sensitive to impurities. The other aminophosphines show some inhibition on increasing the L/Rh ratio, but not to the marked extent of the triphenylphosphine system, Hydride and ethylene complexes were isolated for the phenyldimorpholino and phenyl-dipiperidylphosphine systems; the ethylene complex with the latter, RhCl[PhP(NC$_5$H$_{10}$)$_2$]$_3$C$_2$H$_4$, was active for the hydrogenation of cyclo-pentene.

Aminoarsine and aminostibine complexes are reported by Stern and Sajus (810) to be less effective than the phosphine derivatives.

We have seen how inhibition of hydrogenation by added phosphine can arise generally from the effects of equilibrium 327 and/or 339, and that such effects may also manifest themselves in the dependence on catalyst concen-tration (612, 687). For example, with increasing amounts of RhClL$_3$, the concentration of free phosphine will increase (see eq. 336). Figure 5 shows the dependence on concentration of RhCl(PPh$_3$)$_2$ and RhCl(PPh$_3$)$_3$ for reduction of hex-1-ene (687). A linear dependence is observed for both up to 5×10^{-3} M, except below about 0.5×10^{-3} M. The higher rate here for the bisphosphine complex (and presumably the tris complex) was attributed to further dissociation to an even more active monophosphine complex:

$$\text{RhCl(PPh}_3)_2 \rightleftharpoons \text{RhCl(PPh}_3) + \text{PPh}_3 \qquad (340)$$

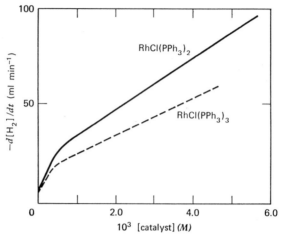

Fig. 5   Hydrogenation of hex-1-ene, 0.6 M in benzene at 25°, 50 cm H_2, 80 ml solution (687).

The dependence is consistent with a rate $R = k_a[RhCl(PPh_3)] + k_b[RhCl(PPh_3)_2]$ where $k_a > k_b$ (cf. the $HRh(CO)(PPh_3)_3$ catalyzed systems, Section XI-D-1), although forming the monophosphine complex *in situ* (L/Rh = 1) did not give a highly active species (687). The interpretation of the dependence on $RhCl(PPh_3)_3$ is not obvious particularly in view of the uncertainty of equilibrium 327 and its sensitivity toward oxygen. The linear region, which again describes activity of the bisphosphine, might in fact be part of a gradual curve that would be expected if equilibrium 327 or 339 plays a role at these concentrations. One of these must be involved in order to account for the difference in slopes of the two linear dependences. The observed linear regions indicate that dimerization to inactive species is not important, although one paper (638) reported some dimerization at 1.25 × 10^{-3} M for the bisphosphine complex in a cyclohexene system. More quantitative data on the various dissociation reactions (eqs. 327, 339, and 340) would be required for a detailed understanding of catalyst dependence. It is possible that addition of 2 or 3 moles of phosphine to the starting rhodium(I) chloride dimers may not always cleave the chloride bridges.

b.   Triphenylarsine, Triphenylstibine, and Phosphites.   Some aspects of the hydrogenating ability of $RhCl(AsPh_3)_3$ and $RhCl(SbPh_3)_3$ have been noted (668, 682, 684, 687). The complexes behave similarly to the triphenyl-phosphine complex (682). Dihydride formation with the arsine complex, however, seems irreversible at ambient temperatures. The hydrogen is not transferred to hex-1-ene, and very low activity has been reported for the hydrogenation of cyclohexene in benzene, THF and benzene-ethanol

(Tables 50, 51, and 53). However, the complex is reported more efficient than the phosphine complex for the reduction of styrene in a styrene-benzene (4:1) system (Table 52). The tris(triphenylstibine) complex is the least stable of the three tris complexes (682); solutions of the dihydride are inefficient for catalytic reduction of cyclohexene (Table 51) and styrene (Table 52).

The bis(arsine) and bis(stibine) complexes are more active than the corresponding tris complexes (Tables 51 and 52), which has been attributed (638, 687) to the absence of reassociation in the hydride (eq. 339).

The complex $H_2RhCl(AsPh_3)_2$ readily exchanges with triphenylphosphine in solution, the equilibrium lying well to the side of the phosphine hydride (682).

The triphenylphosphite complex $RhCl[P(OPh)_3]_3$ does not activate hydrogen (612); this was attributed partly to the higher π-acidity of the phosphite compared to the phosphine increasing the promotional energy of the metal (Section B-5a). Furthermore, the square-planar complex does not dissociate in solution. This is unexpected considering the trans effect, and it was rationalized in terms of less steric hindrance between *cis*-phosphite ligands compared to the phosphine ligands (612). But the effect of trace oxygen promoting phosphine dissociation (Section B-1) could account for the difference.

The importance of the π-acceptor properties is shown by the fact that the bisphosphite complexes formed *in situ* from $P(OMe)_3$, $P(OEt)_3$, $P(OPh)_3$, and $P(OCH_2CH_2Cl)_3$ did not form hydrides or give active catalyst solutions for cyclohexene reduction (638); extremely low activity has been observed with the bis(triphenylphosphite) complex for hex-1-ene reduction (Table 51; 687).

Fluorophosphine-rhodium complexes of the type $RhClL_3$, $(RhClL_2)_2$, and $RhCl(PPh_3)_2L$, where $L =$ fluorophosphine (794, 795, 845), would also be expected to be inactive.

8. Effect of Oxygen, and Oxidized Catalysts.

van Bekkum and coworkers (672) reported that addition of oxygen or hydrogen peroxide (0.2–1.0 molar ratio to Rh) accelerates the catalytic hydrogenation of cyclohexene in benzene by a factor of up to about 1.7. Factors up to 4 were observed for other unspecified olefinic substrates. The oxidants were thought to promote phosphine dissociation by removing the phosphine as triphenylphosphine oxide which was detected (672, 687); this would be consistent with the findings of Shriver and coworkers (859; Section B-1).* The oxidation itself

---

* It is instructive to estimate the possible effect of oxygen. If K for reaction 327 is 10^{-6} M (cf. Section XII-A) then at 10^{-3} M the catalyst will be $\sim 3\%$ dissociated. In the presence of 10^{-4} M oxygen, assuming production of 2×10^{-4} M phosphine oxide, the dissociation is increased to $\sim 20\%$.

was thought to be catalytic at the rhodium center. $RhCl(PPh_3)_3$ has been reported to absorb 1 mole of oxygen in solution (612, 673), and a variety of such 1:1 molecular oxygen complexes, both solvated and nonsolvated, have been formulated (673–677). Augustine and Van Peppen (709–712, 1474), however, consider that the enhancement of hydrogenation rate is not due to oxidation of dissociated phosphine but to new oxidized catalysts, which in benzene-ethanol also promote isomerization (Section B-4). These workers (711, 712) isolated oxidized complexes not showing 1:1 stoichiometry, and they used them for catalytic hydrogenation. Oxidation of $RhCl(PPh_3)_3$ in ethanol yielded a complex $[Rh(H_2O)(PPh_3)Cl(O)]_n$, 108, thus:

$$RhCl(PPh_3)_3 + 3O_2 + C_2H_5OH \longrightarrow$$
$$[Rh(H_2O)(PPh_3)Cl(O)] + CO_2 + CH_3OH + 2OPPh_3 \quad (341)$$

From the reaction in benzene, the complex $Rh_2(PPh_3)_2Cl_2O_5(C_6H_6)_2$, 109, was isolated. Complex 108 in benzene-ethanol and 109 in benzene absorbed hydrogen under mild conditions with formation of cyclohexane from benzene hydrogenation, although the system became inactive after about 20 equivalents of gas had been taken up. The residues from the resulting solutions were active, however, for hydrogenation of hept-1-ene, with accompanying isomerization. In the presence of 1 mole of triphenylphosphine, 108 and 109 were not effective for benzene reduction under hydrogen. But hydrides were formed which were fifteen times more active for hept-1-ene reduction than the nonphosphine-treated residues; furthermore, no isomerization was observed. These data were considered to be consistent with the findings that isomerizations occur during hydrogenations when using $RhCl(PPh_3)_3$ in alcoholic solutions, where no dissociation to free phosphine occurs, but not in benzene even in the presence of oxygen since the dissociated phosphine is present (Section B-4).

Wilkinson and coworkers (612) have stated that oxygen inhibits hydrogenation catalyzed by $RhCl(PPh_3)_3$, and Candlin and coworkers (245) have stated that an oxygen complex itself is not a hydrogenation catalyst.

The effect of oxygen is clearly quite complex and will be solvent dependent; the extent of oxidation and the oxygen/hydrogen pressure ratio will be important factors.

Enhancement of hydrogenating ability by the presence of oxygen has also been observed for $RhCl(CO)(PPh_3)_2$ (Section XI-E) and $IrCl(CO)(PPh_3)_2$ (Section XII-A-2).

9.   Olefin Reduction by RhL_3X_3 Catalysts (L = tertiary phosphine, arsine; X = halide). Intimately related to the rhodium(I) catalyst systems $RhXL_3$, discussed in Sections B-1 to B-8, are those in which the starting catalytic materials have been the rhodium(III) compounds RhL_3X_3 (688, 695–697, 699, 700) or which have involved formation of the rhodium(I) catalysts *in situ* from rhodium trichloride with added L (686, 694).

Wilkinson's group (699, 700) have used benzene-ethanol solutions of 1,2,3-Rh(PPh$_3$)$_3$Cl$_3$ and the dimethylphenylarsine analog under mild conditions to reduce hex-1-ene, and to reduce, at 110° and 60 atm hydrogen, the saturated aldehyde heptanal to the alcohol. However, the latter system may involve the carbonyl catalyst RhCl(CO)(PPh$_3$)$_2$ (Section B-5b). Dewhirst (695) has used similar complexes with tertiary phosphine and arsine ligands containing alkyl and/or aryl groups with up to 20 carbon atoms for the hydrogenation of monoenes such as hex-1-ene and acrylonitrile; pressures up to 90 atm at 90° in inert solvents (e.g., propionitrile) were employed. Similar conditions in benzene-ethanol (1:1) have been used by Knowles and Sabacky (688) for asymmetric hydrogenation (Section XI-C), but a base such as triethylamine was also added to the extent of 3.5 moles per mole of rhodium(III) complex.

The role of the base presumably is to remove acid, similar to that in the production of the ruthenium catalyst HRuCl(PPh$_3$)$_3$ (eq. 132, Section IX-B-2b). This could, for example, involve the following reactions:

$$RhL_3Cl_3 + H_2 \rightleftharpoons HRhL_3Cl_2 + HCl \qquad (342)$$
$$HRhL_3Cl_2 \rightleftharpoons RhClL_3 + HCl \qquad (343)$$

Reactions such as 342 and 343 in aqueous solution leading to the production of labile rhodium(I) have been discussed in Section XI-A-1; tertiary L ligands and one halide could remain coordinated to give the usual rhodium(I) catalyst.

Harmon and coworkers (696) have used trichlorotris-(4-biphenyl-1-naphthylphenyl-phosphine) rhodium(III) in benzene-ethanol (1:1), with no added base at 6 atm hydrogen and 50°, to reduce α,β-unsaturated acids (cinnamic, p-methylcinnamic, α-acetoamidocinnamic, α-phenylacrylic, and itaconic); the catalyst is as effective as RhCl(PPh$_3$)$_3$ (Table 47). Even the highly substituted olefinic bond in 2,5-dimethoxy-β-methyl-β-nitrostyrene was effectively reduced (697). Harmon's group (696) favors a suggestion by Horner and coworkers (686) that the rhodium(I) intermediate is formed after *two* hydride substitution-hydrogen chloride elimination reactions (eqs. 342, 344) in contrast to the scheme of eqs. 342 and 343.

$$HRhL_3Cl_2 + H_2 \rightleftharpoons H_2RhL_3Cl + HCl \qquad (344)$$
$$H_2RhL_3Cl \rightleftharpoons RhClL_3 + H_2 \qquad (345)$$

Horner and coworkers (686) originally postulated such a scheme for the reduction of pent-1-ene using benzene-ethanol (2:1) solutions of rhodium trichloride trihydrate, tertiary arylphosphine, and triethylamine at 24°, 1 atm. A maximum activity was observed when these constituents were present in the ratio 1:2.5:2.5. Other bases such as ethoxide and dimethylphenylamine were equally effective. Tri(p-tolyl) and tri(p-methoxyphenyl) phosphines were more effective than triphenylphosphine (Section B-7a). In the reaction scheme, eq. 345 was written involving the olefin, where L could be phosphine or solvent (S):

$$H_2RhCl(PPh_3)_2(S) + \text{olefin} \longrightarrow RhCl(PPh_3)_2(S) + \text{paraffin} \quad (346)$$

Considerable isomerization to pent-2-enes occurred under these conditions, in contrast to the $RhCl(PPh_3)_3$ catalyzed reduction of alk-1-enes under similar conditions (612, 643). This was ascribed to the presence of the rhodium(III) monohydride, $HRhL_3Cl_2$. Indeed, the observed isomerization gives evidence for the presence of rhodium(III) hydrides in these rhodium(III) catalyst systems.

Production of rhodium(I) via reactions 342 and 343 seems more likely than via reactions 342 and 344. Added base does not always seem essential, and the part played by the ethanol is not clear. In the basic solvent dimethyl-acetamide, $RhCl_3, 3H_2O$ is rapidly reduced by hydrogen in the presence of triphenylphosphine to give the $RhCl(PPh_3)_3$ catalyst; no added base is required (694).

Solutions of the rhodium(III) hydride, $HRhCl_2(PPh_3)_2$, readily prepared by oxidative addition of hydrogen chloride to $RhCl(PPh_3)_3$, have been used for reduction of a hexyne-alkene mixture (Section B-11), but no data are reported for single substrate reductions. The hydride does, however, react with olefins and acetylenes to give rhodium(III)-alkyl and -vinyl derivatives, respectively (683).

Sacco and coworkers (772) have shown that the series of conversions shown in eq. 347, where $X = $ halide and $L = $ tertiary phosphine, is extremely sensitive to small changes in the σ-donor and π-acceptor properties of L.

$$RhX_3 \xrightarrow{+L} RhX_3L_3 \xrightarrow[-X^-]{+H^-} RhHX_2L_3 \underset{+HX}{\overset{-HX}{\rightleftharpoons}} RhXL_3 \quad (347)$$

The more basic phosphines systems (trialkyl) stop at the first step; the less basic triphenylphosphine one goes readily to the third step. Both of the PEt_2Ph and $PEtPh_2$ systems give rhodium(III) hydrides, but the PEt_2Ph system does so only with great difficulty. The rhodium(III) hydride with $PEtPh_2$ shows considerable thermal stability, but it can give the rhodium(I) complex. The relative activity of these RhX_3L_3 systems will be governed in part by the conversions shown in eq. 347. The hydride complexes for a range of tertiary phosphine and arsine ligands have been prepared (772, 848–850).

10. Acetylene Reduction by Rhodium(I) and Rhodium(III) Triphenylphosphine Catalysts. $RhCl(PPh_3)_3$ has been used in benzene, toluene, benzene-ethanol, and benzene-hexane at 25° and 1 atm hydrogen to reduce a number of alkynes (43, 612, 636, 647, 657, 1453) (Table 55).

Table 55   Hydrogenation of Acetylenic Bonds Catalyzed by RhCl(PPh$_3$)$_3$ Systems[a]

Substrate	Products (via Olefin)	Reference
Dehydrolinalool +1 mole H$_2$ (see **102**)	Linalool (50%), dihydrolinalool (25%), dehydrolinalool (25%)	647, 657
Hept-1-yne	Heptane	636
Oct-1-yne	Octane	636
Hex-1-yne	Hexane	43, 612, 636, 699, 1453
Hex-2-yne	Hexane	612
Hex-3-yne	Hexane	636
Acetylene	Ethane	43, 699
Phenylacetylene	Ethylbenzene	43, 636, 1453
Diphenylacetylene	Dibenzyl	43
3-Methylpent-1-yn-3ol	3-Methylpentan-3-ol	43, 699
3-Methylbut-1-yn-3-ol	3-Methylbutan-3-ol	43, 699
Pent-2-yn-1,5-diol	1,5-Pentanediol	699
Pent-2-yn-1,4-diol	1,4-Pentanediol	43
1-Ethynylcyclohexan-1-ol	1-Ethylcyclohexan-1-ol	43, 699
3-Chloropropyne	No reduction	43
Acetylene dicarboxylic acid	No reduction	43

[a] Reference 699 uses RhCl$_3$(PPh$_3$)$_3$ as the added rhodium compound.

Some quantitative data for the reduction of hex-1-yne in benzene-hexane have been reported by Wilkinson and coworkers (612); the dependence of the rate on acetylene concentration was analyzed in the same manner as for cyclohexene (Section B-1), and the same "hydride route" mechanism was postulated. As mentioned in Section B-1, however, Candlin and Oldham (636) consider that the "unsaturate route" may also operate. Partial reduction of terminal alkynes yields the monoene and saturated product, suggesting a two-stage process in which the olefin reduction is more rapid (43, 612, 636). In benzene the rate of reduction of C$_6$—C$_8$ straight chain terminal acetylenes is about 0.85 that of C$_6$—C$_{12}$ straight-chain terminal olefins (636) (see Table 43). Phenylacetylene is reduced about eight times slower than styrene in toluene solution (1453).

Internal acetylenes such as hex-2-yne are reduced slowly at 1 atm pressure, but rapidly at 50 atm (612); partial reduction gave hex-2-enes with a *cis* to *trans* product ratio >20, indicating *cis* addition as for olefin reduction. As for alkenes, steric effects likely determine the hydrogenation rate; for example, diphenylacetylene is initially reduced six times slower than phenyl-acetylene (43, 636; Table 43). 3-Chloropropyne and acetylene dicarboxylic acid were not reduced under mild conditions. Hexafluorobut-2-yne, which forms a complex with $RhCl(PPh_3)_3$ (351), is also unlikely to be hydrogenated.

In the absence of hydrogen, diphenylacetylene forms a complex $RhCl(PPh_3)_2(PhC_2Ph)$; 3-methylbut-1-yn-3-ol forms a complex $RhCl(PPh_3)_2$-$(C_{10}H_{16}O_2)$ with coordinated dimer, and the acetylene can be dimerized catalytically; phenylacetylene is also readily polymerized (707). But these reactions seem unimportant under hydrogenation conditions.

Benzene-ethanol solutions of $Rh(PPh_3)_3Cl_3$ have also been used at 50° and 1 atm to reduce acetylenes to the corresponding saturated compounds (Table 55; 699): the reduction rates decreased in the order: 3-methylbut-1-yn-3-ol > 3-methylpent-11-yn-3-ol > hex-1-yne > 1-ethynylcyclohexan-1-ol > pent-2-yn-1,4-diol > acetylene. These systems presumably involve the rhodium(I) catalyst (Section B-9).

11.   Reduction of Mixed Substrates Using Rhodium(I, III) Triphenylphosphine Complexes—Effect of Solvent.

Candlin and Oldham (636) have presented data for the hydrogenation of mixed substrates when using $RhCl(PPh_3)_3$ in benzene (Table 56) and other solvent mixtures (Table 57). One of the

Table 56   Competition Figure for Hydrogenation of Mixed Substrates Containing an Alk-1-ene (636)[a]

Substrate	Competition Figure (R^*)	Substrate	Competition Figure (R^*)
Acrylonitrile	14.7	C_6–C_{12} alk-1-enes	1.0
But-1-yne-3-ol	9.1	Cycloheptene	0.95
Diethyl maleate	6.6	Vinylcyclohexane	0.93
Allyl nitrile	4.9	Cyclohexene	0.92
Phenylacetylene	4.1	Cycloocta-1,3-diene	0.75
Methyl acrylate	3.5	Cyclooctene	0.72
Allyl alcohol	3.4	Hex-3-yne	0.71(3.6[b])
Styrene	2.6	2-Methylpent-1-ene	0.69
Ethyl vinyl ether	1.8	*Cis*-oct-2-ene	0.54
C_6–C_8 alk-1-ynes	1.7	Neo-hexene	0.48
Vinyl acetate	1.6	2,4,4-Trimethylpent-1-ene	0.41
Cis-penta-1,3-diene[c]	1.5	*Trans*-oct-2-ene	0.17
Trans-penta-1,3-diene[c]	1.3	Diethyl fumarate	0.0
Penta-1,4-diene	1.3		
Cyclopentene	1.2		
Indene	1.1		

[a] 10^{-2} M $RhCl(PPh_3)_3$ in benzene, 1 atm H_2, 22°, 0.5 M in each substrate.
[b] In 2,2,2-trifluoroethanol.
[c] The dienes are initially reduced at the terminal bond.

Table 57   Competition Figure for Hydrogenation in Mixed Solvents (636)[a,b]

Solvent	Relative Rates		Competition Figure
	Hex-1-yne	Oct-1-ene	
Benzene only	1.7	1.0	1.7(1.9[c,d])
Acetone	1.0	0.6	1.5
Amyl acetate	1.4	0.8	1.7
THF	2.1	1.2	1.8
DMF	0.35	0.2	1.8
Cyclohexanol	1.3	0.7	1.9
Methyl ethyl ketone	1.3	0.7	2.0
Octanol	3.0	1.4	2.2
DMSO	0.15	0.07	2.2
Isopropanol	2.5	1.1	2.3
Ethanol	3.0	1.1	2.6(1.7[e])
Methanol	1.6	0.6	2.6
Cyclohexanone	0.8	0.3	2.7
Ethyl acetate	1.1	0.35	3.1
Ethyl acetoacetate	0.9	0.3	3.2
Chlorobenzene	1.5	<0.2	>5.0
Phenol	3.9	f	f
2,2,2-Trifluoroethanol	~7.0	f	f(>20[c,g])
Nitromethane, acetonitrile, and acetic acid gave very slow rates. [f]			

[a] Benzene + other solvent (1:1).
[b] 10^{-2} M RhCl(PPh$_3$)$_3$, 0.5 M in each substrate, 22°, 1 atm H$_2$; in benzene alone $t_{0.5}$ for hex-1-yne and oct-1-ene = 16 and 27 min respectively.
[c] Using HRhCl$_2$(PPh$_3$)$_3$(THF) in the same conditions.
[d] $t_{0.5}$(hex-1-yne) = 90 min, $t_{0.5}$(oct-1-ene) = 160 min.
[e] 2.0 M in each substrate.
[f] Hex-1-yne completely reduced before oct-1-ene or hex-1-ene (product) is hydrogenated.
[g] $t_{0.5}$(hex-1-yne) = 2 min.

substrates was a terminal alkene, and they define a competition figure (R^*) as the ratio of hydrogenation rate of substrate to that of the terminal alkene, both being measured by product analysis. Olefinic and acetylenic substrates containing functional groups (nitrile, carboalkoxy, ether, phenyl, hydroxy) have R^* values of >1.0, which are greater than would be expected from consideration of the individual rates discussed earlier (Table 43). Alkenes that are sterically hindered (internal, branched, and most cyclic ones) have $R^* < 1.0$, while conjugated dienes have $R^* > 1.0$, although individually they are reduced slower than terminal olefins (Table 43). Similarly, all alk-1-ynes show a common rate of reduction with $R^* = 1.7$, in apparent contrast to the data of Table 43 which shows that such alkynes reduce more slowly than alk-1-enes. However, this merely indicates that individual rates of hydrogenation are the wrong criteria to use for predicting which substrate will hydrogenate the most rapidly in a mixture, and this is clear from examination of a rate law such as eq. 329. Of prime importance is the relative magnitude of the substrate complexity constants K_2 and the k' and k'' values. For example, if K_2 is larger for an acetylene, this will complex the available rhodium and could prevent olefin reduction by either the hydride or the unsaturate route. A natural conclusion is that acetylene reduction then probably proceeds via the unsaturate route, although this is not necessarily so if a hydrogen transfer step becomes rate-determining. A similar unpredictable selectivity for alk-1-yne over alk-1-ene reduction has been considered in Section IX-B-2b for a HRuCl(PPh$_3$)$_3$ system.

The rhodium(III) catalyst HRhCl$_2$(PPh$_3$)$_2$(THF) has also been used to reduce a hexyne-alkene mixture in benzene with a similar R^* value (Table 57), although the rates are six times lower than when using RhCl(PPh$_3$)$_3$. Rhodium(I) species are probably involved particularly in view of the reversible nature of the following oxidative addition reaction (683):

$$RhCl(PPh_3)_2 + HCl \rightleftharpoons HRhCl_2(PPh_3)_2 \qquad (348)$$

The data of Table 57 together with those of Table 58, which gives the relative reduction rates of oct-1-ene and hex-1-yne separately in various solvent mixtures (1:1), bear further on the question of selective reduction in a mixture of alk-1-ene and alk-1-yne. Table 58 shows that the rate of oct-1-ene reduction alone varies little on adding alcohols, although hex-1-yne reduction is greatly enhanced in the acidic alcohols. Table 57 shows that in the phenol and trifluoroethanol systems, hex-1-yne is reduced selectively to hex-1-ene before any alkene reduction occurs. Furthermore, the competition

Table 58   Relative Rates of Hydrogenation in Mixed Solvents (636)[a]

Solvent	Oct-1-ene	Hex-1-yne
Benzene only	1.0	0.9
Ethanol[b]	0.9–1.7	0.9
Phenol[b]	0.9–1.0	1.7–2.4
2,2,2-Trifluoroethanol[b]	0.9	>12

[a] 10^{-2} M RhCl(PPh$_3$)$_3$, 0.5 M substrate, 22°, 1 atm H$_2$; $t_{0.5}$(octene) in benzene = 10 min.
[b] Benzene : solvent = 1:1.

figure for hex-3-yne/oct-1-ene in the presence of trifluoroethanol was 3.6, which shows a different hydrogenation sequence to the system in benzene alone (Table 56). Only the coordinating solvents (dimethylformamide, dimethylsulfoxide, nitromethane, acetonitrile, and acetic acid) strongly inhibit hex-1-yne reduction in the substrate mixture.

As mentioned in Section B-1, Candlin and Oldham (636) have given evidence for the existence of both hydrogenation paths in Scheme 9, and they have suggested that the extent of participation of each route could explain the increase in selectivity in various solvents. The evidence for the hydride route involved stoichiometric hydrogenations using the dihydride, H$_2$RhCl(PPh$_3$)$_3$; such a reduction of a hex-1-yne/oct-1-ene mixture gave an $R*$ value of 1.9 in benzene, which compares reasonably to that of 1.7 for catalytic hydrogenation. The $R*$ values for benzene-ethanol (or methanol) systems were > 10 for the dihydride reduction and 2.6 for the catalytic reduction (Table 57). This suggests that, if it is possible to direct the catalytic reduction solely through the hydride route in the presence of alcohols, a high competition figure may result; the 2.6 value implies that both routes are occurring. These workers thus suggest that the selective reduction of hex-1-yne in the acidic alcohol systems (Table 57) may occur via the hydride route. The hydride route will be favored by use of higher hydrogen pressures or lower substrate concentrations; in agreement with this, a higher $R*$ value is observed at a lower substrate concentration (Table 57).

By analogy with data on the catalytic dimerization of hydroxyacetylenes when using RhCl(PPh$_3$)$_3$ (707) and addition reactions of acetylenes with d^8 complexes (708), Candlin and Oldham (636) suggest that the acidic alcohols might promote oxidative addition of acetylene to give a rhodium(III) acetylide by intermolecular hydrogen bonding with the chloride ligand; such an intermediate could accelerate hydrogenation of the alkyne. The uptake rate of acetylene itself by RhCl(PPh$_3$)$_3$ in benzene-solvent mixtures increased markedly in the presence of acidic alcohols, while that of ethylene decreased (636).

It is not too clear how much the finding by Augustine and Van Peppen (709, 710), that RhCl(PPh$_3$)$_3$ does not dissociate in alcoholic solvents (Sections B-1 and B-4), affects the rationalizations of Candlin and Oldham, who assumed the same solvated species RhCl(PPh$_3$)$_2$(S) in the different media. The catalyst had been pretreated with hydrogen, however, and in benzene or benzene-ethanol the dihydride, H$_2$RhCl(PPh$_3$)$_2$, was formed (710). It seems certain that the selectivity results from competition between different reaction paths, but these may involve somewhat different catalysts. The unsaturate path in the alcoholic media could, for example, involve the undissociated tris(triphenylphosphine) complex (709, 710).

12. Solvent Effects and Inhibition by Strong Donors. The solvated intermediates $RhCl(PPh_3)_2$ and $H_2RhCl(PPh_3)_2$ play key roles in the hydrogenation mechanism; the substrate likely displaces coordinated solvent from the former and, in the case of the hydride route, the latter; it is not surprising that marked solvent effects have been observed. Also, the finding that the dissociation of $RhCl(PPh_3)_3$ is inhibited by small amounts of alcohol (Section B-1) suggests that other nonhydrocarbon solvents will give similar effects. Less dissociation would be expected to lead to lower activity, but this is not necessarily so. The tris(triphenylphosphine) complex can still apparently form olefin and hydride complexes which could subsequently lose a phosphine ligand (Sections B-1, B-4, and B-11).

The work of Candlin and Oldham (636) on solvent effects has been discussed in the preceding section. Other reports by Wilkinson's group (612, 637), Hussey and Takeuchi (667, 668), Voelter and Djerassi (664), and Horner and coworkers (686) have given information on comparative rates in various media. The limited discussion presented on these obviously complex effects has not included the possibility of activity by initially undissociated species.

Some kinetic data for hydrogenation of cyclohexenes catalyzed by $RhCl(PPh_3)_3$ in different solvent systems are given in Tables 41, 44, and 59.

Table 59   Rates of Hydrogenation of 1-Methylcyclohexene Using $RhCl(PPh_3)_3$ (667, 668)[a]

Solvent	k (min^{-1})	Solvent	k (min^{-1})
Acetophenone	~0.5	Dichloromethane	0.075
Benzene-ethanol (3:1)	0.49	1,2-Dichloroethane	0.038
Nitrobenzene	0.45	Chlorobenzene	0.025
Cyclohexanone	0.39	Chloroform	0.0
Benzene	0.13	Benzonitrile	0.0

[a] 0.94 M 1-methylcyclohexene, 25°, 1 atm H_2. Initial rate $= k$[catalyst]; see Section B-3.

An ethyl methyl ketone-hexane and a pure benzene system were about twice as active as a benzene-hexane system for cyclohexene itself, because of lower activation energies (Table 41). Also, addition of polar solvents such as ethanol or ethyl methyl ketone to benzene (1:1) increases activity by a factor of about 2 (612); the inverse plots of (rate)$^{-1}$:(cyclohexene)$^{-1}$ suggest that for a hydride route, not only does k' increase but K_2/K_1 decreases (eq. 329). Wilkinson and coworkers (612) report that other effective polar cosolvents include methanol, t-butyl alcohol, isopropanol, phenol, glacial acetic acid, ethyl acetate, dimethylformamide, and dioxane and dimethyl-sulfoxide. Some of these solvents may be used alone, but low catalyst solubility is a limitation. Also, slow abstraction of carbon monoxide, particularly from alcohols, can lead to catalyst deactivation. For the same cyclohexene system, Horner and coworkers (686) have reported that alcohols, phenol, acetone, and nitrobenzene are effective cosolvents, but state that acetic acid and dimethylsulfoxide almost completely inhibit reduction. Dimethylformamide, open and cyclic ethers, nitromethane, and malonates were said to retard the reaction. Benzene-ethanol mixtures (3:1) are more effective than benzene for reduction of cycloalkenes by factors of 1.2 to 3.8, depending on the particular substrate (Table 44). Acetophenone, benzene-ethanol (3:1), nitrobenzene, and cyclohexanone were all effective for the 1-methylcyclohexene system—they were three to four times more efficient than benzene itself (Tables 44 and 59). The solvent effects may well be dependent on the substrate, since the data of Table 58 discussed in the preceding section show that the hydrogenation rate of oct-1-ene is little altered by changes in solvent.

The rate increase does not correlate with the dielectric constant of the solvent; however, the increase in rate with polar solvents could indicate an activated complex more polar than the reacting species, for example, in reaction 330. The coordinating power of the solvent is clearly important; it should be sufficient to promote the dissociation in reaction 327 or 329a (with or without oxygen), but not so strongly as to prevent later displacement by alkene (eq. 330). Solvation of a 5-coordinate alkyl hydride intermediate (**87** in eq. 333) could also play a role; and besides participation in the inner coordination sphere, solvation of halide and hydride ligands could also be important. Hydrogen solubility in the solvent is a further factor that affects the rate.

The hydrogenation (deuteration) rates and isotope product distribution from 2-cholestene in a number of solvent systems (664) have been considered in Section B-5c.

Chlorinated solvents generally do not seem satisfactory. Wilkinson and coworkers (43a, 612, 637) have shown that the dihydride is formed in high concentrations in chloroform and dichloromethane, and that transfer of hydrogen to olefins does occur. However, under catalytic hydrogenation conditions, at low concentration, hydrogen transfer to the solvent occurs with liberation of hydrogen chloride. Depending on the conditions, particularly the hydrogen pressure, the relatively inactive HRhCl$_2$(PPh$_3$)$_2$ may be formed (636, 673, 683) (see Section B-11, eq. 348, Table 57). Some data in chlorinated solvents are given in Table 59 for reduction of 1-methylcyclohexene; other workers (612, 686) have reported that chloroform and chlorobenzene are almost inactive for reduction of cyclohexenes. Hex-1-yne, however, has been effectively and quite selectively reduced in a mixture with oct-1-ene when using a benzene-chlorobenzene medium (Table 57). Studies on styrene reduction in chloroform and dichloromethane (671) have been mentioned in Section B-4.

Coordinating solvents, such as pyridine, benzonitrile, and acetonitrile, inhibit hydrogenation (612, 636, 650, 668, 686), even though dihydride formation may still occur. The strongly bound solvent in a complex such as **84** (Section B-1) could prevent activation (complexing) of the substrate (612). Oct-1-ene is still reduced to the extent of $\sim 3\%$ in benzene-pyridine (1:1) at 10^{-2} M RhCl(PPh$_3$)$_3$ under mild conditions over a period of 2 days; hex-1-yne reduction is not so strongly inhibited, and is reduced to the extent of 40% under the same conditions (636).

Other materials reported to inhibit hydrogenation include thiophene, 8-hydroxyquinoline, conjugated diolefins (612), and thiophenol (657). These may complex with the RhCl(PPh$_3$)$_{2(3)}$ catalyst to give species that do not activate hydrogen, although they could behave similarly to pyridine and acetonitrile and give rise to inactive hydride complexes. Birch and Walker (657) report that the addition of thiophenol to the catalyst system (2.5:1) in benzene reduces the hydrogenation rates of oct-1-ene, ergosterol and the acetylene, dehydrolinalool (cf. **102**), by factors up to 20, but the rates are still usable. A large excess of thiophenol severely inhibits the reductions. Sulphides do not poison the catalyst system (Section B-5d). Unsaturated thiophenes have been reduced (Section B-5d).

The inhibition by added triphenylphosphine has been considered in Section B-7a.

C. ASYMMETRIC HYDROGENATION OF OLEFINS

Knowles and coworkers (688, 689) and Horner and coworkers (690) have described an interesting development in the use of optically active rhodium complexes of the type RhClL$_n$ (where L = optically active tertiary phosphine) for asymmetric hydrogenation. The catalysts have been prepared *in situ* from [RhCl(1,5-C$_6$H$_{10}$)]$_2$ and the phosphine (689, 690) or from rhodium(III) complexes of the type RhL$_3$Cl$_3$, by treating with hydrogen in the presence of a base (686, 688, Section B-9).

The systems studied are summarized in Table 60; the reduction products all have asymmetric carbon centers. The data indicate that asymmetry in the alkyl group has a negligible effect. Assuming the usual H$_2$RhClL$_2$(olefin) intermediate and for L = methylphenyl-*n*-propylphosphine, Horner and coworkers (690) consider that with methyl and *n*-propyl groups of both phosphines in skew positions, the olefin coordinates in such a way that *cis*

Table 60 Asymmetric Hydrogenation of Olefins Using Optically Active RhClL$_n$

Phosphine[a]	Substrate	Optical Purity of Product (%)	Reference[b]
Methylphenyl-n-propyl	α-Phenylacrylic acid	21	689
	α-Benzylacrylic acid	4	689
	Itaconic acid	3	689
	α-Acetoxystyrene	1	689
	3-Methyl-2-cyclohexene-1-one	1	689
	α-Ethylstyrene	8	690
	α-Methoxystyrene	4	690
Methylphenylisopropyl	α-Phenylacrylic acid	17.5, 15[c]	688, 689
	Itaconic acid	3[c]	688
Cyclohexylmethylphenyl	α-Phenylacrylic acid	3	689
Methylphenyl-2-methylbutyl[d]	α-Phenylacrylic acid	15	689
Phenyl-bis(2-methylbutyl)[e]	α-Phenylacrylic acid	1[c]	688

[a] Optically active tertiary phosphine; L/Rh = 2.
[b] References 688, 689: ∼30 atm H$_2$, 60°, up to 0.3 mole % catalyst in benzene: methanol (1:1) with a trace of base, triethylamine. Reference 690: 1 atm, room temperature, 5 × 10^{-3} M catalyst.
[c] Catalyst prepared from RhL$_3$Cl$_3$ in benzene-ethanol (1:1).
[d] Asymmetric alkyl group as well as P atom.
[e] Asymmetric alkyl group only.

addition of hydrogen results in preferential formation of one enantiomorph.

The hydrogenation of α-phenylacrylic acid by the methylphenyl-n-propyl-phosphine system has been studied in some detail in benzene-ethanol at 30 atm H$_2$, 60° (689). This work shows the unusual effects of increase in rate, as well as product optical purity, with increasing amounts of the phosphine ligand (see Section B-7a); a maximum rate is observed for L/Rh = 8 or 16, when the optical purity reaches 23%. The excess ligand, in fact, reacts with the substrate to give the phosphobetaines **110** and **111**, and the effects are due to the presence of the carboxylate anion. Use of a salt of the olefinic acid

$$L^+CH_2CH(C_6H_5)CO_2^- \qquad [L^+CH_2CH(C_6H_5)CO_2H][CH_2{=}C(C_6H_5)CO_2^-]$$
$$\textbf{110} \qquad\qquad\qquad \textbf{111}$$

$$L = (CH_3)(n\text{-}C_3H_7)(C_6H_5)P$$

gave a thirtyfold rate increase and an optical purity up to 28%. More rapid coordination of the substrate through the nucleophilic carboxylate group was suggested. The olefinic bond would be in close proximity to the *cis* hydrogens to assist hydrogen transfer and, if coordination through this bond occurs, the resulting chelate ring could enhance steric control.

Catalysts of the type $(BH_4)RhCl_2py_2$(amide), reported by McQuillin and coworkers (691, 692) (see Section XI-H), give asymmetric hydrogenation if the solvent amide ligand is optically active. Reduction of methyl 3-phenylbut-2-enoate in (+)- or (−)-1-phenylethylformamide gave (+) or (−)-methyl 3-phenylbutanoate in better than 50% optical yield. Configurational correlations led to **112** as a representation of the alkyl intermediate. Use of (−)-lactdimethylamide, (−)-$MeCH(OH)CONMe_2$ gave a smaller induced

asymmetry.

D.   HYDRIDOCARBONYLTRIS(TRIPHENYLPHOSPHINE)RHODIUM(I), $HRh(CO)(PPh_3)_3$

1.   The Catalytic Hydrogenation Reaction and Its Mechanism.   Bath and Vaska (717) first prepared the complex $HRh(CO)(PPh_3)_3$, and Vaska (261) first reported that it catalyzed ethylene hydrogenation and hydrogen-deuterium exchange under mild conditions in benzene or toluene. The catalytic properties of the complex have since been studied in some detail by Wilkinson and coworkers (55, 234, 718–720, 722–726, 767) who have shown it effective for hydroformylation and isomerization as well as hydrogenation. The hydroformylation and reducing properties under H_2/CO atmospheres are discussed in Section F-2.

The solid, which exists in two crystalline forms (854), has a trigonal bipyramidal structure with equatorial phosphines (264, 727), and in benzene at 38°, dissociation gives a yellow square-planar species with *trans* phosphines which on dilution can lose a second phosphine (251, 261, 720):

$$HRh(CO)P_3 \overset{-P}{\underset{+P}{\rightleftharpoons}} HRh(CO)P_2 \overset{-P}{\underset{+P}{\rightleftharpoons}} HRh(CO)P \quad (P=PPh_3) \quad (349)$$

At 38° the second dissociation is apparent at $\sim 10^{-4}$ M; the bisphosphine complex predominates at 10^{-3} M, while at 10^{-2} M there is considerable reassociation to the tris complex. Reassociation also occurs at lower temperatures (720).

The H$_2$–D$_2$ exchange is thought to proceed via a trihydride species (261, 719), although there is no hydrogen absorption by solutions of the catalyst and thus equilibrium 350 must lie far to the left:

$$HRh(CO)(PPh_3)_2 + H_2 \rightleftharpoons H_3Rh(CO)(PPh_3)_2 \qquad (350)$$

HRh(CO)(PPh$_3$)$_3$ is an efficient catalyst at ambient conditions for the hydrogenation of alk-1-enes and other compounds with the grouping RCH = CH$_2$; it is more selective for terminal olefins than the RhCl(PPh$_3$)$_3$ catalyst and is about half as active. HRh(CO)(PPh$_3$)$_3$ resembles the ruthenium catalyst, HRuCl(PPh$_3$)$_3$ in many ways: general properties, selectivity, kinetics, and mechanism (Section IX-B-2b). Although the rhodium compound is <10% as active as the ruthenium compound, it has the advantage of being a substantially more stable species in the solid state (718, 719); however, it does decay to an inactive species in solution (724, Section D-2).

The kinetics of the reduction of hex-1-ene and dec-1-ene in benzene under mild conditions (718, 719) gave first-order in hydrogen and between zero- and first-order in olefin. These data are consistent with the following reactions:

$$HRh(CO)(PPh_3)_2 + alkene(A) \overset{K}{\rightleftharpoons} Rh(CO)(PPh_3)_2(alkyl) \qquad (351)$$

$$Rh(CO)(PPh_3)_2(alkyl) + H_2 \overset{k}{\rightleftharpoons} HRh(CO)(PPh_3)_2 + alkane \qquad (352)$$

The rate is then given as follows:

$$rate = kK[Rh][A][H_2]/(1 + K[A]) \qquad (353)$$

This expression is readily analyzed by plotting rate^{-1} versus [A]$^{-1}$ (see eqs. 130 and 137). The initial report (718) incorrectly suggested that eq. 351 might be rate-determining. The dependence on catalyst concentration is of the form shown in Fig. 5 for RhCl(PPh$_3$)$_n$ systems (Section B-7a). Over the linear region where the detailed kinetics were studied, the bisphosphine complex predominates and the [Rh] term in the rate law refers to this species which was approximated to the total rhodium. At very low catalyst concentration there is dissociation to the more active monophosphine species (eq. 349); the rate law is thus better written as

$$rate = k_a[HRh(CO)(PPh_3)] + k_b[HRh(CO)(PPh_3)_2] \qquad (354)$$

where $k_a > k_b$, and these are pseudo constants incorporating all the variables of eq. 353 except the rhodium term. A more exact but complex expression could be written involving the equilibrium constants of eq. 349, and this would demonstrate an inverse dependence on phosphine which was observed (cf. eq. 137 in Section IX-B-2b and eq. 336 in Section XI-B-4).

Table 61 summarizes the kinetic data obtained by Wilkinson's group. The magnitude of the K values (eq. 351) suggests that significant amounts of alkyl complex should be present under the hydrogenation conditions, but as in the case of the $HRuCl(PPh_3)_3$ systems there is no evidence of such an intermediate. Of all olefins, ethylene has the largest complexity constant with metal centers, and since $K[A]$ is likely to be >1, the high rate recorded is probably the limiting one given by $k[Rh][H_2]$; this means the ethyl complex is fully formed. The nondetection of the intermediate by n.m.r. (55, 719) may be due to the rapidity of equilibrium 351 or, less likely, the presence of a more stable ethylene hydride complex. But in either case it should be possible to estimate K for ethylene and other olefins by some spectrophotometric method; this would be useful for comparison with the kinetically determined value. The suggestion that reaction 351 with ethylene lies well to the left (55) seems inconsistent with the K values presented in Table 61.

That an alkyl intermediate is involved is evidenced by hydrogen atom exchange, studied by observing the growth of the proton resonance or hydride stretch when $DRh(CO)(PPh_3)_3$ is treated with olefins (55, 234, 718, 723). The half-life for pent-1-ene was ~ 20 s, and for cis-pent-2-ene >10 min,

Table 61 Hydrogenation of Terminal Olefinic Bonds Using $HRh(CO)(PPh_3)_3$ in Benzene (719)[a]

Substrate	Initial Rate (ml min^{-1})	kK (M^{-2} s^{-1})	K (M^{-1})
Allyl alcohol	20.3		
Hex-1-ene	16.7	100	~ 0.6
Hexa-1,5-diene	15.2		
Dec-1-ene	14.8	102	~ 0.9
Undec-1-ene	14.3		
Allylbenzene	11.1		
4-Vinylcyclohexene[b]	7.43		
Allyl cyanide	5.57		
Styrene	1.39		
Ethylene[c]	35		(See text)

<0.1 ml min^{-1} with cis-pent-2-ene, cis-hept-2-ene, cis-4-methylpent-2-ene, cyclohexene, penta-1,3-diene, 2-methylpent-1-ene, limonene, allyl phenyl ether, acrylic acid, cinnamaldehyde, 2-chloroprop-1-ene, 1-chloroprop-1-ene, and hex-1-yne.

[a] Initial rate measured when using $1.25 \times 10^{-3}\ M$ catalyst, 0.6 M substrate in 80 ml, and 50 cm H_2; all data at 25°.
[b] Vinyl group reduced.
[c] Using 30 cm of both C_2H_4(0.06 M, (612)) and H_2.

while for *trans*-pent-2-ene there was no exchange over several hours. Isomerization of alkenes was also observed for alk-1-enes and alk-2-enes (55, 234, 718, 723), and the rate was much the same under nitrogen or hydrogen. Isomerization is much slower than exchange and depending on the conditions, in particular the catalyst concentration, may be negligible compared to the hydrogenation rate, or may approach it. Excess phosphine slowed the exchange and isomerization markedly, confirming that a dissociated species is involved. Equation 351 is written in more detail below:

$$trans\text{-}HRh(CO)(PPh_3)_2 + RCH_2CH{=}CH_2$$

$$
\begin{array}{ccccc}
\text{CH}_2\text{R} & & & & \\
| & & & & \\
-\text{Rh}-\text{CH} & \rightleftharpoons & -\text{Rh}-\text{H} & \rightleftharpoons & \text{Rh}-\text{CH}_2\text{CH}_2\text{CH}_2\text{R} \\
| & & | & & \textbf{114} \\
\text{CH}_3 & & \text{CH}_2{=}\text{CH}-\text{CH}_2\text{R} & & \\
\textbf{115} & & \textbf{113} & &
\end{array}
\qquad (355)
$$

Hydride transfer to coordinated olefin is thought to involve the usual four-center transition state (219, eq. 129), hence hydrogen abstraction can only occur from a β-carbon in the metal alkyl. Isomerization must arise by abstraction from the β-methylene group of the alkyl **115**, formed by Markownikov addition of the metal hydride. The complex question of the polarity of the M—H bond and the direction of addition to olefins has been considered for cobalt carbonyl catalysts (Sections X-E-3 and F-1), where it appears that increasing acidity of the hydride leads to more Markownikov addition (more branching).

Further evidence for reactions 351 and 352 has been obtained by studying the tetrafluoroethylene system (725, 726). The hydride reacts with the olefin (at 5 atm) to give the stable, square alkyl, $trans\text{-}Rh(C_2F_4H)(CO)(PPh_3)_2$. This complex reacts with hydrogen (50°, 70 atm) in the presence of excess phosphine to give $C_2F_4H_2$ and regeneration of $HRh(CO)(PPh_3)_3$; the expected dihydride adduct could not be detected. The fluoroalkyl complex undergoes oxidative addition reactions. With hydrogen chloride, for example, the complex reversibly forms an unstable rhodium(III) hydride, which decomposes with hydrogen transfer to give the fluoroalkane and $RhCl(CO)(PPh_3)_2$. Unlike this chloro complex (Section XI-E), the fluoroalkyl readily dissociates a phosphine in solution and itself becomes an active hydrogenation catalyst for alk-1-enes at 25°, 1 atm; the system is thus comparable to that of $RhCl(PPh_3)_3$. Other complexes of the carbonyl hydride with activated olefins have been referred to, but have not yet been described (853).

Table 61 shows the high selectivity by $HRh(CO)(PPh_3)_3$ for the reduction of terminal olefins including nonconjugated diolefins such as hexa-1,5-diene. Conjugated, internal, and cyclic olefins are not hydrogenated at atmospheric pressure, and functional groups such as CHO, OH, CN, CO₂H, or O are not affected. Chloride is apparently not affected in 2-chloro- and 1-chloroprop-1-ene (719), but allyl, methallyl, and vinyl chlorides react with the catalyst to give alkene and the relatively inactive trans-$RhCl(CO)(PPh_3)_2$ (722):

$$HRh(CO)(PPh_3)_2 + CH_2\!=\!\!CH\!\!-\!\!Cl \longrightarrow$$
$$RhCl(CO)(PPh_3)_2 + C_2H_4 \quad (355a)$$

A corresponding reaction occurs with methyl iodide to give methane (722); these hydrogenolysis reactions presumably involve formation of 5-coordinate rhodium(I) or 6-coordinate rhodium(III) intermediates.

The specificity in hydrogenation was related to the exchange reactions (eq. 355), and was considered (55, 719) to arise from the steric difficulty of hydrogen transfer to form the alkyl group which is mutually cis to two trans triphenylphosphine groups (113 → 114 or 115); transfer was thought to occur readily only for a terminal olefin with resulting anti-Markownikov addition (113 → 114). Markownikov addition to RCH = CH_2 would be hindered, thus explaining the slow isomerization. Hydride addition to alk-2-enes would form either Rh—$CH(R^1)$- or Rh—$C(R^1)(R^2)$-; both would be hindered by the phosphine groups, which was consistent with the slower exchange of pent-2-ene. The polarity of the M—H bond and the presence of bulky ligands then are the main factors, which are not independent, determining the direction of addition. Unlike the $HRh(CO)(PPh_3)_2$ catalyst, the $RhCl(PPh_3)_3$ system readily hydrogenates cyclohexene (Section XI-B-1). This was attributed to the fact that in the pretransfer stage the alkene is coordinated in the species $H_2RhCl(alkene)(PPh_3)_2$ which has the phosphine ligands mutually cis and thus steric interaction is unlikely (55, 719). Clearly, the lifetime of the alkyl species must be sufficient for subsequent reaction with hydrogen (eq. 352). In agreement with these rationalizations, the specificity was decreased somewhat at high dilution where a monophosphine species is likely involved (eq. 349). Use of less hindered phosphines, for example, PMe_2Ph, is predicted to give increased rates but lower specificity, and conversely, ortho substituents on the phenyl groups should make hydrogenation and exchange more difficult (719).

The arguments above are based on variation of the "K" factor in the rate expression 353, and more specifically on a hydrogen transfer rate that is incorporated into K, together with the complexity constant for the olefin hydride complex 113, which was thought to be fairly similar for alk-1-enes, alk-2-enes, and cyclic olefins (Section X-B-3). However, it is difficult to account for the nonexchange with trans-pent-2-ene, since it will give the same alkyl species as the cis isomer which did exchange, unless there is some significant difference in the alkene complexity constant (723). There will also, of course, be variations with different alkyls in the value of k for the oxidative addition reaction 352. As in the case of the $HRuCl(PPh_3)_3$ system (Section

IX-B-2b, Scheme 4), a kinetic isotope effect $k'_H/k'_D = 1.47$ ($k' = kK$) and activation parameters ($\Delta H^{\ddagger} = 10.6$ kcal mole^{-1}, $\Delta S^{\ddagger} = -8$ eu) estimated for the hex-1-ene system (719) are considered to be consistent with a rate-determining dihydride formation followed by faster hydrogen transfer steps. A direct hydrogenolysis of the Rh—C bond was considered unlikely. Electron-withdrawing groups on the alkyl would probably decrease the oxidative addition (Section X-B-3), and could certainly contribute to an overall low hydrogenation rate as, for example, with 2-chloroprop-1-ene.

The difference in hydrogenation rates of hex-1-ene and dec-1-ene has been attributed to a difference in the nature of the olefin-solvent mixture; some of the other noted rate differences (Table 61) may be due to solvent effects. The complex is insoluble in alcohols but dissolves in halogenated solvents; a dichloromethane solution, however, is inactive for hydrogenation.

Since the complexes HRh(CO)(PR$_3$)$_3$, where R = p-FC$_6$H$_4$ and p-MeC$_6$H$_4$, probably dissociate to a lesser extent than the triphenylphosphine complex (723), they are likely to give less active systems; indeed they isomerized pent-1-ene at about half the rate of HRh(CO)(PPh$_3$)$_3$ (723).

Strohmeier and Hohmann (1459) have presented some initial rate data for reductions in toluene at 25° and 1 atm hydrogen when using 2×10^{-3} M HRh(CO)(PPh$_3$)$_3$ and 0.8 M substrate. The substrates together with the rates in M hr^{-1} × 10^3 were: ethyl acrylate, 445; hept-1-ene, 171; styrene, 134; dimethyl maleate, 59; *cis*-hept-3-ene, 13.5; *cis*-hept-2-ene, 11.0; *trans*-hept-2-ene, 8.5; *trans*-hept-3-ene, 8.0; cycloheptene, 7.0; cycloocta-1,3-diene, 6.5; dimethyl fumarate, 6.3; β-bromostyrene, 1.5. The trends are similar to those discussed by Wilkinson's group.

Ugo and coworkers (854) report that alcohol-alkoxide solutions of some rhodium trichloride-tertiary phosphine systems catalyze the hydrogenation of oct-1-ene and cyclohexene at 130° under nitrogen. The systems are thought to involve the hydridocarbonyl catalysts HRh(CO)(PR$_3$)$_3$ formed by carbonyl and hydride abstraction reactions according to some reactions shown in Scheme 11 (S = solvent):

$$RhCl_3/PR_3 \xrightarrow[base]{R'OH} RhCl(CO)(PR_3)_2 \underset{HCl}{\overset{R'ONa + H_2O}{\rightleftharpoons}} Rh(OH)(CO)(PR_3)_2$$

$$116$$

$$\downarrow R'ONa$$

$$HRh(CO)(PR_3)_3 \xleftarrow[R'ONa]{R'OH} [Rh(CO)(PR_3)_3]_2 \underset{-PR_3}{\overset{PR_3}{\rightleftharpoons}} [Rh(CO)(PR_3)_2S]_2$$

$$118 \qquad\qquad\qquad\qquad 117$$

Scheme 11

The intermediates **116** to **118** were all isolated for the triphenylphosphine system, while with the more basic phosphine systems, direct conversion to a mixture of **118** and hydride was observed. The hydride catalysts themselves (R = Et, *t*-Bu, Et$_2$Ph, MePh$_2$) were used in benzyl alcohol-benzylate solution; a maximum 35 % reduction of oct-1-ene was noted with the triethylphosphine system, and the recovered olefin was mainly *trans*-oct-2-ene. The benzylate was converted to benzoate. These workers also report that the HRh(CO)(PR$_3$)$_3$ complexes with the more basic phosphines also readily dissociate to the bisphosphine complexes in solution. The dimer **118** will be an active hydrogenation catalyst since it is also readily converted into the hydride using molecular hydrogen (854).

Acetylenes such as hex-1-yne give a brown coloration to benzene solutions of HRh(CO)(PPh$_3$)$_3$, but according to Wilkinson's group (719), no reduction occurs under hydrogen. Strohmeier and Hohmann (1459), however, have listed initial rate data for the hydrogenation of both hex-1-yne and phenylacetylene in toluene which indicate that they are reduced at about half the rate of hept-1-ene under corresponding conditions (see above).

Complexes with activated acetylenes, as yet undescribed, have been mentioned (853).

No exchange has been observed between molecular hydrogen and the orthohydrogen of a phenyl ring of the phosphine ligand when using HRh(CO)(PPh$_3$)$_3$ (719), but the necessary intermediate involving rhodium coordinated to an orthocarbon of one phenyl group (Section IX-B-2b, Scheme 3, and Section X-G) may have been isolated (see **120** in Section D-2); presumably this does not react readily with hydrogen.

Some comparisons of the HRh(CO)(PPh$_3$)$_3$ catalyst with the corresponding iridium analog are made in Section XII-B-2.

2. Decomposition of HRh(CO)(PPh$_3$)$_3$ in Solution and Its Reaction with Carbon Monoxide.

The catalytic hydrogenation and isomerization reactions of the title complex slow down, especially in dilute solutions, because of decomposition of the catalyst at 25° to an orange dimer according to eq. 356 (724):

$$2HRh(CO)P_2 \rightleftharpoons \underset{\underset{P}{|}}{\overset{\overset{P}{|}}{OC-Rh}}-\underset{\underset{P}{|}}{\overset{\overset{P}{|}}{Rh-CO}} + H_2 \qquad (356)$$

(P = PPh$_3$)   **119**

This reaction, which may involve the monophosphine complex, is retarded by the addition of the phosphine and the presence of hydrogen; it can be reversed at 80 atm hydrogen. Although **119** can act as a hydroformylation catalyst, it is not a hydrogenation catalyst.

In refluxing benzene, HRh(CO)(PPh$_3$)$_3$ decomposes to an inactive brown solid which may be **120** where the rhodium is bound to the α-carbon of one phenyl group. The corresponding complex with the carbonyl replaced by triphenylphosphine, Rh(PPh$_3$)$_2$(PPh$_2$C$_6$H$_4$), has been studied by Keim (728, 729) (Section XI-G).

120

On treating HRh(CO)(PPh$_3$)$_3$ in solution with carbon monoxide, a dicarbonyl species HRh(CO)$_2$(PPh$_3$)$_2$, 121, is formed rapidly (eq. 357). This species can lose hydrogen more slowly to give a yellow dimer 122 according to eq. 357a (718, 720, 724, 854):

$$HRh(CO)(PPh_3)_2 + CO \rightleftharpoons HRh(CO)_2(PPh_3)_2, \textbf{121} \qquad (357)$$

$$2HRh(CO)_2(PPh_3)_2 \rightleftharpoons [Rh(CO)_2(PPh_3)_2]_2 + H_2 \qquad (357a)$$
$$\textbf{122}$$

These reactions are reversible and, depending on the partial pressures of carbon monoxide and hydrogen, 122 itself loses carbon monoxide reversibly to give a red dimeric solvate [Rh(CO)(PPh$_3$)$_2$S]$_2$ (S = ethanol, dichloromethane), 117 (see also Scheme 11 in Section D-1). In the presence of one molar excess of phosphine, 117 irreversibly absorbs hydrogen to form HRh(CO)(PPh$_3$)$_3$:

$$[Rh(CO)(PPh_3)_2S]_2 + H_2 + 2PPh_3 = 2HRh(CO)(PPh_3)_3 + 2S \quad (358)$$

Thus the red solvate dimers 117, the dicarbonyl dimer 122, and the dicarbonyl hydride 121 may all become hydrogenation catalysts depending on the conditions. In the absence of excess triphenylphosphine, hydrogen reacts with 117 to give initially the transient bisphosphine catalyst HRh(CO)(PPh$_3$)$_2$, although this subsequently decomposes to the inactive dimer, 119 (eq. 356). Also, solutions of the solvate dimers slowly decompose under nitrogen to give 119. To complete the series of complex reactions, which are of more importance in the hydroformylation systems (Section XI-F-2), both the orange dimer 119 and the yellow dimer 122 react with carbon monoxide to give a tricarbonyl, believed to be [Rh(CO)$_3$(PPh$_3$)]$_2$.

E. CHLOROCARBONYLBIS(TRIPHENYLPHOSPHINE)RHODIUM(I), RhCl(CO)(PPh$_3$)$_2$, AND ANALOGS

The well-known square-planar complex trans-RhCl(CO)(PPh$_3$)$_2$ (730, 732, 733, 740) was reported by Vaska and Rhodes (730) to catàlyze the hydrogenation of ethylene, propylene, and acetylene in benzene or toluene at 40 to 60°

and 1 atm. But the reductions were very slow and inefficient; about 10^{-2} M catalyst at 60° produced 24% ethane from 1 atm ethylene-hydrogen in 22 hr; at higher temperatures slow reduction to metal, a heterogeneous catalyst, was observed. About the same time Wilkinson and coworkers (43a) also noted that the complex was a very slow catalyst.

Subsequent studies on the catalytic properties of this complex as such has been carried out by Wilkinson and coworkers (55, 612), Strohmeier and coworkers (734–736, 1464, 1465), and James and Rempel (596). Its ready formation from $RhCl(PPh_3)_3$ by a variety of decarbonylation reactions suggests that the carbonyl complex could be involved in hydrogenations catalyzed by $RhCl(PPh_3)_3$ under more severe conditions, such as for saturated aldehyde to the alcohol (Section XI-B-5). Indeed, aldehydes can be reduced to alcohols by using $RhCl(CO)(PPh_3)_2$ at 50 atm hydrogen and 100° (55). Furthermore, at higher hydrogenation pressures, $HRh(CO)(PPh_3)_2$ can be formed, and under carbon monoxide this can be converted to the dicarbonyl, $HRh(CO)_2(PPh_3)_2$ (55). These complexes are important for hydrogenation (Section XI-D) and hydroformylation (Section XI-F-2); thus the $RhCl(CO)$-$(PPh_3)_2$ catalyst is intimately related to these other systems.

In contrast to the corresponding iridium catalytic hydrogenation system using $trans$-$IrCl(CO)(PPh_3)_2$, which has been studied in more detail (Section XII-A), and the $RhCl(PPh_3)_3$ system, $RhCl(CO)(PPh_3)_2$ gives no evidence of dihydride formation at 20° and 1 atm hydrogen (612, 730). The π-acid carbon monoxide ligand is thought to increase the promotion energy in comparison to the $RhCl(PPh_3)_3$ system (Section XI-B-5a). The difference with the iridium analog is ascribed to the lower promotion energy in the larger third transition series and the somewhat increased strength of the Ir—H bond compared to Rh—H (612). Similar factors presumably explain the lack of uptake of ethylene (and, carbon monoxide) by the complex at ambient conditions (730, 737). Moreover, the $RhCl(CO)(PPh_3)_2$ complex, although highly labile (738, 739), does not dissociate appreciably in hydrocarbon solution around room temperature (732, 741), even presumably in the presence of traces of oxygen. Hence, even if small amounts of a dihydride species were formed, the complex would initially be coordinatively saturated and have no vacant site for olefin coordination, which could account for the relative inactivity of the system; the activity then would depend on the tendency of the dihydride to dissociate a ligand. Endres (1466) has determined a value of about 10^{-4} M for the dissociation constant of the following reaction in toluene:

$$RhCl(CO)(PPh_3)_2 \rightleftharpoons RhCl(CO)(PPh_3) + PPh_3 \qquad (358a)$$

Strohmeier and Rehder-Stirnweiss (734–736) reported that by using 5×10^{-3} M $RhCl(CO)(PPh_3)_2$ and 1 M substrate in toluene at 70° and 1 atm total pressure, hept-1-ene is reduced to the extent of 80% in 3 hr. Accompanying isomerization to *cis*- and *trans*-hept-2-enes was also observed. The isomerization does not occur under nitrogen, suggesting that some hydride intermediate is formed from the molecular hydrogen. Wilkinson and coworkers (55) reported similar findings for the isomerization of pent-1-ene in benzene at 1 atm and 70° and also that there is accompanying hydrogenation at 70 atm hydrogen. The hydrogenation uptake curve for hept-1-ene (734–736) shows an induction period of about 30 min, unless the catalyst is pretreated in the solvent-olefin solution for 5 hr at 90° under nitrogen. Although not suggested, slow displacement of phosphine by olefin might be occurring, and the 4-coordinate olefin complex could subsequently undergo reaction with hydrogen via the "unsaturate route," as has been postulated for the IrCl(CO)(PPh₃)₂ system (Section XII-A-2) and other rhodium(I) systems (Sections XI-B-1 and XI-H). Alternatively, Wilkinson and coworkers (55) consider that the hydrogenations involve the HRh(CO)(PPh₃)₂ catalyst that is formed as follows:

$$RhCl(CO)(PPh_3)_2 + H_2 \rightleftharpoons HRh(CO)(PPh_3)_2 + HCl \qquad (359)$$

In the presence of a base such as triethylamine, reaction 359 occurs readily at 50 atm and 70° (but not at 1 atm), and an induction period for hydroformylation reactions, for which HRh(CO)(PPh₃)₂ is a catalyst (Section XI-F-2), is removed (55). Reaction 359 in the absence of base was not observed by these workers; one difficulty is that the reverse reaction is also rapid (55). Strohmeier and coworkers (1464) have shown that the induction period for hydrogenation of hept-1-ene was removed on adding triethylamine. Furthermore, the system was about five times more active than the system with the nitrogen pretreatment, and they concluded that the reduction involves HRh(CO)(PPh₃)₂. This group (736) also presented data for the very slow hydrogen uptake by complexes of the type RhCl(CO)(PPh₃)₂ in toluene at 1 atm and 70°; the initial yellow solutions turn brown, but unfortunately the products were not identified.

It thus appears that the HRh(CO)(PPh₃)₂ complex is involved, but the evidence is not strong; paths involving dihydride plus olefin or olefin complex plus hydrogen cannot be entirely ruled out. Different mechanisms could operate at low and high hydrogen pressures.

Addition of base to the IrCl(CO)(PPh₃)₂ catalyst decreased activity, but the HIr(CO)(PPh₃)₂ intermediate was thought not to be involved (1464).

It is again worth noting that the overall "heterolytic" hydrogen splitting in eq. 359 likely involves dihydride formation followed by dehydrochlorination (see eq. 148, Section IX-B-2e).

Table 62 (736) summarizes some hydrogen uptake data for olefins when using different RhX(CO)L₂ complexes. The iodide seems to be the most

Table 62   Hydrogen Uptake for Olefins Using RhX(CO)L$_2$ Complexes (736)^a

Olefin	X L	Cl P(C$_6$H$_{11}$)$_3$	Cl PPh$_3$	Cl P(OPh)$_3$	Cl AsPh$_3$	I PPh$_3$
Hept-1-ene		430	6720	20	1400	6700
Norbornene		860	2260	20	4420	6900
Cyclohexa-1,3-diene		190	190	0	640	5180
Cycloocta-1,3-diene		0	0	10	0	20
Cyclooctene		0	0	0	0	0
2,3-Dimethylbutadiene		20	60	0	0	240

^{a} With 5×10^{-3} M catalyst (no pretreatment), 1 M olefin, 56 cm H$_2$ in toluene at 70°. Uptake given in ml l^{-1} for a 3 hr period.

active catalyst over a 3 hr period, but the uptake curves presented (735) then fall away more rapidly than for the chloride and bromide systems; some papers by the Strohmeier group (734, 735) indicate decreasing activity in the order Cl > Br > I for hept-1-ene reduction. Activity with L decreases in the order PPh$_3$ > P(C$_6$H$_{11}$)$_3$ > P(OPh)$_3$. The arsine complex is more, or less, active than the triphenylphosphine complex, depending on the substrate. Although cyclooctene at 1 M was not reduced at 70°, there was a slow hydrogenation at 80° when using only one-fourth the catalyst concentration (L = PPh$_3$) and 0.6 M olefin, showing again the importance of the preheating treatment (736). Also reduced under the conditions of no pretreatment were, in order of decreasing rate, cis-hept-2-ene > hept-1-ene > styrene > dimethyl fumarate > n-butyl vinyl ether, trans-hept-2-ene > dimethyl maleate (736).

The Strohmeier group (1465) have also listed some initial rate data for the reduction of olefins and acetylenes when using the RhCl(CO)(PPh$_3$)$_2$ catalyst preactivated under nitrogen (Table 63). There are discrepancies in the order of reactivities mentioned above. The complex is usually less active than the iridium analog (Section XII-A-2), although acetylenes and dimethyl fumarate in particular are more readily reduced with the rhodium catalyst. It is also worth noting that the comparative rates for the rhodium system are generally not in good agreement with the rates given by the same group for the same substrates when using the HRh(CO)(PPh$_3$)$_3$ catalyst (Section XI-D-1).

By use of more strongly coordinating solvents, James and Rempel (596) had hoped to promote ligand dissociation from RhCl(CO)(PPh$_3$)$_2$ to give an active hydrogenation catalyst. However, a study in DMA showed that reduction of unsaturated carboxylic acids was extremely slow at 80° and 1 atm hydrogen. On the other hand, inorganic substrates such as iron(III) were readily reduced under the same conditions. Whether this involves a hydride intermediate or, more probably, oxidation to rhodium(III) followed by hydrogen reduction back to rhodium(I), remains to be established. Interestingly the DMA solution of RhCl(CO)(PPh₃)₂ becomes quite active for the hydrogenation of olefinic acids (80°, 1 atm), after exposure to oxygen at 80° causes some oxidation of the triphenylphosphine and/or solvent. This could involve phosphine dissociation with possibly the production of a new oxidized catalyst (Section XI-B-8).

It should be noted that alcohol-alkoxide solutions of the RhCl(CO)(PR₃)₂ complexes may be converted to the active HRh(CO)(PR₃)₃ catalyst under nitrogen or hydrogen atmospheres (854, Section XI-D-1).

Trans-RhCl(CO)(PPh₃)₂ forms complexes with tetracyanoethylene (851) and acetylenic carboxylic acids and esters (852) without displacement of a phosphine ligand; these complexes are unlikely to readily activate hydrogen.

Hartwell and Clark (858) found that the unsaturated tertiary phosphine ligands in the complexes RhCl(CO)[Ph₂P(CH₂)ₙCH=CH₂]₂ (*n* = 0 to 3) are readily reduced in alcohol at 1 atm hydrogen and 25°. The *n* = 2 system with reduction to the butyldiphenylphosphine ligand was studied in detail. Loss of chloride occurs to give RhCOL₂⁺ with coordinated olefins; dihydride formation followed by hydrogen transfer was suggested. Deuterium tracer studies

Table 63 Hydrogenation of Olefins and Acetylenes Using MCl(CO)-(PPh₃)₂ (1465)[a]

Substrate	Initial Rate × 10³ (*M* min⁻¹)	
	RhCl(CO)(PPh₃)₂	IrCl(CO)(PPh₃)₂
Hept-1-ene	3.0	8.9
Dimethyl fumarate	3.0	0.0
Ethyl acrylate	1.7	10.0
Styrene	1.1	6.7
Cis-hept-2-ene	1.0	1.0
Cycloheptene	1.0	4.5
Dimethyl maleate	0.6	0.4
Trans-hept-3-ene	0.4	0.7
Trans-hept-2-ene	0.2	0.55
β-Bromostyrene	0.14	0.0
Hex-1-yne	0.3[b]	0.0[b]
Phenylacetylene	1.1	0.7

[a] M = Rh and Ir; 2 × 10⁻³ *M* catalyst, 0.8 *M* substrate, 1 atm pressure in toluene at 80°.
[b] At 70°.

(Section XI-B-4) indicated stepwise transfer via hydridoalkyl intermediates which would contain a 5- or 6-membered ring, depending on whether the first hydrometalation step was Markownikov or anti-Markownikov.

$Rh(NO_2)(CO)(PPh_3)_2$ (889) and cis-$RhCl(CO)(PPh_3)_2$ (796) have been prepared but not studied for activity. Fluorophosphine complexes, $trans$-$RhCl(CO)L_2$, where L = fluorophosphine (794, 845, 846) and the thio-carbonyl complex $trans$-$RhCl(CS)(PPh_3)_2$ (798), are expected to be less active than $trans$-$RhCl(CO)(PPh_3)_2$, because of the somewhat greater π-acidity of the ligand systems.

F. HYDROGENATION UNDER HYDROFORMYLATION CONDITIONS

Before considering further rhodium-catalyzed hydrogenation systems, it is best to examine hydrogenation under hydroformylation conditions, since the catalysts involved are carbonyls and as such are related to the systems considered in the preceding two sections (Sections XI-D and XI-E).

Hydroformylation and hydrogenation reactions involving cobalt carbonyl catalysts have been discussed from the point of view of hydrogen activation in Sections X-E and X-F. A similar brief presentation is given here for the less documented rhodium systems. The mechanisms for hydroformylation and for reduction of olefin and aldehyde groups by the rhodium systems are very similar to those of the cobalt systems; the rhodium systems have been reviewed to some extent (53, 746, 750).

1. Systems Involving $HRh(CO)_n$. Patent literature (742, 743, 757) first disclosed that use of rhodium salts, such as $RhCl_3 \cdot 3H_2O$, enables the per-formance of the Oxo reaction at less severe conditions than used with cobalt salts; one of these patents (743) also mentioned a small amount of alcohol production from but-1-ene. Since then a number of investigators (in particular 468; 536, 540; 744–746; 747–752; 769; 753, 754; 755; 762–764, 1430; 765; 771), have studied the catalytic properties of such systems or of rhodium carbonyls themselves. The active catalyst in these systems is generally assumed to be $HRh(CO)_4$ or $HRh(CO)_3$, which may be up to 1000 times as active as the cobalt catalyst (746, 755) and which, in some conditions, can produce much more branched-chain aldehydes. This may result from extensive accompanying olefin isomerization (754, 755), although other conditions have given no isomerization (752). The higher activity of the rhodium carbonyls has been attributed (55) to the greater ease of oxidative addition of hydrogen to a rhodium(I) acyl complex (eq. 250, Section X-E), as well as to reduced ligand crowding at the larger rhodium atom (51).

A number of the papers above report hydrogenation as well as hydro-formylation products, and some deal specifically with reduction of aldehydes to alcohols (747–750, 753, 762). These hydrogenations are summarized in Table 64.

Table 64 Hydrogenation of Olefins and Saturated Aldehydes Using Rhodium Carbonyls[a] under Hydroformylation Conditions

Substrate	Reduction Product	Reference
Hept-1-ene	Octanols	747
Cyclohexene	Cyclohexylcarbinol	747
But-1-ene	Amyl alcohols	743
α-Methylstyrene	Isopropylbenzene	744
Butadiene	Monoaldehyde, butenes	468, 744, 745
1,3-Pentadiene	Monoaldehyde	468
1,4-Pentadiene	Cyclohexane carboxaldehyde	468
1,5-Cyclooctadiene	Cyclooctylcarbinol, bis(hydroxymethyl)-cyclooctane	763, 1430
Benzaldehyde	Benzyl alcohol	747, 749
Propionaldehyde	1-Propanol	747, 749, 753
n-Butyraldehyde	1-Butanol	747–749
1-Cyclohexen-4-al	1,3- and 1,4-bis(hydroxymethyl)cyclo-hexane, hexahydrobenzylalcohol	762

[a] Viewed as $HRh(CO)_n$.

Alcohol production from an olefin will occur via the saturated aldehyde. Dienes and unsaturated aldehydes can give a saturated aldehyde, a saturated alcohol or a diol, depending on the extent of reduction and whether hydro-formylation occurs at one or both double bonds (Sections X-E-2 and X-E-3).

Yamaguchi and Onoda (753) reported that ethanol or ketone solutions are more effective than toluene solutions in their overall yield and selectivity for reduction of propionaldehyde to 1-propanol.

Heil and Marko (748) have studied the kinetics of butyraldehyde reduction in hexane solution (25%) at 160°, when using rhodium carbonyls with hydrogen at 115 atm and carbon monoxide from 40 to 60 atm, as well as carbon monoxide at 120 atm with varying hydrogen pressure (17–177 atm). The catalyst was added as $Rh_4(CO)_{12}$ or as $RhCl_3 \cdot 3H_2O$, which can be reduced to rhodium(0) carbonyls or $[Rh(CO)_2Cl]_2$ under such conditions (758–760). Above carbon monoxide pressures of 80 atm and catalyst concentrations of about 8×10^{-5} g atom Rh/g mole of aldehyde, the hydrogenation obeyed the following rate law:

$$\frac{d[\text{alcohol}]}{dt} = k[\text{aldehyde}][\text{Rh}]^{1/6}(p_{\text{H}_2})^{0.5}(p_{\text{CO}})^{-0.3} \qquad (360)$$

Of the data listed, the minimum half-lives for the reaction were about 3 hr. Below 80 atm carbon monoxide, the rate fell off markedly. The reaction mechanism presented was the same as that given by Marko and coworkers for the corresponding cobalt system (Section X-E-2, eqs. 263 to 266), except that the rate-determining step was written as the initial addition of $\text{HRh}(\text{CO})_3$ to the aldehyde to give the alkoxide intermediate; the very small amounts of butyl formate formed were again attributed to reactions such as 265 and 266. With the equilibria 361 to 363 determining the $\text{HRh}(\text{CO})_3$ concentration, and the fact that $\text{Rh}_6(\text{CO})_{16}$ is the thermodynamically stable compound under the reaction conditions, the rate law is accounted for:

$$2\text{Rh}_6(\text{CO})_{16} + 4\text{CO} \rightleftharpoons 3\text{Rh}_4(\text{CO})_{12} \qquad (361)$$

$$\text{Rh}_4(\text{CO})_{12} + 2\text{H}_2 \rightleftharpoons 4\text{HRh}(\text{CO})_3 \qquad (362)$$

$$\text{HRh}(\text{CO})_3 + \text{CO} \rightleftharpoons \text{HRh}(\text{CO})_4 \qquad (363)$$

Infrared studies detected $\text{Rh}_6(\text{CO})_{16}$, $\text{Rh}_4(\text{CO})_{12}$, and $[\text{Rh}(\text{CO})_2\text{Cl}]_2$ as intermediates. The species $\text{Rh}_2(\text{CO})_8$ and $\text{HRh}(\text{CO})_4$, akin to the cobalt catalysts, have not been detected generally (748, 761), but $\text{Rh}_2(\text{CO})_8$ has been detected at $-30°$ and 450 atm carbon monoxide pressure (826). Rhodium acyl intermediates have also not been observed, but the production of a carboxy derivative according to eq. 363a is thought to arise from oxidation of an acyl intermediate by oxygen impurity (769).

$$\text{Rh}_4(\text{CO})_{12} + \text{C}_2\text{H}_4 + \text{H}_2 + \text{CO} \xrightarrow[\text{100 atm}]{75°} [\text{Rh}(\text{CO})_2(\text{OOCC}_2\text{H}_5)]_2 \quad (363a)$$

A patent (1456) describing a rhodium-catalyzed hydroformylation of styrene to α- and β-phenylpropanals also claims that the aldehydes can be subsequently reduced to the alcohols by reaction under a purely hydrogen atmosphere at 200°.

Takesada and Wakamatsu (1437) noted that during hydroformylation of butenes in nitrobenzene solution, the solvent is effectively reduced to aniline; this conversion is less effective than with the $\text{Co}_2(\text{CO})_8$ catalyst.

2. Tertiary Organophosphine-Rhodium Carbonyl Catalysts and Their Analogs. The use of organophosphine rhodium carbonyl complexes for hydroformylation and hydrogenation reactions was first recorded in patents by Slaugh and Mullineaux (524, 525). The catalysts and analogs, involving arsines and phosphites, were prepared *in situ* by adding the tertiary ligand to a carbonyl or a precursor rhodium salt. Further work on these *in situ* systems has been reported by these workers (527, 528), Fell and coworkers (536, 540), Falbe and Huppes (764), and Pruett and Smith (765).

Wilkinson and coworkers (699, 700) discovered that $1,2,3\text{-RhCl}_3(\text{PPh}_3)_3$ was an effective hydroformylation and hydrogenation catalyst under 100 atm of hydrogen/carbon monoxide at 55 to 100° in ethanol-benzene. Under hydrogen, this rhodium(III) compound is readily reduced to $\text{RhCl}(\text{PPh}_3)_3$ (Section XI-B-9), which itself is readily carbonylated to $\text{RhCl}(\text{CO})(\text{PPh}_3)_2$ (612, Section XI-B-5b). Wilkinson's group (43, 55, 612) have since used both these complexes for hydroformylation, even of alk-1-ynes. Craddock and coworkers (766, 773) have also reported on the use of the carbonyl complex. Solutions of $\text{RhCl}(\text{CO})(\text{PPh}_3)_2$, L = P, As, dispersed in porous solids, have been used for hydroformylation of propylene (509, 702).

When using $\text{RhCl}(\text{CO})(\text{PPh}_3)_2$ for hydroformylation of simple alkenes, an observed induction period could be removed by adding a base; this led to the conclusion that the true catalyst was the hydride species $\text{HRh}(\text{CO})(\text{PPh}_3)_2$ formed according to eq. 359 (Section XI-E). Indeed, solutions of the tris(triphenylphosphine) complex, $\text{HRh}(\text{CO})(\text{PPh}_3)_3$, which dissociates to the bis complex (Section XI-D-1), are active for hydroformylation even at 25° and 1 atm (55, 722, 725, 726, 765, 767). The complicated series of reactions involved in the system $\text{HRh}(\text{CO})(\text{PPh}_3)_3$—$\text{H}_2$—CO have been presented in Section XI-D-2; under carbon monoxide the dicarbonyl species $\text{HRh}(\text{CO})_2$-$(\text{PPh}_3)_2$ predominates (55). Accordingly, Wilkinson and coworkers (55, 767) have presented a possible hydroformylation mechanism based on eq. 357 followed by eqs. 364 to 366, which correspond to those of the cobalt systems (Section X-E).

$$\text{HRh}(\text{CO})_2\text{P}_2 + \text{RCH}{=}\text{CH}_2 \longrightarrow \begin{bmatrix} \text{HRh}(\text{CO})_2\text{P}_2 \\ | \\ \text{RCH}{=}\text{CH}_2 \end{bmatrix} \longrightarrow$$

$$\text{RCH}_2\text{CH}_2\text{Rh}(\text{CO})_2\text{P}_2 \quad (364)$$

$$\text{RCH}_2\text{CH}_2\text{Rh}(\text{CO})_2\text{P}_2 \rightleftharpoons \text{RCH}_2\text{CH}_2\text{CORh}(\text{CO})\text{P}_2 \quad (365)$$

$$\text{RCH}_2\text{CH}_2\text{CORh}(\text{CO})\text{P}_2 + \text{H}_2 \longrightarrow [\text{dihydride}] \longrightarrow$$

$$\text{RCH}_2\text{CH}_2\text{CHO} + \text{HRh}(\text{CO})\text{P}_2\,(\text{P} = \text{PPh}_3) \quad (366)$$

Acyl complexes, such as $\text{Rh}(\text{COR})(\text{CO})_2(\text{PPh}_3)_2$, have been isolated (725, 726, 729). The propionyl complex, made by reacting $\text{HRh}(\text{CO})_2(\text{PPh}_3)_3$ with ethylene and carbon monoxide, reacts with hydrogen to give propionaldehyde with loss of carbon monoxide (725, 726):

$$\text{Rh}(\text{COC}_2\text{H}_5)(\text{CO})_2(\text{PPh}_3)_2 + \text{H}_2 \xrightarrow{\text{PPh}_3}$$

$$\text{HRh}(\text{CO})(\text{PPh}_3)_3 + \text{C}_2\text{H}_5\text{CHO} + \text{CO} \quad (367)$$

The reaction is inhibited by high partial pressure of carbon monoxide, suggesting dissociation to give a square acyl which then undergoes oxidative addition (eq. 366). Hydrogenolysis of benzoyl rhodium complexes of the type $PhCORh(CO)_2(PPh_3)_2$ (768), according to an equation like 366, with 1 atm hydrogen, was also reported (725, 726). Tricarbonyl species, such as $Rh(COC_2H_5)(CO)_3PPh_3$ and $Rh(alkyl)(CO)_3PPh_3$, and dissociated species, such as $Rh(alkyl)(CO)_2PPh_3$, have also been isolated; these could also be present in hydroformylation conditions (725, 726, 767).

The hydroformylation systems will not be presented in detail, but some points relating to the hydride addition step, eq. 364, which has been considered for the $HRh(CO)(PPh_3)_2$ catalyst (Section XI-D-1) and the cobalt catalysts (Sections X-E-3 and X-F-1), will be summarized. The rhodium carbonyls considered in Section XI-F-1 can give rise to considerable branched-chain aldehydes from the hydroformylation of alk-1-enes. This may be due to isomerization followed by hydroformylation of the isomer. Use of $RhCl(CO)(PPh_3)_2$ at up to 125° and 100 atm gave a higher straight- to branched-aldehyde ratio of up to 3 (55, 766), but the formation of the branched product, at least at 70°, was not due to isomerization (55). Furthermore, hydroformylation of alk-2-enes can give exclusively branched aldehydes (55). Similarly at 100 atm and 25°, $HRh(CO)(PPh_3)_3$ gave a ratio of about 3 for alk-1-enes, but at 1 atm for a stoichiometric hydroformylation the ratio increased to about 20 at 25° and to about 9 at 50°; again there was no isomerization. These data indicated that the ratio of anti-Markownikov to Markownikov addition, which must determine the product ratio, decreased with increasing temperature. This was consistent with the expected somewhat lower free energy of activation for anti-Markownikov addition of L_nRh—H to an alk-1-ene (55). But other data from using $HRh(CO)(PPh_3)_3$ under catalytic conditions at 1 atm and 25 to 50° have indicated that the product ratio for the hex-1-ene system increases with temperature and also with increasing catalyst concentration, hydrogen pressure, and triphenylphosphine concentration (722, 767). The absence of isomerization in these systems under certain conditions, particularly at lower temperatures, indicates that acyl formation from the alkyl (eq. 365) must be fast. When using $RhCl(CO)(PPh_3)_2$ at about 40 atm and 100 to 200°, isomerization is significant, but decreases with decreasing temperature and increasing added phosphine (766). The large number of variables, including solvent effects, make an interpretation difficult at this stage, but conditions can be found to give quite selective production of linear aldehydes (55, 722, 766).

In view of the ready dissociation or displacement by carbon monoxide of triphenylphosphine, the high phosphine/rhodium ratios generally needed to ensure high specificity are thought (726, 767) to suppress dissociation and maintain phosphine complexes where the steric interaction is greatest. It is not too clear, however, why steric crowding should give anti-Markownikov addition via a 6-coordinate species as exemplified in eq. 364. The detection of $Rh(alkyl)(CO)_2PPh_3$ suggested that the dissociated species $HRh(CO)_2PPh_3$ gives rise to low selectivity and high yields of branched aldehydes via initial Markownikov addition, and such higher yields at the lower catalyst concentrations are also consistent with this (726, 767). A quite similar reaction sequence to that outlined in eqs. 364 to 366, which has been presented by Wilkinson and coworkers (55, 767), involves the initial predissociation of the bisphosphine complex to the monophosphine.

Of interest is the finding that the selectivity in hydroformylation rates between alk-1-ene and alk-2-ene is about 20:1 (55, 722) compared to the hydrogenation rates catalyzed by $HRh(CO)(PPh_3)_2$ that are >200. The latter was attributed to steric hindrance in hydride transfer to form the alkyl group which would be *cis* to two *trans*-phosphine ligands of the $HRh(CO)(PPh_3)_2$ catalyst (Section XI-D-1). In hydroformylation, the lack of selectivity is thought to arise from decrease of steric hindrance while forming the alkyl from $HRh(CO)_2(PPh_3)_2$ where the phosphine groups are *cis* (55). The difference between various alkenes is of the same order as found for hydrogenations when using $H_2RhCl(PPh_3)_2$ which also has *cis*-phosphines (Sections XI-B-1 and XI-B-3).

$RhCl(CO)(PPh_3)_2$, which can be recovered unchanged at the end of a hydroformylation reaction, gives aldehyde selectively, with no hydrogen reduction to alcohol or paraffin under conditions of 100 atm ($H_2:CO = 1$) and $100°$ (55, 766, 773). Such hydrogenation occurs only under a purely hydrogen atmosphere (Section XI-E). Use of the corresponding iodide and bromide complexes, and other complexes $RhCl(CO)L_2$ (where L may be other tertiary phosphines, or triphenylarsine) has given quite similar results, including the relatively high proportion of straight-chain aldehydes (55). When using $RhCl_3(PPh_3)_3$ in ethanol-benzene at $100°$ and 100 atm ($H_2:CO > 1$), some alcohol can be formed from alk-1-enes, and with increasing hydrogen partial pressure, alkanes may also be formed (699, 700).

Catalysts formed *in situ* by adding tri-*n*-butylphosphine or other phosphines, arsines, and phosphites to rhodium trichloride, rhodium carbonyls, or supported rhodium metal also give from alkenes (such as pent-1-ene and oct-1-ene) a highly selective production of aldehydes. Again, these are mainly nonbranched, which certainly result in part from the small amount of accompanying isomerization of the olefin. Longer reaction times can give increasing amounts of alcohol (524, 525, 527, 528, 536, 540, 700, 765, 770). Increases in the partial pressure of hydrogen when using $HRh(CO)(PPh_3)_3$ at $25°$ and 1 atm can lead to some alkane production (722, 767). These reductions to alkanes and aldehydes will involve catalysts of the type $HRh(CO)_nL_m$, particularly with $n = 1$ (Section XI-D-1) and $n = 2$ (cf. eq. 364).

Triaryl and trialkylphosphites are particularly effective additives for straight-chain aldehyde production from alk-1-enes, even at pressures of only 7 atm at 100° (765, 770). The straight- to branched-aldehyde ratio again increased with phosphite concentration and also with decreasing total pressure and carbon monoxide partial pressure; substitution on the ortho positions of triphenylphosphite decreased the ratio to about unity. Pruett and Smith (765) rationalized these data in terms of several reactive catalysts in equilibrium (eq. 368). **123**, which would not form with the highly hindered

$$HRh(CO)_3(PR_3) \underset{CO}{\overset{PR_3}{\rightleftharpoons}} HRh(CO)_2(PR_3)_2 \underset{CO}{\overset{PR_3}{\rightleftharpoons}}$$

$$HRh(CO)(PR_3)_3 \underset{PR_3}{\rightleftharpoons} HRh(CO)(PR_3)_2 \quad (368)$$

$$\underset{\textstyle \mathbf{123}}{} \qquad\qquad \underset{\textstyle \mathbf{124}}{}$$

phosphites, was considered to give a predominance of straight-chain aldehyde via a mechanism involving dissociation to **124** followed by reactions such as 364 to 366 but with a monocarbonyl complex. A greater efficiency of tributylphosphites for straight-chain formation compared to tributylphosphine was attributed to the better π-acidity of the former, which favors formation of **123**.

The hydroformylation of unsaturated esters can give the expected aldehydes (eq. 369), but further reduction to the hydroxyester with possible lactone formation has been observed (764, 765) (eq. 252a, Section X-E-2). For example, methyl methacrylate gives α-methyl-γ-butyrolactone (via **126**) and methyl 2,2-dimethyl-3-hydroxypropionate (via **125**). Similarly methyl crotonate gives β-methyl-γ-butyrolactone, and ethyl cinnamate gives β-phenyl-γ-butyrolactone. These systems are of interest in that hydroformylation of such esters, which is normally electronically controlled by hydridic addition of a carbonyl hydride (Section X-E-3, eq. 287), can become sterically controlled in the presence of excess triphenylphosphite (765):

$$\underset{\delta+}{CH_2}{=}\overset{\overset{\displaystyle CH_3}{|}}{C}CO_2CH_3 \longrightarrow \overset{\overset{\displaystyle CH_3}{|}}{\underset{\underset{\displaystyle CHO}{|}}{CH_3C}}CO_2CH_3 + \overset{\overset{\displaystyle CH_3}{|}}{\underset{\underset{\displaystyle CHO}{|}}{CH_2CH}}CO_2CH_3 \quad (369)$$

$$\qquad\qquad\qquad\qquad \underset{\textstyle \mathbf{125}}{} \qquad\qquad \underset{\textstyle \mathbf{126}}{}$$

Addition of tributylphosphine at high pressure favors production of **125** (764), which is consistent with replacement of a CO group by a tertiary phosphine reducing the acidity of the rhodium hydride (Section X-F-1). Addition of triphenylphosphite at 200 atm gives a product ratio for **126:125** of 0.3 but this increases to 24 on lowering the pressure to 8 atm; again this is considered to be due to steric control via a species such as **123** (765).

The question of what factors determine the aldehyde product ratio remains a complex one. Of importance are isomerization of the olefin, steric effects, and electronic effects reflected both in the acidity of the carbonyl hydride and the nature of the olefin. The greater activity of the rhodium carbonyl phosphine catalysts compared to their cobalt analogs has been attributed to the lack of dissociation of the pentacoordinate cobalt(I) species (55).

3. Systems Involving Other Rhodium Complexes. $[(C_5H_5)_2RhCl]_2$ and $[Rh(CO)_2Cl]_2$ have been used as the starting catalysts (55, 766); the formation of the latter *in situ* has been mentioned in Section XI-F-1. Both catalysts give more than 50% branched aldehydes, probably as a result of extensive olefin isomerization (766). Less than 1% hydrogenated products were observed (766).

The rhodium(III) complex, 1,2,6-$RhCl_3(py)_3$, and the solvated rhodium(I) complex, $Rh_2Cl_2(SnCl_2, C_2H_5OH)_4$, have been used in ethanol to give mainly *n*-heptaldehyde from hex-1-ene at 55° and 90 atm (with $H_2:CO = 1$); at 100° (with $H_2:CO > 1$) alcohols and alkanes can be formed (699, 700). Alderson (756) has described the production of aldehydes, alcohols, and ketones from olefins when using aqueous solutions of rhodium trichloride and pyridine (py:Rh = 1.5) at 200° under 1000 atm carbon monoxide alone. The aldehyde and alcohol products are those expected from hydroformylation; for example, butadiene gives amyl alcohols and 1,5-hexadiene gives heptanol, and so the water probably provides the source of hydrogen.

Foster and Lawrenson (1441) have claimed that complexes such as $Rh(CO)L[P(OPh)_3]$, where L is a chelate ligand such as salicylaldoximate, can be used in toluene under hydroformylation conditions for the production of alcohols from olefins.

The rhodium(III) complexes $RhCl_3(PEt_2Ph)_3$ and $RhCl_3(CO)(PPh_3)_2$ are ineffective for hydroformylation (55), presumably because they are not readily reduced by hydrogen to rhodium(I) (Section XI-B-9).

G. OTHER RHODIUM CATALYSTS CONTAINING PHOSPHINES, ARSINES, AND PHOSPHITES

Sacco and coworkers (566, 579) and Taylor and coworkers (897, 1407) have reported that the cationic complex $Rh(diphos)_2^+$ does not react with hydrogen under ambient conditions. However, the aliphatic analog $[Rh(Me_2PCH_2CH_2PMe_2)_2]Cl$, with higher electron density at the metal, is reported by Chatt and Butter (781, 1392) to undergo reversible oxidative addition to give a *cis* dihydride in THF; the corresponding 5-coordinate complex containing carbonyl gives no reaction with hydrogen (Section XI-E).

The 6-coordinate dihydride would not be expected to be an active catalyst because of the difficulty of dissociation to give a vacant site for olefin. However, the [Rh(diphos)$_2$]Cl complex shows slight activity at 1 atm hydrogen in a refluxing mixture of oct-1-ene and toluene (897). But under severe conditions of 350 atm hydrogen at 80° in the presence of sodium carbonate, this complex has been used to reduce 1,4-dicyanobutenes to adiponitrile (698). A similar ditertiary arsine complex [Rh(As–As)$_2$]Cl, where (As–As) is *cis*-Ph$_2$AsCH=CHAsPh$_2$, is reported by Mague and Mitchener (782) to form reversibly a *cis* dihydride, inactive at ambient conditions. A similar cationic complex with a fluorocarbon-bridged ditertiary phosphine, Ph$_2$PC=CPPh$_2$(CF$_2$)$_2$CF$_2$, reported by Cullen and Thompson (783), does not form a dihydride because of the electron-withdrawing nature of the ligand.

Analogous cationic square planar complexes with nonchelating Group V ligands have been reported by Osborn and coworkers (786, 787), Johnson and coworkers (855), Green and coworkers (856), and Haines (788). The complexes Rh(diene)L$_2^+$ (L is a tertiary phosphine or AsPh$_3$, and diene refers to 1,5-cyclooctadiene, norbornadiene, or 1,5-hexadiene) may be made from the dimeric complex [Rh(diene)X]$_2$ (X = halide) in alcohols or nitromethane. In other organic solvents the halide is not displaced, and neutral Rh(diene)LX species result. Treatment of the cationic complex with hydrogen at ambient conditions in a solvent S gives rhodium(III) dihydrides, for example, H$_2$Rh(PPh$_3$)$_2$S$_2^+$, which very efficiently catalyze the hydrogenation of olefinic and acetylenic bonds (786). The S may be acetonitrile, 2-butanone, dimethylacetamide, acetone, or alcohol. The hydrogenations have also been effectively carried out in dioxane, 2-methoxyethanol, dimethylacetamide, and acetone, but not in acetonitrile. Some rate data were given for a corresponding iridium complex (Section XII-B-1). The rhodium complex with acetonitrile has *cis* hydrides and *cis* solvent molecules. The triphenylphosphine catalyst hydrogenates monoolefins (hex-1-ene > cyclohexene ∼*cis*-hex-2-ene > *trans*-hex-2-ene ≫ 1-methylcyclohexene), dienes (norbornadiene > 1,5- and 1,3-cyclooctadiene), acetylenes (hex-1-yne > hex-2-yne > hex-1-ene), and unsaturated ketones and esters without reduction of the carbonyl. The dienes are reduced rapidly to the monoene which is then reduced more slowly; some accompanying isomerization of the dienes suggest that the hydride transfer to the olefin is stepwise (Section XI-B-4). Excess triphenylphosphine inhibits hydrogenation. Interesting information could result if Osborn and coworkers complete their plans to investigate the possible relationship between solvent exchange and catalytic efficiency of these solvated dihydrides.

The cationic dihydrides with more basic phosphines (e.g., PPh$_2$Me, PPhMe$_2$, PMe$_3$), in contrast to the triphenylphosphine complex, were found to reduce saturated ketones to alcohols (787); for example, a $3 \times 10^{-3}\,M$ solution of $H_2Rh(PPhMe_2)_2S_2^+$ in 1% aqueous acetone absorbs hydrogen at 25° and 1 atm with an initial rate of about 4 ml min^{-1}. Cyclohexanones, acetophenone, butan-2-one, and tetramethylcyclobutan-1,3-dione were also reduced; 4-*t*-butylcyclohexanone gave 86% *trans*-alcohol product. Since benzophenone forms a stable complex, it is not readily hydrogenated. Aldehydes were also reduced initially, but the rate of hydrogen uptake decreased rapidly. The ketone reductions were extremely slow in the absence of water, and maximum rates were observed using about 1% water by volume; on the other hand, the reduction of olefins was inhibited by water. Deuteration gave alcohols labeled at the α-carbon only, showing that the enol form of the ketone was not involved; no primary isotope effect was apparent. Scheme 12 shows the mechanism postulated:

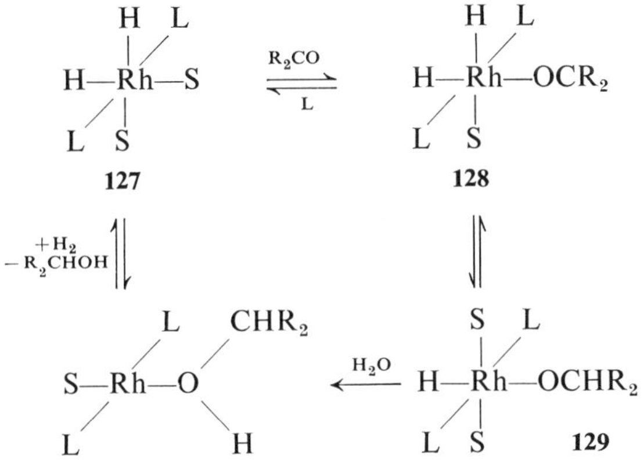

L = phosphine
S = solvent
All species are cationic

Scheme 12

The first hydride transfers by nucleophilic attack at the α-carbon of the coordinated ketone (**128** → **129**); the second hydrogen transfer is thought to be promoted by water as follows:

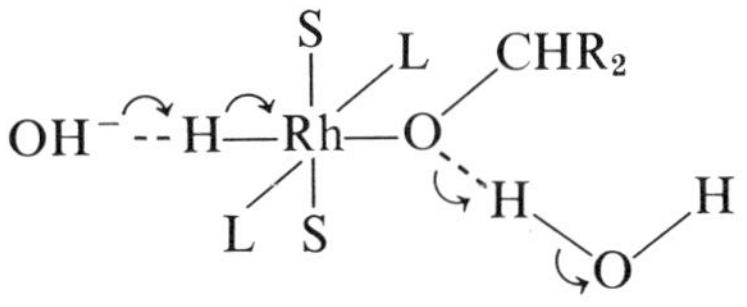

Scheme 13

The reduction of dry ketones was autocatalytic due to the alcohol product which could also promote the deprotonation-protonation process depicted in Scheme 13. Species **127** and **128** themselves may be reversibly deprotonated to yield neutral monohydride species which do not absorb hydrogen;

for example:

$$H_2RhL_2S_2^+ \overset{base}{\rightleftharpoons} HRhL_2S_2 + H^+ \tag{370}$$

$$(base = Et_3N, H_2O, alcohols)$$

Equilibrium 370 readily accounted for the D_2–H_2O exchange observed in the absence or presence of ketone.

Haines (788) has described cationic phosphites, $Rh[P(OR)_3]_4^+$ ($R = CH_3$, C_2H_5, n-C_4H_9) but their reactivity under hydrogen has not been reported.

Deprotonation of $H_2Rh(PPh_3)_2S_2^+$ according to eq. 370 in the presence of excess phosphine (787) gives the hydridotetrakis(triphenylphosphine)-rhodium(I) complex, $HRh(PPh_3)_4$. This compound has also been studied by Takesada and coworkers (789, 811), Dewhirst, Keim, and coworkers (251, 728, 790), Yamamoto and coworkers (252, 1368), Levison and Robinson (254, 1410), Ilmaier and Nyholm (791), Baker and Pauling (1177), and Ugo and coworkers (854). Only one report (790) describes its use as a hydrogenation catalyst: in the presence of excess triphenylphosphine at 80° and using the substrate as solvent, the olefinic bond in acrylonitrile, mesityl oxide and hex-1-ene is reduced quantitatively. Deuterium exchange occurs with the hydride ligand and phenyl hydrogens (1368; Scheme 3 in Section IX-B-2b).

$HRh(PPh_3)_4$, which can also be obtained from $RhL(PPh_3)_3(L = Cl$ or $OH)$ by hydrogen abstraction procedure from alcohol-alkoxide solutions (854), is converted by carbon monoxide into the active catalyst $HRh(CO)(PPh_3)_3$ (854). Two other preparations of $HRh(PPh_3)_4$, which has the four phosphorus atoms tetrahedral and the hydrogen atom likely on a threefold axis (1177), are of interest in that they involve activation of hydrogen by rhodium(I) complexes. One involves the mild hydrogenolysis of a phenyl complex with resulting liberation of benzene (789):

$$(diene)Rh(C_6H_5)PPh_3 + H_2 \xrightarrow{PPh_3} HRh(PPh_3)_4 + C_6H_6 \tag{371}$$

The other involves dehydrohalogenation by base via a dihydride (251, 790):

$$RhCl(PPh_3)_3 + H_2 \longrightarrow H_2RhCl(PPh_3)_3 \xrightarrow[PPh_3]{base}$$
$$HRh(PPh_3)_4 + base\ HCl \tag{371a}$$

Other rhodium(I) monohydrides such as $HRh(Ph_2PMe)_4$ (251), $HRh(PPh_3)$-$(AsPh_3)_3$ (789, 857), $HRh(PPh_3)_3$ (728, 791, 792), $HRh(diene)(PPh_3)$ (793), and $HRhL_2S_2$ (eq. 370; 787, 791) have been formed similarly; σ-bonded methyl derivatives have also been used for reaction 371. A reaction such as 371 is almost certainly involved in hydrogenations catalyzed by rhodium(I) monohydrides; it likely involves initial oxidative addition of the hydrogen (Section XI-D-1).

The alkyl and aryl complexes (eq. 371), which have been prepared using Grignard reagents on rhodium(I) halide complexes such as $RhCl(PPh_3)_3$ or

RhCl(diene)PPh$_3$ (789, 792, 793), have themselves been used as hydrogenation catalysts. The phenyl complex C$_8$H$_{12}$RhPh(PPh$_3$), **130**, is reported by Hagihara and coworkers (793, 811) to be active for reduction of olefins, dienes, and diphenylacetylene at 20° and 1 atm hydrogen. Although initial hydrogenolysis to HRh(C$_8$H$_{12}$)(PPh$_3$) (eq. 371) followed by reaction with unsaturate seems feasible, these workers also reported that **130** forms complexes with reducible olefins such as butadiene and isoprene, for example, (C$_4$H$_6$)RhPh(PPh$_3$), and that this complex could be hydrogenated presumably to C$_4$ products, suggesting an "unsaturate route" (Sections XI-B-1 and XI-H, eq. 376). Ethylene reacts with **130** in the presence of triphenylphosphine to form the hydridotetrakis(phosphine) catalyst and styrene:

$$C_8H_{12}RhPh(PPh_3) + C_2H_4 \xrightarrow{PPh_3} HRh(PPh_3)_4 + PhCH{=}CH_2 \quad (371b)$$

Compare this with eq. 371. Whether such reactions occur in the absence of added phosphine to form HRh(C$_8$H$_{12}$)(PPh$_3$) is not clear; the mechanism of the hydrogenations remains to be established. Methyl acrylate, however, was hydrogenated stoichiometrically to methyl β-phenylpropionate, showing that the unsaturated moiety after coordination inserts into the rhodium-phenyl bond before hydrogenolysis:

$$\textbf{130} + \overset{\delta+}{CH_2}{=}CHCO_2CH_3 \longrightarrow \underset{/}{\overset{\backslash}{}}Rh{-}CHCO_2CH_3 \xrightarrow{H_2}$$

$$H_5C_6{-}\overset{|}{C}H_2$$

$$\underset{/}{\overset{\backslash}{}}Rh{-}H + C_6H_5CH_2CH_2CO_2CH_3 \quad (372)$$

It was not reported whether the hydride product, HRh(C$_8$H$_{12}$)(PPh$_3$), was a hydrogenation catalyst or not, but the corresponding chloride was said to be inactive. Keim (792) has concluded that a rhodium(I)-isopropyl complex would spontaneously decompose into hydride and propylene at 25°:

$$(Ph_3P)_3{-}Rh{-}CH(CH_3)_2 \rightarrow (Ph_3P)_3RhH + CH_3CH{=}CH_2 \quad (373)$$

This may be contrasted to reactions 371 and 372 involving the more stable rhodium-phenyl bonded systems. Clearly the chemistry of metal alkyls and aryls is as important as that of metal-hydrides for catalytic hydrogenation reactions. Of interest here is the thermal decomposition of the complex (Ph$_3$P)$_3$Rh—CH$_3$, by hydrogen abstraction at an ortho position of a phenyl ring, to give methane and Rh(PPh$_3$)$_2$(PPh$_2$C$_6$H$_4$) (728, 729, cf. **120**, Section XI-D-2). This complex also reacts with hydrogen to give, in the presence of phosphine, HRh(PPh$_3$)$_4$ (728).

Nixon and Clement (1329) have shown that the π-allyl complex

$Rh(C_3H_5)(PPh_3)_2$ (1330, 1331) is an effective catalyst for the hydrogenation of oct-1-ene in toluene at 25° and 1 atm.

Rhodium(I) monohydrides involving chelating phosphines, for example, $HRh(diphos)_2$ (252, 566, 789), are expected to be inefficient hydrogenation catalysts. The tetrakis(triarylphosphite) complexes $HRh[P(OAr)_3]_4$ (784) and the trifluorophosphine complex $HRh(PF_3)_4$ (797), because of the relatively high π-acidity of the ligands (Section XI-B-7), are also unlikely catalysts. The $HRh[P(OPh)_3]_4$ complex undergoes intramolecular oxidative addition of the *ortho*-phenyl C—H bond with subsequent hydrogen elimination to give $Rh[P(OPh)_2(OC_6H_4)][P(OPh)_3]_3$ (242, 959). (cf. Scheme 3, Section IX-B-2b.)

A phenoxide complex $Rh(OPh)(PPh_3)_3$ (782, 792), a hydroxide complex $Rh(OH)(PPh_3)_3$ (854), and rhodium(0) derivatives such as $Rh(PPh_3)_4$ (952) have been prepared, but their reactivity under hydrogen has not been reported.

Collman and coworkers (799) report that the nitrosyl complex $Rh(NO)(PPh_3)_3$ catalyzes the hydrogenation of hex-1-ene and cyclohexene in dichloromethane at 25° and 1 atm; deuteration of cyclohexene gives stereo-specifically $C_6H_{10}D_2$. The rhodium nitrosyl shows no reactivity toward hydrogen in the absence of substrate, but a series of other oxidative addition reactions were observed. The role of the nitrosyl group remains to be elucidated; its ability to function as a three- or one-electron donor (with linear or bent M—N—O geometries, respectively), when change to the latter creates a vacant coordination site at the metal center, could be critical in inducing catalytic activity generally in nitrosyls in contrast to similar carbonyls (799, 800). Added triphenylphosphine retarded the hydrogenations. Dewhirst (801) has patented the use of this complex and the arsine analog for reduction of nonaromatics such as hex-1-ene in toluene at 20 to 130° and 1 to 200 atm hydrogen; excess of the Group V ligand was said, however, to increase catalyst efficiency.

The complexes, *trans*-$RhX(CO)_2(PPh_3)$ (X = halide), undergo a wide variety of oxidative addition reactions (843, 844), but reactivity toward hydrogen would be expected to be less than that of the $RhX(CO)(PPh_3)_2$ complexes (Section XI-E). Benzene solutions of the metal-metal bonded complexes, $Cl_2(CO)(PPh_3)_2Rh–CuPPh_3$, (847) and $X_2(Ph_2AsMe)_3Rh–HgY$, where X is Cl, Br, and Y is halide, acetate (887), are reported unreactive toward hydrogen.

Khidekel and coworkers (812, 813) have prepared the rhodium-bis(di-methylglyoximate) complexes containing triphenylphosphine, $(Ph_3P)Rh-(DMGH)_2Cl$ and $[(Ph_3P)Rh(DMGH)_2]_2$, where $DMGH_2$ is dimethyl-glyoxime. Each of these complexes absorbs 1 mole of hydrogen in aqueous ethanol to give the rhodium(III) monohydride, $HRh(DMGH)_2PPh_3$, **131**, presumably according to eqs. 373a and 373b, respectively.

$$Rh^{III}Cl + H_2 \longrightarrow HRh^{III} + HCl \tag{373a}$$

$$(Rh^{II})_2 + H_2 \longrightarrow 2Rh^{III}H \tag{373b}$$

Complex **131**, the analog of a cobaloxime complex mentioned in Section X-H, behaves similarly as a hydrogen carrier for the reduction of nitrobenzene (812, 1472). Equation 373b corresponds to activation of hydrogen by dimeric cobaloxime(II) complexes. In these systems, decomposition of the hydride to anionic cobaloxime(I) species occurs readily in mildly alkaline solutions (eq. 312b), but such a reaction for the rhodium(III) hydride has not been reported. Complex **131** (or the chloride) in aqueous ethanol catalyzes the hydrogenation of butadiene to *cis*- and *trans*-but-2-enes at 5° and 1 atm pressure. An isolated intermediate, $(C_4H_7)Rh(DMGH)_2PPh_3$, was thought to be a mixture of the two possible σ-butenyl derivatives. This species liberated butenes on treatment with borohydride or hot butanol but not hydrogen, so that reactions similar to those in the corresponding $HCo(CN)_5{}^{3-}$ system are probably involved in the catalytic hydrogenation (cf. Section X-B-2a, eqs. 181 and 182).

Complexes such as $RhCl[P(CH_2CH_2CH{=}CH_2)_3]$ with a tetradentate phosphine ligand have been prepared by Hartwell and Clark (858, 1381) for use in studies of homogeneous hydrogenation (see also Section XI-E).

H.   OTHER RHODIUM(I, III)-CATALYZED HYDROGENATIONS IN NONAQUEOUS SOLVENTS

Besides the systems discussed in Sections B to G, which involved neutral phosphine, carbonyl, or carbonyl-phosphine complexes and their analogs, a significant number of other rhodium systems have been used for hydrogenations.

The activity of rhodium(I) complexes formed *in situ* from the cyclooctene complex $[RhCl(C_8H_{14})_2]_2$ by addition of a ligand L has been studied. Wilkinson and coworkers (638) reported that in benzene at 25° and 1 atm hydrogen, addition of pyridine, bipyridyl, or diethylsulfide to L:Rh ratios of up to 2:1 gave essentially inactive systems for the reduction of cyclohexene; addition of diphenyl sulfide resulted in some activity, but decomposition of the catalyst, probably to metal, was observed. James and Ng (774, 775) find, however, that addition of diethyl or dibenzyl sulfide (L:Rh = 2) or even chloride (L:Rh ⩾ 100) to the cyclooctene complex in DMA gives active solutions at 80° and 1 atm hydrogen for the reduction of unsaturated acids. The kinetics were independent of the substrate concentration and gave this simple rate law:

$$\frac{-d[H_2]}{dt} = k_2[Rh][H_2] \tag{374}$$

Table 65   Kinetic Data for the Hydrogenation of Maleic Acid and Other Compounds by Rhodium Complexes in DMA (694, 775, 776)[a,b]

Initial Complex	$k_2^{80°}$ $(M^{-1}\,s^{-1})$	$\Delta H_2^{\ddagger}$ (kcal mole^{-1})	$\Delta S_2^{\ddagger}$ (eu)	k_2^H/k_2^D
RhCl$_3$,3H$_2$O + LiCl(L:Rh $\geqslant$ 100)	1.5	18.4	−6.1	1.08
Rh(I) + LiCl(L:Rh $\geqslant$ 100)	2.1	18.0	−6.8	—
RhCl$_3$(Et$_2$S)$_3$	0.33	21.4	−1.0	1.05
Rh(I) + Et$_2$S(L:Rh = 2 or 3)	2.85	14.1	−17.4	1.2
Rh(I) + 2Et$_2$S + LiCl(Cl:Rh = 2)	0.56	—	—	—

[a] Rh(I) added as [Rh(C$_8$H$_{14}$)$_2$Cl]$_2$.
[b] Also hydrogenated: ethylene, fumaric acid, cinnamic acid, maleic anhydride, diethyl maleate, and maleic acid monoamide.

Some kinetic data for the reduction of maleic acid are summarized in Table 65. Conductometric data showed that the rhodium diethyl sulfide catalyst was a neutral species; the chloride had not been displaced (cf. the Rh(diene)L$_2^+$ catalysts considered in Section XI-G). In the absence of substrate, all the rhodium(I) complexes are rapidly reduced by hydrogen to the metal; the sulfur ligands, for example, are weaker π-acceptors than triphenylphosphine and no stable hydrides are formed. For this reason, James and Ng suggested that these reductions occur via the "unsaturate route" (Section XI-B-1). Spectrophotometric data showed rapid complexing of the olefin with the rhodium complexes believed to be present as monomers.

$$Rh^I + olefin \xrightarrow{\text{fast}} Rh^I(olefin) \tag{375}$$

$$Rh^I(olefin) + H_2 \xrightarrow{k_2} H_2Rh^{III}(olefin) \longrightarrow$$
$$Rh^I + saturated\ product \tag{376}$$

Reaction 376 is thought to proceed by oxidative addition of hydrogen to a square planar RhClL$_2$(olefin) complex followed by faster hydrogen transfer steps as in the phosphine type systems (cf. Section XI-B-4). The cyclooctene, liberated on dissolution of the starting complex in DMA containing the ligand L, is not hydrogenated. Terminal olefins such as hex-1-ene do not form sufficiently stable rhodium(I) olefin complexes to prevent decomposition to metal.

James and coworkers (596, 606, 694, 775, 776, 778) have also used DMA solutions of the rhodium(III) compounds 1,2,6-RhCl$_3$(Et$_2$S)$_3$, 1,2,6-RhCl$_3$[(PhCH$_2$)$_2$S]$_3$, and RhCl$_3$, 3H$_2$O (with excess chloride) for the hydrogenation of α,β-unsaturated acids and esters, and ethylene (as well as the reduction of inorganic substrates) at 1 atm and 80°. The uptake plots for the olefinic systems at constant pressure consist of an initial region involving reduction of rhodium(III) to rhodium(I), and then a long linear region involving reduction of the substrate by the reactions shown in eqs. 375 and 376. Both regions analyze according to a simple rate law of the form of eq. 374. Some data for the k_2 step (eq. 376) for the maleic acid reduction are given in Table 65. The good agreement between the parameters for the hydrogenations catalyzed by the rhodium(I) and rhodium(III) chloride systems show that the same reactive rhodium(I) intermediate and rate-determining step are involved in both systems. The faster hydrogenation rate in the Rh(I)–Et$_2$S system compared with the RhCl$_3$(Et$_2$S)$_3$ system is due to a lower activation energy. However, different rhodium(I) species are involved in these two systems, since the ratio of Rh:Cl is different. On addition of another 2 moles of chloride per rhodium to the rhodium(I) system, the rate constant becomes comparable to that for the rhodium(III) system. It was noted in Section XI-A-1 that rhodium(I) chlorides in aqueous solution were not effective for these hydrogenations although reaction 375 occurred. The promotion of reaction 376 in nonaqueous media is thought to be due to factors such as increased hydrogen solubility, the donor power, and dielectric constant of the DMA (776). Reaction 376, which will involve anionic chlororhodate(I) species, should be favored in a medium of lower dielectric due to greater attraction between the anion and the induced dipole of the hydrogen molecule.

Rates of hydrogenations catalyzed by RhCl$_3$(Et$_2$S)$_3$ are in the order *trans*-cinnamic acid > fumaric acid > maleic acid, which is probably the reverse order of the stability of the rhodium(I) olefin complex (775). The trend is opposite to that observed in the RhCl(PPh$_3$)$_3$ systems where the sterically hindered and *trans*-substituted olefins are less prone to hydrogenation, and where it was postulated that olefin complexation to the rhodium(III) dihydride was rate-determining (Section XI-B-3). The catalyzed deuteration of maleic acid, which was not isomerized under the conditions, when using the diethyl sulfide system, yielded a mixture of *meso*-2,3-dideuterosuccinic acid and unsymmetrically dideuterated succinic acid (775). The former product indicates overall *cis* addition, while the unsymmetrical product must arise from alkyl intermediates, thus showing that the hydrogen molecule transfers in two consecutive steps (cf. Section XI-B-4). The reaction scheme corresponds to that of Scheme 5 for a ruthenium(I) system (Section IX-B-2c).

The initial hydrogen reduction of the rhodium(III) complexes proceeds via rhodium(III) hydride intermediates as discussed in Sections XI-A-1 and XI-B-9. For the sulphide systems, the second-order rate constant (cf. eq. 374) showed an inverse dependence on added sulfide ligand, and the kinetic results were analyzed in terms of a predissociation of this ligand (775):

$$RhCl_3L_3 \xrightleftharpoons{K} RhCl_3L_2(DMA) + L \qquad (377)$$

$$RhCl_3L_2(DMA) + H_2 \xrightarrow{k_1} HRhCl_3L_2^- + H^+ + DMA \qquad (378)$$

These reactions lead to rate law 379.

$$\frac{-d[H_2]}{dt} = \frac{Kk_1[Rh][H_2]}{(K + [L])} \qquad (379)$$

Data for the reduction in the presence of maleic acid are given in Table 66.

Table 66   Kinetic Data for Reaction between Hydrogen and Rhodium(III) Complexes in DMA and DMSO (694, 775, 776, 808)

Initial Complex	$k_1^{80°}$	$\Delta H^{\ddagger}$ (kcal mole^{-1})	$\Delta S^{\ddagger}$ (eu)	K (M)	k_1^H/k_1^D
$1,2,6\text{-}RhCl_3(Et_2S)_3{}^a$	1.92	12.9	−21.4	0.05	1.1
$1,2,6\text{-}RhCl_3(Et_2S)_3{}^b$	3.35	16.8	−9	—	—
$1,2,6\text{-}RhCl_3[(PhCH_2)_2S]_3{}^a$	3.55	—	—	0.02	1.1
$RhCl_3,3H_2O^a$	1.67	17.3	−9.2	—	1.0
$RhCl_3,3H_2O^b$	2.00	27.0	+18	—	1.0
$RhCl_3,3H_2O + LiCl^a$ (L:Rh $\geqslant$ 100)	0.74	17.3	−10.8	—	1.0
$RhCl_3,3H_2O^c$	0.55	24.6	+9.0	—	—

^{a} DMA.
^{b} DMSO.
^{c} In 3 M HCl solution (605, 606); see Section XI-A-1.

The sulphide systems also showed a rate inhibition by added chloride; supporting spectral data suggested that this was probably due to the presence of $RhCl_4L_2^-$ which would be less efficient for reaction 378 where hydride has to displace only a solvent molecule. The inhibition of maleic acid hydrogenation observed on adding diethyl sulfide to the $RhCl_3(Et_2S)_3$ system was originally attributed incorrectly to repression of dissociation from $RhCl(Et_2S)_3$ by analogy with the corresponding triphenylphosphine system (694). There is no such inhibition when the *in situ* rhodium(I) catalyst is used. Such dissociations, however, seem important in the hydrogenation of cinnamic acid by this system, and in the hydrogenation of maleic acid by the benzyl sulfide system, since these reductions are inhibited by excess sulfide ligands (775). The dissociations are thought to relieve steric crowding in the $RhClL_2$(olefin) complex before oxidative addition of hydrogen (eq. 376). When $RhCl_3,3H_2O$ is used, the initial reduction process to rhodium(I) is similarly inhibited by added L (chloride), but the data do not analyze according to eq. 379. The k_1 value likely refers to a mixture of active species which with added chloride give less reactive higher chloro species (776).

The hydrogenation rate of maleic acid catalyzed by $RhCl_3,3H_2O$ in DMA in the absence of excess chloride showed a second-order dependence on rhodium, between first- and second-order on hydrogen (increasing with decreasing concentration), and a complex inverse dependence on substrate (i.e., after the initial rhodium(I) production according to the kinetics of eq. 374) (776). A qualitative explanation was given in terms of an active dimeric catalyst $[Rh^I(MA)]_2$ formed from a $Rh^I(MA)_2$ complex (MA = maleic acid; chloride and solvent ligands omitted):

$$2Rh^I(MA)_2 \underset{k_{-3}}{\overset{k_3}{\rightleftharpoons}} [Rh^I(MA)]_2 + 2MA \tag{380}$$

$$[Rh^I(MA)]_2 + 2H_2 \xrightarrow{k_4} 2Rh^I + 2 \text{ (succinic acid)} \tag{381}$$

The rate law obtained is of the correct form, assuming a steady state treatment for the dimer:

$$-\frac{d[H_2]}{dt} = \frac{k_3 k_4 [H_2]^2 [Rh_T]^2}{k_4 [H_2]^2 + k_{-3}[MA]^2} \tag{382}$$

At lower $[H_2]$, the rate becomes second-order in H_2, resulting from the termolecular reaction 381. A similar intermediate to the dimeric hydride **85** (Section XI-B-1), with MA, Cl, or solvent replacing PPh_3, was postulated. A dimethyl maleate complex $(C_6H_8O_4)RhCl$ has been described (777). Excess chloride is thought to prevent formation of dimers, which leads to the simpler kinetics discussed above. In the absence of chloride, the dimeric complex $[RhCl-(C_2H_4)_2]_2$ is inactive for ethylene hydrogenation in DMA (776, 778). Rylander and coworkers (248) have reported that rhodium trichloride in DMF is effective for the hydrogenation of dicyclopentadiene at room temperature and 1 atm.

$RhCl_3(Et_2S)_3$ is not an active catalyst in benzene, since the initial hydrogen reduction to rhodium(I) does not occur (694, 775). The basicity of DMA is thought to play an important role, particularly in stabilization of the released proton in reaction 378 (cf. Section XI-B-9 for similar rhodium(III) phosphine systems, and Section IX-B-1 for ruthenium chloride systems). As in the ruthenium systems, the enhanced reactions in DMA compared to the hydrochloric acid system are reflected in lower activation energies (Table 66).

Ethanolic solutions of $RhCl_3$, $3H_2O$, or 1,2,6-Rhpy$_3$Cl$_3$ catalyze the reduction of hex-1-ene at 25° and 1 atm hydrogen; accompanying isomerization to *trans*-hex-2-ene occurs and the uptake was considered due to this isomer (618). The pyridine complex is said to react with hydrogen to give HRhpy$_3$Cl$_2$ with elimination of hydrogen chloride but no details were given (612). The ionic, chelated 2,5-dithiahexane complex $[RhCl_2(CH_3SCH_2CH_2SCH_3)_2]Cl$ is inactive for hydrogenation in DMA at 80° and 1 atm hydrogen, presumably because of the difficulty of securing dissociation that leaves a vacant site for hydrogen and/or olefin (775).

On studying the catalytic activity of $RhCl_3,3H_2O$ in DMSO, James and coworkers (808) observed that a hydrogenation occurred in the absence of any organic substrate at 80° and 1 atm. The linear uptake plots at constant pressure pertained to a catalytic reduction of the solvent to dimethyl sulfide and water:

$$(CH_3)_2SO + H_2 \longrightarrow (CH_3)_2S + H_2O \tag{383}$$

The reaction was first-order in rhodium and hydrogen; this was ascribed to monohydride formation from a DMSO complex probably $RhCl_3(DMSO)_3$ (cf. eq. 378). The hydride then decomposed rapidly via a mechanism presented in eq. 384, which involves protonation of the basic oxygen atom followed

$$\tag{384}$$

by nucleophilic attack by the coordinated hydride. 1,2,6-RhCl$_3$(Et$_2$S)$_3$ was about twice as efficient for the same reaction. The kinetic data obtained are given in Table 66. A decline in rate after a few hours was due to formation of inactive rhodium(I) complexes via the rhodium(III) hydride intermediate (eq. 314, Section XI-A-1). Use of a hydrogen/oxygen mixture prevented rhodium(I) formation, but then a catalytic oxidation to dimethyl sulfone occurred. This observation is of interest, since neither the initial rhodium(III) complexes nor the rhodium(I) complexes formed during the hydrogenation studies performed this oxidation under oxygen alone, which suggests that rhodium hydrides are the oxygen carriers. Trocha-Grimshaw and Henbest (809) have reached similar conclusions for the same oxidation. The possibility that hydrogen peroxide is being produced catalytically is worthy of study.

McQuillin and coworkers (691, 692, 802–805) have used DMF solutions of the rhodium(III) complex, py$_2$(DMF)RhCl$_2$(BH$_4$), for the catalytic hydrogenation of a wide range of organic substrates (Table 67). The reductions

Table 67   Catalytic Hydrogenation and Deuteration Using py$_2$(DMF)RhCl$_2$(BH$_4$) in DMF at Room Temperature and 1 atm H$_2$^{a}

Substrate	Product	Reference
Oct-1-ene	Octane	691
Cyclopentene	Cyclopentane	802
Cyclohexene	Cyclohexane	802
Cycloheptene	Cycloheptane	802
Cyclooctene	Cyclooctane	802
Norbornene	Norbornane	802
Maleic acid (+D$_2$)	*Meso*-2,3-d_2 succinic acid	804
Fumaric acid (+D$_2$)	DL-2,3-d_2 succinic acid	804
Dimethyl acetylenedicarboxylate	Dimethyl maleate	804
Butyne-1,4-diol	*Cis*-but-2-ene-1,4-diol	804
Diphenylacetylene	*Trans*-stilbene	804
Azobenzene	Hydrazobenzene, aniline	805
Nitrobenzene	Aniline	805
Benzalaniline	Benzylaniline	805
Pyridine	Piperidine	805
Quinoline	1,2,3,4-Tetrahydroquinoline	805
3-Oxo-Δ^4-steroids^b	5α- and 5β-Hydrogenated products (See text)	803

^{a} [Rhpy$_3$Cl$_3$] = (3.7–7.5) × 10^{-3} M, [NaBH$_4$] = 2.2 × 10^{-2} M, [substrate] up to 1 M.
^{b} Cholestenone, testosterone, progesterone, and methyltestosterone.

have been carried out at room temperature and 1 atm pressure (806). The catalyst can be formed *in situ* by using Rhpy$_3$Cl$_3$ and borohydride. The hydrogen and not the borohydride is involved in the reduction process, but few kinetic details have been reported. The DMF seems to remain as a ligand during hydrogenation, since use of optically active amides led to asymmetric hydrogenation (692, Section XI-C). Added pyridine retards the reactions. Acetylenes appear to be reduced in a stepwise manner via the olefin. Besides unsaturated carbon-carbon bonds, —N=N—, —CH=N—, and —NO$_2$ groups have been reduced; only the hetero ring is hydrogenated in quinoline. 3-Oxo-Δ^4 steroids, which are not easily reduced by the RhCl(PPh$_3$)$_3$ system (Section XI-B-5c), are readily hydrogenated with the carbonyl bond remaining unaffected.

Cis-hydrogen addition seems general in the systems (804) but diphenylacetylene exceptionally gave *trans*-stilbene. On deuteration, deuterium was incorporated at the ortho position of the phenyl group of the product; it was suggested that the phenyl group was the primary hydrogen acceptor in a *cis* addition with the product arising by a 1,3-hydride shift (Scheme 14, path A). An alternative mechanism for a monodeuteride catalyst which involved migration prior to orthodeuteration was also presented (path B):

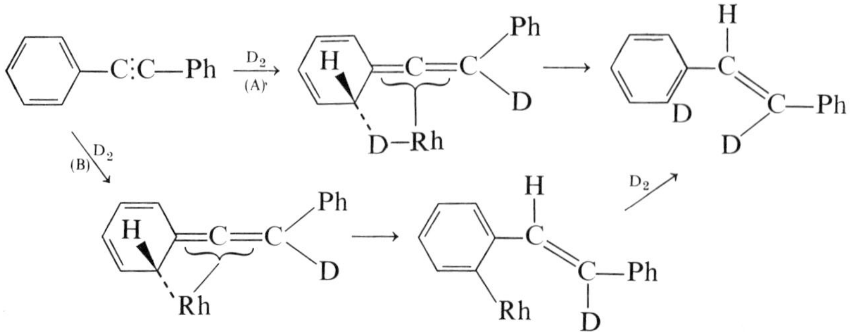

Scheme 14

Ortho-hydrogen exchange independent of addition to the triple bond (cf. Section IX-B-2b, Scheme 3) was considered unlikely.

The hydrogenation rate increased with substrate concentration for oct-1-ene (691) and the cyclic olefins (802), and with the latter it reached a limiting value which decreased in the order norbornene > cyclohexene > cycloheptene > cyclopentene > cyclooctene. Furthermore, the log of this limiting rate was proportional to the heat of hydrogenation of the substrate. Analogous data had been obtained by Jardine and McQuillin (807) for heterogeneous hydrogenation of the cyclic olefins; these workers concluded that the same rate-determining step of hydrogen transfer to a coordinated (or chemisorbed) olefin was involved (802). It should be noted, however, that in homogeneous systems a limiting rate with increasing olefin concentration does not necessarily imply that hydrogen transfer is rate-determining (cf. Section XI-B-1, and eq. 376 of this section). The product isomer distribution from the 3-oxo-Δ^4-steroids also showed the same trend in stereoselectivity as heterogeneous reduction on rhodium (803). In the homogeneous system, progesterone and testosterone gave about 80% of 5α product, while methyl-testosterone and cholestenone gave about 25% of the 5α isomer. Since these steroids substrates differ only in substitution at the C_{17}-position, the selectivity is thought to be governed by the bulk of the remote substituents.

Wilkinson and coworkers (700) have reported that ethanol solutions of the dimeric rhodium(I) chlorine-bridged complex $[Rh_2Cl_2(SnCl_3)_4]^{4-}$ are active at 25° and 1 atm hydrogen for the reduction of hex-1-ene, and at 110° and 50 atm for the reduction of n-heptaldehyde to n-heptanol. James and Pavlis (779) have used the same complex in acetone for the reduction of α,β-unsaturated acids under mild conditions; a dihydride is formed rapidly and the kinetics are similar to those discussed for the $RhCl(PPh_3)_3$ system (cf. eq. 329). It is not yet certain whether the hydride or unsaturate route is involved, but more interestingly whether the catalysts are monomeric or dimeric. The order in hydrogen is never greater than 1, however (cf. eq. 382). The catalysts are initially present as solvated neutral dimers, $Rh_2Cl_2(SnCl_2,$ solvent)$_4$ (681). The system is inactive in acetonitrile (779). It is possible that quite strong donors are required to cleave the chlorine bridge to produce active monomers.

The cationic diene complexes $Rh(diene)_2^+$ reported by Wilkinson and coworkers (681) and Green and coworkers (856) appear useful starting materials for production of hydrogenation catalysts. They react with a variety of donor ligands, including acetonitrile, to give, for example, $[Rh(1,5-C_8H_{12})(CH_3CN)_2]^+$ which effectively catalyzes the hydrogenation of hex-1-ene, and cycloocta-1,5-diene to cyclooctene; but cycloocta-1,3-diene is not reduced (856). The hydrogenation rate is solvent dependent, that is, acetone $>$ THF $\sim$ acetic acid. The catalyst resembles the phosphine-containing catalysts $Rh(diene)(PR_3)_2^+$ discussed in Section XI-G, which are readily formed from the acetonitrile complex on addition of the tertiary phosphine (856).

Khidekel and coworkers (188, 814, 815) have described semisandwich-type rhodium(I) complexes which are active hydrogenation catalysts for olefins in DMF, but little detail is available. The black, amorphous unstable compounds have been formed as follows: (a) reduction of rhodium trichloride in the presence of strong π-donors, such as 1,3,5-triphenylbenzene; (b) reduction or hydrogenation of the complex $[RhCl(duroquinone)]_n$ (816, 817), when the quinoid ligand becomes benzenoid; (c) heating rhodium trichloride in ethanol with 1,3-cyclohexadiene, when the coordinated ligand is again aromatized; (d) reduction of rhodium complexes with anthracene, naphthacene, or acridine. Reduction involves treatment with borohydride. A review (57a) has referred to this work of Khidekel, and states that similar catalysts can be formed from complexes of ruthenium, iridium, and molybdenum, although no details are available. The catalysts are thought to be of the same type as the ruthenium complex, $[C_6H_6RuCl_2]_n$, prepared by Winkhaus and Singer (818); the catalytic capability of which has been described (Section IX-B-2c).

Related to these complexes are some active catalysts based on rhodium(I) π-complexes with carboxylates containing a phenyl ring, which have also been studied by Khidekel and coworkers (819–822, 1447–1449). The complexes are formulated $H[Rh_2A_2Cl]$, where A is phenylacetate or the anion of the amino acids, N-phenylanthranilic acid, or L-tyrosine; their DMF solutions are active for hydrogenation of olefins, acetylenes, and aromatics at 20° and 1 atm (Table 68). The conditions are probably the mildest yet reported for the hydrogenation of aromatic rings. Keto groups were not reduced. The reduction of stilbene when using the phenylanthranilate complex (820)

Table 68   Catalytic Hydrogenation and Deuteration Using about $10^{-3} M$ $H[Rh_2A_2Cl]$ in DMF at 20° and 1 atm[a]

Substrate	Product	Reference
Ethylene	Ethane	819, 820
Hex-1-ene	Hexane	820
Fumaric acid $(+D_2)$	D,L-2,3-d_2 succinic acid	820
Stilbene	Ethylbenzene	820
Cyclohexene	Cyclohexane	819
Mesityl oxide	Iso-butyl methyl ketone	821
Methyl vinyl ketone	Ethyl methyl ketone	821
Acetylene $(+D_2)$	Cis-1,2-d_2-ethylene	820
Benzene	Cyclohexane	822
Naphthalene	Tetrahydro-, decahydronaphthalene	822, 1447
Anthracene	1,2,3,4-Tetrahydroanthracene	822, 1447
Chrysene, phenanthrene, pyrene, fluorene, fluoranthene	Readily hydrogenated	822
Diphenylacetylene, 2-propyn-1-ol, ethynylbenzene	Very slow reduction	820

[a] A = carboxylate, particularly N-phenylanthranilate.

followed the third-order rate law:

$$\frac{-d[H_2]}{dt} = k[Rh][H_2][\text{stilbene}] \qquad (385)$$

where $k = 10^5 \ M^{-2} \ min^{-1}$ and [Rh] is thought to refer to a monohydride produced rapidly according to the following reactions:

$$Rh_2A_2Cl^- + H_2 \longrightarrow HRh_2A_2Cl^{2-} + H^+ \qquad (386)$$

$$HRh_2A_2Cl^{2-} \longrightarrow HRhA^- + RhACl^- \qquad (387)$$

(DMF ligands omitted).

Conductivity data suggested the dissociation reaction 387 (820); hydride i.r. bonds were detected but these vanished on reaction with ethylene or cyclohexene (819). The mechanism postulated involved a reversible reaction of $HRhA^-$ with olefin (acetylene) to give an alkyl(vinyl) followed by (a) a rate-determining hydrogenolysis (cf. eqs. 351, 352 in Section XI-D-1) or (b) decomposition by proton, in which case the rhodium is regenerated as RhA and the rate-determining step was thought to be

$$RhA + H_2 \longrightarrow HRhA^- + H^+ \qquad (388)$$

However, it is not clear how an olefin dependence arises if eq. 388 is rate-determining; therefore, (a) seems more likely. The complete rate law for such a mechanism is given in eq. 130 (Section IX-B-2a); the dependence on olefin should decrease with increasing concentration as more alkyl is formed.

The *cis* deuterium addition to acetylene is readily accounted for by a four-centered transition state for the initial hydride transfer, while *cis* addition to fumaric acid requires that the final hydrogenolysis or proton attack occurs with retention of configuration at the metal-bonded carbon (see eq. 129 in Section IX-B-2a, and Scheme 2). With deuterium under 20 to 40 atm in the absence of substrate, exchange was effected at the tertiary carbon atom of DMF (819), presumably coordinated to the DRhA$^-$ species. In the presence of 10% water, a very slow exchange was also observed with protons. Deuteration of C$_2$H$_4$, however, at similar high-pressure conditions in the aqueous DMF quite rapidly gave C$_2$H$_6$ as well as C$_2$H$_5$D and C$_2$H$_4$D$_2$, suggesting that deuterium exchange is much faster in the intermediate DRhA(C$_2$H$_4$)$^-$.

Plots of autocatalytic hydrogen uptake were obtained for the reduction of mesityl oxide and methyl vinyl ketone by using the amino acid complexes. These uptake plots were not discussed, but a lower maximum rate for the former substrate was attributed to greater substitution at the olefinic center (821). The reduction rate of acetylene decreases with increasing acetylene partial pressure due to formation of an inactive complex (820).

In the Rh$_2$A$_2$Cl$^-$ anion with N-phenylanthranilate, the monosubstituted phenyl ring is thought to form a stabilizing semisandwich π-bonded structure through one rhodium atom while the carboxylate of the other ring co-ordinates to the second rhodium (819, 820); the chloride may be bridging. The catalysts were likened to the enzyme catalyst hydrogenase (cf. Section XVI-I), in view of the ligand substituents and its mechanism of hydrogen activation (eq. 386) in a polar solvent under mild conditions (820). Further-more, the catalysts can be used in the same manner as HRh(DMGH)$_2$PPh$_3$ (Section XI-G) and cobaloxime catalysts (Section X-H) to promote hydrogen transfer to nitrobenzene from another hydrogenation catalyst, that is, as a hydrogen carrier in a catalytic chain. Reaction 386 can take place in aqueous solution at pH 12, but the resulting hydride decomposes rapidly (820).

I. CATALYSIS BY RHODIUM(II) COMPLEXES

As mentioned in Section XI-G the rhodium(II) compound [(Ph$_3$P)Rh-(DMGH)$_2$]$_2$ is an effective hydrogenation catalyst due to its ready hy-drogenolysis to a rhodium(III) hydride (eq. 373b). Ng (775) finds that the DMA solutions of rhodium(I) chloride containing carboxylic acids (Sec-tion XI-H, eq. 375) can be oxidized stoichiometrically to green solutions containing rhodium(II). These solutions are also active for hydrogenation of unsaturated acids but the systems seem identical to the rhodium(I) catalyzed systems (Section XI-H). Initial hydrogen reduction to rhodium(I) via a rhodium(III) hydride formed according to eq. 373b seems likely. Rhodium(II) complexes appear to remain dimeric in solution and are commonly green (187, 823, 824).

Wilkinson and coworkers (187a, 825) have protonated the dimeric rhodium(II) acetate $Rh_2(CO_2Me)_4$ in fluoroboric acid-methanol media to displace the acetate and yield a green, air-stable diamagnetic ion, Rh_2^{4+}. On addition of triphenylphosphine (phosphine/rhodium $= 2:1$), the solutions become red and active for hydrogenation at 25° and 1 atm. Using 2.5 × 10^{-3} M solutions of dimer under these conditions with 1 M substrate, the following initial rates (ml min^{-1}) were recorded per 50 ml: 3-methylbut-1-yn-3-ol, 44.8; hexa-1,5-diene, 36.6; hex-1-yne, 34.2; allyl phenyl ether, 14.1; hex-1-ene, 10.9; diethyl maleate, 4.8; hex-2-enes, 3.3; *cis*-hept-2-ene, 2.2; cyclohexene, 1.1. Cycloocta-1,5-diene, propargyl alcohol, and acetylene-dicarboxylic acids were also hydrogenated at somewhat different conditions. This variation in sequence with substrate is similar to that reported by Osborn and coworkers (786) with their $H_2Rh(PPh_3)_2(solvent)_2^+$ catalysts (Section XI-G); such a catalyst could well be produced after an initial hydrogen reduction of the rhodium(II) dimer to rhodium(I), as mentioned above. The hex-1-ene reduction was approximately first-order in each of rhodium, hydrogen, and olefin, at lower concentrations; but the orders decreased at higher concentrations. The red catalyst solutions, which reacted with unsaturated carboxylic acids and alcohols to give yellow-green solutions, also yielded a solid $Rh(PPh_3)_3^+BF_4^-$ on treatment with excess phosphine. This complex could formally be rhodium(II) if hydrogen abstraction occurs from the ortho position of a phenyl ring to give a metal-carbon bond (cf. Scheme 3 in Section IX-B-2b), but there was no evidence of such an interaction.

Hui and Rempel (824) have used DMF solutions of $Rh_2(CO_2Me)_4$ by itself at 30 to 50° and 1 atm pressure for the hydrogenation of terminal olefins (including ethylene), maleate, and cyclooctene. The initial rates ranged from 17 × 10^{-2} ml min^{-1} for hex-1-ene to 1.3 × 10^{-2} ml min^{-1} for cyclooctene, on using 5 ml solution with 5 × 10^{-3} M in dimer and 1 M in substrate. Hexadiene, *trans*-hex-2-ene, and fumarate were not reduced. The rhodium(II) dimer in this case seems to be the true catalyst, since it can be recovered unchanged. The systems are not sensitive to oxygen, since this is catalytically reduced to water (cf. Sections II-B and IX-B-1). With dec-1-ene as substrate, the reaction is first-order in dimer and hydrogen, and the olefin dependence changes from first- to zero-order as its concentration increases; there was no observable reaction of the dimer with either hydrogen or olefin. The following mechanism accounted for these data (cf. Section IX-B-2c, eqs. 143 to 143b):

$$Rh_2(CO_2Me)_4 + H_2 \underset{k_{-1}}{\overset{k_1}{\rightleftharpoons}} Rh_2(CO_2Me)_4H_2 \quad (k_{-1} > k_1) \qquad (389)$$

$$Rh_2(CO_2Me)_4H_2 + olefin \overset{k_2}{\longrightarrow} Rh_2(CO_2Me)_4 + paraffin \qquad (389a)$$

Since the dimer is known to rapidly form solvent (S) adducts of the type $SRh(CO_2Me)_4RhS$, with S occupying the terminal positions of the bridged complex (823, 827–829), displacement by hydrogen and olefin seems likely. Hydrogenation was thought to proceed at only one of the rhodium atoms (824). Comparable activity was observed in THF. Lower activity was observed in DMA, ethanol, and dioxane, as well as in DMSO where a long induction period was noted.

XII

Group VIII Metal Ions and Complexes

IRIDIUM

A. CHLOROCARBONYLBIS(TRIPHENYLPHOSPHINE)IRIDIUM(I), IrCl(CO)(PPh₃)₂, AND ANALOGS

Renewed interest in iridium chemistry surged after Vaska and Diluzio (34) reported on the simple reversible hydrogen-activating system involving the title compound at 25° and 1 atm hydrogen. The importance of this discovery was mentioned in the Introduction (Section I-A). The general reaction occurring in solution is shown in eq. 390, where X is a halide or other anionic ligand and PR₃ is a tertiary phosphine or related ligand:

$$\text{trans-IrX(CO)(PR}_3)_2 + \text{H}_2 \underset{k_{-1}}{\overset{k_1}{\rightleftharpoons}} \text{H}_2\text{IrX(CO)(PR}_3)_2 \qquad (390)$$

The iridium(III) dihydride complexes are just one example of a very wide range of adducts that may be formed with these remarkable iridium(I) compounds (e.g., Ref. 815). Some comparisons with the rhodium analogs have already been made in Section XI-E. The iridium complexes have been used as hydrogenation catalysts, and, although not very efficient, the systems are highly suitable for such studies, particularly because the carbonyl stretching frequency responds to changes in the environment of the central atom and is a measure of the "inherent basicity" of these complexes (36, 41a, 860, 861). Increasing electron density at the iridium results in decreasing

288

ν_{CO} values for a series of complexes with the same halide ligand (cf. Table 70). Variation of the halide ligand with the same phosphine ligand generally has little effect on ν_{CO}, except for triphenylphosphine when ν_{CO} increases in the order Cl < Br < I (cf. Table 70 and Ref. 247) which is the inverse of that expected from the basicity trend.

1. Activation of Hydrogen. The square-planar *trans*-IrX(CO)(PPh₃)₂ complexes (X = halide) were first prepared by Angoletta (884) and were subsequently synthesized more easily and characterized more fully by Vaska and coworkers (35, 730). The dihydride, particularly of the chloride, with *cis* hydrogens and *trans* phosphines has been characterized especially by using n.m.r. and i.r. methods (34, 84, 264, 730, 860, 931). Deeming and Shaw (932) have demonstrated the same configuration for H₂IrCl(CO)(PMe₂Ph)₂ and H₂IrBr(CO)(PEt₂Ph)₂. Incorporation of phosphine (or arsine) ligands of the type PMe₂Ph can be useful, since the methyl n.m.r. pattern may be used for generally studying oxidative addition reactions as exemplified by eq. 390 (934). Vaska and coworkers (36, 860, 869–871) and Strohmeier and coworkers (41a, 861–866) have studied the IrX(CO)(PR₃)₂ complexes and equilibrium 390 in some detail; Chock and Halpern (247) and James and coworkers (867, 868) have also obtained data for the triphenylphosphine complexes. Some kinetic data for the second- and first-order rate constants, k_1 and k_{-1}, respectively, and thermodynamic parameters for the equilibrium constant $K_1 (= k_1/k_{-1})$ are given in Tables 69 to 72. Reaction 390 has been followed by gas-uptake techniques and spectroscopic methods (u.v. and i.r.).

The addition reactions are exothermic, spontaneous, and generally, as expected, have a negative entropy change. The $\Delta S°$ values in Table 69 depended little on X and were also stated to depend little on L. These were thought to reflect mainly the loss of the translational entropy of the hydrogen molecule upon coordination (860). Vaska and Werneke (860) have used the data of Table 69 with eq. 391 (40, 61, 612) to estimate the average iridium-hydrogen bond dissociation energy, $D(\text{IrH})$.

$$-\Delta H° = 2D(\text{IrH}) - D(\text{HH}) - R \qquad (391)$$

$D(\text{HH})$ is the dissociation energy of H₂, and R is the reorganization or rehybridization energy which is comprised of the electronic promotion energy, E [for the formal iridium(I) to iridium(III) change] and a stereochemical component, S. Vibrational spectra (264, 860) suggested that $D(\text{IrH})$ was essentially invariant in the complexes H₂IrX(CO)(PR₃)₂; for a particular complex the two iridium-hydrogen stretching frequencies generally differed by about 100 cm⁻¹. Neglecting R in eq. 391 gave $D(\text{IrH})$ as 59, 60, and 61 kcal for the chloro, bromo, and iodo complexes, respectively (Table 69). These values are close to that of $D(\text{CoH})$ in HCo(CN)₅³⁻,

Table 69   Data for Reaction 390 with IrX(CO)(PPh$_3$)$_2$ in Chlorobenzene at 30° (860)

X	k_1 (M^{-1} s^{-1})	$\Delta H_1^{\ddagger}$ (kcal)	$\Delta S_1^{\ddagger}$ (eu)	$10^5 k_{-1}$ (s^{-1})	$\Delta H_{-1}^{\ddagger}$ (kcal)	$\Delta S_{-1}^{\ddagger}$ (eu)	$\Delta H°$ (kcal)	$\Delta S°$ (eu)	$\Delta G°$ (kcal)
Cl	1.2^a	12	−20	4.0	26	+6	−14	−26	−6.3
Br	8.2^b	7.8^c	−28^c	—	—	—	−17	−31	−7.6
I	430	6	−26	1.1	25	+1	−19	−27	−11
NCO	0.022	13	−23	1.4	26	+5	—	—	—

^{a} The k_1 values for the p-tolyl and p-anisole analogs were 1.7 and 2.1 M^{-1} s^{-1}, respectively.
^{b} Estimated from Fig. 5 of Ref. 860.
^{c} Calculated from eq. 391 (with $R = \Delta H_1^{\ddagger}$) using $D(\text{IrH}) = 64$ kcal (see text).

58 kcal, estimated similarly (Section X-A-2), and of $D(\text{IrH})$ for chemisorbed hydrogen on supported iridium metal, 65 kcal (7). Such considerations indicated that R was relatively small compared to $D(\text{IrH})$ and was probably an endothermic term in the overall energetics. Furthermore, E is expected to decrease with increasing basicity of the complex or ligands (40, 860), that is, $R:\text{Cl} > \text{Br} > \text{I}$; thus the true $D(\text{IrH})$ values were thought to be somewhat higher and possibly closer together than those given above. The activation energy for dehydrogenation, $\Delta H_{-1}^{\ddagger}$, was said to be essentially independent of X and PR$_3$, while $\Delta H_1^{\ddagger}$ was inversely proportional to $-\Delta H°$ (860, Table 69). Vaska and Werneke have tentatively suggested an intrinsic relationship between $\Delta H_{-1}^{\ddagger}$ and $D(\text{IrH})$ and, more importantly, between $\Delta H_1^{\ddagger}$ and R, since the former follows the expected trend of the latter; for example, the trend in $\Delta H_1^{\ddagger}$, I < Br < Cl, follows the electronegativities. Assuming $R = \Delta H_1^{\ddagger}$, the $D(\text{IrH})$ values for the IrX(CO)(PPh$_3$)$_2$ complexes (X = halide) were all 64 kcal, very close to the chemisorbed hydrogen value. The assumption was partly rationalized in that the transition state configuration closely resembles the product, that is, the reorganization and formation of hydride bonds are nearly complete in the activated complex (see eq. 392 below). The tris(cyclohexyl)phosphine complex, IrCl(CO)[P(C$_6$H$_{11}$)$_3$]$_2$, although more basic than the triphenylphosphine analog (cf. Table 70), has a somewhat higher $\Delta H_1^{\ddagger}$ value of 14 kcal; yet, the calculated $D(\text{IrH})$ value is again 64 kcal (860). Since $\Delta H_1^{\ddagger} = R = E + S$, the activation energy increase was thought to reflect the few kilocalories necessary for reaction of the sterically hindered complex (see below). Data from Strohmeier's group, however, (Table 70) indicate a low activity due to an unfavorable $\Delta S_1^{\ddagger}$ value.

According to Vaska and Werneke (860), $-\Delta H°$ is inversely proportional to the electronegativity of X in the complexes IrX(CO)(PPh$_3$)$_2$, where X = halide (Table 69), N$_3^-$, NCO$^-$, NCS$^-$, and NO$_3^-$; also $-\Delta H°$, the stabilities of the hydrides ($-\Delta G°$), and k_1 increase with polarizability of X.

The K_1 (and k_1) values for $IrCl(CO)(PR_3)_2$ increase somewhat with basicity of the phosphine ligands having para-substituents on the phenyl ring: $OCH_3 > CH_3 > H$ (Tables 69 and 70). The tris(perfluorophenyl)phosphine complex is a very weak base ($\nu_{CO} = 1994$ cm^{-1}, cf. Table 70) and shows no reaction with hydrogen (860). These data are all consistent with reaction 390 being thought of as interaction of the Lewis acid hydrogen molecule with a basic iridium complex (860). The tri-n-butyl- and triethylphosphine analogs are somewhat less active than might be expected from their basicities (860; Table 70). This has been attributed to differences in detailed stereochemistry of these complexes compared to the phenyl and substituted-phenyl phosphine complexes (860). Table 70 indicates that the low activity of the isopropyl complex, presumably sterically hindered, is due to an unfavorable $\Delta S_i^{\ddagger}$ value. On descending the series given in Table 71 the decreasing ν_{CO} values indicate increasing basicity in the $IrCl(CO)(PR_3)_2$ complex; however, the orthomethyl and the cyclohexyl groups sterically hinder the attack by hydrogen.

The interaction with hydrogen has been depicted (860) as follows (the *trans* triphenylphosphine ligands, normal to the plane of the paper, are not shown):

$$\tag{392}$$

The $\Delta S_i^{\ddagger}$ values in chlorobenzene are of the same magnitude as ΔS° (Table 69), implying that the geometry of the activated complex **132** resembles that of the dihydride. Furthermore, linear free energy relationships (LFER) were apparent on plotting $\ln k_1$ (or $\ln k_{-1}$) versus $\log K_1$ for various $IrX(CO)$-$(PPh_3)_2$ complexes, indicating a common mechanism (860). The slope of the k_1 plot was about 0.8 which represents a "resemblance factor" of the activated complex to the product; the Cl–Ir–C angle in **132** was predicted to be about 110° (36, 860). Similar LFER plots were obtained for the complexes $IrCl(CO)(PR_3)_2$ with R = ethyl, butyl, phenyl, m- and p-tolyl, and p-anisyl; the data for the sterically hindered cyclohexyl complex deviated markedly (860).

The k_1 values for $IrCl(CO)(PPh_3)_2$ increase somewhat with solvent polarity; for example, in ratios of approximately $1:2:3$ for the series toluene < chlorobenzene < DMF (Table 72; 247). But according to Vaska and Werneke, the activation parameters (Table 72), ΔH°, and calculated $D(IrH)$ values are little affected by using these different solvents, and the mechanism is thought to be the same. Rates in benzene are somewhat faster than those in toluene (247). From data given in Ref. 867 for DMA solution, values of ΔH° and

Table 70 Data for Reaction 390 in Toluene Mostly (862, 863, 865)

Complex		ν_{CO}	$k_1(30°)$	$k_1(80°)$	$\Delta H_1^‡$	$\Delta S_1^‡$	$10^4\,k_{-1}(80°)$	$\log K_1' = -A/T + B$[b]	
X	R	(cm^{-1})	$(M^{-1}\,s^{-1})$	$(M^{-1}\,s^{-1})$	(kcal)	(eu)[a]	(s^{-1})	$-A$	B
Cl	C_6H_{11}	1932	0.001	0.016	11.4	−35	1.68	2604	−4.45
Cl	$i\text{-}C_3H_7$	1935	0.008	0.104	10.9	−32	1.71	2330	−2.87
Cl	C_4H_9	1940	0.36	—	11.4	−23	—	—	—
Cl	$R_3 = C_4H_9Ph_2$	—	—	—	—	—	—	2506	−3.23
Cl	CH_2Ph	1956	0.64	7.67	10.6	−25	1013	3259	−6.41
Cl	$p\text{-}CH_3C_6H_4$	1963	0.89	11.5	10.9	−23	489	2879	−4.78
Cl	Ph	1967	0.59	8.38	11.3	−22	500	3462	−6.64
			0.93^c		10.8^c	-23^c			
Cl	OPh	2001	0.07	1.21	12.0	−24	20.8	2692	−3.91
Br	C_6H_{11}	1932	0.016	0.39	13.5	−22	3.41	2710	−3.67
Br	$i\text{-}C_3H_7$	1935	0.082	2.00	13.6	−19	12.3	2867	−3.97
Br	Ph	1969	11.2	183	11.9	−14	300	1810	−0.40
			14.3^c		12.0^c	-14^c			
Br	OPh	2001	3.4	41	10.6	−21	198	2131	−1.77
I	C_6H_{11}	1932	0.69	10.0	11.4	−22	320	3954	−7.76
I	$i\text{-}C_3H_7$	1938	2.49	26.0	10.0	−24	15.3	894	+2.64
I	Ph	1970	~400	~4300	—	—	~8600	1143	+1.38
			$>100^c$						
I	OPh	2003	~100	~1000	~10	−16	~1200	600.4	+3.16

[a] Calculated from data in Refs. 862 and 863.

[b] K_1' written in terms of mole fractions; Ref. 865 gives $K_1(=k_1/k_{-1})$ as about $K_1' \times 10^{-1}$.

[c] In benzene (247).

Table 71   Data for Reaction 390 with IrCl(CO)(PR₃)₂ in Chlorobenzene at 30° (860)

R	ν_{CO} (Nujol) (cm^{-1})	k_1 (M^{-1} s^{-1})	$10^{-3} K_1$ (M^{-1})
m-CH₃C₆H₄	1967	0.70	24
p-CH₃C₆H₄	1956	1.7	44
o-CH₃C₆H₄	1946	No reaction	
C₆H₁₁	1934	0.0034	6

$\Delta S°$ are estimated to be about -12 kcal and -22 eu, while from Table 70 (toluene) the values are about -15.5 kcal and -26 eu respectively; the values are close to those reported for the chlorobenzene medium (Table 69). Preliminary data in sulfolane indicate corresponding values of about -11 kcal and -18 eu (868). The higher dipole moment of the dihydride, $\mu \approx 6D$ (860) compared to that of the starting complex, $3.9D$ (35) suggests an increase in polarity on going to the activated complex, which is consistent with the higher rates in the more polar DMF (247, 860). In agreement with the close resemblance of **132** to the dihydride, k_{-1} was independent of the solvent (Table 72).

James and Memon (867) report that the following dissociation equilibrium is slowly established in DMA solution:

$$IrCl(CO)(PPh_3)_2 \underset{k_{-2}}{\overset{k_2}{\rightleftharpoons}} IrCl(CO)(PPh_3)(DMA) + PPh_3 \qquad (393)$$

Table 72   Kinetic and Equilibrium Data for Hydrogenation of IrCl(CO)(PPh₃)₂ in Different Solvents

Solvent	k_1 (M^{-1}s^{-1})^a	$\Delta H_1^{\ddagger}$ (kcal)	$\Delta S_1^{\ddagger}$ (eu)	$10^5 k_{-1}$ (s^{-1})^a	$\Delta H_{-1}^{\ddagger}$ (kcal)	$\Delta S_{-1}^{\ddagger}$ (eu)	$10^{-4} K_1$ (M^{-1})^b	Reference
C₆H₆	0.67	10.8	-23	—	—	—	1.5	247
C₆H₅CH₃	0.48	11	-22	1.7	25	$+4$	—	860
	0.59^b	11.3	-22	—	—	—	0.6	862, 863
C₆H₅Cl	0.88	12	-20	1.7	26	$+6$	3.2	860
DMF	1.4	11	-22	1.7	24	-1	—	860
DMA	—	—	—	—	—	—	0.4	867
Sulfolane	—	—	—	—	—	—	2.5	868

^{a} At 25°.
^{b} At 30°; the solubility of H₂ in the different solvents appears much the same, about (2–3) × 10^{-3} M atm^{-1} (cf. 612, 776, 863).

At 25°, $k_2 \simeq 10^{-5}$ s^{-1} and $k_{-2} \simeq 3.9$ M^{-1} s^{-1} which corresponds to about 15% dissociation at 10^{-4} M complex; at 80°, $k_2 \approx 10^{-4}$ s^{-1} and $k_{-2} \approx$ 75 M^{-1} s^{-1}. No dissociation was observed in benzene; Refs. 41a and 862 incorrectly cite a measurable dissociation (see also Section A-2). Strohmeier and Onoda (866) have studied the solvation of IrX(CO)(PR$_3$)$_2$ complexes, without loss of the phosphine ligand. The tendency to solvate (measured by changes in ν_{CO}) increases as follows: toluene $<$ benzene $\simeq$ chloroform $<$ carbon disulfide: chloride $<$ bromide $<$ iodide; for a given halide, R $=$ cyclohexyl $\ll$ phosphite $<$ phenyl. Endres (1466), however, has apparently determined an equilibrium constant of about 10^{-5} M for dissociation of a phosphine ligand in toluene.

The data for reaction 390 in toluene observed by Strohmeier and coworkers (Table 70) show serious differences from those of Vaska and Werneke (Table 69) measured in chlorobenzene. However, as indicated above, at least for the IrCl(CO)(PPh$_3$)$_2$ complex, the solvent used seems to be of little importance, but the possibility of slower dissociation reactions (cf. eq. 393) should not be ignored. Although the two sets of data agree well for the k_1 values for the IrXCO(PPh$_3$)$_2$ (X = halide) complexes at 25 to 30° and the activation parameters agree for the chloro complex, the $\Delta H_1^{\ddagger}$ and $\Delta S_1^{\ddagger}$ values for the bromo complex are at variance. The same reactivity trend Cl $<$ Br $<$ I is true for all the different phosphine ligands in Table 70, but this is not due to a decreasing $\Delta H_1^{\ddagger}$ value. Indeed, Strohmeier and Onoda (862) have stressed the essential constancy of $\Delta H_1^{\ddagger}$ and the importance of the $\Delta S_1^{\ddagger}$ term, in contrast to the discussions of Vaska and Werneke (860). Chock and Halpern's (247) activation parameters for the chloro and bromo complexes in benzene agree well with those of Strohmeier's group (Table 70). The overall enthalpy and entropy changes for the bromo and iodo triphenyl-phosphine complexes in Table 70 are again quite different from those given in Table 69. These changes indicate that the increasing K_1 values (Cl $<$ Br $<$ I) result from more favorable $\Delta S°$ values—again, quite contradictory to the data of Vaska and Werneke. Solvation effects at the more polarizable larger halide ligands could possibly explain these apparent discrepancies.

Camia and coworkers (883) have reported that a complex formulated as "Ir(SnCl$_3$)(CO)(PPh$_3$)$_2$," but believed to be the pentacoordinate species IrCl(CO)(PPh$_3$)$_2$SnCl$_2$, forms a dihydride. Although details are not available, the product is likely to be the 6-coordinate H$_2$Ir(SnCl$_3$)(CO)(PPh$_3$)$_2$ in comparison with the formulation of isolated ethylene and acetylene adducts (Section A-2). This dihydride also has been synthesized by another method (931).

Malatesta and coworkers (890) have reported the *cis* complex HIr(CO)-(PPh$_3$)$_2$ which readily undergoes reversible hydrogenation to give *cis*-H$_3$Ir(CO)(PPh$_3$)$_2$. *Cis*-HIr(CO)(AsPh$_3$)$_2$ behaves similarly (910). However,

Harrod and coworkers (214) have questioned the existence of HIr(CO)(PPh$_3$)$_2$ and indicated that it disproportionates to HIr(CO)(PPh$_3$)$_3$ (see Section XII-B-2).

The dihydride H$_2$IrBr(CO)(PPh$_3$)$_2$, with *cis* hydrogens and *cis* phosphines, and complexes such as *cis*-IrCl(CO)(PEt$_3$)$_2$ are known (891). The arsine complex IrCl(CO)(AsPh$_3$)$_2$ and a 5-coordinate stibine complex IrCl(CO)-(SbPh$_3$)$_3$ have been isolated, but their reactivity toward hydrogen was not reported (896). Dihydrides of the type H$_2$Ir(SiR$_3$)(CO)(PPh$_3$)$_2$, R = alkyl, have been synthesized (935–937) as have the corresponding trialkylgermanium analogs (891, 1396). The complexes H$_2$Ir(GeR$_3$)(CO)(PPh$_3$)$_2$ (R = Me, Et) have been prepared by Glockling and Wilbey (1396) by reacting hydrogen with Ir(GeR$_3$)(CO)(PPh$_3$)$_2$ under mild conditions. The square phenyl complex IrPh(CO)(PPh$_3$)$_2$ is known (725, 726).

The nitrogen analog of Vaska's compound, IrClN$_2$(PPh$_3$)$_2$, studied by Collman and workers (913, 914), is unreactive toward hydrogen up to 4 atm.

The reaction of IrCl(CO)(PPh$_3$)$_2$ with hydrogen in benzene exhibited only a small deuterium isotope effect, $k_{H_2}/k_{D_2} = 1.22$ (247). Reaction with HD gave, in addition to the two isomeric HD products, an equimolar mixture of the dihydride and dideuteride (871). This resulted from slow H$_2$–HD–D$_2$ exchange reactions which were demonstrated by using H$_2$–D$_2$ mixtures in toluene at 25° (869). The exchange reactions were written involving 8-coordinate intermediates (eq. 394), but the possibility of a mechanism involving concerted steps (e.g., eq. 395) or a partially dissociated starting complex was not ruled out. Hydrogen deuteride adds to IrBr(CO)(PEt$_2$Ph)$_2$

$$\text{Ir} \underset{-H_2}{\overset{H_2}{\rightleftharpoons}} \text{IrH}_2 \underset{-D_2}{\overset{D_2}{\rightleftharpoons}} \text{IrH}_2\text{D}_2 \underset{+HD}{\overset{-HD}{\rightleftharpoons}} \text{IrHD} \underset{-H_2}{\overset{H_2}{\rightleftharpoons}} \text{IrH}_3\text{D}, \ldots \quad (394)$$

$$\text{IrH}_2 + \text{D}_2 \rightleftharpoons \text{IrHD} + \text{HD}, \qquad \text{Ir} = \text{IrCl(CO)(PPh}_3)_2 \quad (395)$$

to give the expected isomeric mixture (932). IrCl(CO)(PPh$_3$)$_2$ also catalyzes parahydrogen conversion (870).

Vaska and Werneke (860) consider that the reaction with hydrogen involves the feeding-in of electron density of a filled nonbonding metal orbital (the base) to an antibonding molecular orbital of hydrogen with a resultant reduction of the H—H bond order to zero as well as a simultaneous formation of two metal-hydrogen covalent bonds. Such an electronic model was originally suggested by Nyholm (61); Wilkinson's group (612) had adopted a similar model for their RhCl(PPh$_3$)$_3$ catalyst, although they had suggested that the higher energy p orbitals of hydrogen could act as acceptors. In earlier years, Halpern (25, 26) had suggested that the bonding electrons of hydrogen could attack a vacant metal orbital. More recently, Carra and Ugo (40) have discussed the data of Chock and Halpern (Table 70; 247) for reaction 390 in such terms and suggested that the free energy of activation, which is lower for the bromide than for the chloride, is a rough measure of the availability of the vacant iridium $6p_z$ orbital (see also Chapter XVII).

2.   Hydrogenation of Olefins, Acetylenes, and Inorganic Substrates.   Studies on the reduction of olefins and acetylenes have been reported by Vaska and coworkers (730, 869), James and coworkers (867, 868), Strohmeier and coworkers (872–874, 1465), and Yamaguchi (902).

Vaska and Rhodes (730) first reported the catalyzed hydrogenation of ethylene, propylene and acetylene by using $IrX(CO)(PPh_3)_2$, X = halide, in benzene or toluene at 40 to 60° and 1 atm. The yields with the chloride were somewhat better than when using the rhodium analog (Section XI-E): at 60° in 18 hr, ethylene gave 40% ethane, propylene gave 10% propane, and acetylene gave 10% ethylene and 5% ethane. Slow reduction to metal occurred at higher temperatures. Unlike the rhodium analogs, the iridium complexes reacted reversibly with the substrates at ambient conditions:

$$
\begin{array}{ccc}
\overset{\displaystyle X}{\underset{\displaystyle CO}{\overset{|}{\underset{|}{Ir}}}} + \overset{\displaystyle CH_2}{\underset{\displaystyle CH_2}{\|}} \rightleftharpoons & \overset{\displaystyle X}{\underset{\displaystyle OC}{}}\overset{}{Ir}\overset{\displaystyle CH_2}{\underset{\displaystyle CH_2}{}} & \text{or} \quad \overset{\displaystyle X}{\underset{\displaystyle OC}{}}Ir{-}\overset{\displaystyle CH_2}{\underset{\displaystyle CH_2}{\|}}
\end{array}
\qquad (396)
$$

$$\textbf{133} \qquad\qquad\qquad \textbf{133a}$$

The products, which were not isolated, can be formulated as either iridium(III) complexes with two σ Ir—C bonds (**133**) or as a π-bonded iridium(I) complex (**133a**). Such formulations have been discussed, particularly for the corresponding oxygen complexes (e.g., 36). In reaction 396, the iodo complex was much more reactive than the chloride.

Eberhardt and Vaska (869) later studied the deuteration of ethylene in toluene at 60° and 1 atm total pressure, using $IrCl(CO)(PPh_3)_2$; the reaction yields d_0 to d_4 ethanes (mainly d_2), and d_1 and d_2 ethylenes, as well as HD and H_2. Since there was no D_2–C_2H_6 exchange in this system, the reaction must proceed at least in part via a mechanism in which both reactants are associated with the catalyst, with hydrogen transfer occurring at the same

$$
\begin{array}{ccccc}
D_2IrC_2H_4 & \rightleftharpoons & DIrC_2H_4D & \rightleftharpoons & HDIrC_2H_3D & \rightleftharpoons \\
\Updownarrow & & \downarrow & & \Updownarrow \\
D_2 + C_2H_4 & & C_2H_4D_2 & & HD + C_2H_3D
\end{array}
$$

$$
\begin{array}{ccc}
& HIrC_2H_3D_2 & \rightleftharpoons & H_2IrC_2H_2D_2 \\
& \downarrow & & \Updownarrow \\
& C_2H_4D_2 & & H_2 + C_2H_2D_2
\end{array}
\qquad (397)
$$

metal atom. Addition and elimination of the hydride (deuteride) through reversible ethylene-ethyl complexes accounts for the observed products; for example, as shown in eq. 397, followed by recycling. Whether deuterium or ethylene coordinated first was not considered. Concurrent mechanisms involving attack by gas-phase ethylene (eq. 398) or two metal centers (eq. 399) could not be excluded. "Ir" was written to represent $IrCl(CO)(PPh_3)_2$ and in eq. 397 implied 8-coordinate intermediates (cf. eqs. 394, 396), although

$$IrH_2 + C_2H_4 \longrightarrow Ir + C_2H_6 \tag{398}$$

$$IrH_2 + IrC_2H_4 \longrightarrow 2Ir + C_2H_6 \tag{399}$$

this may not necessarily be so (see below). The accompanying isomerization of but-1-ene to but-2-enes, observed during hydrogenation to butane, was readily accounted for by the addition-elimination of hydride (869).

James and Memon (867) observed very slow hydrogenation of hex-1-ene, styrene, and cyclohexene in benzene at 50° and 1 atm, but noted enhanced activity for these substrates and ethylene in DMA solution. The reductions of unsaturated carboxylic acids were studied in more detail. For maleic acid, hydrogen uptake plots at constant pressure exhibited a linear region after an initial autocatalytic-type region. The kinetics in the linear region indicated between zero- and first-order in $IrCl(CO)(PPh_3)_2$, hydrogen and the substrate (the order decreasing with increasing concentration) and an inverse dependence on added triphenylphosphine. An important finding was the slow dissociation of a phosphine ligand according to eq. 393 (Section A-1) which accounted for the autocatalytic region and the phosphine inhibition, since maleic acid (MA) complexes with the solvated species rapidly, but not with the bisphosphine complex:

$$IrCl(CO)(PPh_3)(DMA) + MA \overset{K_m}{\rightleftharpoons} IrCl(CO)(PPh_3)MA \; (+DMA) \tag{400}$$

A rate-determining oxidative addition of hydrogen followed by hydrogen transfer steps was then postulated, that is, the unsaturate route (cf. eqs. 375, 376 in Section XI-H, and Section XI-B-1).

$$\tag{401}$$

IrCl(CO)(PPh₃)DMA + Succinic acid

The 6-coordinated dihydride $H_2IrCl(CO)(PPh_3)_2$ is formed in the usual way (eq. 390) with its concentration determined by K_1, but it appears to play no role in the hydrogenation. Once the slow dissociation equilibrium 393 has been established ($K_2 = k_2/k_{-2}$), reactions 390, 400, and 401 lead to the rate law 402, consistent with the observed kinetics:

$$\text{rate} = \frac{k_3 K_m [H_2][MA][Ir]_T}{1 + K_m[MA] + [PPh_3](1 + K_1[H_2])/K_2} \tag{402}$$

The reaction becomes less than first-order in iridium because of the increasing association back to $IrCl(CO)(PPh_3)_2$ with increasing total iridium; this is reflected in the $[PPh_3]/K_2$ denominator term. At 80°, K_m was ~ 1200 M^{-1}; k_3, ~ 0.3 M^{-1} s^{-1}; K_1, 2.6×10^2 M^{-1}; and K_2, $\sim 10^{-6}$ M.

No evidence was obtained for a 5-coordinate hydride, $H_2IrCl(CO)(PPh_3)$, but the possibility of this reacting with the substrate via the hydride route (Section XI-B-1) could not be completely ruled out. A slower reduction rate (one-seventh) for fumaric acid, which forms a weaker complex (lower K_m), is consistent with the unsaturate route. The reactivity of the halides increased in the order $Cl < Br < I$ (1:2.5:4); a first-order dependence noted for the iodo complex was attributed to easier dissociation of a phosphine ligand (greater K_2).

Chan and James (868) have extended these studies to other solvent systems using unsaturated acids and esters as substrates. The activity increases in the order DMSO < DMA < sulfolane (about 1:2:5), and preliminary studies indicate that the chloride ligand may also dissobiate to some extent, particularly in sulfolane (compare with cationic iridium catalysts in Section XII-B-1). No activity was observed in nitromethane, acetonitrile, acetone, 2-butanone, and pyridine.

The presence of a few per cent oxygen in the gas phase increased the hydrogenation rate of maleic acid in DMA by a factor of about 100 for the $IrCl(CO)(PPh_3)_2$ system (867). This seems likely to involve promotion of the phosphine dissociation reaction 393 via oxidation of the free phosphine (672), or via phosphine oxidation at the metal center through the $IrCl(CO)(PPh_3)_2O_2$ complex (876) (cf. Sections XI-B-1, XI-B-8, and XI-E).

Strohmeier and coworkers (872, 874) have studied the hydrogenation of dimethyl maleate in toluene at 80° and 1 atm pressure, using a range of $IrX(CO)(PR_3)_2$ complexes. Only the initial linear rates for about 5% reduction were measured (Table 73); there was less than 1% isomerization to fumarate which in any case was reduced about three times slower. Contrary to the data for dihydride formation (Table 70), the rates are all similar in magnitude. For all the phosphine complexes listed, the reactivity decreases in the order $Cl > Br > I$ (opposite to the trend observed in DMA). The reduction rates were generally much slower than the initial hydrogen uptake rates in the

Table 73 **Initial Hydrogenation Rates of Dimethyl Maleate using IrX(CO)(PR$_3$)$_2$ complexes (872, 874)**[a]

X	R	Rate × 10^3 (ml s^{-1})	X	R	Rate × 10^3 (ml s^{-1})
Cl	C$_6$H$_{11}$	3.2	Cl	i-C$_3$H$_7$	1.4
Br	C$_6$H$_{11}$	0.8	Br	i-C$_3$H$_7$	0.4
I	C$_6$H$_{11}$	0.6	I	i-C$_3$H$_7$	~0
Cl[b]	C$_6$H$_5$	4.8	Cl	OC$_6$H$_5$	6.7
Br	C$_6$H$_5$	1.3	Br	OC$_6$H$_5$	2.0
I	C$_6$H$_5$	0.8	I	OC$_6$H$_5$	0.1

[a] Catalyst, 2 × 10^{-3} M; 0.4 M maleate, 50 ml toluene, 1 atm pressure.
[b] The same rate was recorded for the p-tolyl analog; the benzyl and n-butyldiphenyl analogs gave rates of 2.8 × 10^{-3} ml s^{-1}.

absence of olefin, except for the cyclohexyl and isopropyl phosphine chloride complexes. It was concluded that the H$_2$IrX(CO)(PR$_3$)$_2$ complex was not the active complex and that the rate-determining step probably involved attack of hydrogen on a IrX(CO)(PR$_3$)$_2$(olefin) complex (872). More detailed kinetic studies on the IrCl(CO)(PPh$_3$)$_2$ system (874) revealed an inverse dependence on added phosphine. These workers reached the same conclusion as James and Memon—that a 4-coordinate iridium olefin complex was involved (eq. 401). However, Strohmeier and Onoda (874) consider that the phosphine dissociates from an initially formed IrCl(CO)(PPh$_3$)$_2$(olefin) complex (eq. 404) since, although phosphine exchange is very rapid for IrCl(CO)(PPh$_3$)$_2$ in toluene (612, 877), no measurable dissociation was said to occur in this medium (869, 874). It should, however, be noted that the value of the dissociation constant K_2 in DMA at 80° (~10^{-6} M) corresponds to only ~3% dissociation for a 10^{-3} M solution of the complex, and a K_2 value in toluene of 10^{-5} M (temperature not stated) has been reported (1466). However, complexes of the type IrX(CO)(PPh$_3$)$_2$(olefin) (X = halide, SnCl$_3^-$) are known; for example, with tetrafluoroethylene (36, 878, 879), cyanoethylenes (36, 847, 851, 852, 880–882, 885, 893, 1401), maleic and fumaric acids, esters, and anhydrides (852, 868, 873), ethylene (36, 730, 883), and cyclohexene (852). Molecular weight data in solution for some isolated cyanoethylene complexes indicate that no phosphine dissociation occurs (880). The different trends in halide activity noted in DMA and toluene may indicate differences in mechanism. Wilkinson and coworkers (612) have suggested possible catalyst production by phosphine dissociation from the dihydride, H$_2$IrX(CO)(PPh$_3$)$_2$.

The reaction scheme of Strohmeier and Onoda (874) gives a rate law similar in type to that shown in eq. 402; between zero- and first-order dependences were observed for both iridium and maleate. In the presence of added phosphine, the lower rates can become first-order in iridium in agreement with eq. 402 (cf. Section IX-B-2b). Using a measured association constant K_3 of 0.2 M^{-1} for reaction 403 at 80° (873), analysis of the kinetic

$$IrCl(CO)(PPh_3)_2 + \text{maleate} \xrightarrow{K_3} IrCl(CO)(PPh_3)_2(\text{maleate}) \quad (403)$$

$$IrCl(CO)(PPh_3)_2(\text{maleate}) \rightleftharpoons IrCl(CO)(PPh_3)(\text{maleate}) + PPh_3 \quad (404)$$

data at 80° gave k_3 for reaction 401 as ~2.5 M^{-1} s^{-1} and the dissociation constant for reaction 404 as 1.5×10^{-2} M.

Strohmeier and Fleischmann (873) have presented K_3 values and thermodynamic data for exothermic reactions such as 403 in toluene for dimethyl maleate and fumarate, maleic anhydride, ethyl acrylate, styrene, and hept-1-ene with various $IrX(CO)(PR_3)_2$ complexes (X = halide, R = C_6H_{11}, C_6H_5, OC_6H_5). For each substrate K_3 increases in the order I < Br < Cl which is consistent with the activity trend shown in Table 73, considering the rate law (874) (cf. K_m term in eq. 402). The K_3 also increases in the order R = C_6H_{11} < C_6H_5 < OC_6H_5, again generally consistent with activity trend in Table 73. Both trends indicate increasing stability of the complexes with decreasing electron density at the iridium (cf. Table 70). The stability also increases with stronger π-acceptor properties of the olefin, that is, with decreasing electron density at the carbon-carbon double bond (708, 873, 880, 881). At 20°, K_3 varies from about 10^4 M^{-1} for the maleic anhydride adduct with $IrCl(CO)[P(OC_6H_5)_3]_2$ to the smallest measured (~3 × 10^{-2} M^{-1}) for the styrene adduct with the same complex. Unactivated olefins are reduced extremely slowly by these complexes.

It seems that the olefin stability constant has to be sufficiently large for a reasonable hydrogenation rate (cf. K_m in eq. 402) but not too large, since k_3 will be decreased due to an increased electronic promotion energy for the subsequent dihydride formation. For example, it is unlikely that the strongly complexed tetrafluoro- or tetracyanoethylene (36) will be readily hydrogenated. It is worth noting the contrast between the data of Table 73 and that for the $RhX(PR_3)_n$ catalysts in similar media where the activity trends increase in the order Cl < Br < I (Section XI-B-6) and R = OC_6H_5 ≪ C_6H_5 (cf. Table 51) which appear consistent with a hydride route.

Strohmeier and coworkers (1465) have listed some initial rate data for the reduction of olefins and phenylacetylene by using $IrCl(CO)(PPh_3)_2$ (see Table 63 in Section XI-E). The iridium catalyst seems more active than the rhodium analog, at least for the more readily reduced olefinic substrates: ethyl acrylate, hept-1-ene, styrene, and cycloheptene.

Yamaguchi (902) has reduced terminal olefins by using $IrCl(CO)(PPh_3)_2$ in toluene and 30 atm hydrogen at 60 to 110°; the reaction was inhibited by added phosphines, arsines, pyridine, or carbon monoxide. Fotis and McCollum in a patent (970) have described similar reductions, using the complex in benzene at 90° with 5 atm hydrogen.

Lyons (918) has reported that the complexes $IrX(CO)(PPh_3)_2$ (X = halide) catalyze the disproportionation of cyclohexa-1,4-diene to benzene and cyclohexene at 80° (60% disproportionation after 180 hr). Two possible mechanisms involving separate stages of dehydrogenation and hydrogenation (eqs. 405 and 406) or a direct hydrogen transfer through a bis-diene complex were suggested:

$$C_6H_8 + Ir \longrightarrow C_6H_6 + IrH_2 \qquad (405)$$

$$IrH_2 + C_6H_8 \longrightarrow C_6H_{10} + Ir \qquad (406)$$

The $Ir(GeR_3)(CO)(PPh_3)_2$, R = Me,Et, complexes may possibly act as catalysts for ethylene hydrogenation, since Glockling and Wilbey (1396) report that the corresponding iridium(III) dihydride complexes, which are readily formed with molecular hydrogen (1 atm), react stoichiometrically with ethylene at 60° in benzene to give ethane:

$$H_2Ir(GeR_3)(CO)(PPh_3)_2 + C_2H_4 \longrightarrow Ir(GeR_3)(CO)(PPh_3)_2 + C_2H_6 \quad (406a)$$

With 80 atm hydrogen, the dihydrides undergo hydrogeonlysis:

$$H_2Ir(GeR_3)(CO)(PPh_3)_2 + H_2 \longrightarrow H_3Ir(CO)(PPh_3)_2 + R_3GeH \quad (406b)$$

The possibility of an 8-coordinate intermediate was considered, as well as other conceivable mechanisms (cf. eqs. 394 and 395).

As noted above, the reduction of acetylene when using $IrCl(CO)(PPh_3)_2$ in benzene or toluene is extremely inefficient (730). In the "more active" sulfolane medium at 80°, no measurable reduction occurs even for activated acetylenes, such as acetylene dicarboxylic acid when using $IrCl(CO)(PPh_3)_2$ (868). Adducts with acetylenes have been formed using $IrX(CO)(PR_3)_2$ complexes, X = Cl, SnCl₃; R₃ = Ph₃, Ph₂Me (730, 852, 883).

Inorganic substrates such as iron(III) are readily hydrogenated by using DMA solutions of the $IrX(CO)(PPh_3)_2$ complexes, the activity in this medium again increasing in the order Cl < Br < I (886). As for the rhodium analogs (XI-E), the mechanism remains to be established. The iridium complex $Cl_2(CO)(PPh_3)_2Ir–HgCl$ in benzene reacts with molecular hydrogen to give metallic mercury and the hydride $Cl_2(CO)(PPh_3)_2IrH$ (888); the complexes $Cl_2(CO)(PPh_3)_2Ir–M(PPh_3)$, where M is copper or gold, were reported unreactive at 1.5 atm and 25° (847), although the existence of the gold complex has since been disputed (1388).

Some comparisons of the $IrCl(CO)(PPh_3)_2$ complex with the rhodium analog were considered in Section XI-E.

B.   OTHER IRIDIUM CATALYSTS CONTAINING PHOSPHINES, PHOSPHITES, ARSINES, AND STIBINES

1.   Catalysts without Carbonyl or Nitrosyl Ligands.   Hieber and Frey (896), Sacco and coworkers (579), Taylor (897), and Vaska and Catone (898) have reported that the cationic complex $Ir(diphos)_2^+$ adds hydrogen in the solid state or in solution, the reaction being irreversible at 25°. The related complex $IrCO(Ph_2PCH_2PPh_2)_2^+$ (899), and the nonionic $IrBr(Ph_2PCH_2CH_2SPh)_2$ do not react with 1 atm hydrogen (897). The presence of chelating phosphines inhibits catalytic activity at ambient conditions (cf. Section XI-G), but at 150 to 175° with 100 atm hydrogen, reduction of hex-1-yne/oct-1-ene mixtures occurs (897). The order of increasing activity is

$$IrBr(Ph_2PCH_2CH_2SPh)_2 < [Ir(diphos)_2]Cl < [IrCO(Ph_2PCH_2PPh_2)_2]Br$$

In all cases the hex-1-yne is reduced faster than the oct-1-ene (to hexene and hexane).

The cationic catalyst complex with nonchelating phosphines, $Ir(diene)$-$(PPh_3)_2^+$, was reported along with the more effective rhodium analogs (Section XI-G; 786, 856, 900). The catalysts behave similarly, with di-hydrides $H_2Ir(PPh_3)_2S_2^+$ (S = solvent) being readily formed. Data were presented for the complex in acetone which at $5 \times 10^{-3} M$ (25°, 1 atm) hydrogenates 1,5-cyclooctadiene (0.5 M) to the monoene at an initial rate of $\sim 0.1 \ M \ hr^{-1}$; slow hydrogenation of butyraldehyde to 1-butanol was observed at 50° in dioxane. The complex with the more basic diphenylmethyl-phosphine is ineffective for reducing hex-1-ene and 1,5-cyclooctadiene because of accompanying rapid isomerization (856). The dihydride species $H_2Ir(PPh_3)_{3,4}^+$ have been made (786, 901), but their potential activity was not reported.

The hydride complexes $H_3Ir(PPh_3)_{2,3}$ (890, 904, 1410) have been used as hydrogenation catalysts. Coffey (905) has used a solution of $H_3Ir(PPh_3)_3$ ($5 \times 10^{-3} M$) in acetic acid at 50° and 1 atm hydrogen to reduce n-butyralde-hyde (0.56 M) to 1-butanol at an initial rate of 0.34 $M \ hr^{-1}$ (cf. above). The reaction was first-order in catalyst and aldehyde; no uptake occurs in toluene or unmixed aldehyde. Octenes were not reduced under the conditions used in acetic acid, but acrylic acid and methyl acrylate were reduced at a rate of ca. 0.1 $M \ hr^{-1}$. The true catalyst was thought to be a mixture of hydridoacetates such as $H_2Ir(OCOCH_3)(PPh_3)_3$, since this is formed from acetic acid and $H_3Ir(PPh_3)_3$ (906).

Formic acid, which is catalytically decomposed to hydrogen and carbon dioxide by the $H_3Ir(PPh_3)_3$-acetic acid system, may be used as the hydrogen source for these reductions (908). A patent by Coffey (907) reports that 1:1

benzene–oct-1-ene solutions of $H_3Ir(PPh_3)_3$, $H_3Ir(PPh_3)_2$, or $H_3Ir(PEt_2Ph)_2$ absorb hydrogen (1 atm) under reflux to give *n*-octane. Similarly, 1,5-cyclooctadiene gave cyclooctene, and phenylacetylene gave styrene and ethylbenzene. Ketones could be reduced to alcohols at 100 atm. The bis(triphenylphosphine) complex, but not the tris compound, effected hydrogenation of oct-2-ene. Fotis and McCollum (970) have used benzene solutions of $H_3Ir(PPh_3)_3$ to reduce hex-1-ene with 5 atm hydrogen at 90°; hex-2-enes were also formed. The hydrogenation rates were increased by u.v. radiation presumably by a photochemically promoted phosphine dissociation.

Dolcetti and coworkers (909) have used a suspension of $H_3Ir(PPh_3)_2$ in dichloromethane at room temperature to reduce ethylene and hex-1-ene according to eq. 407:

$$H_3Ir(PPh_3)_2 + RCH:CH_2 \longrightarrow HIr(PPh_3)_2(\text{solvent}) + RCH_2CH_3 \quad (407)$$

$$HIr(PPh_3)_2(\text{solvent}) + H_2 \rightleftharpoons H_3Ir(PPh_3)_2 \quad (407a)$$

Since the green monohydride can absorb hydrogen reversibly to re-form the trihydride (eq. 407a), this system has been used at 1 atm to catalytically reduce the olefins. Addition of hex-1-ene to the monohydride solution gives a blue-green coloration, presumably due to an alkyl, which reacts with hydrogen after an induction period to give hexane and regeneration of the monohydride. The mechanism could involve this unsaturate route (as suggested) or a hydride route involving reactions 407 and 407a. $H_3Ir(PPh_3)_3$ was ineffective under these conditions (909). Coffey (941) has reported that benzene solutions of $D_3Ir(PPh_3)_3$ give no exchange with 1 atm hydrogen at 80° showing the absence of a ready equilibrium such as 407a for the trisphosphine system.

Lappert and coworkers (915–917) have prepared diglyme solutions of the hydride $HIr(PPh_3)_3$. Hydrogen abstraction occurs from toluene to give a trihydride complex (916, 917); if, as suggested, a bisphosphine complex is involved, the system might also possibly be used for hydrogenation in the absence of molecular hydrogen (eqs. 407, 407a). The catalyzed decomposition of formic acid mentioned above could well involve an equilibrium such as 407a followed by hydrogen abstraction via the monohydride.

Shaw and coworkers (929, 958) have recently shown that the complex initially formulated $H_3Ir(PEt_2Ph)_2$ (904) is a pentahydride $H_5Ir(PEt_2Ph)_2$, and they suggest that the corresponding triphenylphosphine complex discussed above should also be reformulated as such. Equations 407, 407a and the discussion above will have to be modified if a pentahydride exists in solution. Parshall and coworkers (1361) also consider this complex to be a pentahydride and have reported that the complexes H_5IrL_2 (L = PEt_2Ph, PEt_3, PMe_3) lose hydrogen in solution (cf. eq. 407a) and also catalyze exchange between deuterium and benzene (cf. eq. 78 in Chapter IV). Reaction of the pentahydride with diethylphenylphosphine yielded the trihydride $H_3Ir(PEt_2Ph)_3$, showing that the formulation for the trisphosphine complexes is correct (958). Seven-coordinate monohydride complexes involving a quadridentate phosphine ligand have been described (957).

$IrCl(PPh_3)_3$, the analog of the well-studied rhodium hydrogenation catalyst, was first isolated by Bennett and Milner (911, 912) and later by Collman and coworkers (913); its presence in solution was suggested earlier by Vaska (250) and by Collman and Kang (914). Lappert and coworkers (915–917) have also reported on this species. The complex and its analogs $IrXL_3$ [X = Cl, Br, and L = PPh_3, $AsPh_3$, $SbPh_3$, and other ring-substituted triphenylphosphines (912)] are extremely reactive at ambient conditions to give a wide range of adducts including an irreversibly formed dihydride, for example, $H_2IrCl(PPh_3)_3$ (911–913, 916). However, the complexes are ineffective hydrogenation catalysts (911–913) mainly because the dihydride, unlike the rhodium analog, remains 6-coordinate in solution; the somewhat stronger iridium-hydrogen bonds will also be a contributing factor. The $IrXL_3$ complexes, which are probably undissociated in oxygen-free solutions (912, 913, Section XI-B-1), undergo hydrogen abstraction from the ortho position of an aromatic ring to give octahedral hydridoaryl iridium(III) species, for example $HIrCl[(o-C_6H_4)Ph_2P](PPh_3)_2$ (911, 912, 917; cf. Scheme 3, Section XI-B-2b). $IrCl(PPh_3)_3$ even forms the dihydride by hydrogen abstraction from hydrocarbon solvents such as toluene (916, 917). This dehydrogenation is apparently catalytic over long periods (917), which suggests that the dihydride might slowly lose hydrogen.

Van der Ent and coworkers (924) have formed $IrClL_n$ ($n = 1, 2$; L = PPh_3, $AsPh_3$, and $SbPh_3$) in situ in benzene from the cyclooctene complex $[IrCl(C_8H_{14})_2]_2$ (925, 926) (compare with the rhodium systems in Sections XI-B-7 and XI-H). The species $IrCl(PPh_3)_2$ (2.3×10^{-4} M) catalyzes the hydrogenation of 0.27 M hex-1-ene (25°, 1 atm) at an initial rate of 6×10^{-4} M s^{-1}, which is some ten times greater than that for the corresponding rhodium system (cf. Table 51). The rate quickly falls because of an accompanying rapid isomerization (not found with the rhodium system) to hex-2-enes, which are reduced about 180 times more slowly. The monophosphine system and the cyclooctene complex itself were about six times less active than the bisphosphine system. $IrCl(AsPh_3)_2$ was also said to be active. An isolated ethylene complex, trans-$IrCl(PPh_3)_2C_2H_4$, exhibited fast exchange between free and coordinated ethylene. The trisphosphine complex, $IrCl(PPh_3)_3$, formed in situ, was inactive in agreement with the findings on the isolated complex.

Besides the hydridophosphine-type complexes mentioned above, iridium(III) forms an extensive range of such complexes, including $H_3Ir(LR_3)_3$, $H_3Ir(LR_3)_2$ (but see above), $H_2IrX(LR_3)_3$, $H_2Ir(AsPh_3)_4{}^+$,

$H_2Ir(LPh_3)_3{}^+$, $HIrX_2(LPh_3)_3$, $HIrX_2(PPh_3)_2$, and $HIr(PPh_3)_2{}^{2+}$, where L is P, As and more rarely Sb, and X is commonly halide (75, 850, 958, 961); this list is not exhaustive. Some activity has been reported for some of the 6-coordinate species.

Vaska (250) has reported that reaction 408 occurs in toluene at ambient conditions, and thought that it occurs via addition of deuterium to $IrCl(PPh_3)_3$ formed by a prior dissociation of hydrogen chloride. $HIrBr_2(AsPh_3)_3$,

$$HIrCl_2(PPh_3)_3 + D_2 \longrightarrow D_2IrCl(PPh_3)_3 + HCl \qquad (408)$$

$HIrCl_2(SbPh_3)_3$, and $HIrX_2(PPh_3)_3$ (X = halide) have been reported by Yamaguchi (902) to hydrogenate terminal olefins in toluene at 30 to 100° with 30 atm hydrogen; the stibine and arsine complexes were the most active. A patent (970) has described the use of $HIrCl_2(LR_3)_3$ (L = P, As, Sb) in benzene with 5 atm hydrogen for the reduction of hex-1-ene; at 90° accompanying isomerization is observed which decreases in the series L = Sb > As > P. For the phosphine system the extent of isomerization increases with decreasing temperature: at 150°, hydrogenation occurs exclusively; at 90°, there are 15% isomerization products; and at 25°, there is exclusive isomerization. Isomerizations of cycloocta-1,5-diene and oct-1-ene catalyzed by $HIrCl_2(PEt_2Ph)_3$ have also been reported (940, 941). Jardine and McQuillin (240) have reported that the complexes $HIrCl_2(PPh_3)_3$, $H_2IrCl(PPh_3)_3$, and $H_2Ir(SnCl_3)(PPh_3)_3$ do not catalyze the hydrogenation of norbornadiene at ambient conditions; the stannochloride complex forms a stable adduct with the diene. $H_2IrCl(PPh_3)_3$ can be used in the same manner as $H_3Ir(PPh_3)_3$ in acetic acid–formic acid mixtures for reducing aldehydes to alcohols (see above; 908). Horner and coworkers (686) have used benzene-ethanol (2:1) solutions of chloroiridate, tertiary arylphosphine, and triethylamine for hydrogenation of pent-1-ene, but no details were given (compare with the similar rhodium systems in Section XI-B-9).

Henbest and coworkers (919), followed by Browne and Kirk (920) and Orr and coworkers (713), have described the use of refluxing aqueous propan-2-ol solutions of chloroiridate ($IrCl_6{}^{2-}$) and trimethylphosphite for producing axial alcohols from cyclic ketones, particularly steroid 3-ketones. Acetone production indicates that the alcohol is the source of hydrogen; deuterium studies show that the C-2 hydrogen is transferred to the ketone (919). Since alcohols can reduce iridium(IV) to iridium(III) (Ref. 75, p. 241), a hydridoiridium(III) phosphite complex, possibly with a coordinated carbonyl, seems to be the likely catalyst (compare with some corresponding iridium-DMSO systems in Section XII-C, and the rhodium-catalyzed hydrogen transfer reactions discussed in Section XI-D-1, Scheme 11). During the reduction, the phosphite is hydrolyzed to phosphorous acid which itself can be used instead of the ester but the axial selectivity is somewhat less. A number of alkyl-substituted cyclohexanones (919), and a range of 5α- and 5β-androstan-3-ones, 5α- and 5β-pregnan-3-ones, and 5α-cholestan-3-ones were reduced with >90% selectivity to the axial product in 24 hr (713, 919, 920); 4-, 6-, 7-, 11-, 12-, and 20-keto groups were unaffected while 2- and 19-keto groups were reduced more slowly (113, 920). Some isomerization at the 17-position was observed during selective 3-reduction of 5α-pregnane-3,20-dione (713, 920). Addition of sodium hydroxide gave a threefold increase in rate of stereospecificity, although slight reduction of the 17-keto group was now apparent (713); testosterone, with a 4-ene-3-one grouping, gives androsta-3,5-diene-17β-ol under these conditions (713). An attempt to reverse these reactions by starting with a steroid alcohol and using aqueous acetone solutions of the chloroiridate-phosphite mixture was not very successful (713).

Ainscough and Robinson (959) have reported that the iridium phosphite complexes $HIrCl_2[P(OPh)_3]_3$ (256), $HIr[P(OPh)_3]_4$ (903), as well as the carbonyl-containing complexes $HIr(CO)[P(OPh)_3]_3$ (784) and $IrCl(CO)$-$[P(OPh)_3]_2$ (see Section XII-A) all abstract hydrogen from the ortho positions of an aromatic ring in boiling decalin. The hydrido-dichloride eliminates hydrogen chloride to give $IrCl[P(OPh)_3]_3$, which immediately undergoes intramolecular oxidative addition to give $HIrCl[P(OPh)_2(OC_6H_4)][P(OPh)_3]_2$, and then $IrCl[P(OPh)_2(OC_6H_4)]_2[P(OPh)_3]$ with hydrogen elimination (cf. Scheme 3, Section IX-B-2b). The tetrakisphosphite, and the carbonyls (which lose their carbon monoxide) react in a corresponding manner. $HIr(PF_3)_4$ has been made (797).

2. Hydridocarbonyltris(triphenylphosphine)iridium(I), $HIr(CO)(PPh_3)_3$, and Analogs.

The title complex in the solid state closely resembles the rhodium analog (Section XI-D; 264, 717, 727, 892) and was first reported by Bath and Vaska (717); Malatesta and coworkers had simultaneously prepared the compound by a different method (890, 892). Vaska (250, 250a, 261) has reported on its catalytic activity in benzene or toluene, which was said to be greater than that of the rhodium complex (250a). Unlike the rhodium complex, it does not measurably dissociate in solution (717); indeed, O'Connor and Wilkinson (719) have stated that this gives rise to a lower activity for the iridium complex.

Strohmeier and Hohmann (1459) have presented initial rate data for catalyzed hydrogenation with the iridium and rhodium complexes (Section XI-D-1) in toluene at 25° and 1 atm. From using 2×10^{-3} M $HIr(CO)(PPh_3)_3$ and 0.8 M substrate, the rates in M hr^{-1} $\times 10^3$ were listed as follows: styrene, 21.0; ethyl acrylate, 16.0; hept-1-ene, 16.0; dimethyl maleate, 13.5; phenyl-acetylene, 9.0; hex-1-yne, 8.0; cis-hept-3-ene, 8.0; cis-hept-2-ene, 8.0;

trans-hept-2-ene, 8.0; cycloocta-1,3-diene, 5.5; cycloheptene, 5.0; *trans*-hept-3-ene, 4.5; dimethyl fumarate, 3.0; β-bromostyrene, 0.5. The rhodium complex is more active for each substrate (see Section XI-D-1), and is much more selective for terminal olefin reduction.

Hydrogen and ethylene are absorbed reversibly at ambient conditions in toluene, again in contrast to the rhodium species, and were thought to give $H_3Ir(CO)(PPh_3)_3$ (see below) and $HIr(CO)(PPh_3)_3(C_2H_4)$ (or the ethyl complex), respectively (261) (see also Section XII-D). An observed hydrogen-deuterium exchange was written involving the 7-coordinate trihydride complex (250a). A hydrogen-ethylene mixture (1 atm) is completely converted overnight to ethane at 30° (261); the mechanism suggested corresponded to that shown in eqs. 351 and 352 in Section XI-D for the rhodium complex, but involving the 5-coordinate ethyl complex. A catalyzed isomerization, reported by Lyons (918), is indicative of such intermediates.

The evidence that the adducts above and intermediates in solution contain three coordinated phosphines is not strong; Baddley and coworkers (853, 895) and Strohmeier and Hohmann (1459) have isolated adducts of the type $HIr(CO)(PPh_3)_2$(olefin), which contain only two such ligands. These compounds were obtained for a range of activated ethylenes with substituent cyano, carboxyl, and carbmethoxy groups. Acrylonitrile gives such an adduct, but may also form a complex, $CNCH_2CH_2Ir(CO)(PPh_3)_2(CH_2{=}CHCN)$, in which the hydrogen has transferred to the π-bonded olefin and a second acrylonitrile has coordinated (853). A similar complex is formed with tetracyanoethylene (895), although the Ir–H adds in a 1,4 manner to the olefin, instead of the usual 1,2-addition, to give $(CN)_2CHC(CN){=}C{=}N{-}Ir(CO)(PPh_3)_2(C_6N_4)$.

Baddley and coworkers (895) have also reacted the bisphosphine complexes, $H_3Ir(CO)(PPh_3)_2$ and $HIr(CO)_2(PPh_3)_2$ with activated olefins to give $HIr(CO)(PPh_3)_2$(olefin) complexes with loss of hydrogen and carbon monoxide, respectively.

Wilkinson and coworkers (725, 726, 930), Collman and coworkers (969), and Shaw and coworkers (958) have also reported on $HIr(CO)_2(PPh_3)_2$, $H_3Ir(CO)(PPh_3)_2$ and their analogs (see also Section XII-D). The dicarbonyl can be formed from carbon monoxide and $HIr(CO)(PPh_3)_3$; here a slow predissociation of phosphine ligand seems likely. Yagupsky and Wilkinson (930) doubt the existence of the 7-coordinate trihydride $H_3Ir(CO)(PPh_3)_3$ postulated by Vaska. The $HIr(CO)_2(PPh_3)_2$ exists in solution as two isomeric pentacoordinate species and reacts with hydrogen under pressure to give an isomeric mixture of the hydrides $H_3Ir(CO)(PPh_3)_2$ (930) (see also Section XII-A-1); reaction with ethylene is considered in Section XII-D.

$HIr(CO)(PPh_3)_3$ has been used to hydrogenate acetylene to ethylene and ethane, but the efficiency of the process is lowered by an irreversible reaction of the complex with acetylene (261). The acetylene complexes RCH$=$CR—Ir-(CO)(PPh$_3$)$_2$(RC $\vdots$ CR), with R = CF$_3$ and CO$_2$CH$_3$, similar in type to the acrylonitrile complex, have been isolated (894).

The activity of the triarylphosphite complexes HIr(CO)[P(OAr)$_3$]$_3$ (784) has not been reported; an intramolecular hydrogen abstraction in this complex was noted in Section B-1.

3. Other Hydrido-Carbonyl and Carbonyl Complexes. Other neutral and cationic mono- and dihydro iridium(III) complexes containing carbonyl ligands, besides those mentioned in Sections A and B-2, have been prepared (75, 510, 706, 871, 933, 939, 953, 954, 969, 1385, 1386, 1408, 1414, 1415) but they are all coordinatively saturated and are unlikely to be active catalysts. Yamaguchi (902), however, has used toluene solutions of HIrCl$_2$(CO)(PPh$_3$)$_2$ at 60 to 90° and 30 atm hydrogen for the reduction of terminal olefins.

Mays and coworkers (954, 1409) have made 5-coordinate cationic car-bonyls of the type Ir(CO)$_3$L$_2{}^+$ and Ir(CO)L$_2$[P(OMe)$_3$]$_2{}^+$ (L = tertiary phosphine or arsine). They readily undergo oxidative addition with loss of one ligand; thus Ir(CO)$_3$(PPh$_2$Me)$_2{}^+$ reacts with 1 atm hydrogen to give H$_2$Ir(CO)$_2$(PPh$_2$Me)$_2{}^+$; this reaction can be reversed by adding carbon monoxide. Mays and Stefanini (1348) have subsequently studied the kinetics of the reductive elimination of hydrogen from the H$_2$Ir(CO)$_2$L$_2{}^+$ complexes (L = tertiary phosphine or arsine) by reaction with the ligands L' (L' = CO, PPh$_2$Me, and P(OMe)$_3$) in chloroform. The reactions proceed by a dissociative mechanism, in which the first-order rate constant is essentially independent of L':

$$H_2Ir(CO)_2L_2{}^+ \overset{k}{\rightleftharpoons} Ir(CO)_2L_2{}^+ + H_2 \tag{408a}$$

$$Ir(CO)_2L_2{}^+ + L' \overset{fast}{\longrightarrow} Ir(CO)_2L_2L'^+ \tag{408b}$$

H$_2$Ir(CO)$_2$(PPh$_2$Me)$_2{}^+$ undergoes exchange with deuterium according to 408a and no HD is formed under H$_2$/D$_2$ atmospheres; also, 4-coordinate species with L = P(C$_6$H$_{11}$)$_3$ and PPr$_3{}^i$ were isolated. With L = PPh$_2$Me, an isotope effect (k_{H_2}/k_{D_2}) of 2.1 was measured at 20°; the activation parameters were $\Delta H^{\ddagger} = 20$ kcal mole^{-1} and $\Delta S^{\ddagger} \approx -5$ eu. The small isotope effect suggests partial bond formation between the hydrogen atoms in the activated state (compare with the addition of hydrogen to Vaska's compound; eq. 392 in Section A-1). In these reactions, k decreased along the series L = PPh$_3$ > AsPh$_3$ > PPh$_2$Me > PPhEt$_2$ > PEt$_3$ > P(C$_6$H$_{11}$)$_3$ $\approx$ PPr$_3{}^i$, and k gave values of $(25-0.35) \times 10^{-4}$ s^{-1}. This is thought to reflect increasing iridium-hydrogen bond strength; a similar trend was observed by Deeming and Shaw (933) during some protonation reactions of $IrCl(CO)L_2$; except for the anomalous arsine ligand, the basicity increases in the same order to a maximum in the isopropyl phosphine. These findings are consistent with the general activity trends discussed in Section A-1, the more basic phosphines decreasing the rate of reductive elimination of hydrogen from $H_2IrCl(CO)L_2$ complexes (Table 70, k_{-1} values).

Deeming and Shaw (512, 1415) have also studied the complexes $Ir(CO)_xL_{5-x}{}^+$, where $x = 1$–3 and $L = PPhMe_2$ or $AsMe_2Ph$.

Solutions containing the 4-coordinate anionic species $Ir(CO)_3(PPh_3)^-$ and some related 5-coordinate metal-metal bonded complexes of the type $(PPh_3)(CO)_3Ir\text{-}Au(PPh_3)$ (cf. Section A-2) have been characterized by Collman and coworkers (969) but, as with other d^{10} systems, the species are probably unreactive toward hydrogen (38, Section XIII-A).

Malatesta and coworkers (1347) have reported that a THF suspension of the complex $Ir_2(CO)_6(PPh_3)_2$ reacts with molecular hydrogen at ambient conditions according to eq. 408c:

$$\tfrac{1}{2}Ir_2(CO)_6L_2 \xrightleftharpoons[-\frac{1}{2}H_2]{\frac{1}{2}H_2} HIr(CO)_3L \xrightleftharpoons[+CO-H_2]{-CO+H_2} H_3Ir(CO)_2L \qquad (408c)$$

4. Nitrosyls. The versatility of the nitrosyl group as a donor ligand and its possible potential in inducing catalytic activity (Section XI-G) has renewed interest in iridium nitrosyl complexes.

The formally d^{10} complex $Ir(NO)(PPh_3)_3$, first reported by Malatesta and coworkers (943), has been prepared by other methods by Reed and Roper (944, 1412) and Collman and coworkers (799) who have studied its oxidative addition reactions and hydrogenating activity. It is not reported to give a dihydride, but it catalyzes the reduction of hex-1-ene (with isomerization) in benzene at 85° under a few atmospheres of hydrogen (799).

The coordinatively unsaturated species $IrX(NO)(PPh_3)_2{}^+$ (X = Cl, OH, OR) reported by Reed and Roper (945) readily undergo addition reactions to yield 5-coordinate species. But the cationic complexes are less active than the neutral carbonyl analogs, for example, $IrCl(CO)(PPh_3)_2$; reaction with hydrogen was not reported.

The complexes $IrX(CO)(NO)(PPh_3)_2{}^+$ (X = Cl, I) with a bent M—N—O linkage (946, 947), seem worthy of study for potential activity, as do other complexes such as $Ir(NO)(CO)(PPh_3)_n$ ($n = 1, 2$), $IrX_2(NO)(PPh_3)_2$ (X = halide), $Ir(H_2O)(NO)(PPh_3)_2{}^{2+}$, $HIr(NO)(PPh_3)_3{}^+$, $HIrCl(NO)(PPh_3)_2$, $Ir(NO)_2(PPh_3)$, $Ir(NO)_2(PPh_3)_2{}^+$, $IrX(NO)_2(PPh_3)_2$ (X = halide) (894, 943–945, 1395, 1410, 1412), and the oxygen-bridged species $[Ir(NO)(PPh_3)]_2O$ (948).

C.   OTHER IRIDIUM-CATALYZED HYDROGENATIONS

The cationic diene complexes $Ir(diene)_2^+$ and derived catalysts such as $[Ir(1,5-C_8H_{12})(CH_3CN)_2]^+$, **132**, were reported along with the rhodium analogs (Section XI-H; 681, 856). Complex **132** similarly catalyzes hydrogenation of 1,5- but not 1,3-cyclooctadiene, but the hex-1-ene reduction is accompanied by some isomerization. This catalyst is more effective than $Ir(diene)(PPh_3)_2^+$ (Section XII-B-1) which is readily formed on addition of triphenylphosphine (856).

The high activity of the cyclooctene complex $[IrCl(C_8H_{14})_2]_2$ in benzene (924) was discussed with that of the phosphine systems $IrCl(PPh_3)_n$ in Section XII-B-1; the activity presumably results from lability of the cyclooctene ligand. A number of other related olefin complexes, such as $[IrCl(diene)]_2$, $IrCl(diene)_2$, $[IrCl(C_2H_4)_2]_2$, and $IrCl(C_2H_4)_4$, have been prepared (e.g., 75, 681, 926, 927), but they have not been studied for activity. The last-mentioned complex, however, undergoes ethylene exchange relatively slowly (926); it seems that iridium(I)-olefin and -chlorine bonds are generally stronger than the corresponding rhodium(I) bonds (926–928). One diene ligand at least in the diene complexes is not easily removed, for example, even by phosphines (681, 927); high activity for these complexes seems unlikely. Mention should be made of the complex $[HIrCl_2(C_8H_{12})]_2$, first reported by Robinson and Shaw (942, 955), which appears to be the earliest example of an isolated hydridoolefin species. The complex can be formed from $[IrCl(C_8H_{12})]_2$ in HCl-EtOH; but in DCl-EtOH some deuteration of the ring occurs, which was explained by *cis* addition of Ir-D to a C=C bond followed by proton elimination (956).

An iridium catalyst stabilized by quinonoid and aromatic ligands is said to be effective in DMF solution (57d, 188) but no details are available (see Section XI-H for the corresponding rhodium catalyst).

Jardine and McQuillin (240, 240a) have used solutions of chloroiridate and stannous chloride to slowly hydrogenate norbornadiene; an intermediate $Ir(SnCl_3)_3$ complex was suggested. The dependence of the rate on diene concentration was of the form shown in Fig. 4 (see Section IX-B-2b).

Closely related to the chloroiridate-trimethylphosphite-catalyzed hydrogen transfer reactions considered in Section XII-B-1 are those involving chloro-iridate-DMSO systems. Henbest and coworkers (919) reported that addition of DMSO to aqueous propan-2-ol solutions of chloroiridate also gave an effective system (on reflux) for reduction of 4-*t*-butylcyclohexanone and cholestan-3-one, although a somewhat lower ratio ($\sim 3:1$) of axial/equatorial alcohols were formed than in the phosphite systems. The 1,2,3- and 1,2,6-isomers of $IrCl_3(DMSO)_3$ and two forms of the acid $H[IrCl_4(DMSO)_2]$ were isolated from these mixtures; all these complexes, particularly the acids, catalyzed the reductions (919). Trocha-Grimshaw and Henbest (921) later reported that the acid, in boiling propan-2-ol with 1 to 2% water, effectively reduced the olefinic bond in the α,β-unsaturated ketones RCOCH=CHPh (R = Ph, C(CH$_3$)$_3$, and PhCH = CH); both olefinic bonds in PhCO-(CH=CH)$_2$Ph were reduced. The acid-propan-2-ol reaction also yielded the hydride HIrCl$_2$(DMSO)$_3$, which is believed to be the true catalyst, since this was also effective in the presence of a trace of hydrochloric acid. Furthermore, a likely intermediate, (C$_{15}$H$_{13}$O)IrCl$_2$(DMSO)$_2$, with a chelated benzylacetophenone grouping, was formed according to eq. 409 (921, 922):

$$\text{PhCOCH=CHPh} + \text{HIrCl}_2(\text{DMSO})_3 \longrightarrow$$

$$
\begin{array}{c}
\text{CH}_2 \\
\diagup \quad \diagdown \\
\text{PhC} \qquad \text{CHPh} \qquad\qquad (409)\\
\parallel \qquad\quad\; | \\
\text{O} \longrightarrow \text{IrCl}_2(\text{DMSO})_2 + \text{DMSO} \\
\mathbf{134}
\end{array}
$$

McPartlin and Mason (1406) have determined the structure of both the hydride and the chelate. An alternative π-enol intermediate (cf. Section X-E-3, eq. 282), was thus ruled out. The final protonolysis step was given as follows:

$$\mathbf{134} + 2\text{HCl} \longrightarrow \text{PhCH}_2\text{CH}_2\text{COPh} + \text{H[IrCl}_4(\text{DMSO})_2] \quad (409a)$$

The HIrCl$_2$(DMSO)$_3$-propan-2-ol-2% water-HCl system has also been used to reduce diphenylacetylene to cis-stilbene (923):

$$
\begin{array}{cccc}
\text{PhC}\!\equiv\!\text{CPh} & \qquad & \text{Ph} \quad\;\; \text{Ph} & \qquad \text{Ph} \quad\;\; \text{Ph} \\
+ & \longrightarrow & \diagdown \quad \diagup & \xrightarrow{\text{H}^+} \quad \diagdown \quad \diagup \\
\text{HIrCl}_2(\text{DMSO})_3 & & \text{C=C} & \qquad \text{C=C} \quad (410)\\
& & \diagup \quad\quad \diagdown & \quad \diagup \quad\quad \diagdown \\
& & \text{H} \quad\;\; \text{IrCl}_2(\text{DMSO})_3 & \quad \text{H} \quad\quad \text{H}
\end{array}
$$

The intermediate was isolated and formulated cis; the second protonolysis step could be carried out separately. This type of study is important in that it could show that protonolysis at an alkenyl-metal bond occurs with retention of configuration. However, Shaw and coworkers (1355), based on their work and that of Tripathy and Roundhill (1356) on protonation of platinum(0) acetylene and alkenyl complexes, suggest that this hypothesis may not be correct. For example, reaction of hydrogen chloride with PtCl(CPh:CHPh)(PPh$_3$)$_2$, which has the hydrogen cis to platinum as in the iridium complex above, yielded trans-stilbene, which implies that either platinum-carbon bond fission goes with inversion of configuration or there is subsequent isomerization of an initial cis-stilbene product.

D. HYDROGENATION UNDER HYDROFORMYLATION CONDITIONS

Hydroformylation reactions involving iridium carbonyls have been little studied relative to the analogous cobalt (Sections X-E and X-F) and rhodium (Section XI-F) systems. Few details have been reported, but the studies of Imyanitov and Rudkovskii (744, 745, 963, 964) and a patent (742), show that carbonyls such as $[Ir(CO)_4]_n$ and $[Ir(CO)_3]_n$ are very inefficient for hydroformylation. However, they seem more active than the cobalt and rhodium carbonyls for hydrogenation under hydroformylation conditions—accompanying hydrogenation of cyclohexene (963, 964), butene, butadiene, and α-methylstyrene (744, 745) has been reported.

Unsubstituted iridium carbonyl hydrides are known (75, 965, 966) and evidence has been presented for $HIr(CO)_4$, the analog of the cobalt hydroformylation catalyst (965). A hydridodichloro-dicarbonyl complex $[HIrCl_2(CO)_2]_n$ has also been reported (925).

The use of phosphine, phosphite, and arsine carbonyl complexes, formed *in situ* by adding the tertiary ligand to a carbonyl or a precursor iridium salt, was first noted in patents by Slaugh and Mullineaux (524, 527, 968). In the presence of bases such as acetate, hydroformylation of pent-1-ene can give exclusively C_6 alcohol products (968). Benzoni and coworkers (967) have used the complexes $IrCl(CO)(PR_3)_2$ (R = Bu, Ph) for hydroformylation of propylene; aldehydes (~65% linear) were almost exclusively formed. Hydroformylation when using $IrCl(CO)(PPh_3)_2$ is presumably close to that involving the rhodium analog, which is thought to involve the $HRh(CO)_2$-$(PPh_3)_2$ complex (Section XI-F-2; 55, 725, 726, 767). The relatively low activity of iridium complexes has been attributed (55) to the greater stability of pentacoordinate iridium(I) and octahedral iridium(III) species (cf. Section XI-F-2; eqs. 364 to 366) and indeed the types of intermediates expected are more readily isolated than for rhodium.

The dicarbonyl $HIr(CO)_2(PPh_3)_2$ is readily synthesized (930, 958, 969). Reaction with ethylene gives a mixture of ethyl and propionyl species; $Ir(COC_2H_5)(CO)(PPh_3)_2$ and, in the presence of carbon monoxide, $Ir(COC_2H_5)(CO)_2(PPh_3)_2$, **134a**, were isolated (725, 726). A corresponding acetyl complex $Ir(COCH_3)(CO)_2(PPh_3)_2$ has also been characterized (969). Complex **134a** reacts with hydrogen according to the following equation (cf. eq. 367):

$$Ir(COC_2H_5)(CO)_2(PPh_3)_2 + H_2 \longrightarrow HIr(CO)_2(PPh_3)_2 + C_2H_5CHO \quad (411)$$

An expected dihydride intermediate, such as $H_2Ir(COC_2H_5)(CO)(PPh_3)_2$ or $H_2Ir(COC_2H_5)(CO)_2(PPh_3)$, could not be isolated (725, 726). However, oxidative addition with hydrogen chloride yielded $HIr(COC_2H_5)Cl(CO)-(PPh_3)_2$, which readily decomposed to propionaldehyde and $IrCl(CO)(PPh_3)_2$. As with the rhodium system, **134a** is partially converted to $Ir(COC_2H_5)(CO)_3-(PPh_3)$ under carbon monoxide (725). As pointed out by Wilkinson and coworkers (726), the analogy between the two elements should be treated carefully. For example, although the rhodium benzoyl linkage undergoes hydrogenolysis to give benzaldehyde (Section XI-F-2), the iridium benzoyl $Ir(COPh)(CO)_2(PPh_3)_2$, unlike the propionyl (eq. 411), gives benzene via hydrogenolysis of iridium-phenyl complexes present in equilibrium with the benzoyl:

$$IrPh(CO)(PPh_3)_2 + CO \rightleftharpoons IrPh(CO)_2(PPh_3)_2 \xrightarrow{CO} \atop \rightleftharpoons$$
$$Ir(COPh)(CO)_2(PPh_3)_2 \quad (412)$$

E. ACTIVATION OF HYDROGEN IN AQUEOUS SOLUTION

Ipatieff and Tronev (993) reported that chloroiridate solutions were slowly hydrogenated to iridium(III) at temperatures below 50° (2 days at 100 atm), and that at higher temperatures metal precipitated. Acidic chloride solutions containing iridium(III) and iridium(IV) were reported by Halpern and Harrod (135) to be inactive toward hydrogen, but the conditions were not stated. James and Memon (886) also observed an extremely slow and autocatalytic hydrogen uptake by acid solutions of chloroiridate at 80° and 1 atm pressure to finally give metal. The system could involve activation of hydrogen via an iridium(III) chloride complex in an analogous manner to the chlororuthenate(III) complexes (Section IX-B-1, eqs. 107 and 109), followed by decomposition of an iridium(III) hydride to metal via iridium(I), as in the chlororhodate system (Section XI-A-1, eqs. 314 and 319) (see also Section XVI-B).

The synthesis of inert iridium(III) complexes containing nitrogen donor ligands by using hydridic reducing agents such as ethanol and hypophosphorous acid as catalysts (e.g., 950, 951) likely involve labile iridium(I) species formed via an iridium(III) hydride as in the analogous rhodium systems (Section XI-A-2).

XIII

Group VIII Metal Ions and Complexes

THE NICKEL TRIAD

A. NICKEL SPECIES

Relatively little work has been reported on the activation of hydrogen by nickel complexes in solution.

Corresponding to studies on the efficient cobalt cyanide complexes [Sections X-A to X-D], a few reports have appeared concerning nickel cyanides. Spencer and Dowden (971) first noted the use of aqueous solutions of hexacyanodinickelate(I), $Ni_2(CN)_6^{4-}$, at 0 to 60° for the reduction of acetylene to ethylene; molecular hydrogen was generated during this reaction and so was not a necessary additive for the hydrogenation. Griffith and Wilkinson (350) had reported an unstable acetylene complex with $Ni_2(CN)_6^{4-}$, possibly $Ni_2(CN)_6C_2H_2^{4-}$. Goerrig (362) also noted the use of tetracyano-nickelate(II) solutions for catalytic reductions by borohydride but, in the abstract available, no substrates were mentioned.

Burnett (972) subjected butadiene-hydrogen mixtures to aqueous solutions of excess $Ni_2(CN)_6^{4-}$ at 20° and produced but-1-ene, *cis*-but-2-ene, and *trans*-but-2-ene in about a $1:2:3$ ratio. It is not too clear how essential is the presence of the hydrogen, since it can be generated by the oxidation of nickel(I) by water (e.g., 973, 1404) (cf. eqs. 150 and 152, Section X-A-1):

$$Ni_2(CN)_6^{4-} + 2H_2O + 2CN^- \longrightarrow 2Ni(CN)_4^{2-} + 2OH^- + H_2 \quad (413)$$

The butadiene appears to form a reasonably stable complex with $Ni_2(CN)_6^{4-}$ (see below), which could react with water (cf. eq. 413) or with an intermediate hydride species (undetected) formed during reaction 413. The latter was favored, since on using a lower nickel concentration, the rate of butene production decreased ten times more than that of diene absorption (972). $Ni(CN)_4^{2-}$ itself is unreactive toward water, hydrogen, or butadiene (972). Acid solutions containing $Ni(H_2O)_6^{2+}$ or nickel heptanoate are also unreactive toward hydrogen at 150° (25, 92).

Kwan and coworkers (974–976) have used $Ni(CN)_4^{2-}$-borohydride solutions for the hydrogenation and hydrogenolysis of a number of monoenes and dienes; again the complex has to be present in excess (Table 74). The initial red solution, which contains the $Ni_2(CN)_6^{4-}$ anion, becomes yellow o n

Table 74 Hydrogenation Using Hexacyanodinickelate(I) Solutions at 0 to 60°[a]

Substrate	Product	Reference
Acetylene	Ethylene	971
Isoprene	2-Me-but-1-ene, 2-Me-but-2-ene	974
Butadiene	But-1-ene, but-2-enes	972, 974
Allene	Propane (20%), propylene (80%)	974
Dimethyl maleate	Succinate (12%), fumarate (88%)	976
1,5-Cyclooctadiene	Cyclooctene, 1,3- and 1,4- COD	976
But-1-ene	Butane (10%) and but-2-enes	974, 975
1-Chloro-but-2-ene	Butane (6%), butenes	974
1-Hydroxy-but-2-ene	Butane (6%), butenes	974
1-Hydroxy-but-1-ene	But-2-enes	974
3-X-propene (X = halide, NCS, CO_2CH_3)	Propane (25%), propylene (75%)	974
3-Hydroxy (or -amino) -propene	Propane (5%), propylene (95%)	974
$(CH_2=CHCH_2)_2S$	Propane (23%), propylene (77%)	974
$(CH_2=CHCH_2)_2O$	Propane (3%), propylene (97%)	974
Maleic acid	Succinic acid (95%)	518
Citraconic acid	Methylsuccinic acid (68%)	518
Itaconic acid	Methylsuccinic acid (70%)	518
Crotonic acid	Butyric acid (95%)	518
Tiglic acid	2-Methylbutyric acid (89%)	518
Methacrylic acid	Isobutyric acid (90%)	518
Sorbic acid	Caproic acid (95%)	518
2,3-Dimethylmaleic acid	No reaction	518
α-Methylstyrene[b]	Cumene (8%)	518
Triethylbenzyl-ammonium chloride[b]	Toluene (10%)	518
Benzonitrile[b]	Dibenzylamine (24%)	518

[a] Except for Refs. 971 and 972, borohydride is also present; reaction time is 0.5 to 4 hr.

[b] Reaction time: 16 hr.

adding isoprene, butadiene, or allene (974). The selective formation of propane and propylene from allene and allyl derivatives depends on the substituted group. The nickel system differs from $Co(CN)_5^{3-}$ which under hydrogen gives exclusively propylene from allyl halides (Table 30, Section X-B-3). The reduction of monoenes is relatively slow and is accompanied by a more rapid isomerization reaction, for example, with maleate, 1,5-cyclooctadiene, and but-1- and but-2-enes. Under similar conditions, the complex $Ni(CN)_2(phen)$ enhanced the hydrogenation rates of these substrates in comparison to $Ni(CN)_4^{2-}$, while the triphenylphosphine complex, $Ni(CN)_2$-$(PPh_3)_2$ gave over 95% isomerization products (975, 976).

The isomeric butene mixture obtained from isoprene and butadiene in these systems suggested (974) that π-allyl intermediates were formed as in the cobalt cyanide systems, via a borohydride-produced nickel hydride; decomposition of the allyl complex could involve attack by further nickel hydride or borohydride (cf. Section X-B-2a). A similar mechanism was presented (976) for the maleate reduction, but it involved an intermediate σ-complex.

Dennis and coworkers (518) have reduced α,β-unsaturated acids by using the aqueous $Ni(CN)_4^{2-}$-borohydride solutions at 25° (Table 74). The reactions were catalytic in nickel, with the borohydride regenerating the nickel(I) catalyst (cf. eq. 413):

$$Ni_2(CN)_6^{4-} + substrate(S) \xrightarrow[2CN^-]{2H_2O} 2Ni(CN)_4^{2-} + SH_2 \qquad (413a)$$

Since crotonic acid could be reduced stoichiometrically when using $Ni_2(CN)_6^{4-}$ without borohydride, and borohydride reductions in D_2O gave dideuteride products, the water must be the hydrogen source (no exchange occurs between D_2O and borohydride); the rate also increased with acidity. These workers thus favored reduction via an electron transfer mechanism involving nickel(I) followed by protonation, as sometimes postulated for the $Co(CN)_5^{3-}$-borohydride reductions (cf. Section X-B-5). Excess cyanide was necessary for effective reduction and a paramagnetic $Ni(CN)_4^{3-}$, which has been observed by Nast and von Krakkay (520), was considered to be the active species:

$$Ni_2(CN)_6^{4-} + 2CN^- \rightleftharpoons 2Ni(CN)_4^{3-} \qquad (413b)$$

$$2Ni(CN)_4^{3-} + S \rightleftharpoons [Ni(CN)_3]_2S^{4-} + 2CN^- \qquad (413c)$$

$$[Ni^I(CN)_3]_2S^{4-} \rightleftharpoons Ni^{II}(CN)_3^- + Ni^I(CN)_3S:^{3-} \qquad (413d)$$

$$Ni^I(CN)_3S:^{3-} \xrightarrow{H^+} Ni^I(CN)_3SH^{2-} \rightleftharpoons Ni^{II}(CN)_3^- + :SH^- \qquad (413e)$$

$$:SH^- + H^+ \longrightarrow SH_2 \qquad (413f)$$

The hydrogenation of α-methylstyrene and benzonitrile and hydro-genolysis of triethylbenzyl-ammonium chloride in aqueous ethanol were much slower; there was no effective reduction of benzyl cyanide, oleic acid, cyclohexene, and, in contrast to Kwan's studies using excess nickel complex, isoprene.

Vlcek (973), Llopis and Sanchez Robles (1405), and Bingham and Burnett (1404) have shown that reaction 413 is first-order in complex and cyanide at pH >13, although at lower pH an acid-dependent path becomes important (1404). Bingham and Burnett wrote the paths in terms of eqs. 413g and

$$Ni_2(CN)_6^{4-} + CN^- \longrightarrow Ni^0(CN)_3^{3-} + Ni(CN)_4^{2-} \qquad (413g)$$

$$Ni_2(CN)_6^{4-} + HCN \longrightarrow HNi(CN)_3^{2-} + Ni(CN)_4^{2-} \qquad (413h)$$

413h; rate constants were evaluated at 25° (1404). The faster steps involving hydrogen evolution were written as follows:

$$Ni^0(CN)_3^{3-} + H_2O \xrightarrow{CN^-} Ni^I(CN)_4^{3-} + \tfrac{1}{2}H_2 + OH^- \qquad (413i)$$

$$2Ni(CN)_4^{3-} \longrightarrow Ni_2(CN)_6^{4-} + 2CN^- \qquad (413j)$$

$$2HNi(CN)_3^{2-} \longrightarrow Ni_2(CN)_6^{4-} + H_2 \qquad (413k)$$

These equations indicate that in acid solutions nickel hydrides may be possible intermediates for hydrogenation reactions; in less acid solutions, a nickel(0) species appears important (eq. 413i). The initial reaction (eq. 413g) differs from that postulated by Dennis and coworkers (eq. 413b). Further studies would seem useful on these cyanide catalysts.

Turco and coworkers (987) have observed reduction to nickel(I) on treating the tertiary phosphine complexes $Ni(CN)_2(PR_3)_2$ with borohydride. Nickel(I) triphenylphosphine complexes $NiX(PPh_3)_3$ (X = halide) have also been made; they dissociate a phosphine in solution, but their chemistry has not been reported (988, 989). It should be noted, however, that hydrido-phosphine complexes such as $HNiX(PR_3)_2$ and $HNi(BH_4)(PR_3)_2$ have been isolated from borohydride reduction of $NiX_2(PR_3)_2$, where X is chloride or bromide and R is cyclohexyl or isopropyl (981, 982, 1013).

An unstable $HNiBr(PEt_3)_2$ has also been reported (1008), and the complexes $HNiX[P(C_6H_{11})_3]_2$, where X = $OCOCH_3$, OPh, CH_3, Ph, and so on, have been made by other methods (519).

Cationic hydrides such as $HNi(diphos)_2^+$ and $HNi[P(OEt)_3]_4^+$ have been formed by protonation of the nickel(0) complexes (1364, 1365), and Tolman (1366) has studied the kinetics and mechanism of formation, and decay (eq. 413l) of the phosphite complex in acidic methanol (cf. eq. 442 in Section XIII-C-3a). Tolman (1369, 1370) has also studied the kinetics at 0 to 25° of

$$HNiL_4^+ \underset{}{\overset{-L}{\rightleftharpoons}} HNiL_3^+ \xrightarrow{H^+} H_2 + Ni^{2+} + 3L, \qquad L = P(OEt)_3 \quad (413l)$$

reaction of the hydride with conjugated dienes to give π-allylnickel complexes; for example, butadiene gives the π-crotyl complex $C_4H_7NiL_3^+$. The major products were the anti-isomers, indicating that the dienes assume a cisoid configuration as the nickel hydride is added. Dissociation of a phosphite ligand is the rate-determining step in these reactions; the chelated complex $HNi(diphos)_2^+$ reacts much more slowly.

Abley and McQuillin (243) have reported that $NiX_2(PPh_3)_2$ complexes (X = Br, I) in benzene very slowly reduce oct-1-ene at 20° with 1 atm hydrogen. For a 5×10^{-3} M catalyst solution with 4.5×10^{-2} M oct-1-ene, the linear uptakes measured at constant pressure were about 5×10^{-3} ml min^{-1} per 20 ml of solution for both complexes. The systems were less active than those of the corresponding palladium and platinum complexes (see Sections XIII-B and XIII-C).

Also patents by Heck (575) and Gosser (977) have described hydrogenations using $NiX_2(PR_3)_2$ complexes, where X is halide and R is alkyl or aryl. 1,5,9-Cyclododecatriene is reduced to cyclododecene; for example, by using $NiI_2(PPh_3)_2$ at 175° and 85 atm hydrogen with the substrate as solvent (977). α-Olefins have been reduced in heptane containing borohydride with the complexes; for example, $NiCl_2(PBu_3)_2$, at 50° and 2 atm hydrogen (575). Bisphosphine-nickel halides have also been used by Bailar and coworkers (978–980) for the hydrogenation of esters of fatty acids to the monoene stage (see Section XIV-F-2). The hydrogenating activity of similar complexes with chelated phosphines, for example, $NiCl_2(diphos)$, has not been reported and, although it would be expected to be less than that with the monodentate systems (cf. Section XI-G), a report by Kumada and coworkers (990) shows that the chelates are much more effective at 120° for hydrosilylation of olefins.

Solutions probably containing the unstable cation $NiX(CO)(PR_3)_2^+$, electronically related to $IrCl(CO)(PR_3)_2$, have been mentioned by Clark and coworkers (992); $NiCl(PEt_3)_3^+$ probably exists (1074).

There is now an extensive literature on the addition reactions of phosphine complexes of nickel(0), palladium(0), and platinum(0) (e.g., 38, 986). These d^{10} systems, however, apparently do not readily add hydrogen, which lends support to the idea (25, 26, 40) that interaction of a transition metal with hydrogen involves attack of a vacant metal orbital by the bonding electrons of the hydrogen molecule (38). This contrasts with the electronic model suggested for hydrogen activation by Wilkinson's rhodium catalyst and Vaska's compound (Section XII-A-1). Palladium(0) and platinum(0) phosphine complexes, however, have been used at elevated temperatures and pressures for hydrogenation (see Sections XIII-B and XIII-C).

Nickel phosphite complexes of the type $Ni[P(OPh)_3]_n$ (n = 2, 3, 4) are known (e.g., 785, 989, 1383), but although activity under hydrogen is not reported, Brown and Rick (992a) have stated that these complexes and the corresponding palladium complexes have been used in the first useful homogeneous process for hydrocyanation of nonactivated olefins.

The report by Barnett and coworkers (983) that nickelocene, $(\pi\text{-}C_5H_5)_2Ni$, reacts with 30 atm hydrogen at 50° in THF to give about 50% of $\pi\text{-}C_5H_5Ni$-$\pi\text{-}C_5H_7$ is of interest. The reaction is apparently heterogeneous at nickel metal formed during the reaction, but it suggests that nickelocene reacts with hydrogen to give an unstable hydride. Benzene solutions of nickelocene have been added to ethylene polymerization systems in the presence of about 11 atm hydrogen at 150° to reduce unsaturation in the polyethylene product (984). Hydrogenation of the coordinated ligands in bimetallic complexes such as 2,2'-bis(π-allyl)dicyclopentadienyl-nickelpalladium, to give cyclopentane and 2,3-dimethylbutane (985) with the metals, is presumably heterogeneous.

Nickel tetracarbonyl has been used at 200° with about 30 atm hydrogen to reduced acetonitrile to ethylamine and diethylamine, and acrylonitrile to propylamine and dipropylamine; the nitriles were used for the solvent media (211). A complex was isolated from the acrylonitrile system. The tetra-carbonyl is inactive as a hydroformylation catalyst (963). A patent (1441) has described the use of a nickel carbonyl/salicylaldoxime/triphenylphosphite system for the production of alcohols from olefins under hydroformylation conditions.

Soluble Ziegler catalysts containing nickel (see Chapter XV) have been used extensively for the reduction of carbon-carbon double and triple bonds (56, 147, 163, 167, 168, 173, 175, 195, 570–572, 1157). Such catalysts and nickel acetylacetonate have also been used to hydrogenate fats (144, 145, 206, 207; Section XIV-E).

B. PALLADIUM SPECIES

Ipatieff and Tronev (994) in 1935 reported the ready precipitation of metal from palladous chloride solutions by reaction with 1 atm hydrogen; the reaction was autocatalytic and heterogeneously catalyzed by the metal. In the presence of an oxidizing substrate, such as ferric chloride, no metal appeared until reduction of the substrate was complete. This suggested a homogene-ously catalyzed hydrogenation; Halpern and coworkers (995) later studied the kinetics of this reaction at 65 to 90° in 3 M hydrochloric acid when the predominant palladium species is $PdCl_4^{2-}$. The description of the system follows closely that described for the chlororhodate(III) system (Section XI-A-1; eqs. 315 and 316). The reaction was first-order in both hydrogen and catalyst and independent of substrate, and was described in terms of an initial rate-determining formation of an intermediate hydride, $HPdCl_3^{2-}$, followed by a rapid reaction with ferric; $\Delta H^{\ddagger}$ was 19.3 kcal mole^{-1} and $\Delta S^{\ddagger}$ was -7 eu. As later pointed out by Halpern (59, 60), on analogy with dihydride formation by iridium(I) complexes (Section XII-A-1), the hydride could be formed via $H_2PdCl_4^{2-}$ with subsequent loss of hydrogen chloride. Metal production presumably occurs via reaction 414:

$$Pd^{II}H \rightleftharpoons Pd^0 + H^+ \qquad (414)$$

With <3 M hydrochloric acid solutions, metal tended to precipitate during the ferric reduction. A similar complication occurred when using $Pd(ClO_4)_2$ in perchloric acid, although it was concluded qualitatively that the aquo complex and the lower chloride complexes were less active than $PdCl_4^{2-}$ (cf. Section XI-A-1); the ethylenediamine complex was also relatively inactive (995).

Fasman and coworkers (997) have studied the kinetics of the hydrogen reduction of palladium(II) complexes ($PdCl_4^{2-}$, $PdBr_4^{2-}$, $Pd(ClO_4)_2$) to the metal at 10 to 50° and 1 atm hydrogen. Prior to the autocatalytic region, the initial region is concerned with production of an intermediate hydride but, in the case of the chloride complex, this was thought to involve substitution of hydride for an axial water ligand and not chloride.

Maxted and Ismail (998) have hydrogenated ethyl crotonate at 30° and 1 atm hydrogen by using aqueous ethanolic solutions of $PdCl_2$ in the presence of promoting ions whose effectiveness falls somewhat in the sequence: sodium(I), zinc(II), mercury(II), cobalt(II), copper(II), magnesium(II), cadmium(II), silver(I), chromium(III), aluminium(III), nickel(II), and calcium(II). The quite rapid initial uptake rates (up to 9 ml min^{-1} for about 7 ml solution containing 3.5×10^{-2} moles of $PdCl_2$, 1 ml crotonate and 2×10^{-8} moles Na$^+$) decreased with time even though metal was produced; this was taken as evidence that the reductions were homogeneous. This seems somewhat indefinite, however, particularly since the uptakes tend to level off at a stage corresponding to about 25% reduction of the substrate *and* the fact that the $PdCl_2$ is present in some three times excess over the substrate.

Sibata and Matsumoto (1088) in 1939 reported on quinone hydrogenation catalyzed by aqueous solutions of $Pd(en)Cl_2$; in any one experiment, the total gas uptake was said to be proportional to the square root of time.

A number of anionic palladium(II)–tin(II) species of the type $[PdCl_x(SnCl_3)_{4-x}]^{2-}$ and a cycloocta-1,5-diene metal cluster complex, $(C_8H_{12})_3Pd_3(SnCl_3)_2$ have been reported (e.g., 1006) but unlike their platinum analogs (Section XIII-C-2) their activity toward hydrogen has not been reported. The anionic carbonyl $Pd(CO)Cl(SnCl_3)_2^-$ is also known (1012).

Rylander and coworkers (248) have used DMF or DMA solutions of palladous chloride (~0.025 M) to hydrogenate dicyclopentadiene (~1.5 M) at 20° and 1 atm hydrogen. One mole of hydrogen per substrate was absorbed at a constant rate ($\sim$10 ml min^{-1} for 120 ml solution) to give **135**; at this stage, metal appeared and a much slower reduction to the saturated hydro-

135

carbon followed. Hydrogenation to **135** was thought to be homogeneous; addition of thiophene (an inhibitor for heterogeneous systems) enhanced activity by a factor of 4. Palladous chloride was much more effective than chlorides of ruthenium (Section IX-B-2c), rhodium (Section XI-H), and platinum (Section XIII-C-1). Quinone and naphthaquinones were effectively reduced homogeneously by the palladium chloride–DMF system, but furan and other cyclic dienes, and crotonaldehyde, mesityl oxide, and unsaturated and saturated nitriles did not prevent immediate metal production.

Palladium(II)–tertiary phosphine complexes alone are generally not very effective catalysts. Tayim and Bailar (999) note less than 5% conversion of cyclooctadienes to monoene when using $Pd(CN)_2(PPh_3)_2$ in dichloromethane for 10 hr at 90° and 40 atm hydrogen. However, this complex and the arsine analog have been used by Bailar and coworkers (980, 1000, 1001) to reduce fats to the monoene stage (Section XIV-F-2). The $PdCl_2(MPh_3)_3$ complexes (M = P, As, Sb), and systems in which divalent chlorides $M'Cl_2$ (M' = Sn, Ge, Hg) have been added, have also been used to hydrogenate fats (978, 980, 1000, 1001; Section XIV-F-2). Abley and McQuillin (243) have used benzene–methanol (3:2) solutions of the $PdX_2(PPh_3)_2$ complexes (X = halide, CN) in combination with 2 moles of stannous chloride to hydrogenate (and isomerize) oct-1-ene at 20° and 1 atm. Complexes of the type $(PPh_3)_2Pd$-$(SnCl_3)Cl$ and $(PPh_3)_2Pd(SnCl_3)_2$ have been synthesized (1004). The systems involving the iodo and cyano complexes (5×10^{-3} M) with 4.5×10^{-2} M substrate gave linear uptake plots (at constant pressure) of 28×10^{-3} and 2×10^{-3} ml min^{-1} per 20 ml solution, respectively. The chloro and bromo analogs gave initial rates of 125×10^{-3} and 50×10^{-3} ml min^{-1}, respectively, but these rapidly decreased partly because of an accompanying isomerization to oct-2-ene, which was reduced more slowly. The chloro system forms a hydride (reversibly) and an olefin complex, possibly according to eqs. 415 and 415a; however, the exact nature of the remaining ligands is not known (see below):

$$(PPh_3)_2Pd(SnCl_3)Cl + H_2 \rightleftharpoons HPd(SnCl_3)(PPh_3)_2 + HCl \quad (415)$$

$$(PPh_3)_2Pd(SnCl_3)Cl + \text{oct-1-ene} \longrightarrow$$
$$(PPh_3)Pd(SnCl_3)Cl(\text{octene}) + PPh_3 \quad (415a)$$

Reaction 415 was thought to form a monohydride, since the corresponding system with platinum was shown to be reversible when using hydrogen chloride (Section XIII-C). The suggested mechanism of the reaction via this hydride, for both hydrogenation and isomerization, resembles that for the HRuCl(PPh$_3$)$_3$ and HRh(CO)(PPh$_3$)$_3$ systems (e.g., eqs. 351 and 352 in Section XI-D-1). Reaction 415a, however, may tie up available catalyst in the palladium (and platinum) systems, which becomes more significant as the reaction proceeds via the reversibility of reaction 415, since the amount of hydrochloric acid builds up. The drop in rate sometimes observed was attributed to this, since the presence of acetate buffer in some closely related platinum systems did prevent this "catalyst poisoning" (Section XIII-C-3b). Initial rate data as a function of oct-1-ene concentration were of the form shown in Fig. 4 (Section IX-B-2b) for the chloro and iodo systems, and only indicate a linear dependence above a 2:1 substrate/catalyst ratio; again the nature of ligand displacements involving olefin(s) is obscure (Section IX-B-2b).

Heck (575) has used complexes of the type PdX$_2$(PR$_3$)$_2$ (X = halide, R = alkyl or aryl) with borohydride in heptane for reduction of terminal olefins at 50° and 2 atm hydrogen. Similar complexes have been used with alcoxyalanates (1187) and aluminum alkyls (56) for the formation of Ziegler type catalysts (Section XV-A).

$$\begin{array}{ccccc}
Y & & X & & PR_3 \\
\diagdown & & \diagup\diagdown & & \diagup \\
& Pd & & Pt & \\
\diagup & & \diagdown\diagup & & \diagdown \\
R_3P & & X & & Y
\end{array}$$

136 R = Bu, Pr; X = Cl, Br, PPh$_2$; Y = Cl, Br.

A patent (1002) has described the use of binuclear palladium(II)–platinum(II) complexes of type **136** for the hydrogenation of hex-1-yne/oct-1-ene mixtures in benzene at 175° and 100 atm. The acetylene is preferentially reduced to hex-1-ene but hexane and octane are formed in smaller amounts. Palladium(II)–copper(II) complexes of the type PdCu(O$_2$CR)$_4$ (R = Me, Et, CH$_2$Ph) were also used as catalysts. Thompson (1007) has reported that complexes of type **136a** (1007, 1010) are also hydrogenation catalysts.

$$2M(PPh_2H) + [(\pi\text{-}C_3H_5)PdCl]_2 \longrightarrow \begin{array}{ccccc}
Ph_2P & & Cl & & M \\
\big| \diagdown & & \diagup\diagdown & & \diagup \big| \\
\big| & Pd & & Pd & \big| \\
\big| \diagup & & \diagdown\diagup & & \diagdown \big| \\
M & & Cl & & PPh_2
\end{array} + 2C_3H_6$$

$$M = Cr(CO)_5, \ Fe(CO)_4 \qquad \textbf{136a} \qquad \qquad (416)$$

The preparation is of interest in that it involves hydrogen transfer to a coordinated allyl group from a diphenylphosphine moiety.

Scheben and coworkers (1003) have patented the use of some aryl-palladium(II) chlorides for a variety of homogeneously catalyzed reactions including hydrogenation. A complex formed by treating palladous chloride with *p*-(dimethylamino)-*N*-benzyldieneaniline, probably **136b**, was used in

136b

ethyl acetate to reduce oct-1-ene at 3 to 4 atm of hydrogen.

Stabilized palladium(II) hydrides such as $HPdCl(PEt_3)_2$ (982, 996, 1005, 1008, 1014), $HPd(GePh_3)(PEt_3)_2$ (1014), and $HPd(CO)Cl_2^-$ (1011) have been described, but they decompose rapidly in solution to metal (cf. eq. 414). The more stable hydrides $HPdCl(PR_3)_2$ and $HPd(BH_4)(PR_3)_2$, where R is phenyl, cyclohexyl, or isopropyl, have also been synthesized (1013, 1362). Brooks and Glockling (1014) report one route for the germanium derivative by a reaction involving molecular hydrogen (100 atm, 20°), although some decomposition to metal occurred:

$$Pd(GePh_3)_2(PEt_3)_2 + H_2 \longrightarrow HPd(GePh_3)(PEt_3)_2 + HGePh_3 \quad (417)$$

At 0° under 1 atm hydrogen $Pd(GePh_3)_2(PEt_3)_2$ decomposes readily to metal; Deganello and coworkers (1015) have reported a similar decomposition with the lead-bonded analog, $Pd(PbPh_3)_2(PEt_3)_2$. Reaction 417 could involve oxidative addition to a palladium(IV) dihydride, although such a process is not so favorable as for corresponding platinum(II) complexes (37a, 39, 888); see also Section XIII-C-3b.

The cationic complexes, $PdCl(CO)(PEt_3)_2^+$ and $PdCl(PEt_3)_3^+$, electronically related to $IrCl(CO)(PPh_3)_2$ and $RhCl(PPh_3)_3$, have been synthesized (992, 1074). In view of the interest in the potential activity of nitrosyls (Section XI-G), the synthesis (1009) of $NO(PPh_3)_2Pd–Pd(PPh_3)_2NO$ is perhaps worth noting.

White and Parshall (1382) have studied the reductive cleavage of some palladium(II) phenyl complexes with methanol, and have noted the resemblance of such a reaction to hydrogenolysis of a palladium-carbon bond. Such a hydrogenolysis was, in fact, carried out on complex **136c** using deuterium. Benzyldimethylamine containing one deuterium atom was liberated; metal is also produced although in the presence of triphenylphosphine the complex $Pd(PPh_3)_4$ is formed.

Palladium(0) phosphine complexes (e.g., 986) do not readily add hydrogen (see Section XIII-A), but Dunning and coworkers (1026) have patented their

136c

use at 150 to 250° and 70 to 300 atm hydrogen for the hydrogenation of hex-1-yne/oct-1-ene mixtures in benzene. Hex-1-yne was converted completely into hex-1-ene and hexane, and oct-1-ene was partially reduced over an hour by using 10^{-4} M Pd(PPh$_3$)$_4$ at 175° and 100 atm. A palladium(0) carbonyl phosphine complex Pd(CO)(PPh$_3$)$_3$ has been synthesized (1363).

Takahashi and coworkers (1351) have noted that reaction of hydrogen with the bis(dibenzylideneacetone)palladium(0) complex, Pd(PhCH=CHCOCH=CHPh)$_2$, in methanol yields 1,5-diphenylpentan-3-one and 1,5-diphenylpentan-3-ol, but this may be a heterogeneous reduction, since the complex slowly decomposes to metal.

The use of palladium(0) phosphite complexes for catalytic hydrocyanation was mentioned in Section XIII-A (992).

Reviews dealing with palladium olefin complexes with mention of hydrogenation catalyzed by palladium species have appeared (149, 150).

C. PLATINUM SPECIES

Reviews of platinum olefin and acetylene complexes and some aspects of homogeneous hydrogenation with this metal have appeared (149–151). The systems are generally not very active, but the chemistry of the large number of hydride and alkyl species known contributes significantly to possible mechanisms generally for homogeneous hydrogenation.

1. Platinum Chloride Catalysts.   Ipatieff and Tronev (993) reported a slow precipitation of metal from aqueous solutions of chloroplatinic acid, H$_2$PtCl$_6$6H$_2$O, on exposure to hydrogen. Acid solutions containing chloro complexes of platinum(IV) were, however, reported by Halpern and Harrod (135) to be inactive toward hydrogen. Lack of details prevents a meaningful comparison. The reaction is catalyzed by the metal, however (993) (see also Section XVI-B).

Rylander and coworkers (248) have used DMF or DMA solutions of tetrachloroplatinate(II) (0.1 M) to reduce dicyclopentadiene ($\sim$1.5 M) at 20° and 1 atm hydrogen; a linear hydrogen uptake ($\sim$0.2 ml min^{-1} for 120 ml solution) resulted and gave saturated product before metal was precipitated. This contrasts with the palladium chloride system where metal is formed at the monoene stage (Section XIII-B; **135**); the monoene product (**135**) may stabilize the platinum ion. Addition of thiophene had no effect on the reduction rate, indicating a homogeneous system. A variety of other unsaturated substrates (Section XIII-B) did not prevent immediate metal production.

The ethylenediamine complex Pt(en)Cl$_2$ is reported by Bailar and Itatani (978) to be inactive for hydrogenation of linoleate in benzene-methanol even at 90° and 40 atm hydrogen in the presence of stannous chloride (Section, C-2 and XIV-F-1).

Flynn and Hulbert (1018) first studied the reaction of Zeise's dimer, [PtCl$_2$(C$_2$H$_4$)]$_2$, with hydrogen in solution, but Anderson (1019) had earlier studied the quantitative reduction to metal in the solid state at 20° according to eq. 418:

$$[PtCl_2(C_2H_4)]_2 + 4H_2 \longrightarrow 2Pt + 4HCl + 2C_2H_6 \qquad (418)$$

In solution (chloroform, acetone, toluene), reaction 418 occurs even at $-40°$ (1018). Using a 1:6 mixture of hydrogen and ethylene, metal precipitation could be avoided yet ethane was still formed, indicating a homogeneous hydrogenation. In acetone solution, the rate of ethane formation increased with temperature up to about $-15°$ but decreased at higher temperatures. Since Chatt and Wilkins (1019) had shown that a bis-ethylene complex PtCl$_2$(C$_2$H$_4$)$_2$ was stable under ethylene below about $-60°$, the low temperature reduction was written in terms of eqs. 419 and 420 via this species:

$$[PtCl_2(C_2H_4)]_2 + 2C_2H_4 \rightleftharpoons 2PtCl_2(C_2H_4)_2 \qquad (419)$$

$$2PtCl_2(C_2H_4)_2 + 2H_2 \longrightarrow [PtCl_2(C_2H_4)]_2 + 2C_2H_6 \qquad (420)$$

An activation energy of 2.5 kcal ascribed to reaction 420 should be treated cautiously, since of the three used rate measurements one referred to a much higher partial pressure of hydrogen and another occurred at a sharp maximum in the rate versus temperature plot. Studies by Rylander and coworkers (248) and Hayes (1020) confirm that the hydrogenation is homogeneous at the lower temperatures and at hydrogen to ethylene ratios of <1. Gow and Heinemann (1021) and Cramer (1022) have shown that ethylene exchange in complexes of the type shown in reaction 419 and Zeise's anion, PtCl$_3$(C$_2$H$_4$)$^-$, is rapid, but at low hydrogenation temperatures no hydrogen-deuterium exchange occurs between C$_2$D$_4$ and [PtCl$_2$(C$_2$H$_4$)]$_2$; such deuterium exchange is common in heterogeneous systems (1021). Olefin rotation in platinum(II) olefin complexes has been demonstrated (e.g., 1117).

The kinetics and mechanism of the reaction between tetrachloroplatinate(II) and ethylene and other olefins have been studied (e.g., 1377, 1378).

Tayim and Bailar (999) have noted that cycloocta-1,5-diene complexes of platinum, PtX$_2$(C$_8$H$_{12}$) (X = chloride, iodide) are ineffective for the hydrogenation of the diene even at 90° and 40 atm hydrogen, but they are active in the presence of stannous chloride (see footnote c of Table 76).

Garnett and coworkers (1080–1086, 1098, 1358, 1359) have discovered a useful procedure for labeling (deuteration and tritiation) a wide range of aromatic compounds, including benzene, substituted benzenes, polyphenyls, polycyclic hydrocarbons, and some steroids by using tetrachloroplatinate(II) solutions. The systems are homogeneous using, for example, CH_3COOD/D_2O solvent at 75° for deuteration; acid is necessary at higher temperatures to prevent decomposition to metal. Mechanistic evidence (e.g., 1085) suggests that platinum(II) π-bonded intermediates undergo a reversible oxidation to a hydride platinum(IV) phenyl complex, which subsequently exchanges by reversible loss of hydrogen chloride; for example:

$$Cl_4Pt^{II}\!\!-\!\!\langle\ \rangle \rightleftharpoons Cl_4\overset{\overset{H}{|}}{Pt^{IV}}\!\!-\!\!\langle\ \rangle \underset{+DCl}{\overset{-HCl}{\rightleftharpoons}} Cl_3Pt^{II}\!\!-\!\!\langle\ \rangle \qquad (421)$$

The second equilibrium seems well established for platinum(II) systems (Section C-2, eq. 437) and hydrogen abstraction from a bonded aromatic moiety has been demonstrated for a ruthenium system (Section IX-B-2b, eq. 140b). Loss of hydrogen from the aromatic ring of azobenzene has also been observed in the formation of platinum(II) complexes from $PtCl_4^{2-}$ and azobenzene (1096). An observed exchange of alkyl hydrogens in substituted benzenes could occur similarly but involving a π-allylic intermediate as shown in eq. 422. The mechanism initially favored (1085) involved a reversible dealkylation as shown in eq. 423 (for a methyl substituent, the mechanism

$$(422)$$

was modified to involve a coordinated —CH_2Cl group in the platinum(IV)

$$(423)$$

hydride). There are no data on the reversible reaction of platinum(IV) alkyls to hydridoolefin species, although this is well known for platinum(II) systems (see Section C-3). In any case, some observed dealkylation was attributed to heterogeneous catalysis by traces of metal present, and the latest exchange data support the π-allylic species (1103).

Closely related to these exchange studies is the work of Gol'dshleger and coworkers (577) and Hodges and coworkers (1087), who have used similar deuterated-acetic acid/chloroplatinate(II) solutions to measure exchange in saturated hydrocarbons such as methane, ethane, pentanes, and cyclohexane; at about 100°, up to 90 % of a monodeuterated species has been obtained with ethane after about 6 days. The reactions presumably involve alkyl hydride intermediates (cf. eq. 421, and eq. 312 in Section X-G) but the interesting unanswered question is, what is the nature of initial interaction between the saturated substrate and the metal center? Comparing with molecular hydrogen (cf. Section I-B), it could correspond with "dihydride" formation, heterolytic splitting (via dihydride formation?), both exemplified in eq. 424, or homolytic splitting involving two metal centers (eq. 425). Both mechanisms

$$M^{II} + RH \rightleftharpoons HM^{IV}R \rightleftharpoons [M^{II}-R]^- + H^+ \qquad (424)$$

$$[M^{II}-M^{II}] + RH \rightleftharpoons [HM^{III}-M^{III}R] \rightleftharpoons [HM^{IV}-M^{II}R] \qquad (425)$$

seem feasible for platinum(II) systems, in view of evidence for platinum(IV) hydrides and the existence of platinum(II) alkyls (see Sections C-2 and C-3); the reverse of reaction 424 has been demonstrated (eq. 438 in Section C-2).

Platinum(IV) dihydride complexes of formulation $H_2Pt(CO)X_3^-$, where X = halide, CNS, have recently been described by Matveev and coworkers (1341).

The use of platinum salts in Ziegler catalyst systems has been described in a patent (1166; see Section XV-A).

2. Platinum-Tin Chloride Complexes without Phosphines.

Cramer and coworkers (30) made the important discovery that methanol solutions containing chloroplatinic acid and stannous chloride dihydrate were effective for converting a 1:1 ethylene-hydrogen mixture quantitatively to ethane at 20° and 1 atm pressure. Acetylene was similarly converted to ethane and ethylene. The homogeneous systems showed maximum activity at tin/platinum ratios above 5:1, and a first-order dependence on platinum ions was noted for the ethylene hydrogenation. The stannous chloride stabilizes the platinum against reduction to metal. Further studies on the catalytic activity of this system and the chemistry of the platinum-tin chloride complexes have appeared since, including work by the same group (1022–1025), Wilkinson and coworkers (616, 681), Bond and Hellier (1027, 1028), van Bekkum and coworkers (1029, 1030), van't Hof and Linsen (1031–1034), Shilov and coworkers (1035–1037), Belluco and coworkers (1038, 1376), Adams and Chandler (1039), Baranovskii and coworkers (1042), Bailar and coworkers (978, 980, 1040, 1041), and Kheifits and coworkers (1458).

Table 75 summarizes the catalytic hydrogenations reported. In methanol, higher olefins than ethylene are reduced more slowly but they are readily isomerized (30, 1027, 1028); the slower hydrogenation has been attributed to weaker complexing with the higher olefins (30). In higher alcohols (especially propan-2-ol), carboxylic acids, esters, ethers, or ketones, however, the hydrogenation rate is greatly enhanced (1029–1034). Internal olefins are less readily reduced and generally the order of reactivity is terminal > 1,2-disubstituted > trisubstituted (1028, 1029, 1031); 2-alkyl-2,5-cyclohexadienes have been selectively reduced to the 2-alkyl-2-cyclohexene derivative (1029). Addition of hydrochloric or hydrobromic acid, alkali metal chlorides, or bromides increased significantly the hydrogenation rates of hex-1-ene and cyclohexene (1029, 1031). The bromine promoters were particularly effective, and such a system has been used to reduce hexamethyl-Dewar-benzene, with tetrasubstituted olefinic bonds, to hexamethylbicyclo[2.2.0]hex-2-enes (1030; see below). The bromide-promoted cyclohexene reduction in propan-2-ol was further promoted by addition of water to a maximum of 1 M, although added water decreased the efficiency of the chloride-promoted reduction (1029). The reductions observed for cyclohexene at 0.15 M (1029), and pent-1-ene at $\sim$2.0 M (1028) were noted to be first-order in substrate at 4×10^{-3} M platinum. Both first- and zero-order reductions in substrate have been recorded for hex-1-ene, depending on the catalyst concentration (1031). Maximum activity has been observed generally at tin/platinum ratios of about 5 to 6 (30, 1029–1037), although an optimum ratio of 12 was noted for the pent-1-ene and styrene systems (1028); at ratios $\leqslant$3, metal may precipitate (1031). One tin per platinum reduces the platinum to the divalent state (30, 1037).

The platinum-tin chloride system has also been used to hydrogenate unsaturated fats (978, 980, 1031–1034, 1040, 1041; Section XIV-F-1).

The trigonal bipyramidal anionic complex $Pt(SnCl_3)_5^{3-}$ and cis- and trans-$PtCl_2(SnCl_3)_2^{2-}$ have been isolated, and in solution a series of equilibria involving $[PtCl_x(SnCl_3)_{4-x}]^{2-}$ exist (30, 616, 681, 1037, 1038). An anionic cluster species, $Pt_3Sn_8Cl_{20}^{4-}$ and a derived cycloocta-1,5-diene complex $(C_8H_{12})_3Pt_3(SnCl_3)_2$, with structures based on a trigonal bipyramidal metal atom cluster, have also been made (1055, 1056). The trichlorostannate ligand is a strong π-acceptor (1024, 1038), which reduces electron density on the platinum atom making it susceptible to attack by nucleophiles such as hydride or an olefinic bond (1064). The hydride $HPt(SnCl_3)_4^{3-}$ was isolated from $Pt(SnCl_3)_5^{3-}$ at 30° with 500 atm hydrogen, although its formation could be followed when using 3 atm pressure (1024). The true catalytic

Table 75 Hydrogenations of Olefins and Acetylenes Catalyzed by Platinum–Tin Chloride Complexes Without Phosphine Ligands

Substrate[a]	Solvent	Product	Reference
Ethylene	Methanol	Ethane	30, 1035, 1036
But-1-ene	Methanol	But-2-enes, butane(~2%)	1025
Pent-1-ene	Methanol	Pent-2-enes, pentane(4%)	1027, 1028
Hex-1-ene	Acetic acid, esters, ethers, ketones	Hexane, hex-2- and -3-enes	1031–1034
Oct-1-ene	Acetic acid	—	1034
Octadec-1-ene	Acetic acid	—	1034
Cis-pent-2-ene	Methanol	No reduction	1028
Hex-2-enes	Acetic acid	Slow reduction	1031
Cyclohexene	Alcohols	Cyclohexane, 1-methylcyclohexene	1029
Styrene	Methanol	Ethylbenzene	1028
2-Alkyl-2,5-cyclohexadiene-1-carboxylic acid	Alcohols	2-Alkyl-2-cyclohexene-1-carboxylic acid	1029
Hexamethyl-Dewar-benzene	2-Propanol, methyl ethyl ketone	Hexamethylbicyclo[2,2,0]hexenes	1030
Substituted norbornadienes	—	—	1458
Acetylene	Methanol	Ethane, ethylene	30
Acetylenes (unspecified)	2-Propanol	—	1029

[a] At 20–30°, 1 atm H_2.

species is not definitely known, however. Bond and Hellier (1028) consider that $trans$-$PtCl_2(SnCl_3)_2{}^{2-}$ is the active catalyst (eq. 426), while formation of the more thermodynamically stable cis isomer at lower tin/platinum ratios and formation of $Pt(SnCl_3)_5{}^{3-}$ at higher ratios results in lower activity. The

$$PtCl_2(SnCl_3)_2{}^{2-} + H_2 \rightarrow HPtCl_2(SnCl_3)^{2-} + H^+ + SnCl_3{}^- \quad (426)$$

existence of $Pt(SnCl_3)_5{}^{3-}$ in solution has been questioned (681). Mössbauer studies (1037, 1042) have confirmed the π-acceptor character of $SnCl_3{}^-$, and indicate that complexes containing more than three tin ligands dissociate and readily exchange these ligands in methanol, while complexes with two or three $SnCl_3{}^-$ ligands do not readily do so.

Stannous chloride promotes the coordination of ethylene and other olefins to tetrachloroplatinate(II) to give complexes such as Zeise's salt (30, 1022, 1376). A kinetic study of these reactions (1038, 1376) indicates that the cis- and $trans$-$PtCl_2(SnCl_3)_2{}^{2-}$ complexes, particularly the cis-, are the active absorbing intermediates; the results were rationalized in terms of the high trans effect of the $SnCl_3{}^-$ ligand. Exchange between coordinated and free ethylene in the platinum–tin system is rapid (30). No ethylene absorption was observed for solutions of $Pt(SnCl_3)_5{}^{3-}$ (1038). Much earlier, Nyholm (61) had suggested reactions 427, 428 for ethylene hydrogenation via the unsaturate route (cf. Section XI-B-1), but this can be readily modified using

$$Pt(SnCl_3)_5{}^{3-} + C_2H_4 \rightarrow (SnCl_3)_xPt(C_2H_4), \quad x = 4 \text{ or } 3 \quad (427)$$

$$(SnCl_3)_xPt(C_2H_4) + H_2 \rightarrow (SnCl_3)_xPtH_2(C_2H_4) \xrightarrow{SnCl_3{}^-} Pt(SnCl_3)_5{}^{3-} + C_2H_6 \quad (428)$$

the square-planar, 4-coordinate complexes as the active species. Alternatively, initial formation of an alkyl via a monohydride (eq. 426) followed by hydrogenolysis, was suggested for hydrogenation of pent-1-ene and styrene (1027, 1028) (cf. Section IX-B-2b):

$$HPtCl_2(SnCl_3)^{2-} + C_5H_{10} \rightleftharpoons Pt(C_5H_{11})Cl_2(SnCl_3)^{2-} \xrightarrow{H_2} HPtCl_2(SnCl_3)^{2-}$$
$$+ C_5H_{12} \quad (429)$$

Shilov and coworkers (1035, 1036) have reported that above 200 mm hydrogen at ambient temperatures using a 5:1 tin–platinum ratio, the ethylene hydrogenation rate in methanol is given simply by $k[Pt][C_2H_4]$. Also, reduction of ethylene with D_2 in CH_3OH or H_2 in CH_3OD gave mainly C_2H_5D, which shows that one of the added hydrogens comes from the solvent and one from the molecular hydrogen. The following reaction scheme was initially presented (1035):

$$Pt + C_2H_4 \xrightleftharpoons{k} Pt(C_2H_4) \tag{430}$$

$$Pt(C_2H_4) + H_2 \rightleftharpoons [H_2Pt(C_2H_4)] \rightleftharpoons HPt(C_2H_4) + H^+ \tag{431}$$

$$HPt(C_2H_4) \rightleftharpoons Pt(C_2H_5) \tag{432}$$

$$Pt(C_2H_5) + H^+ \longrightarrow Pt + C_2H_6 \tag{433}$$

Pt represents the active complex which, in view of the later Mössbauer studies (see above), is thought to contain at least three $SnCl_3^-$ ligands (1037). The mechanism differs from the usually postulated unsaturate route (cf. eq. 428) in that *both* hydrogens in the dihydride intermediate are not transferred (eqs. 431, 432). The hydrolytic decomposition of platinum alkyls has been demonstrated for certain phosphine-substituted species (see below). The scheme above was based on the observation of the absence of hydrogen exchange with the solvent in the presence of the ethylene, that is, $HPt(C_2H_4)$ did not exchange. However, in the absence of olefin, exchange did occur, that is, Pt exchanges, and since the slow step involved ethylene coordination to Pt, the scheme was subsequently modified (1036) in that the Pt complex was thought to already have one coordinated ethylene, and was a species that did not exchange. The later paper (1036) also preferred hydride formation (cf. eq. 426) prior to reaction with ethylene. Kinetic measurements on this system were also made in ethanol, acetone, and THF; the activation energies were in the range 14.5 to 16.0 kcal mole^{-1}.

N.m.r. evidence for a conversion of π-bonded ethylene to a σ-alkyl derivative has been obtained by Kaplan and coworkers (1092), although the required nucleophilic attack involved deuterated pyridine and not hydride:

$$\tag{434}$$

Van Bekkum and coworkers (1030) used a bromide promoted tin-platinum chloride system to reduce hexamethyl-Dewar-benzene to bicyclo[2.2.0]hex-2-enes, namely, the 1,2,3,4,5 *endo*, 6 *endo*-(**137**), the 1,2,3,4,5 *endo*, 6 *exo*-(**138**), and the 1,2,3,4,5 *exo*, 6 *exo*-(**139**)-hexamethyl derivatives (eq. 435). Only one double bond is reduced mainly with *trans*-hydrogen addition to give **138**. The mechanism was discussed in terms of alkyl formation, via hydride formation (eq. 426), followed by substrate coordination on the exo side and hydride migration. An attempt was then made to determine the nature of the final reaction step which could involve (*a*) reaction with

exo

endo

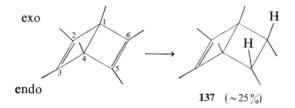

137 (~25%)

$+$ **138** (~70%) $+$ **139** (<5%)   (435)

molecular hydrogen (cf. eq. 429) or (*b*) some decomposition involving acid (cf. eq. 433) (see Scheme 15).

Mechanism (*a*) was written involving hydrogenolysis of the Pt—C bond, although it could equally well be written involving a platinum(IV) dihydride intermediate (cf. Scheme 4, Section IX-B-2b); both will give rise to *cis* addition and **137**. However (*a*) was ruled out, since deuteration in propan-2-ol

$\longrightarrow$ **137** + HPt—   (a)

$\xrightarrow{\text{H}^+\text{X}^-}$ $\longrightarrow$ **137** + XPt—   (b)

$\downarrow$ X⁻(Cl, Br)

$\longrightarrow$ $\longrightarrow$ **138** + XPt—   (c)

Scheme 15

gave 92% nondeuterated products, whereas d_1 products would be largely expected for the hydrogenolysis mechanism. The nondeuterated products were accounted for in terms of rapid hydride exchange in the initially formed monohydride (cf. eq. 426). It should be pointed out, however, that rapid exchange in a platinum(IV) dihydride intermediate via mechanism (a) would also be consistent with mainly d_0 products. Chatt and Shaw (1043) and Halpern and Falk (1044) have studied exchange of *trans*-HPtCl(PEt$_3$)$_2$ with D$_2$O/DCl mixtures and in acetone (1044). The kinetics fit rate law 436, and the suggested mechanism is shown in eq. 437; this does involve a platinum(IV) dihydride type intermediate, and compound H$_2$PtCl$_2$(PEt$_3$)$_2$ can decompose with loss of hydrogen chloride to regenerate *trans*-HPtCl(PEt$_3$)$_2$ (1043).

$$-d[\text{HPtCl(PEt}_3)_2]/dt = (k_1 + k_2[\text{Cl}^-])[\text{D}^+][\text{HPtCl(PEt}_3)_2] \qquad (436)$$

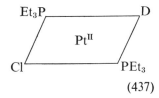

$$(437)$$

(Replacement of the added Cl$^-$ by solvent accounts for the k_1 path.)

Mechanism (b), involving addition of HX to the complex leads to a platinum(IV) alkyl hydride intermediate which eliminates alkane with resulting *cis* addition (Scheme 15). This is closely related to the mechanism of eq. 437; Belluco and coworkers (1045) have, in fact, studied the acid decomposition of the methyl complexes *trans*-PtX(PEt$_3$)$_2$(CH$_3$) (X = Cl, I) which goes according to a rate law such as 436 (see also Section C-3a):

$$\text{PtX(PEt}_3)_2(\text{CH}_3) + \text{HCl} \longrightarrow [\text{HPtX(PEt}_3)_2(\text{CH}_3)\text{Y}] \longrightarrow$$
$$\text{PtX(PEt}_3)_2\text{Cl} + \text{CH}_4 \quad (438)$$

(Y = Cl$^-$ or solvent)

Trans addition to give **138** could arise via (b) but involving direct protonolysis through back-side attack (Scheme 15): precoordination of a nucleophile X$^-$ would probably assist the reaction; front-side protonolysis would probably be indistinguishable from addition of HX. The situation, however, is more complex than just involving mechanisms (a)—ruled out by van Bekkum and coworkers—and (b), since deuteration of the Dewar-benzene in C_2H_5OD gave **137** and **138** containing up to d_5 fractions showing the presence of isotope exchange; this was thought to involve one methyl group, for example, as shown in eq. 439. *Cis* addition could now occur also via **140**, and *trans* addition via the exocyclic olefin, **141**. Related studies have

(439)

been carried out on reduction of substituted norbornadienes by Kheifits and coworkers (1458), but details are not yet available.

Anionic complexes such as $Pt(CO)X_3^-$ (X = halide) and $Pt(CO)Cl(SnCl_3)_2^-$ have recently been described (513, 1012).

3. Platinum Complexes with Tertiary Phosphine and Arsine Ligands. *a. In the Absence of Stannous Chloride.*

Platinum(II) hydrides of the type $HPtX(PR_3)_2$ (X = univalent anion and R = alkyl or aryl) are well known through the studies of Chatt and coworkers (996, 1043). They are not readily formed from molecular hydrogen; for example, reaction 440 takes place at 95° and 50 atm in ethanol (1043). The slow reverse reaction has been demon-

$$cis\text{-}PtCl_2(PEt_3)_2 + H_2 \rightleftharpoons trans\text{-}HPtCl(PEt_3)_2 + HCl \qquad (440)$$

strated by Belluco and coworkers (1045); the equilibrium involves a platinum(IV) dihydride intermediate, $H_2PtCl_2(PEt_3)_2$, which more readily loses hydrogen chloride than hydrogen to regenerate the monohydride (see above, eq. 437). Bailar and Itatani (978) have made $HPtCl(PPh_3)_2$ according to eq. 440 under high-pressure conditions (see also 1397). A monohydride has been generated (1043) using 1 atm hydrogen at 20° from a phenyl complex (eq. 441), but in any case these hydrides appear too stable to be efficient hydrogenation catalysts. The formation of platinum(II) hydrides according to

$$cis\text{-}PtClPh(PEt_3)_2 + H_2 \longrightarrow trans\text{-}HPtCl(PEt_3)_2 + C_6H_6 \qquad (441)$$

reactions such as 426, 440, and 441 appears to involve platinum(IV) dihydride intermediates. Another possibility pointed out by Tayim and Bailar (1064), besides a "direct" heterolytic mechanism, is initial reduction to platinum(0), which could be followed by reaction with acid to yield the platinum(II) hydride; such oxidation has been demonstrated by Cariati and coworkers (1067, 1076) for tris-phosphine and -arsine complexes:

$$PtL_3 \underset{KOH}{\overset{HX}{\rightleftharpoons}} [HPtL_3]X \underset{+L}{\overset{-L}{\rightleftharpoons}} HPtXL_2 \qquad (442)$$

(L = PPh_3, $AsPh_3$; X = Cl^-, NO_3^-, CN^-, ClO_4^-, etc.) The position of the equilibrium depended on the nature of X.

A platinum(IV) acetylide dihydride complex,

$$H_2Pt(PPh_3)_2\left(C\equiv C-\overset{HO}{\diagdown}\bigcirc \right)_2$$

has been synthesized by Roundhill and Jonassen (1097).

Chatt and coworkers (1043, 1049) have shown that the hydrides HPtX-$(PEt_3)_2$ (X = Cl, Br, CN) react reversibly with olefins to give alkyl derivatives; for example,

$$trans\text{-}HPtCl(PEt_3)_2 + CH_2{=}CHR \rightleftharpoons$$

$$\overset{CH_2}{\underset{RCH}{\|}}{-}PtHCl(PEt_3)_2 \rightleftharpoons Pt(CH_2CH_2R)Cl(PEt_3)_2 \quad (443)$$

Deuterium studies on the ethylene system, which involves high pressures, showed that either the α- or β-carbon atoms of the ethyl group can provide the hydride ligand, which is consistent with a hydride-π-olefin intermediate. With higher olefins, equilibrium 443 lies more to the left. Anti-Markownikoff addition only to give n-alkyl species was thought to be involved, since no isomerization was noted in the oct-1-ene system. Cramer and Lindsey (1025), however, have isomerized hex-1-ene with the hydridochloride. Platinum(II) alkyls and aryls appear too stable (1050) to be effective intermediates for efficient hydrogenation. A bond energy E(Pt–C) of at least 60 kcal mole^{-1} has been estimated for the $(PEt_3)_2PtPh_2$ complex (1110). Cationic platinum(II) alkyls have been reported (e.g., 1379, 1402).

Baddley and coworkers (1051, 1052) isolated the hydridoolefin complex $HPtCN(PEt_3)_2(C_6N_4)$ from the reaction of tetracyanoethylene with $trans$-$HPtCN(PEt_3)_2$, but with the hydridochloro complex the product was

$Pt(PEt_3)_2(C_6N_4)$, the initial intermediate adduct eliminating hydrogen chloride (cf. eq. 437).

Clark and coworkers (1069–1071) studied the reaction of fluoroolefins with *trans*-HPtCl(PEt$_3$)$_2$. One product is a fluorovinyl compound, formed by loss of hydrogen fluoride; for example,

$$HPtCl(PEt_3)_2 + C_2F_4 \longrightarrow [HPtCl(PEt_3)_2(C_2F_4)] \xrightarrow{-HF}$$
$$PtCl(PEt_3)_2(CF{=}CF_2) \quad (444)$$

The π-olefin intermediate was thought to have been isolated, but later the isolated compound was shown to be $[PtCl(CO)(PEt_3)_2]^+SiF_5^-$ formed by hydrolysis of the vinyl complex (1072, 1073). By using nonglass apparatus, the tetrafluoroethyl complex $PtCl(PEt_3)_2(CF_2CF_2H)$ was obtained from reaction 444 (1073). Fluoroacetylenes generally added across the platinum-hydrogen bond to give vinyl complexes; for example, perfluorobut-2-yne gives $PtCl(PEt_3)_2[C(CF_3){=}CHCF_3]$ (1070, 1071). Kemmitt and coworkers (1099) have also shown that similar ethyl and vinyl complexes can be formed by protonation of the fluoroolefin and -acetylene complexes such as $Pt(PPh_3)_2(C_2F_4)$.

Brookes and Nyholm (1100) have shown that the orthovinyl group of the chelating phosphine (*o*-vinylphenyl)diphenylphosphine (L) inserts into the platinum-hydrogen bond of *trans*-HPtCl(PPh$_3$)$_2$ to give a σ-alkyl derivative **142**, presumably via a hydride-olefin intermediate:

(445)

The alkyl reacted with hydrochloric acid to yield **143**. The net result is a stoichiometric hydrogenation of the ligand vinyl group (see below, eq. 446).

Lewis and coworkers (1105) have studied similar insertion reactions of platinum hydrides using some dienes. *Trans*-HPt(NO$_3$)(PEt$_3$)$_2$ in methanol substitutes the nitrate ligand with resulting formation of the cationic complexes $PtL(PEt_3)_2^+$ where L = π-allyl (from allene), π-crotyl (from butadiene), cyclooct-4-enyl (from cycloocta-1,5-diene), or norborneyl (from norbornadiene). Interestingly, on using ethylene, a hydridoethylene complex *trans*-HPt(C_2H_4)(PEt_3)$_2^+$ was isolated; the intramolecular nucleophilic attack of hydride on the coordinated ethylene does not occur in the cationic species (cf. eq. 443, R=H). The α,β-unsaturated ketones CH_2=CHCOCH$_3$ and PhCH:CHCOPh behaved similarly to the dienes and gave nonhydridic cationic products.

Giustiniani and coworkers (1053) have reported a closely related stoichiometric hydrogenation of hex-1-ene, oct-1-ene, 2-methyl-2-butene, and cyclohexene according to eq. 446 when using *trans*-HPtCl(PEt_3)$_2$ and acid in ethanol solution at 30°. The reactions were first-order in platinum. The favored mechanism was written as involving platinum alkyl formation,

$$\textit{trans-}HPtCl(PEt_3)_2 + RCH{=}CHR' \xrightarrow{\text{HCl}}$$

$$\textit{trans-}PtCl_2(PEt_3)_2 + RCH_2CH_2R' \quad (446)$$

oxidative addition of the acid and subsequent elimination of hydrocarbon (cf. Scheme 15, eqs. 437 and 438 in Section C-2). No reaction was observed between the hydride and the olefins, that is, eq. 443 lies well to the left, and there was no isomerization of hex-1-ene under the mild conditions. The slow reaction of the hydride with hydrogen chloride was noted above (eq. 440), but the possible mechanism shown in eq. 447 was thought unlikely. $H_2PtCl_2(PEt_3)_2$ did not hydrogenate hex-1-ene in solution.

$$\rightarrow \quad \searrow CH-CH\nearrow + PtCl_2(PEt_3)_2 \quad (447)$$

Of related interest is the reaction of a platinum(II) π-allyl complex with 1 atm hydrogen in chloroform to yield saturated hydrocarbons (eq. 448) as reported by Volger and Vrieze (1054). The mechanism of this reaction, which

$$cis\text{-}PtCl_2(PPh_3)_2 + CH_3CH_2CHR_2 \quad (448)$$

has not been established, would be of interest in view of the low hydrogen pressure required; highly reactive, coordinatively unsaturated σ-allyl complexes, such as $(Ph_3P)_2PtCH_2CH{=}CR_2$, might be involved (1053, 1054).

Jardine and McQuillin (240) report that *trans*-HPtX(PPh$_3$)$_2$ (X = halide) and *cis*-PtCl$_2$(PPh$_3$)$_2$ do not catalyze the hydrogenation of norbornadiene in alcohol at ambient conditions. The hydridochloride is reported by Tayim and Bailar (999) to be inactive for the hydrogenation of some polyolefins even at 90° with 40 atm hydrogen. H$_2$Pt(PPh$_3$)$_2$ (1046, 1047), which again requires high pressure for its direct formation from Pt(PPh$_3$)$_4$ (1047), is similarly inactive as are the Pt(PPh$_3$)$_n$ complexes themselves (240, 1095). Other platinum(0) bis(triphenylphosphine) complexes, Pt(PPh$_3$)$_2$L, where L is acetylene, acrylonitrile, methyl vinyl ketone, norbornadiene, maleic anhydride, and phenylacetylene, and the tetracyanoethylene complex, Pt(PEt$_3$)$_2$(C$_6$N$_4$), have been used by Dunning and coworkers (1026) for the reduction of hex-1-yne/oct-1-ene mixtures at 150 to 200° and 70 to 300 atm hydrogen in benzene. The acetylene is more effectively reduced than the oct-1-ene, or the hex-1-ene product.

A range of platinum(II) and platinum(0) complexes containing triphenylphosphine, -arsine, or -stibine have been used for the hydrogenation of unsaturated fats at 90° and 40 atm pressure, although they are not generally effective in the absence of a cocatalyst (978, 980, 1040, 1041, 1048; see Sections XIII-C-3b and XIV-F-2).

The use of binuclear platinum(II)–palladium(II) phosphine complexes as catalysts for hydrogenation of acetylene/olefin mixtures was described in Section XIII-B **(136)**.

A range of cationic carbonyls PtX(CO)(MR$_3$)$_2$$^+$ (X = halide, hydride, aryl; M = P, As, Sb; R = alkyl or aryl), electronically related to Vaska's compound, have been made (954, 1073–1075, 1104, 1105), but they are generally less reactive than the corresponding iridium compounds. Clark and coworkers (1073) report that PtCl(CO)(PR$_3$)$_2$$^+$, for example, does not react with hydrogen under mild conditions. However, certain of the complexes undergo an interesting reaction with water forming hydrides according to eq. 449 (1075, 1077, 1078, 1380); the corresponding alkoxy carbonyl inter-

$$\text{PtCl(CO)(PEt}_3\text{)}_2{}^+ + \text{H}_2\text{O} \xrightarrow{-\text{H}^+} [\text{PtCl(COOH)(PEt}_3\text{)}_2] \longrightarrow$$
$$\text{HPtCl(PEt}_3\text{)}_2 + \text{CO}_2 \quad (449)$$

mediates were isolated from the corresponding reaction with alcohols (see also 1400). The hydride HPt(CO)(PEt$_3$)$_2$$^+$ reacts with (*o*-vinylphenyl)diphenylphosphine with loss of carbon monoxide to give the cationic σ-alkyl complex **144** (1100, cf. eq. 445)

$$\left[\text{C}_6\text{H}_4 \underset{\text{CH(CH}_3)}{\overset{\text{PPh}_2}{\diagup \diagdown}} \text{Pt(PEt}_3\text{)}_2 \right]^+$$

144

Complexes such as $PtCl(PEt_3)_3^+$, related to $RhCl(PPh_3)_3$, are known (1074), but no reactions under hydrogen have been reported. The hydrides $HPt(PR_3)_3^+$ (R = alkyl or aryl, cf. eq. 442) and some related ones $HPtL(MEt_3)_2^+$ (M = P, As; L = tertiary phosphines, arsines, stibines, and phosphites, nitrogen bases, CO, etc.) are also known (1067, 1074, 1076, 1104, 1106, 1107, 1403). Cationic complexes containing dienes, for example, $PtCl(C_8H_{12})PR_3^+$, corresponding to the active rhodium and iridium catalysts, $M(diene)(PR_3)_2^+$ (Sections XI-G and XII-B-1), have also been made (855).

Formation of species containing platinum-carbon bonds from tertiary phosphine complexes such as $PtCl_2(PPhBu^t_2)_2$, by hydrogen abstraction from an orthophenyl position, is known (e.g., 1352); the hydrogen halide is lost, presumably from a platinum(IV) intermediate, for example, the complex above yields $PtCl(C_6H_4PBu^t_2)(PPhBu^t_2)$.

Use of organophosphine and organoarsine platinum complexes as hydroformylation catalysts has been mentioned in patents (e.g., 527, 699, 968). Formation of acyl complexes from alkylplatinum carbonyls (cf. Section X-E) is known (1093, 1094). Apart from halogenocarbonyls, neutral carbonyl-phosphine complexes of platinum seem relatively rare (1108, 1393, 1394).

b. Stannous Chloride Systems. Following the discovery of the effectiveness of stannous chloride to promote the activity of platinum chlorides (Section C-2), a number of workers have successfully used the same co-catalyst (or Group IV analogs) with certain platinum(II)–tertiary phosphine type complexes.

The stannous chloride becomes coordinated as $SnCl_3^-$; a range of complexes of formulation $PtX(SnCl_3)(PR_3)_2$ are known, where R is alkyl or aryl, and, for example, X is chloride (30, 681, 1026, 1057), $SnCl_3^-$ (30), hydride (978, 1024, 1048, 1057) or aryl (1024). Both *cis*- and *trans*-hydride complexes are known. A methanol solution containing $PtCl_2(PEt_3)_2$ and $SnCl_2 \cdot 2H_2O$ (1:2) absorbs 1 mole of hydrogen rapidly at 25° and 1 atm to yield the 5-coordinate $HPt(SnCl_3)_2(PEt_3)_2^-$; this may also be formed according to eq. 450 (1023). Clearly, hydrides may be much more readily formed in the

$$HPt(SnCl_3)(PEt_3)_2 + SnCl_3^- \longrightarrow HPt(SnCl_3)_2(PEt_3)_2^- \qquad (450)$$

presence of the tin (cf. Section C-3a), although it has not been specifically reported that the 4-coordinate hydrides are formed under mild conditions. $HPt(SnCl_3)(PPh_3)_2$ has been isolated from a $PtCl_2(PPh_3)_2/SnCl_2 \cdot 2H_2O$ mixture (1:7) under 40 atm hydrogen (978). Equilibria such as 451 certainly exist (30), but the reactivity of the chlorostannate complexes toward hydrogen is not known quantitatively, and it is not clear if a reaction such as 450 is an

$$PtCl_2(PPh_3)_2 \underset{\longleftarrow}{\overset{SnCl_2}{\rightleftharpoons}} PtCl(SnCl_3)(PPh_3)_2 \underset{\longleftarrow}{\overset{SnCl_2}{\rightleftharpoons}} Pt(SnCl_3)_2(PPh_3)_2 \qquad (451)$$

Table 76 Reductions Using Platinum(II) Phosphine (Arsine)–Stannous Chloride Systems under Hydrogen

Catalyst	Substrate	Product	Reference
HPtCl(PPh$_3$)$_2$ + SnCl$_2$[a](1:2)	Norbornadiene	Norbornane	240
	2,5-Dimethylhex-3-yn-2,5-diol	2,5-Dimethylhexane-2,5-diol	240
	Oct-1-ene	Octane, oct-2-ene	243
PtX$_2$(PPh$_3$)$_2$ + SnCl$_2$(1:2) (X = halide, CN)	Oct-1-ene	Octane, oct-2-ene	243
PtCl$_2$(MPh$_3$)$_2$ + SnCl$_2$(1:10) (M = P, As)	Acrylonitrile[b]	Propionitrile (94%)	1062
PtCl$_2$(PPh$_3$)$_2$ + SnCl$_2$(1:10)	3-Cyclohexenylcyanide	Cyclohexyl cyanide (10%)	999
	4-Vinylcyclohexene	Monoene (93%)	999
	3-Ethylidenecyclohexene	Monoene (100%)	999
	Hexa-1,5-diene	Hexenes (70%)	999, 1001
	Hepta-1,5-diene	Heptenes (91%)	999, 1001
	Octa-1,7-diene	Octenes (26%)	999
	Cyclohexa-1,4-diene	Cyclohexene (11%)	999
	Cycloocta-1,5-diene	Cyclooctene (15%)[c]	999, 1001
	Cycloheptatriene	Cycloheptene (2%)	999
	Deca-1,4,9-triene	Decene (100%)	999
	Cyclooctatetraene	No monoene	999
HPt(SnCl$_3$)(PR$_3$)$_2$ (R is alkyl or cycloalkyl)	Cycloocta-1,3-diene	Cyclooctene (74%)	1066
	Isoprene	2-Methylbut-2-ene (73%), 2-methylbut-1-ene (23%)	1442

[a] Identical results were obtained using anhydrous SnCl$_2$ or the dihydrate (999).
[b] The α- and β- chloroacrylonitriles and crotonitrile were not reduced (1062).
[c] PtCl$_2$(AsPh$_3$)$_2$ and PtCl$_2$(SePh$_2$)$_2$ gave about 100 and 25% monoene under similar conditions. With cycloocta-1,3-diene the figures were 65 and 17%, respectively, while PtCl$_2$(SPh$_2$)$_2$ reduced 14%. The cycloocta-1,5-diene complex PtI$_2$(C$_8$H$_{12}$) was similar in activity to PtCl$_2$(AsPh$_3$)$_2$.

important equilibrium. Hydrides such as $HPt(SnCl_3)(PPh_3)_2$ appear to be the active catalyst in the catalytic hydrogenations to be discussed below; as Tayim and Bailar (1064) have pointed out this could be formed via $HPtCl(PPh_3)_2$ or via hydrogen activation by $PtCl(SnCl_3)(PPh_3)_2$. The latter seems more likely. Addition of stannous chloride to the hydride $HPtCl(PPh_3)_2$ results in similar catalysis as addition to the dichloro compound (978). The complex $HPt(SnCl_3)(PPh_3)_2$ is not active for the hydrogenation of linoleate in benzene-methanol (Section XIV-C), except in the presence of excess stannous chloride, and this has been taken as indicating the existence of equilibrium 452 (978); indeed, *both* hydrides have been isolated from $PtCl_2(PPh_3)_2-SnCl_2-H_2$ mixtures (978, 999). However, the hydrides HPt-$(SnCl_3)(PR_3)_2$, R = Me, Et, have been used as such in benzene by Kanai and

$$HPt(SnCl_3)(PPh_3)_2 \rightleftharpoons HPtCl(PPh_3)_2 + SnCl_2 \qquad (452)$$

Miyake (1066, 1442) for the hydrogenation of cycloocta-1,3-diene and isoprene (Table 76).

Related to the formation of the hydrides is the hydrogenolysis by 1 atm hydrogen at ambient temperatures of a variety of triaryl and trialkyl Group IV derivatives of platinum phosphine or arsine complexes, reported on particularly by Glockling and coworkers (1058–1060, 1101, 1109), Deganello and coworkers (1061), and Akhtar and Clark (1079); for example:

$$Pt(MPh_3)_2(PR_3)_2 + H_2 \longrightarrow HPt(MPh_3)(PR_3)_2 + HMPh_3 \qquad (453)$$
$$PtCl(MMe_3)(PR_3)_2 + H_2 \longrightarrow HPtCl(PR_3)_2 + HMMe_3 \qquad (454)$$
$$(M = Sn, Si, Ge, Pb; R = Et, n\text{-}Pr, Ph)$$

These reactions proceed via 6-coordinate platinum(IV) dihydride intermediates (1059, 1061); a low activation energy of 9 kcal mole^{-1} estimated for reaction 453 (M = Ge, R = Et) was attributed to easy oxidation to the tetravalent state (1058). The reverse of reactions such as eqs. 453 and 454 has also been observed in certain systems (e.g., 1089, 1090). Some related platinum(IV) monohydride complexes, such as $HPt(SnMe_3)_3(diphos)$, have also been isolated (1101, 1102).

The role of the tin moiety in the stannous chloride systems (see Section C-2) may not be unique, and the study of other complexes such as $(Ph_3P)_2ClPt$-$HgCl$ and $(Ph_3P)_2ClPt$-$M(PPh_3)_n$, where M = Au, Cu (1111), may be worthwhile.

Tables 76 and 77 summarize the reported hydrogenations using platinum(II)–phosphine–stannous chloride systems; the studies by Bailar and coworkers (978, 980, 1001, 1040, 1041, 1062, 1063, 1138, 1139, 1475) on the hydrogenation of fats by using these systems are not included and will be considered in Section XIV-F-2.

The studies of McQuillin and coworkers (240, 243) (Table 76) involved quite slow reactions using 1 atm hydrogen at 20°. The bromo and iodo

Table 77 Reduction of Olefins Using PtCl$_2$(PPh$_3$)$_2$ + SnCl$_2$ (1:10) in Benzene-Methanol (3:2)^a

Substrate	Product	Reference
Ethylene	Ethane (100%)	1062, 1065
Propylene	Propane (34%)	1062, 1065
But-1-ene	Butane (11%), but-2-enes	1062, 1065
2-Methylbut-1-ene	2-Methylbutane (0.2%), methylbutenes	1062, 1065
3-Methylbut-1-ene	2-Methylbutane (27%), methylbutenes	1062, 1065
Cis-but-2-ene	Butane (1.6%), but-2-enes	1062, 1065
Trans-but-2-ene	Butane (0.6%), but-2-enes	1062, 1065
Butadiene	But-2-ene (0.2%)	1062, 1065
Pent-1-ene	Pentane (11.6%), pent-2-enes	1062, 1065
Pent-2-enes	Pentane (1.8%), pentenes	1062, 1065
Hex-1-ene	Hexane (12%), hexenes	1062, 1065
Hex-2- and -3-enes	No reduction	1062, 1065
Penta-1,4-diene	n-Pentane (3.4%), pentenes (66%), pentadienes	1062
Penta-1,3-diene	No pentane, pent-2-ene (3%), penta-1,3-diene	1062
2,3,3-Trimethylpenta-1,4-diene^b	2,3,3-Trimethylpent-1-ene, 3,3,4-trimethylpent-1-ene	1062, 1065
Hexa-1,5-diene	Hexane (2.5%), hexenes (20%), hexadienes	1062, 1065
Hexa-1,4-diene	Hex-2-enes (18%), hexadienes	1062, 1065
Hexa-1,3-diene	Hex-3-ene (8%), hexadienes	1062, 1065
Hexa-2,4-diene	Hex-2-enes (1%), hexadienes	1062, 1065
2,4-Hexadienoate (sorbate)	Monoene (9%)	1062
3-Ethylidenecyclohexene	Monoene (77%)	999
4-Vinylcyclohexene	Monoene (30%)	999
Octa-1,7-diene	Monoene (89%)	999
Octa-1,3,6 (or 2,4,6)-triene	Monoene (90%), dienes	999
Deca-1,4,9-triene	Monoene (24%), dienes	999
Mesityl oxide	Isobutylmethyl ketone (67%)	999
Cyclohexen-3-one	Cyclohexanone (100%)	999
Norbornadiene	Ethylidene cyclopentane	999
Allyl alcohol^c	Acrolein	1091
3-Hydroxy-but-1-ene^c	But-2-ene, butadiene, methyl vinyl ketone	1091
Cholest-4-ene-3β-ol^c	5α-Cholest-3-ene, cholesta-2,4-diene	1091

^{a} With 40 atm hydrogen at 90 to 105° for 3 hr (1062, 1065) or 8 hr (999).
^{b} The PtCl$_2$(AsPh$_3$)$_2$ complex was also used.
^{c} At room temperature, 30 atm H$_2$.

complexes, HPtX(PPh$_3$)$_2$, were ineffective for the reduction of norbornadiene and the hex-1-yne diol in methanol, although all the halide complexes PtX$_2$(PPh$_3$)$_2$ were used to reduce oct-1-ene in benzene-methanol (3:2). The oct-1-ene systems resemble closely those of the palladium systems discussed in Section XIII-B (cf. eqs. 415, 415a) but are less active. The iodo and cyano complexes give linear uptake plots with no accompanying isomerization to oct-2-ene. The initial rates for the chloro and bromo species fell off rapidly partly because of production of oct-2-ene and partly because of tying up the catalyst as an olefin complex (cf. eq. 415a), since the amount of active hydride in the system was reduced as a result of the reverse reaction of 415 becoming important. In the presence of acetate buffer, the chloride system gave rise to a completely linear rate exactly the same as that obtained starting with the hydridochloride complex. The mechanism was written in terms of alkyl formation followed by subsequent reaction with hydrogen to yield product and regenerate the hydride; the variation of initial rate with oct-1-ene concentration for the PtCl$_2$(PPh$_3$)$_2$ system was as shown in Fig. 4 (Section IX-B-2b). As in the palladium systems the activity of the PtX$_2$(PPh$_3$)$_2$ systems decreased in the order Cl > Br > I > CN.

The hydrogenations listed in Tables 76 and 77 by Bailar and coworkers (999, 1001, 1062, 1065) and Ichinohe and coworkers (1091) have generally been carried out with 35 to 55 atm at 80 to 105° for 3 to 8 hr in dichloromethane, (Table 76) or benzene-methanol (3:2) (Table 77). Ichinohe and coworkers (Table 77; 1091) used benzene-methanol at high pressure, but at room temperature. Kanai and Miyake (Table 76; 1066, 1442) used 100 atm hydrogen at 125° in benzene for 24 hr. The hydrogenations were usually accompanied by much faster isomerization reactions which occur even at ambient conditions, and these were studied for cyclooleta-1,5-diene (999, 1064), hexa-1,5-diene (999, 1062), and hepta-1,5-diene (999). These dienes very rapidly formed the thermodynamically more stable conjugated systems, and it was first thought that conjugated systems gave rise to the selective monoene product, which was observed generally from the reduction of polyenes (Table 76; 999). However, more careful study of a range of dienes and monoenes (Table 77; 1062, 1065) shows that this is not necessarily so. But-1-ene, pent-1-ene, and hexa-1,5-diene do give saturated product; the conjugated hexa-2,4-diene gives only traces of hexenes; penta-1,4-diene is more reactive than penta-1,3-diene, and 2,3,3-trimethylpenta-1,4-diene which cannot isomerize to a conjugated molecule yields trimethylpent-1-enes. The results were then better rationalized in terms of decreasing activity in the order terminal olefins ≫ internal olefins > short chain conjugated dienes, with the fact that terminal olefins are isomerized faster than they are hydrogenated (999, 1062). Butadiene slows hydrogenation and isomerization of other olefins presumably by coordinating strongly to the metal (1062). This is not the case with long-chain conjugated dienes, which form weaker bonds and can give rise to unreactive internal monoenes (Section XIV-F). Ethylidenecyclopentane and not norbornene is produced from norbornadiene (999).

The olefinic bond in α,β-unsaturated ketones (mesityl oxide, cyclohexen-3-one) and acrylonitrile is readily reduced; the presence of an ester group in hexa-2,4-diene also promotes activity (sorbate). Substitution at the double bond (crotonitrile, chloroacrylonitriles, 2-methylbut-1-ene) hinders hydrogenation. Substrates with an allylic alcohol system undergo more complex reactions (Table 77; 1091). 3-Hydroxy-but-1-ene and cholest-4-en-3β-ol give some products showing dehydration from the saturated alcohol system.

Bailar and coworkers particularly have examined the effect of several variables on the activity of these platinum systems. Some of the data has been obtained for the reduction of linoleate and linolenate (e.g., 978, 1040, 1041; Section XIV-F), but the same conclusions generally seem to hold also for the substrates listed in Tables 76 and 77.

No hydrogenation occurs at less than 30 atm pressure, and varying the pressure above about 40 atm gives no advantage; the optimum tin/platinum ratio was 10:1; activity was independent of the catalyst concentration above $10^{-2} M$ (999, 1064, 1091). Stannous chloride (either anhydrous or the dihydrate) was the most effective cocatalyst; when using $PtCl_2(PPh_3)_2$ as the platinum source, the activity with other Group IV halides decreased in the order: $SnCl_2 > SnCl_4 > GeCl_2 + HCl > GeCl_4 > SiCl_4 > PbCl_2 + HCl > HSiCl_3$ (978, 980). Antimony and aluminum trichlorides, mercuric acetate, and nickel acetylacetonate were ineffective (1040). The bromides $PtBr_2(PPh_3)_2$-$SnBr_2$ were more effective than the chloride system, but solids were precipitated under the hydrogenation conditions. Use of the catalysts $PtCl_2L_2$-$SnCl_2$, where L is the neutral π-acceptor ligand, for the conversion of cyclooctadienes (see Table 76) and linoleate shows reactivity decreasing in the order $L = P(OPh)_3 > AsPh_3 > SePh_2 > SPh_2 \approx PPh_3 > SbPh_3 > PBu_3$ (978, 999, 1041), and $PMe_2Ph > PMePh_2 > PPh_3 > PMe_3$ (1139, 1475). During hydrogenation of linoleate, the $AsPh_3$ and $SbPh_3$ systems gave a precipitate, probably platinum metal, but this was shown to be essentially inactive (978); no precipitate was recorded using the arsine system to reduce cyclooctadienes (999). The order shows increasing activity with π-acceptor capacity or decreasing σ-donation (weaker base character); there is also thought to be less steric hindrance with the larger central atom ligands (As > P, Se > S). The weaker Pt—H bond in $HPtCl(AsPh_3)_2$ compared to the phosphine analog is also thought to contribute to the more efficient hydrogenation (999).

As in the platinum-tin chloride systems without phosphines (Section XIII-C-2) there are marked solvent effects. Some of the hydrogenations in benzene-methanol were just as effectively carried out under about 25 atm nitrogen, the alcohol being the hydride source (978, 980, 999), for example, for the reduction of octatrienes and octa-1,7-dienes using $PtCl_2(PPh_3)_2$-$SnCl_2$. In dichloromethane solution, the yield of monoene from dienes and trienes, is usually markedly increased (Tables 76 and 77, 4-vinylcyclohexene, 3-ethylidenecyclohexene, hexa-1,5-diene, deca-1,4,9-triene); this is in line with the finding that the isomerization of hexa-1,5-diene to the inert 2,4-isomer (Table 77) is much slower in dichloromethane than in benzene-methanol (1062). However, it should be noted that octa-1,7-diene gives a lower monoene yield in dichloromethane (999) and cycloocta-1,5-diene is isomerized more rapidly under hydrogen in this solvent than in benzene-methanol (1064). Decreasing hydrogenating activity in the series, dichloromethane $\approx$ dichloroethane $>$ acetone $>$ THF $\approx$ methanol $>$ pyridine was paralleled with increasing coordinating ability of the solvents (999). In acetone some aldol condensation to mesityl oxide occurs and this gives rise to some isobutyl methyl ketone (999).

In addition to isolating hydride complexes from the $PtCl_2(PPh_3)_2$-$SnCl_2$ mixtures (see above), Bailar's group (999, 1064) have also isolated the hydridoolefin reaction intermediates $[HPt(SnCl_3)(PPh_3)_2]_2$diene, where diene is cycloocta-1,5-diene (**145**) and norbornadiene; the cyclooctene complex $HPt(SnCl_3)(PPh_3)_2(C_8H_{14})$ was also isolated from the cycloocta-1,5-diene system. $[HPt(PPh_3)_2(C_6H_8)]SnCl_3 \cdot CH_2Cl_2$ was obtained from the cyclohexa-1,4-diene system, and $HPt(SnCl_3)(PPh_3)_2(C_8H_{16})$ from oct-1-ene. The non-

145

hydridic complex $[Pt(PPh_3)_2(C_7H_8)]SnCl_4$ was also isolated from the norbornadiene system. Isoprene (C_5H_8) has yielded $[Pt(PPh_3)_2(C_5H_{10})_2]Sn_2Cl_6$, containing monoene units, and also the isopropenyl complex (1048), $PtCl(PPh_3)_2(C_5H_9)$, in which a hydride has been transferred to the coordinated olefin. This latter synthesis is carried out in the presence of stannous chloride, which presumably assists the reaction in much the same way as it does reaction 455; in the absence of stannous chloride, reaction 455 requires

$$\textit{trans-}HPtCl(PEt_3)_2 + C_2H_4 \rightleftharpoons PtCl(PEt_3)_2(C_2H_5) \qquad (455)$$

high ethylene pressure (cf. eq. 443, Section C-3a), but Cramer and Lindsey (1025) report that the equilibrium is set up at 25° and 1 atm ($K = 35\ M^{-1}$) in the presence of the tin. This could be due to promotion of ethylene coordination as with the $PtCl_4^{2-}$ system (Section C-2) or an increased rate of hydrogen transfer.

Bailar and coworkers (978, 999, 1064) consider that the catalytic reductions involve alkyl formation via a hydride followed by reaction with molecular hydrogen (cf. eq. 429 and discussion in Section C-2); isomerization occurs as usual via hydride elimination from the alkyl. The olefins in the system must also readily exchange. Heterolytic fission was favored (1064) for hydride formation (cf. Section C-3a), which was not considered rate-determining since the hydrogenation (and isomerization) rate was the same with either $PtCl_2(PPh_3)_2$ or $HPtCl(PPh_3)_2$ as catalysts. Similarly, olefin complex formation was ruled out as rate-determining, since $PtCl_2(1,5-C_8H_{12})$ gave comparable rates to $PtCl_2(PPh_3)_2$ for hydrogenation and isomerization of cycloocta-1,5-diene (999, 1064). Reactions involving formation or loss of the hydrido-metal-olefin complexes were considered rate-determining. As mentioned above, Bailar and coworkers had first thought that hydrogenation to monoene occurred only via a conjugated diene and they had proposed (999) a mechanism for this step:

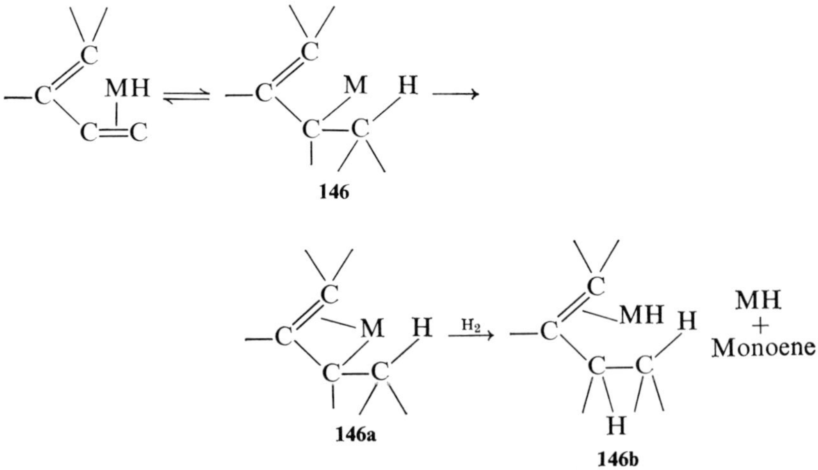

146

146a   146b

(456)

Hydrometalation initially yields an alkyl 146 in which the metal is bonded in a position α to the free double bond; this rearranges to give the σ-π complex 146a which then reacts with hydrogen as shown. 146a could rearrange to a π-allyl complex prior to hydrogenation although none were detected by n.m.r. Complexes such as 146b were isolated (see above). A π-allyl intermediate might also be expected to give rise to a saturated product (cf. eq. 448 in Section C-3a). An alternative mechanism postulated somewhat earlier (978) for selective reduction of linoleate had involved reaction of 146

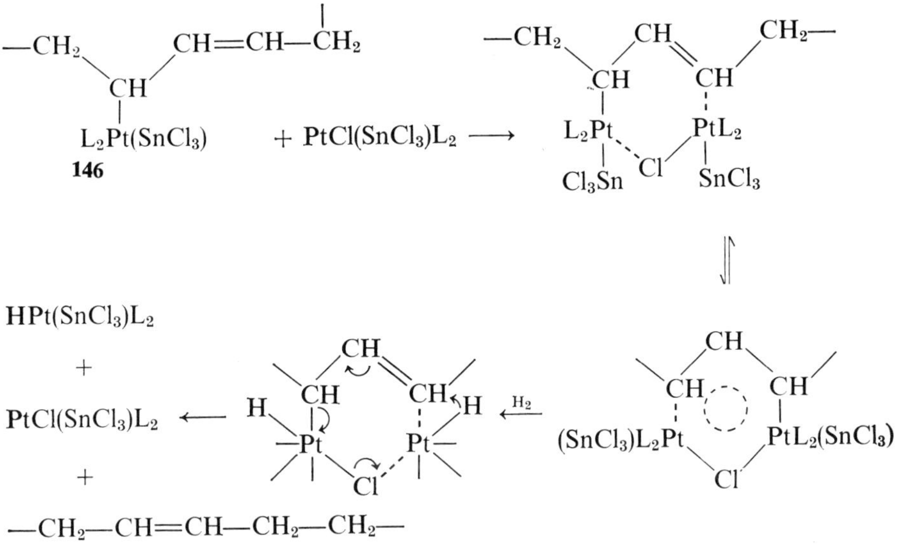

Scheme 16

with a second nonhydridic platinum moiety, for example, as shown in Scheme 16 (L = PPh₃).

The allylic intermediate was said to be isolated, but no details were given. Such reaction schemes are still thought valid for reduction of long-chain olefins (unsaturated fats, Section XIV-F), but are not necessary for the short-chain olefins, which undergo efficient reduction only at terminal positions (see above).

XIV

Homogeneous Hydrogenation of Unsaturated Fats

Soybean and linseed methyl oils consist of mixtures of long-chain unsaturated compounds, such as monoenes, dienes, and trienes. A catalyst has long been sought for the selective reduction of the linolenate constituents (9,12,15-octadecatrienoates, which give an undesirable flavor instability to the oils) to linoleate (*cis*-9, *cis*-12-octadecadienoate).

The subject has been reviewed quite extensively from the point of view of both heterogeneous and homogeneous catalysts. Among the articles that include some discussion of homogeneous systems are those of Frankel and Dutton (81), Leuteritz (1312), Nakamori (1435), Achaya and Raju (1436), and Ucciani (157, 1440). Bailar (1138, 1475) has reviewed studies on some homogeneous platinum metal systems.

A. PENTACYANOCOBALTATE(II)

The successful selective reduction of sorbate to hexenoates by using $Co(CN)_5^{3-}$ (Section X-B-2a) suggested the use of this complex. De Jonge and DeVries (1112) and Mabrouk (319) found that 9,11,13-octadecatrienoic acid was reduced to a diene; Mabrouk (319) found that monoene and conjugated and unconjugated diene fatty acids were not reduced, and concluded that a conjugated triene or a conjugated diene α to a carbonyl (e.g., sorbate) was necessary. It should be noted, however, that α,β-unsaturated acids have been reduced, but more severe conditions are required if the acids lack an α-substituent (Section X-B-2b Table 22). The use of the catalyst is limited because it is not fat soluble and is restricted to conjugated systems.

B. IRON CARBONYLS

Hashimoto and Shiina (949) first reported on the hydrogenation of soybean oil by using $Fe(CO)_5$ at 180° with 20 atm hydrogen to give products including conjugated diene fatty esters. The same workers (1114, 1115) also reported on the isomerization of soybean oil and linoleate under nitrogen atmospheres. Frankel and coworkers (198) studied the reduction of soybean oil and its methyl esters in more detail at 180° with 7 to 70 atm hydrogen in the presence or absence of cyclohexane as solvent. Relatively high reduction of the linolenate and linoleate to monoene occurred, and considerable geometric and positional isomerization was noted; conjugated dienes were formed and new iron carbonyl complexes were detected. The same group (199) also studied the $Fe(CO)_5$-catalyzed reduction of methyl linoleate in cyclohexane, at 180° and 30 atm hydrogen; the major monoene products showed a double bond distribution in good agreement with that calculated by assuming that only conjugated dienes (also observed as products) are reduced. Loss of dienes followed first-order kinetics. A complex π-diene-$Fe(CO)_3$ (diene = conjugated methyl octadecadienoates, **147**) was isolated (199, 1116) as a mixture of isomers and this was a more efficient catalyst than

$$CH_3(CH_2)_y-CH \underset{Fe(CO)_3}{\overset{CH-CH}{\diagup \quad \diagdown}} HC-(CH_2)_x-CO_2CH_3 \qquad \mathbf{147}$$

$x, y = 4, 8; 5, 7; 6, 6; 7, 5; 8, 4; 9, 3; 10, 2.$

$Fe(CO)_5$. The diene complexes could be hydrogenated to give monoenes and the fully saturated stearate, which was also formed in smaller quantities in the pentacarbonyl reductions. Furthermore, decomposition of the diene complexes gave conjugated dienes of about the same composition as those formed during linoleate hydrogenation. Later studies (200), using a mixture of methyl linoleate-1-^{14}C and conjugated octadecadienoate as substrate in the $Fe(CO)_5$ system, showed that most of the activity was found in the monoenes, stearate, and diene-$Fe(CO)_3$ complex with very little in the conjugated diene, which suggested that the diene was not a significant intermediate; linoleate was also reduced at about the same rate as the diene which showed that if hydrogenation followed conjugation the latter is not rate-determining. With unlabeled linoleate and diene-1-[14]C-$Fe(CO)_3$, a large amount of activity was transferred to the reduction products, showing that significant exchange occurs between linoleate and the complex during hydrogenation. Moreover, initial formation of stearate showed that direct reduction paths from both linoleate and diene-$Fe(CO)_3$ to stearate occur without going via a monoene stage. The kinetic data for the reduction of linoleate (composition and radioactivity versus time plots) were finally incorporated into the reaction mechanism shown in Scheme 17, which will be discussed in more detail below.

Misono and coworkers (48, 201) had similarly found that $Fe(CO)_5$ catalyzed the hydrogenation of dehydrated castor oil and cottonseed and linseed oils to give mainly monoene products. These workers (202) also isolated **147** and found it to be an active catalyst (203, 204). Ogata and Misono (205) studied the thermal decomposition of **147** in nitrogen and hydrogen; under hydrogen $H_2Fe(CO)_3$ was thought to be formed via the intermediate $Fe(CO)_3$. The activation energy of the decomposition was $\sim$45 kcal mole^{-1} at 210 to 230°, which was the same as that measured around 200° for the hydrogenation of cottonseed oil (loss of dienes) catalyzed by **147** [a reaction independent of hydrogen from 1 to 25 atm (204)] and comparable with a value of 60 kcal mole^{-1} measured for the same hydrogenation catalyzed by $Fe(CO)_5$ (48). These workers thus suggested (205) that the rate-determining step in these systems involved formation of $Fe(CO)_3$:

$$LFe(CO)_3 \underset{}{\overset{-L}{\rightleftharpoons}} Fe(CO)_3 \overset{H_2}{\longrightarrow} H_2Fe(CO)_3 \qquad (457)$$
$$(L = \text{diene or 2CO})$$

Subsequent coordination of diene, followed by hydrometalation, gives a hydride π-allyl intermediate (cf. Scheme 1, Section IX-A, for the corresponding sorbate system, where the unsaturate route was favored over the hydride route suggested here). Rather than transfer of the second hydrogen, a final step involving reaction with molecular hydrogen to give monoene and regeneration of $H_2Fe(CO)_3$ was suggested:

$$HFe(CO)_3(\pi\text{-allyl}) + H_2 \longrightarrow H_2Fe(CO)_3 + \text{monoene} \qquad (458)$$

Hydrogenation with $Fe(CO)_5$ is less effective in the presence of alcohols, and for reduction of fatty acids compared to the esters; pyridine also inhibited activity (48). Interaction of a hydroxyl group or the nitrogen base with the $Fe(CO)_3$ moiety was postulated (cf. Section IX-A for similar effects on sorbate reduction).

On comparison with cobalt hydrotetracarbonyl chemistry (Section X-E-1), Bird (57) suggested formation of the dihydride from the pentacarbonyl via the sequence:

$$Fe(CO)_5 \underset{+CO}{\overset{-CO}{\rightleftharpoons}} Fe(CO)_4 \overset{H_2}{\rightleftharpoons} H_2Fe(CO)_4 \qquad (459)$$

$$H_2Fe(CO)_4 \rightleftharpoons H_2Fe(CO)_3 + CO \qquad (460)$$

Ogata and Misono (48, 203) and Frankel and coworkers (200) found that monoenes (fatty esters from camelia oil and methyl oleate) were as readily hydrogenated as the diene fatty acids with the carbonyl catalysts. However, in mixtures of linoleate and monoene (oleate or palmitoleate), diene reduction was dominant at low catalyst concentration; the catalyst is tied up as the diene-tricarbonyl complex and monoene reduction is inhibited (200).

A number of intermediates were invoked (200) for the hydrogenation of linoleate shown in Scheme 17. The conjugation of linoleate was explained in terms of an allyl hydride intermediate as shown in eq. 461 (200), although in the presence of hydrogen this could equally well be written in terms of an

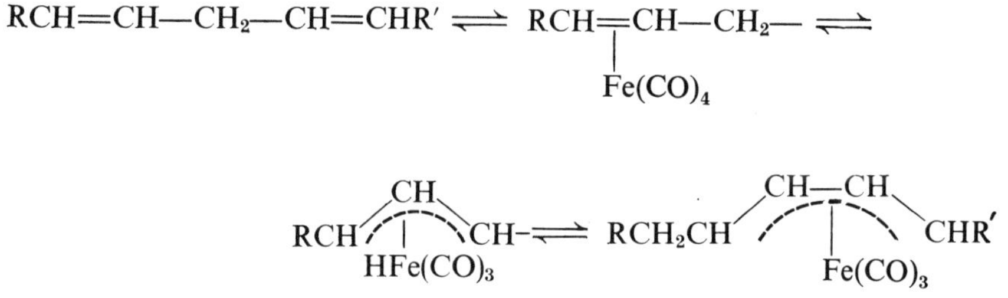

$$RCH{=}CH{-}CH_2{-}CH{=}CHR' \rightleftharpoons RCH{=}CH{-}CH_2{-} \rightleftharpoons$$
$$\underset{Fe(CO)_4}{|}$$

$$RCH\overset{\displaystyle CH}{\underset{HFe(CO)_3}{\diagup\ |\ \diagdown}}CH{-} \rightleftharpoons RCH_2CH\overset{\displaystyle CH{-}CH}{\underset{Fe(CO)_3}{\diagup\ |\ \diagdown}}CHR'$$

$$(461)$$

alkyl species by reversible addition of Fe—H to an olefinic bond.

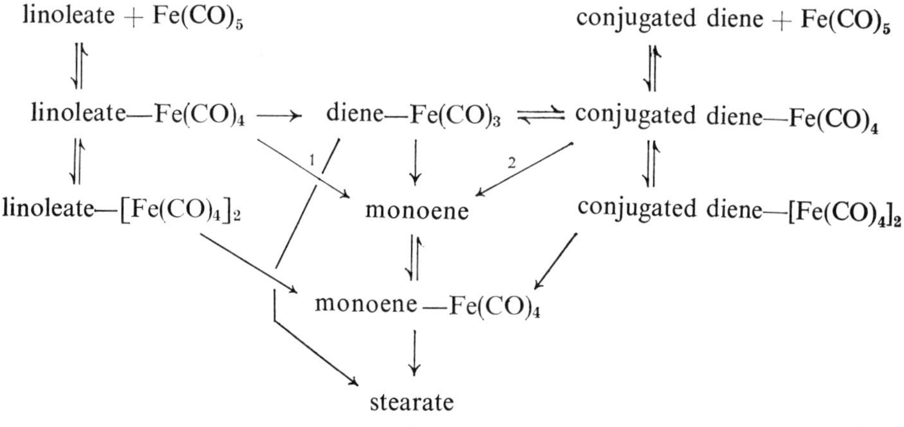

linoleate + Fe(CO)$_5$ ⇅ linoleate—Fe(CO)$_4$ ⇅ linoleate—[Fe(CO)$_4$]$_2$

conjugated diene + Fe(CO)$_5$ ⇅ conjugated diene—Fe(CO)$_4$ ⇅ conjugated diene—[Fe(CO)$_4$]$_2$

linoleate—Fe(CO)$_4$ → diene—Fe(CO)$_3$ ⇌ conjugated diene—Fe(CO)$_4$

monoene

monoene —Fe(CO)$_4$

stearate

Scheme 17

The π-complexes involving one double bond of linoleate can undergo reduction to monoenes (steps 1 and 2) or can form the more stable diene-Fe(CO)$_3$ complexes; these can be reduced to monoene and stearate. Diene-[Fe(CO)$_4$]$_2$ complexes would be reduced to monoene-Fe(CO)$_4$ complexes, which could dissociate or be hydrogenated to stearate. Formation of Fe(CO)$_3$ complexes is favored at low catalyst concentration because they are more stable; at higher concentrations formation of $Fe(CO)_4$ complexes becomes more important, and selectivity for diene hydrogenation is decreased. Ligand exchange, when diene-$Fe(CO)_3$ was used as catalyst, could readily be explained on the basis of Scheme 17 (see also Section IX-A; eqs. 98 to 101), but initial displacement of diene by hydrogen was preferred (eq. 457), followed by displacement of hydrogen.

Closely related to these studies is the later work on sorbate reduction using diene-$Fe(CO)_3$ complexes which was considered earlier (Section IX-A; Scheme 1). As pointed out, it is difficult to determine whether the hydrogen is activated first, as favored by Ogata and Misono (eq. 457), or the olefinic substrate, as favored by Frankel and coworkers (Scheme 17).

Frankel and coworkers (47) also studied the $Fe(CO)_5$-catalyzed reduction of pure methyl linolenate. Monoenoic fatty esters were the main reduction products, but also formed were dienes (half of which were conjugated), trienes (with two or three conjugated double bonds), small amounts of stearate, diene-$Fe(CO)_3$ complexes similar to **147**, and a mixture of triene complexes **148** and **149**, which could be formed by an isomerization procedure (cf. eq. 461). A reaction scheme was presented (47, 81) in which

$$R'CH \overset{\overset{\displaystyle CH\text{---}CH}{\diagup \quad | \quad \diagdown}}{\underset{Fe(CO)_3}{}} CHCH=CHR; \quad R'CH \overset{\overset{\displaystyle CH\text{---}CH}{\diagup \quad | \quad \diagdown}}{\underset{Fe(CO)_3}{}} CH(CH_2)_x CH=CHR''$$

148 $(x > 1)$ **149**

monoenes and stearate are formed only from $Fe(CO)_3$ complexes of 1,4-conjugatable dienes and conjugated dienes, although other more direct paths to monoene (cf. Scheme 17) could also be important, particularly at higher catalyst concentrations:

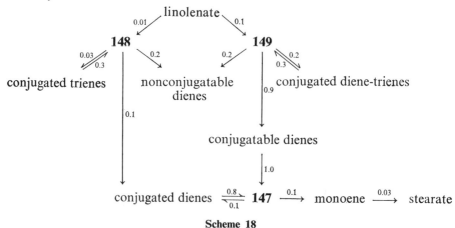

Scheme 18

The relative rates (assumed first-order) shown for all the arrowed steps in Scheme 18 were estimated by simulating composition rate curves with an analog computer (1118). By consideration of the two conjugatable systems in linolenate (C_9–C_{12} and C_{12}–C_{15}), the double bond distribution in monoene product was calculated on the basis of intermediates **147** to **149**, and it agreed well with the experimental distribution.

C. COBALT AND MANGANESE CARBONYLS

Frankel and coworkers (1132) and Ogata and Misono (1133) have used $Co_2(CO)_8$ as a catalyst for the reduction of soybean methyl esters, linoleate, and linolenate. The catalyst is somewhat more efficient than $Fe(CO)_5$ in that it can be used at lower temperatures (75 to 150°). No hydrogenation of monoene occurs and the system is thus very selective for linoleate reduction; less accumulation of conjugated dienes, less double bond migration, but much higher *trans* unsaturation in monoene and diene products are observed in the cobalt systems. The selectivity for reduction of linolenate to monoenes was low. No complexes were isolated.

The simplified scheme shown in 462 was presented for linoleate reduction (1132). Both groups tended to favor conjugation prior to hydrogenation,

$$\text{linoleate} \xrightarrow{k_1} \text{monoenes} \qquad (462)$$

with k_2 going down-left to and k_3 going up-right from **conjugated dienes**

that is, $k_1 = 0$ (see Section XIV-B), and since the diene concentration remained small, k_3 must be larger than k_2. However, the hydrogenation rate of linoleate was found to be nearly the same as that of intermediate conjugated dienes, which required an alternative explanation with $k_1 \approx k_3$ and k_2 small (1132). This problem has not been resolved. The conjugation (k_2) and hydrogenation (k_3) mechanisms were written in terms of the $HCo(CO)_3$ intermediate (1133) (see Section X-E-1). The hydrogenation steps (eq. 462a) correspond to those for some pentacyanocobaltate(II) systems (see Section X-B-2a, eqs. 180, 181, and 185).

$$C{=}C{-}C{=}C \xrightarrow{HCo(CO)_3} \left[\begin{array}{c} C{=}C{-}C{-}C \\ \ \ \ \ | \ \ \ | \\ (CO)_3Co \ \ H \end{array} \rightleftharpoons \begin{array}{c} C \\ \diagup\!\diagdown\!\diagdown \\ C \ \ \ \ \ C{-}C \\ | \ \ \ \ \ \ | \\ (CO)_3Co \ \ \ H \end{array} \right] \xrightarrow{HCo(CO)_3}$$

$$C{=}C{-}CH{-}CH + 2[Co(CO)_3] \quad (462a)$$

Frankel (1134) has reported that the activity of $Mn_2(CO)_{10}$ toward hydrogenation of unsaturated fatty esters is very similar to that of $Co_2(CO)_8$.

D. CHROMIUM CARBONYLS

The use of arenechromium tricarbonyl complexes for the selective reduction of dienes to *cis* monoenes was discussed in Section VII-A (eqs. 82 to 84). Fatty esters seem to be reduced in the same manner, which involves 1,4 hydrogen-addition to a conjugated diene system.

Cais and coworkers (178) first noted that in a mixture of dienoic esters derived from dehydrated methyl ricinoleate, $cis\text{-}CH_3(CH_2)_5CH(OH)CH_2CH=CH(CH_2)_7CO_2CH_3$, the conjugated dienes (mainly *cis*-9, *trans*-11-octadecadienoate) were completely reduced to monoene while the nonconjugated dienes were unaffected. The reduction of β-eleostearate (9 *trans*, 11 *trans*, 13 *trans*-octadecatrienoate) using methyl benzoate-$Cr(CO)_3$ catalyst yielded mainly a mixture of 9,12 and 10,13-unconjugated dienes, which is consistent with 1,4 hydrogen-addition at C-9 and C-12, and at C-11 and C-14 of the eleostearate. Tung oil was said to be similarly reduced.

Frankel and coworkers (183, 183a, 1473) used the same tricarbonyl complexes for the hydrogenation of the unconjugated linoleate and linolenate in soybean esters with no stearate formation; the monoene products were mainly *cis* (see Section VII-A). The dienes in safflower oil are similarly reduced to monoenes high in *cis* unsaturation (183a). 1,4-Reduction of eleostearate (9,11,13-trienes) in tung oil produced an oil that simulated safflower oil; the glyceride structure of cocoa butter was also simulated by selectively hydrogenating linoleate in cottonseed oil stearines (1473). The activity of the arene-$Co(CO)_3$ catalysts fell in the order: methyl benzoate > benzene > toluene > cycloheptatriene > mesitylene, which parallels the decreasing thermal stability of the complexes. Somewhat higher temperatures are needed than for conjugated systems, since isomerization to a conjugated state must precede hydrogenation (see Section VII-A and below). The double-bond distribution in monoene product when using the methyl benzoate catalyst indicated that the monoenes were derived mainly from hydrogenation of linoleate after conjugation into the 9,11- and 10,12-dienes. The most favorable stereochemistry for the reduction of 9,11-octadecadienoates (*trans*, *trans* > *cis*, *trans* > *cis*, *cis*) was considered in Section VII-A (1135).

Frankel and coworkers (1135, 1136) have also studied hydrogenation and deuteration of linoleate and oleate, as well as 9,11-octadecadienoates. The 1,4-diene (linoleate) and 1,3-dienes gave mainly *cis* monoenes in yields of 94 to 100% using the methyl benzoate or benzene-$Cr(CO)_3$ catalysts. An alkali-conjugated linoleate mixture (consisting of 9,11- and 10,12-dienes) gave the same monoene composition as linoleate itself, thus confirming the prior conjugation. However, the linoleate was reduced about twenty times slower than the conjugated mixture, suggesting that the conjugation step is rate-determining. The more unstable cycloheptatriene complex is effective for the reduction of the 1,3-dienes at 125°, but not for linoleate, which requires a temperature of ~170° for the initial isomerization. The 1,6-diene, *cis*-9,*cis*-15-octadecadienoate was not conjugated or therefore reduced with the methyl benzoate catalyst, although it was extensively isomerized to the *trans* form; oleate (*cis*-9-octadecenoate) was similarly isomerized.

Although no deuterium exchange occurred with conjugated dienes (see Section VII-A), such a reaction is important with linoleate and oleate and occurs with hydrogen on α-methylenes (1136); since all monoenes from the reduction of conjugated dienes have a deuterium content close to 2.0 (see Section VII-A), the exchange of monoenes must be inhibited in the presence of dienes. Because of the exchange, linoleate deuteration yields monoenes with an average of more than two deuteriums per molecule. Equation 462b, with the reverse steps involving deuterium, was presented for the exchange and involved initial formation of a linoleate-Cr(CO)$_3$ complex, **150**; the

$$\text{(462b)}$$

150 $Cr = Cr(CO)_3$

possibility of the hydrogenation reaction involving prior activation of the substrate before hydrogen addition cannot therefore be completely ruled out. However, the arene-Cr(CO)$_3$ catalysts are unique in their ability to reduce diene and triene fatty esters into mostly *cis* monoene fatty esters, and contrasts with the corresponding iron complexes have been discussed in terms of the differing mechanisms (81, 180; Section IX-A), the chromium being thought to activate the hydrogen initially, and the iron the substrate.

E. METAL ACETYLACETONATES AND ZIEGLER-TYPE CATALYSTS

Emken and coworkers (144) first reported that transition metal acetyl-acetonates were effective for hydrogenation of linseed and soybean methyl esters at 100 to 180° and 7 to 70 atm hydrogen. Hydrogenation occurred rapidly in methanol but only slowly in DMF and acetic acid; no reduction occurred in cyclohexane, benzene, acetone, pyridine, acetylacetone, chloroform, and heptanoic acid. The relative activity of the complexes was in the order $Ni(acac)_2 > Co(acac)_3 > Cu(acac)_2 > Fe(acac)_3$, which was also the order of the thermal stabilities of the acetylacetonates; black precipitates, formed to differing extents, were catalytically inactive. The selectivity for linolenate hydrogenation in methanol also decreased in the same order; for the most selective nickel catalyst the ratio of the rate of triene reduction to that of diene reduction was between 3 and 5. Stearate production became important only when the trienes had been reduced. Analysis of the monoene products from hydrogenations using the nickel catalyst suggests the following: (1) the high proportion of *cis* monoenes found (65%) are formed by 1,2-addition to linoleate with no movement of double bonds, (2) there is some geometric isomerization (*cis* to *trans*) without migration, and (3) some geometric isomerization via conjugation prior to reduction occurs, as considered for the cobalt carbonyl system (Section XIV-C).

The methanol was thought (81, 144) to form a solvated species $Ni(acac)_2$-CH_3OH which subsequently reacted with hydrogen to form hydrides such as $HNi(acac)_2CH_3OH$ or $HNi(acac)$. The role of the solvated species was likened to that of the alkylated intermediate (RMX_{n-1}) in the initial step of a mechanism suggested for Ziegler homogeneous hydrogenation catalysts (see Chapter XV, eq. 469, and below).

Nanya and coworkers (145) later used acetylacetonates in alcohols under nitrogen pressure at $\sim 200°$ for the hydrogenation of linoleate and its conjugated isomers in the absence of molecular hydrogen. The relative activity of the alcohols as hydrogen donors decreased in the order: 2-propanol > 1-propanol > ethanol > 2-butanol > 1-butanol > 2-pentanol > 1-pentanol > methanol. Acetone was generated from 2-propanol in an amount corresponding to the degree of hydrogenation. The nickel bisacetylacetonate complex $Ni(acac)_2$ was the most effective and was selective for the reduction of dienes to monoenes; monoene reduction to stearate occurred after all the diene had been reduced. $Fe(acac)_3$ was of lower activity, while $Co(acac)_3$, $Cu(acac)_2$, and $Cr(acac)_3$ were essentially inactive. *Cis*-octadecenoate was reduced at the same rate as the *trans* isomer, but due to steric hindrance conjugated *trans,trans*-octadecadienoate was reduced very much slower than the conjugated *cis,trans* isomer. These workers suggested hydrogen abstraction from the alcohol to give a hydride (Section F-2) followed by hydrometalation of the substrate to give an alkyl, which finally underwent hydrogenolysis via another hydride:

$$\text{alkyl—M} + \text{HM} \longrightarrow \text{product} + M_2 \qquad (463)$$

No measurable reduction occurred in methanol over 2 hr when using the nickel catalyst, in marked contrast to the results of Emken and coworkers at lower temperatures but under hydrogen pressure. Hydride formation, or hydrogenolysis of the alkyl, could be easier using hydrogen.

Tajima and Kunioka (206) have used a binary catalyst system (1:5) of transition metal compounds and triethylaluminum (Ziegler catalysts, Chapter XV) for the reduction of soybean fatty esters, linoleate, and linolenate to monoenes in hexane solution. Catalytic activity at 150° and 150 atm hydrogen was in the order $Ni(acac)_3 > Co(acac)_2 \approx CoCl_2 > FeCl_3$, which, as pointed out by Frankel and Dutton (81), is the same as that of the acetylacetonate-methanol-H_2 systems. A hydride is likely to be formed via an alkylated intermediate (Chapter XV):

$$R_3Al + MX_n \longrightarrow R_2AlX + RMX_{n-1} \qquad (464)$$

$$RMX_{n-1} + H_2 \longrightarrow RH + HMX_{n-1} \qquad (465)$$

Small amounts of conjugated diene products were observed, and were considered by Tajima and Kunioka to be intermediates in the hydrogenation (cf. Sections XIV-B, XIV-C, and XIV-D):

trienes $\longrightarrow$ dienes $\longrightarrow$ conjugated dienes $\longrightarrow$ monoenes

Misono and Uchida (207) have used iron, cobalt, and nickel acetylacetonates with triethylaluminum (1:3) to reduce cottonseed esters to oleate at 25° and 1 atm hydrogen.

F. PLATINUM METAL COMPLEXES AND NICKEL COMPLEXES

1. Platinum-Tin Chloride Complexes without Phosphines. The activity of the platinum(II) chloride-tin(II) chloride system in the absence of added phosphines for the reduction of olefins and acetylenes was discussed in Section XIII-C-2.

Bailar and coworkers have used benzene-methanol solutions of platinum(II–IV) chlorides/stannous chloride (1:5 or 1:10) at $\sim$90° with 40 atm hydrogen for the reduction of soybean esters (1040), linolenate (1041), and linoleate and oleate (978, 1040). Soybean ester is reduced to monoenes and stearate; linolenate gave dienes and monoenes with no stearate; linoleate was reduced to monoenes, while oleate underwent reduction and extensive isomerization to the *trans* isomer. $PtCl_2(en)$ is not an effective starting platinum complex. In pure methanol the 1:10 mixture ($H_2PtCl_4 + SnCl_2$) is active at 1 atm hydrogen, and catalyzes the conjugation of linolenate at 30°, and its selective reduction to diene at 40° (1041). The mechanism of linolenate reduction is considered in Section F-2. Van't Hof and Linsen (1031–1034) have used a similar 1:5 catalyst mixture at the same mild conditions in acetic acid, dialkyl ketones, dialkyl ethers, THF, and nitrobenzene for the reduction of soybean oil and esters; higher carboxylic acids up to stearic acid were also suitable solvents. The system showed a remarkable preference for hydrogenation of diene (linoleate) to monoene (particularly oleate), while the triene (linolenate) was not reduced until the diene had been reduced; extensive isomerization to *trans* monoenes was again noted.

These complexes without coordinated phosphines are not effective in the absence of elemental hydrogen for reduction of linoleate or oleate (978).

2. Platinum, Palladium, Nickel, and Rhodium Complexes Containing Tertiary Phosphine-Type Ligands (with and without Added Stannous Chloride). The studies on the platinum, palladium, and nickel systems have all been carried out by Bailar and coworkers. In the absence of stannous chloride, the following platinum complexes were essentially inactive at 90° with 40 atm hydrogen in benzene-methanol for the hydrogenation of soybean esters and linoleate (978, 980, 1001, 1040, 1063): $HPtX(PPh_3)_2$, X = Cl, NO_2, SCN, OCN, CN, $SnCl_3$; $PtCl_2(MPh_3)_2$, M = P, Sb; $Pt(CN)_2(MPh_3)_2$, M = P, As, and $Pt(MPh_3)_4$, M = P, As. The complexes $PtCl_2(AsPh_3)_2$ and $PtCl_2[P(OPh)_3]_2$, however, effected reduction of linoleate to monoene, although catalyst decomposition occurred. In the presence of stannous chloride, the complex $PtCl_2(PPh_3)_2$ was first found to reduce soybean ester to the monoene stage only; linoleate was reduced to monoene, and oleate was isomerized only (1040); $PtCl_2(diphos)_2$ is equally effective (1138, 1475).

The studies were extended with linoleate and oleate as substrates (978, 980, 1139) using a number of $PtX_2(MR_3)_2$ complexes (X = halogen or pseudo-halogen; M = P, As, Sb; R = alkyl or aryl) with several Group IV halides as cocatalysts (Sn, Si, Ge, Pb). These catalyst systems and the likely intermediates, such as $HPt(SnCl_3)(PPh_3)_2$, have already been considered in some detail in Section XIII-C-3b (Scheme 16). These complexes generally catalyze isomerization of oleate to the *trans* isomer with no hydrogenation, and isomerize linoleate to conjugated dienes, which are then reduced selectively to monoenes. The reactions with $PtCl_2(MPh_3)_2$, M = P, As, are less effectively carried out in the benzene-methanol media under nitrogen pressure, the alcohol being the hydride source (cf. Section XIV-E). Such hydrogen abstraction (eq. 466) is well established for platinum complexes (e.g., 1043):

$$CH_3CH_2OH + Cl—Pt \xrightarrow{-HCl} CH_3—CH \overset{O}{\underset{H}{\diamond}} Pt— \longrightarrow$$

$$CH_3CHO + HPt— \quad (466)$$

As discussed in Section XIII-C-3b, short-chain conjugated dienes are not readily reduced by these platinum systems ($PtCl_2(PPh_3)_2$-$SnCl_2$) because they complex strongly with the catalyst. The conjugated 9,11-dienes (both *cis,trans* and *trans,trans*) derived from linoleate can be reduced to monoene, since the steric effect imposed by the chain length weakens the diene-platinum bond (1062, 1065); the resulting internal monoenes are relatively unreactive.

The hydrogenation of linolenate under hydrogen pressure in benzene-methanol using $HPtCl(PPh_3)_2$, $PtCl_2(AsPh_3)_2$, as well as H_2PtCl_4 (see Section F-1), in conjunction with stannous chloride, and $HPt(SnCl_3)(PPh_3)_2$, has also been studied (1041). The hydrogenations were accompanied by extensive conjugation, *cis,trans* isomerization and double-bond migration. The hydrogenating activity, to mainly diene and monoene formation, decreased in the order: $H_2PtCl_4 + SnCl_2 > PtCl_2(AsPh_3)_2 + SnCl_2 > HPtCl(PPh_3)_2 + SnCl_2 > HPt(SnCl_3)(PPh_3)_2$. The H_2PtCl_4 system is highly selective for diene formation, while with $HPt(SnCl_3)(PPh_3)_2$ the linolenate is mainly isomerized to conjugated diene-trienes (trienes with two double bonds conjugated). A mechanism very similar to that for the $Fe(CO)_5$ system (Scheme 18, Section XIV-B) was proposed. The active catalyst was written as $HPtL_x(SnCl_3)_y$, where $L = H$, PPh_3, $AsPh_3$, which can effect conjugation of either the 9,12- or 12,15-diene systems of the linolenate. Conjugated diene-trienes and conjugated trienes were important initial products in these systems. The conjugated diene part of these trienes forms complexes with the hydrido-platinum species; alkyl formation followed by hydrogenolysis results in unreactive nonconjugated diene products. The C_{12} double bond, being involved in two conjugatable diene systems, is more susceptible to hydrogenation, as was observed.

In general the palladium systems (81, 980, 1000, 1001) are more active and selective than their platinum analogs; the palladium complexes are less stable, however, and somewhat lower temperatures around 60° are preferable. No pressure dependence was observed at 14 to 40 atm of hydrogen. The effectiveness of complexes decreases in the order: $PdCl_2(PPh_3)_2 + SnCl_2 > PdCl_2(PPh_3)_2 + GeCl_2 > Pd(CN)_2(PPh_3)_2 > Pd(CN)_2(AsPh_3)_2 > PdCl_2(PPh_3)_2 \gg PdCl_2(AsPh_3)_2$. $PdCl_2(AsPh_3)_2 + SnCl_2$ is a very poor catalyst and $K_2PdCl_4 + SnCl_2$ is inactive, whereas the corresponding platinum systems (Section F-1 and above) are effective. $Pd(CN)_2(PPh_3)_2$, and $Pd(CN)_2(AsPh_3)_2$, in contrast to the platinum analogs, are effective in the absence of stannous chloride; this was attributed (978) to a relatively easier expansion of the coordination sphere of the palladium ion. $PdCl_2(SbPh_3)_2 + SnCl_2$ is inactive, and for an analogous series of palladium complexes the activity trend falls in the order phosphines > arsines > stibines, which again contrasts with platinum systems (Section XIII-C-3b). Addition of mercuric chloride to $PdCl_2(PPh_3)_2$ forms a chlorine-bridged complex (1137) which is ineffective. The active palladium systems catalyze the same reactions as the platinum systems: conjugation of linoleate, selective hydrogenation of polyenes to monoenes, and isomerization of oleate. No hydrido complexes were isolated, but the reaction paths were thought to be similar to those of the platinum systems (see above and Section XIII-C-3b). The reactions again took place under nitrogen pressure in benzene-methanol. Although conjugation of linoleate might occur via addition and elimination of a metal hydride species formed from the alcohol, another possibility involving a hydride shift from the substrate to the metal was also proposed (eq. 467). Indirect evidence for such a shift was the finding that oleate isomerization catalyzed by Pd(CN)$_2$-

$$\tag{467}$$

(PPh$_3$)$_2$ under hydrogen was accompanied by formation of small amounts of dienes and stearate. Some aldehydes were formed even when hydrogen atmospheres were used, suggesting that hydrides can still be formed from the alcohol in these conditions (cf. eq. 466).

The tetrahedral nickel complexes NiX$_2$(PPh$_3$)$_2$, where X = halide, catalyze isomerization and selective hydrogenation of linoleate to monoene in benzene or THF in the presence or, less effectively, in the absence of hydrogen (979, 980); the complexes are unstable in alcohol. The less soluble, square-planar Ni(SCN)$_2$(PPh$_3$)$_2$ complex was inactive. The data show that the activity of the halides falls in the order I > Br $\gg$ Cl, although this conclusion is not clearly presented in the detailed paper on the nickel systems (979), and the reverse order has been stated in a review article (81). Unlike the platinum and palladium analogs, addition of stannous chloride does not enhance activity in the benzene-methanol medium. Under hydrogen, benzene was a more effective solvent than THF, although the reverse was true under nitrogen. *Trans* isomerization of oleate occurred under hydrogen but not nitrogen, and this has been attributed to the lack of formation of hydrides in the nonalcoholic media (81), although linolcate is hydrogenated in these conditions. These systems have not been studied in any detail, but the reductions under hydrogen are thought to follow closely those of the platinum and palladium complexes involving hydrogenation via conjugation.

It has been suggested by Schunn (480) that the hydrogen required for the NiX$_2$(PPh$_3$)$_2$ systems in the absence of molecular hydrogen might be abstracted from the ortho positions of the phenyl groups (cf. Scheme 3 in Section IX-B-2b; see Section XI-G).

The RhCl(PPh$_3$)$_3$ complex catalyzes the deuteration of linoleate and oleate under mild conditions to methyl stearate-d_4 and methyl stearate-d_2, respectively (Table 46; Section XI-B-5a). This complex does not catalyze reduction of simple conjugated dienes because of strong complexing, although more complex conjugated dienes have been reduced (see Sections XI-B-5a and

XI-B-5c). Deuterium location by mass spectrometry indicates that linoleate is reduced by 1,2-addition (647), although the isomerization properties of this catalyst with long-chain unsaturated fats have not yet been reported.

G. MOLYBDENUM AND TUNGSTEN COMPLEXES

Bailar and coworkers (1139, 1475) have briefly reported the use of molybdenum(II) and tungsten(II) complexes of the type $MCl_2(CO)_3(PPh_3)_2$, or the arsine analogs, in the presence of excess stannous chloride for the reduction of soybean esters to the monoene in dichloromethane at $>150°$ and 40 atm hydrogen. The complexes add stannous chloride to give compounds such as $MoCl(SnCl_3)(CO)_2(PPh_3)_2$. Activity decreased in the order: $WCl_2(CO)_3(PPh_3)_2 > MoCl_2(CO)_3(AsPh_3)_2 > MoCl_2(CO)_3(PPh_3)_2$. Simpler dienes were also said to be reduced and, as with the platinum phosphine-tin chloride catalysts (Section XIII-C-3b), hydrogenation continues to a saturated species when a double bond is terminal. The binuclear chloride-bridged complex $[C_3H_5(OC)_2MoCl_3Mo(CO)_2C_3H_5]^-$ is also active even in the absence of stannous chloride. The true catalysts in these systems may be formed by the high-temperature decomposition of the starting complexes (1475).

H. CONCLUSIONS

Some conclusions on the homogeneous hydrogenation of unsaturated fats have been presented by Frankel and Dutton (81). The catalysts commonly require a readily complexed conjugated diene substrate [e.g., $Co(CN)_5^{3-}$] for hydrogenation, or the ability to effect conjugation prior to hydrogenation (metal carbonyls, phosphine complexes of platinum, palladium, and nickel). Ligand dissociation (e.g., metal carbonyls), solvation (e.g., acetylacetonates), and ligand substitution (e.g., platinum-tin chloride systems) may be important for catalyst activation. The substrate may be activated directly (e.g., possibly iron carbonyls) or after formation of a monohydride [e.g., $Co(CN)_5^{3-}$] or dihydride (e.g., possibly chromium carbonyls); reduction may involve 1,2- or 1,4-addition of hydrogen on a π-complexed diene (e.g., metal carbonyls) or 1,2-addition on a σ-complexed diene [e.g., $Co(CN)_5^{3-}$].

Selectivity of hydrogenation is related to the reactivity and stability of, for example, the diene and monoene complexes [e.g., with $Fe(CO)_5$]. Stereoselectivity toward *trans,trans* conjugated dienes has been achieved by 1,4 hydrogen-addition on a cisoid complexed diene (chromium carbonyls).

Linolenate has been reduced to unreactive dienes with unconjugatable double bonds (e.g., the platinum-tin complexes), but the aim of reducing only the C_{15} double bond (to give linoleate) has not yet been achieved. One approach is to form a stable complex with the C_9 and C_{12} double bonds of linolenate, chemically reduce the free C_{15} double bond, and then decompose the complex to give linoleate; such a sequence was used to characterize the triene-$Fe(CO)_3$ complexes of methyl linolenate (Section XIV-B). A direct approach would involve complexing and reduction at the C_{15} double bond only.

XV

Ziegler-Type Catalysts

It is well known that reaction of transition metal halides with alkyls of aluminum, alkaline earth, and alkali metals gives Ziegler catalysts for dimerization and polymerization of olefins. These systems are usually heterogeneous, but homogeneous systems have been developed by modification of the ligands coordinated to the transition metal and have been found effective for catalytic hydrogenation. The isolation or identification of the actual catalysts has not been reported, and it will become evident that in some systems the catalyst may not be entirely homogeneous. These systems seem to be complicated and the mechanisms are not well understood, but the systems appear practical in many cases and have been pursued particularly by petrochemical companies.

A. REDUCTION OF OLEFINS AND AROMATICS

Sloan and coworkers (56) first reported on catalysts derived from (a) acetylacetonates of cobalt(III), chromium(III), iron(III), manganese(II and III), molybdenum(VI), and ruthenium(III); (b) alkoxides of titanium(IV) and vanadium(V); (c) the dichloride complexes $(C_5H_5)_2TiCl_2$, $(C_5H_5)_2ZrCl_2$, $CoCl_2(PPh_3)_2$, $NiCl_2(P^nBu_3)_2$, and $PdCl_2(PBu_3)_2$. In combination with $^iBu_3Al$, triisobutylaluminum, in heptane or toluene at 30 to 50° and $\sim$4 atm hydrogen, one or more of the monoenes (cyclohexene, oct-1-ene, 2-methyl-but-2-ene, pent-2-ene, tetramethylethylene, and stilbene) were successfully reduced (Table 78). Et_3Al and $^iBu_2AlH$ were equally effective organometallic components. ^nBuLi was less effective, while Et_2AlCl, Et_3B, Et_4Sn, $^nBu_3P$, Et_2Zn, and ^nBuMgBr were ineffective. An aluminum-metal ratio corresponding to one aluminum for each oxygen present was generally used,

Table 78 Reduction of Monoenes Using Soluble Ziegler Catalysts

Catalyst	Alkyl/Metal	Substrates Reduced	Reference
Ti(O^{i}Pr)$_4$-^{i}Bu$_3$Al	3.3	Cyclohexene, oct-1-ene, stilbene	56
Ti(O^{i}Pr)$_4$-^{n}BuLi	9.9	Cyclohexene	56
Ti(O^{i}Pr)$_4$-Et$_3$Al	4	Hept-1-ene, hex-1-ene	147, 1157
(C$_5$H$_5$)$_2$TiCl$_2$-BuLi^a	4	Cyclohexene	1172
(C$_5$H$_5$)$_2$TiCl$_2$-Et$_3$Al	1.2	Oct-1-ene	56
(C$_5$H$_5$)$_2$TiCl$_2$-R$_n$AlX$_{3-n}$^{b}	≥ 6	Cyclohexene, hex-1(3)-ene, styrene	147, 163, 165, 166, 195, 1157, 1172, 1173
(C$_5$H$_5$)$_2$ZrCl$_2$-^{i}Bu$_3$Al	2.9	Cyclohexene	56
VO(OEt)$_3$-R$_3$Al	4.0	Cyclohexene	56
VO(O^{n}Bu)$_3$-^{i}Bu$_3$Al	3.9	Cyclohexene, oct-1-ene	56
VO(OEt)$_3$-^{i}Bu$_2$AlH	4.1	Cyclohexene	56
VO(acac)$_2$-R$_n$AlX$_{3-n}$^{b}	6–18	Cyclohexene, hex-1-ene, hept-1-ene	147, 163, 195
Cr(acac)$_3$-R$_n$AlX$_{3-n}$^{b}	6–14	Cyclohexene, tri- and tetramethylethylenes, (di)methylbut-2-enes, trans-pent-2-ene, 3-methyl-pent-2-ene, hex-1-ene, hept-1(3)-ene, oct-1-ene, styrene	56, 147, 163, 166, 176, 177, 195, 1170, 1173, 1178, 1427
Mn(acac)$_2$-Et$_3$Al	~10	Hept-1-ene, hex-1-ene	147, 163
Mn(acac)$_{2,3}$-^{i}Bu$_3$Al	6.0	Cyclohexene	56
Fe(acac)$_3$-R$_n$AlX$_{3-n}$^{b}	4–14	Cyclohexene, methylbut-2-enes, methylpent-2-enes, hex-1-ene, hept-1(3)-ene, styrene	56, 147, 163, 177, 195, 1141, 1157, 1170, 1427, 1429
Fe(acac)$_2$pY$_2$-Et$_3$Al	12	Cyclohexene	147, 1157
Fe(DMGH)$_2$pY$_2$-Et$_3$Al	15	Cyclohexene	147, 1157
FeCl$_3$-Et$_3$Al	~3	Styrene, cyclohexene, indene, safrole, d-limonene	173
Co(acac)$_3$-^{i}Bu$_3$Al	6.0	Cyclohexene, hex-1-ene	56
Co(acac)$_3$-^{n}BuLi	12.6	Cyclohexene	56
Co(acac)$_2$-R$_3$Al	≥ 6	Cyclohexene, methylbut-2-enes, methylpent-2-enes, hex-1-ene, hept-1(3)-ene	147, 163, 1141, 1170

Co(Et.Hex)$_2$-Et$_3$Al	3–14	Pent-1-ene, oct-1-ene, ethyl vinyl ether, mesityl oxide, acrylonitrile	1147, 1176
CoCl$_2$-Et$_3$Al	0.2	Styrene	173
CoCl$_2$(PPh$_3$)$_2$-^{i}Bu$_3$Al	3.5	Cyclohexene	56
Co(stearate)$_2$-Et$_3$Al^c	3–6	Cyclohexene, monoenes	147, 1157, 1174
Co(stearate)$_2$-^{n}BuMgBr	5	Ethylene, propylene, butenes, pent-1-ene^d	570–572
M(stearate)$_2$-RMgX^e	3–8	Hept-1-ene	570–572
Ni(acac)$_2$-Et$_3$Al^f	2–8	Cyclohexene, styrene, methylbut-2-enes, methylpent-2-enes, hex-1-ene, hept-1(3)-ene	147, 163, 195, 1157, 1427
Ni(acac)$_2$-^{i}Bu$_3$Al	6	Cyclohexene	1141
NiCl$_2$-Et$_3$Al	2	Styrene	173
Ni(DMGH)$_2$-R$_n$AlX$_{3-n}$^{b}	15	Cyclohexene, hex-1-ene	147, 195, 1157
Ni(Et.Hex)$_2$-Et$_3$Al	3	2-Methylhept-1-ene, oct-1(2)-ene, cyclooctene, maleic anhydride, cinnamic acid	1147
NiL$_2$-Et$_3$Al^g	6	Cyclohexene	147, 1157
NiCl$_2$(P^{n}Bu$_3$)$_2$-^{i}Bu$_3$Al	4.3	Hex-1-ene	56
NiX$_2$(PPh$_3$)$_2$-Et$_3$Al^h	7	Hex-1-ene	1157
MoO$_2$(acac)$_2$-^{i}Bu$_3$Al	7.1	Cyclohexene, oct-1-ene	56
Ru(acac)$_3$-^{i}Bu$_3$Al	7.1	Oct-1-ene	56
PdCl$_2$(P^{n}Bu$_3$)$_2$-^{i}Bu$_3$Al	4.2	Hex-1-ene	56
Et.Hex = 2-ethylhexanoate			

[a] EtMgCl and PhMgBr also used; Et$_2$Zn ineffective.

[b] R = Et, ^{i}Bu; X = Cl, OEt.

[c] Also studied in presence of 2 moles of EtOH, pyridine and PPh$_3$ per Co.

[d] All stoichiometric hydrogenations (see text).

[e] M = Fe, Co, Ni; R = Et, ^{n}Bu, ^{i}Bu, ^{t}Bu; X = Cl, Br.

[f] Also studied (1157, 1427) in presence of 2 moles of H$_2$O, NH$_3$, Et$_3$N, Et$_2$NH, PPh$_3$, PCl$_3$, and P(OR)$_3$ (where R = Et, ^{i}Pr, Ph) per Ni.

[g] L = o-aminophenolate, salicylaldoximate; 8-hydroxyquinolate.

[h] X = Cl, I, NO$_3$.

although for the $Cr(acac)_3-{}^iBu_3Al$ reduction of cyclohexene the rate increased as the Al/Cr ratio increased from 1 to 10. The most active catalysts were cobalt(III) > iron(III) > chromium(III) acetylacetonates. $Cu(acac)_2$ precipitated copper and was ineffective.

Ziegler and coworkers (1140) have reported that with a large excess of trialkylaluminum, cobalt and nickel acetylacetonates give the colloidal metal, although Kroll (1141) found no evidence for colloids at the dilutions and concentration ratios used in the hydrogenation conditions. Sloan and coworkers (56) stated that the homogeneous catalysts were killed by addition of ethanol or acetone, and they used this as an argument against the presence of colloidal metal. Kroll (1141), however, has reported that addition of butanols to the $1:12$ $Co(acac)_2-{}^iBu_3Al$ catalyst in benzene forms new but less active catalysts.

Since the report of Sloan and coworkers, papers on the use of these Ziegler catalysts for hydrogenation of olefins and aromatics have appeared by many workers. Some of the systems studied are summarized in Tables 78 to 83; the alkyl to metal ratio listed is generally that for optimum reactivity although this can vary somewhat with conditions, substrates, and so on (Section B).

Sloan and coworkers presented some kinetic data for cyclohexene reduction with a $1:6$ $Cr(acac)_3-{}^iBu_3Al$ catalyst in heptane at $30°$ measured by pressure drop. Up to a catalyst concentration of $5 \times 10^{-3}\ M$, the rate is given by $k[H_2][\text{catalyst}]$; for their particular constant-volume apparatus with about 200 ml dead space, their k value was about $0.07\ M^{-1}\ s^{-1}$. Hydrogenation rates decreased with increasing substitution in acyclic ethylenes, although cyclohexene was reported to be reduced at twice the rate of oct-1-ene (56). Kroll (1141), however, reported that hex-1-ene is reduced some three times faster than cyclohexene with the $Co(acac)_2-{}^iBu_3Al$ catalyst in benzene at ~ 2 atm hydrogen, and has selectively reduced the terminal olefin in a mixture of the two. He also observed a first-order dependence on hydrogen for the cyclohexene reduction. Other workers (see below) have found decreasing rates with increased substitution, and lower reduction rates for cyclohexene than for terminal olefins.

A speculative mechanism for these hydrogenations was presented by Sloan and coworkers (56) and this has usually been referred to by later workers.

$$R_3Al + MX_n \longrightarrow R_2AlX + RMX_{n-1} \tag{468}$$

$$RMX_{n-1} + H_2 \longrightarrow RH + HMX_{n-1} \tag{469}$$

$$HMX_{n-1} + \text{olefin} \underset{k_{-1}}{\overset{k_1}{\rightleftharpoons}} \text{alkyl-}MX_{n-1} \tag{470}$$

$$\text{alkyl-}MX_{n-1} + H_2 \underset{}{\overset{k_2}{\rightleftharpoons}} HMX_{n-1} + \text{product} \tag{471}$$

The transition metal complex is first alkylated (eq. 468), and hydride formation was thought to involve the alkylated species (eq. 469). An alternative final hydrogenolysis to product via another hydride molecule (eq. 472), with subsequent regeneration of catalyst via eq. 473, was also presented:

$$\text{alkyl-}MX_{n-1} + HMX_{n-1} \longrightarrow \text{product} + (MX_{n-1})_2 \qquad (472)$$

$$(MX_{n-1})_2 + H_2 \longrightarrow 2HMX_{n-1} \qquad (473)$$

Reaction 474, involving formation of the hydride by elimination from the metal alkyl, was considered to be an alternative to eq. 469:

$$RCH_2CH_2MX_{n-1} \longrightarrow HMX_{n-1} + RCH\!=\!CH_2 \qquad (474)$$

It should be noted that the aluminum alkyls are commonly written for simplicity as monomers; of the R_3Al species, those with R = Me, Et, or ^nPr are dimeric in benzene solution, while those with R = ^iPr, ^iBu are monomeric (1142).

Kroll's studies (Tables 78, 79, and 83; 1141, 1145), which were referred to above, were carried out in benzene at 20° using 2 to 7 atm hydrogen. The aluminum alkyl was added as a Lewis base adduct, for example, as the 1:1 p-dioxane adduct. These adducts have an advantage over the uncomplexed alkyl in that they do not poison the catalyst system at higher aluminum/metal ratios. The activity of these acetylacetonate systems was cobalt(II) > iron-(III) > nickel(II). 4-Vinylcyclohexene and cyclopentadiene were selectively reduced to 4-ethylcyclohexene and cyclopentene, respectively. Kroll (1160, 1166) has patented the use of salts of copper, zinc, vanadium, titanium, chromium, molybdenum, and platinum as well as the iron triad; besides acetylacetonates, the use of chlorides, bromides, acetates, hexafluorosilicates, and naphthenates were also reported in conjunction generally with R^1R^2AlX, where X = H, R^3 OR^3, SR^3, and $NR_2{}^3$, and R^1, R^2, R^3 are alkyl, cycloalkyl, aryl, and alkylaryl. He has also patented the use of supported $Co(acac)_2$-Me_3Al catalysts for the reduction of monoenes, dienes, and benzenes (1161), and the use of iron pentacarbonyl-aluminum alkyl systems at more severe conditions (50°, 70 atm hydrogen in pentane) for the reduction of dienes (Table 79; 1161). Many of these systems were said to be adaptable for homogeneous, heterogeneous or liquid slurry conditions.

Tajima and Kunioka (Table 79; 1146) used 40 atm hydrogen at 90 to 110° in benzene solution for the reduction of butadiene and isoprene using cobalt complexes. The data shown refer to 90° conditions. The hydrogenation of monoenes to saturated product seems retarded when stronger coordinating ligands are present, and the $CoCl_2(PPh_3)_2$, $Co(NO_3)_2(OPPh_3)_2$, and $Co(NO_3)_2(py)_3$ catalysts became highly selective for monoene production; the preferential formation of but-1-ene and cis–but-2-ene from butadiene

Table 79 Reduction of Dienes and Trienes Using Soluble Ziegler Catalysts

Catalyst	Alkyl/Metal	Reaction	Reference
Cp_2TiCl_2-^nBuLi	2.4	Butadiene $\rightarrow$ butane (100%)	167
Cp_2TiCl_2-PhMgBr	3.5	Butadiene $\rightarrow$ butane (97%) + butenes	167
Cp_2VCl_2-^nBuLi	2.4	Butadiene $\rightarrow$ butenes (100%)	167
Cp_2VCl_2-PhMgBr	3.5	Butadiene $\rightarrow$ butenes (100%)	167
Cp_2VCl_2-^nBuLi	2.4	Isoprene $\rightarrow$ butenes (100%)	167
Cp_2VCl_2-PhMgBr	3.5	Isoprene $\rightarrow$ butenes (100%)	167
Cp_2ZrCl_2-^nBuLi	2.4	Butadiene—no reduction	167
Cp_2ZrCl_2-PhMgBr	3.5	Butadiene—no reduction	167
$CpFe(CO)_2Cl$-Et_3Al	2.4	Butadiene $\rightarrow$ butenes (100%)	167
$CpFe(CO)_2Cl$-PhMgBr	3.5	Butadiene $\rightarrow$ butenes (100%)	167
$CpCo(CO)_2$-Et_3Al	2.4	Butadiene $\rightarrow$ butane (7%) + butenes	167
$CpCo(CO)_2$-PhMgBr	3.5	Butadiene $\rightarrow$ butane (3%) + butenes	167
$CpNiC_3H_7$-Et_3Al	2.4	Butadiene[a] $\rightarrow$ butane (20%) + butenes	167
$CpNiC_3H_7$-PhMgBr	3.5	Butadiene $\rightarrow$ butane (10%) + butenes	167
$Co(acac)_2$-Et_3Al	5	Butadiene[a] $\rightarrow$ butane (50%) + butenes, isoprene $\rightarrow$ 2-methylbutane (43%) + methylbutenes	1146
$CoCl_2$-Et_3Al	5	Butadiene[a] $\rightarrow$ butane (75%) + butenes isoprene $\rightarrow$ 2-methylbutane (65%) + methylbutenes	1146
$Co(Sal)_2$-Et_3Al	5	Butadiene[a] $\rightarrow$ butane (38%) + butenes, isoprene $\rightarrow$ 2-methylbutane (38%) + methylbutenes	1146
$CoCl_2/2PPh_3$-Et_3Al	5	Butadiene $\rightarrow$ butane (11%) + butenes isoprene $\rightarrow$ 2-methylbutane (6%) + methylbutenes	1146
$Co(acac)_3/L^b$-Et_2AlCl	40	Butadiene $\rightarrow$ butenes	1425

Co(acac)$_2$/2PPh$_3$-Et$_3$Al	5	Butadiene → butane (10%) + butenes, isoprene → 2-methylbutane (4%) + methylbutenes	1146
CoCl$_2$(PPh$_3$)$_2$-Et$_3$Al	5	Butadiene → butane (3%) + butenes, isoprene → 2-methylbutane (1%) + methylbutenes	1146
Co(NO$_3$)$_2$(OPPh$_3$)$_2$-Et$_3$Al	5	Butadiene → butane (1%) + butenes, isoprene → 2-methylbutane (0.5%) + methylbutenes	1146
Co(NO$_3$)$_2$(py)$_3$-Et$_3$Al	5	Butadiene → butane (0.3%) + butenes, isoprene → 2-methylbutane (0.3%) + methylbutenes	1146
CoCl$_2$(diphos)$_2$-Et$_3$Al	5	Butadiene → butane (1%) + butenes	581
CoCl$_2$/POCl$_3$-Et$_3$Al	5	Isoprene → 2-methylbutane (15%) + methylbutenes	581
Co(SCN)$_2$(PPh$_3$)$_2$-Et$_3$Al	5	Isoprene → 2-methylbutane (2%) + methylbutenes	581
Fe(CO)$_5$-R$_2$AlX^c	15	4-Vinylcyclohexene → 4-ethylcyclohexene	1161
FeCl$_3$/PPh$_3$-Et$_3$Al	3.5	Butadiene → butenes	1423
Co(acac)$_2$-^{i}Bu$_3$Al	≥6	4-Vinylcyclohexene → 4-ethylcyclohexene, cyclopentadiene → cyclopentene	1141, 1145
Co(Et.Hex)$_2$-Et$_3$Al	14	Furan → THF; hexadienoate → hexanoate; penta-1,3-diene → pentane	1176
Ni(Et.Hex)$_2$-Et$_3$Al	3	Butadiene → butane; 4-vinylcyclohexene → ethylcyclohexane; 1,5,9-cyclododecatriene → cyclododecane	1147
Cr(acac)$_3$-^{i}Bu$_3$Al	10	4-Vinylcyclohexene, hexadiene and d-limonene reduced	1178

Sal = salicylaldehyde anion; Cp = π-C$_5$H$_5$; Et.Hex = 2-ethylhexanoate.

[a] Cyclooligomerization also occurred.
[b] L = 1,3-bis(diphenylphosphino)propane.
[c] X = H,R; R = ^{n}Bu, ^{i}Bu, Et.

suggests little isomerization (cf. Section X-B-2a). The selective reduction of unsaturated fats to monoenes using transition metal compounds with triethylaluminum (1:5), also reported by these workers (206) and Misono and Uchida (207), was considered in Section XIV-E, where the following activity trend was noted: Ni(acac)$_3$ > Co(acac)$_2$ $\approx$ CoCl$_2$ > FeCl$_3$.

Iwamoto (Table 79; 581, 1423 1425) similarly used cobalt and iron complexes with more strongly bound phosphorus ligands for selective reduction of butadiene and isoprene to monoenes. The use of hydridocobalt complexes under the same conditions was considered in Section X-G.

Tajima and Kunioka (Table 79; 167) have also used a variety of cyclo-pentadienyl complexes with organometallics in benzene at 40 to 100° with 60 atm hydrogen for the reduction of butadiene and isoprene. The titanium systems gave saturated product, while the vanadium systems gave mono-enes; the zirconium systems were inactive, in contrast to the system of Sloan and coworkers (56) which reduced cyclohexene (Table 78). The CpFe(CO)$_2$Cl systems were selective for butene production, while CpCo(CO)$_2$ and CpNiC$_3$-H$_7$ gave some butane production. Cp$_2$M and Cp$_2$M$^+$X$^-$ complexes (M = Fe, Co, Ni) were not active using the alkylating reagents BuLi, PhMgBr, Et$_3$Al, or ^{i}Bu$_3$Al, and interestingly the Cp$_2$MCl$_2$ complexes (M = Ti, V, Zr) were not activated using Et$_3$Al or ^{i}Bu$_3$Al. Lack of activity was attributed to failure to alkylate the transition metal complex (eq. 468). Coordinated cyclopentadienyl ligands were said to retard the hydrogenating ability in Ziegler-type catalysts.

Lapporte and Schuett have reported on the reduction of monoenes (Table 78; 1147; 20°, 3 atm H$_2$), polyenes (Table 79; 1147; 25 to 100°, 40 to 70 atm H$_2$), and aromatics (Table 80; 175, 1147, 150 to 210°, 70 atm H$_2$) by using triethylaluminum-reduced salts of nickel, cobalt, iron, chromium, and copper, particularly the 2-ethylhexanoates. The more reactive nickel systems were most extensively studied (Ni > Co > Fe > Cr > Cu). Studies on the aromatic reductions showed that ^{i}Bu$_3$Al and (C$_6$H$_{13}$)$_3$Al were equally effective alkyls, while Et$_3$B was inferior. Also an observed anion order of 2-ethylhexanoate > benzoate > acac > acetate > chloride roughly followed their solubility in the hydrocarbon solvents used, and the catalysts prepared from acetates and chlorides were, in fact, suspensions rather than solutions (see below). For the C$_8$ monoenes reduced with the nickel hexanoate catalyst (Table 78), the hydrogenation rates were in the order monosubstituted > unsymmetrically disubstituted > cyclic > symmetrically disubstituted, agree-ing with the findings of Kroll (1141) that internal olefins are reduced more slowly than terminal ones; addition of strong donors such as triphenylphos-phine and nitrobenzene again retarded monoene reduction. In the absence of such donors, dienes and a triene were reduced completely to the saturated product. The aromatic nuclei are reduced under more severe conditions, generally to fully saturated species. The hydrogenations appear to be kineti‐ cally *cis*. Nitro groups inhibit reduction presumably because of strong co‐ ordination to the catalyst, and addition of nitrobenzene also retards the hydrogenation of benzene. The second ring in naphthalene is harder to hydrogenate than the first. The possibility that the reduction of aromatics involved free metal was not excluded, and indeed a patent by Lapporte (1159) has described even the ethylhexanoate systems as being "finely dispersed catalysts." Angelescu and coworkers (1168) have, however, detected the formation of π-benzene complexes in a $Cr(acac)_3$-Et_3Al benzene solution at lower temperatures by e.s.r. measurements.

A similar range of aromatics were effectively reduced by Bressan and Broggi (1176) using the cobalt(II) 2-ethylhexanoate-Et_3Al catalyst (1:14) at $25°$ under 20 to 65 atm hydrogen (Table 80). Also, benzene was selectively reduced in a mixture with *o*-xylene, and naphthalene was reduced preferen‐ tially in a mixture with benzene. The same catalyst was used for monoene reduction (Table 78) and to reduce dienes to the saturated product (Table 79). Acrylonitrile gave exclusively propionitrile, while mesityl oxide gave mainly 4-methylpentan-2-one with smaller amounts of the saturated alcohol; oct-1-ene was selectively reduced in a mixture with benzene. In the presence of added phosphines, thiophene, or pyridine, the hydrogenations of pent-1- ene and penta-1,3-diene were accompanied by some isomerization to pent-2- enes, consistent again with such donors retarding monoene reduction.

Kalechits and coworkers (147, 163–165, 195, 1157, 1170) have made extensive studies on some Ziegler catalysts for the hydrogenation of hex-1-ene and cyclohexene in toluene at 20 to $40°$ and up to 2 atm hydrogen (Table 78), and for benzene reduction in octane under more severe conditions (Table 80; 147, 164).

Reactions 470 and 471 lead to rate law 475, and Kalechits and

$$\frac{-d[H_2]}{dt} = \frac{k_1 k_2 [\text{catalyst}][\text{olefin}][H_2]}{k_{-1} + k_1 [\text{olefin}] + k_2 [H_2]} \qquad (475)$$

coworkers (147, 163, 1157) have demonstrated first-order in catalyst over certain concentration ranges, and between zero- and first-order in both hydrogen and olefinic substrate for a number of systems by monitoring the hydrogen pressure and/or substrate concentration. Various limiting forms of rate law 475, depending on the reactant concentrations and type of catalytic system, were also presented. At lower olefin and higher hydrogen concentration, reaction 470 can be rate-determining with the rate law = k_1[catalyst][olefin]. Several such rate constants were presented (163, 165, 195) to give a reactivity order of $Co > Ni > Fe > Cr > V$ for cyclohexene reduction, and an order $Co > Ni > Fe > Cr \geqslant Ti > Mn > V$ for hept-1- ene reduction, when using Et_3Al (10:1) with the acetylacetonates in all cases

Table 80 Hydrogenation of Aromatics

Substrate (% Conversion)[a]	Products (% Composition)[a]	Reference[b]
Benzene (100)	Cyclohexane (100)	147, 164, 175, 1145, 1176
o-Xylene (98)	Cis- and trans-1,2-dimethylcyclohexane (65, 35)	175, 1145, 1176
m- or p-Xylene (50)	Cis- and trans-dimethylcylcohexanes (63, 37)	1176
Phenol (97)	Cyclohexanol (92), cyclohexanone (5)	175, 1145, 1176
Dimethylterephthalate (100)	Cis-dimethyl hexahydrophthalate (100)	1145
Dimethylphthalate (100)	Cis-dimethyl hexahydrophthalate (100)	175
Diethylphthalate (95)	Cis- and trans-diethyl hexahydrophthalate (83, 17)	1176
Diphenyl (100)	Bicyclohexyl (81), phenylcyclohexane (19)	1176
Nitrobenzene (6)	Aniline (100)	175, 1145
p-Nitrophenol (0)	No reduction	175, 1145
Aniline (73)	Cyclohexylamine (50), dicyclohexylamine (44)	175, 1145
Pyridine (25)	Piperidine (100)	175, 1145
Naphthalene (100)	Tetralin (84), decalin (13)	175, 1145
Naphthalene (100)	Tetralin (37), decalin (54)	175, 1145
Tetralin (100)	Cis- and trans-decalins (66, 34)	1176

[a] The percentages refer to the highest conversion with either the nickel (1145) or the cobalt (1176) catalyst; they were generally quite similar.

[b] References 175 and 1145: Ni(2-ethylhexanoate)$_2$-Et$_3$Al(1:3), 150 to 210°, 70 atm H$_2$ in pentane or cyclohexane; Co, Fe, Cr, and Cu complexes also used.
References 147 and 164: M(acac)$_3$(M = Fe, Cr), M(acac)$_2$(M = Ni, VO, Mn, Co, Cu), Ni(DMGH)$_2$, (C$_5$H$_5$)$_2$TiCl$_2$ complexes + Et$_3$Al, 80 to 150°, 60 atm H$_2$ in hydrocarbons. Al/metal = 5–12.
Reference 1176: Co(2-ethylhexanoate)$_2$-Et$_3$Al (1:14), 25° 20 to 65 atm H$_2$ in heptane.

except for titanium [(C_5H_5)$_2$TiCl$_2$]. In later papers (147, 195) incorporating data on hex-1-ene reduction, the order has been modified slightly to Ni $\geqslant$ Co > Fe > Cr $\approx$ Mn > V. A similar order was presented for benzene reduction (147, 164): Ni $\geqslant$ Co > Cu $\geqslant$ Fe > Mn $\geqslant$ Cr > V > Ti, the exception being copper which is inactive for olefin reduction. A π-crotyl complex was said to be formed from butadiene with the nickel system (1157), thus giving evidence for interaction with a hydride (cf. eq. 185, Section X-B-2a). These workers (1157) also find that (C_5H_5)$_2$TiCl$_2$ systems are more active for olefin reduction than those based on Ti(O^{i}Pr)$_4$. Activation energies of about 8 to 9 kcal mole^{-1} were estimated for the reduction of benzene by the iron(III), chromium(III), and nickel(II) acetylacetonate systems (147). In contrast to the absolute rates, these energies were independent of the aluminum: metal ratio (164); a value of 11 kcal mole^{-1} was given for the copper system (147). For cyclohexene reduction, activation energies decreased in the order (147, 195): vanadium, 9.9 kcal mole^{-1}; chromium, 7.0; iron, 4.8; and Ni, 3.1—the reverse of the activity trend.

Kalechits and coworkers (147, 164, 195, 1157) have also studied ligand effects particularly in some nickel complexes with Et$_3$Al. For benzene reduction, Ni(DMGH)$_2$ was much less active than Ni(acac)$_2$, suggesting that the weaker field ligand had the higher activity. Further studies on cyclohexene reduction confirmed this in the activity order: acetylacetonate > o-amino-phenolate > salicylaldoximate > 8-hydroxyquinolate > dimethylglyoximate, which is also the reverse order of the complex stability constants. An activation energy of 7 kcal mole^{-1} was estimated for the dimethylglyoximate system, some 4 kcal higher than the acetylacetonate system (147). Addition of ligands such as ethanol, pyridine, and triphenylphosphine to some cobalt stearate and iron acetylacetonate systems suggests similar trends here; inhibitory effects increased in this order to a maximum with the phosphine. Addition of amines and phosphorus ligands to a Ni(acac)$_2$-cyclohexene system (Table 78, footnote f) showed that the amines decreased activity only slightly while the phosphorus ligands caused inhibition by factors of at least 7; the activity fell in the order P(OEt)$_3$ $\geqslant$ P(O^{i}Pr)$_3$ > PPh$_3$ > P(OPh)$_3$ > PCl$_3$, which probably results from increasing π-acceptor strength (1158). The kinetics of the P(O^{i}Pr)$_3$ system and the parent acetylacetonate system followed rate law 475; at lower hydrogen pressures the reactions were first-order in hydrogen; the data were analyzed in terms of this limiting form of eq. 475 with k_2[H$_2$] omitted from the denominator. In this particular study (1157), eq. 470 was written to include the intermediate π-complex (eq. 476), hence kinetic analysis

$$\text{HM} + \text{olefin} \overset{K}{\rightleftharpoons} \text{HM(olefin)} \underset{k_{-1}'}{\overset{k_1'}{\rightleftharpoons}} \text{alkyl-M} \qquad (476)$$

gave values of K and k_2k_1'/k_{-1}' (or k_2K_1'). The kinetic and thermodynamic parameters obtained are given in Table 81. The higher K value with the

Table 81   Thermodynamic and Kinetic Parameters for Hydrogenation of Cyclohexene Using Nickel(II) Acetylacetonate Ziegler Catalysts (1157)[a]

Catalyst	$K(30°)$ (M^{-1})	$-\Delta H$ (kcal mole^{-1})[b]	ΔS (eu)[b]	$k_2 K_1'$ $(30°)$ $(M^{-1}\,s^{-1})$	$\Delta H^{\ddagger}$ (kcal mole^{-1})[c]	k_0 $(M^{-1}\,s^{-1})$[d]
Ni(acac)$_2$-Et$_3$Al(I)	1.5	7	-22	16.0	13.0	3.5×10^{10}
I + 2P(O^{i}Pr)$_3$	21	11.5	-31	0.22	19.5	2×10^{13}

[a] Toluene solution, up to 2.5 atm H$_2$, 20 to 40°, $(0.5-4.2) \times 10^{-2}$ M Ni, Al: Ni = 2, up to 0.1 M substrate. (eqs. 471 and 476).
[b] Thermodynamic parameters from reaction isochore for K.
[c] $\Delta H^{\ddagger}$ value from Arrhenius plot for combined constant $k_2 K_1'$.
[d] Preexponential term for combined constant $k_2 K_1'$.

phosphite system is due to an increase in enthalpy, indicating a stronger metal-olefin bond; the lower hydrogenating rate probably results from the higher activation energy. This was attributed to an increase in strength of the metal-carbon and metal-hydrogen bonds in the phosphite system, although this is not at all clear considering the K_1' equilibrium (eq. 476) and k_2 step (eq. 471) involved.

The same group (147, 195) also studied the effectiveness of the organo-aluminum compounds for reduction by acetylacetonates, Cp$_2$TiCl$_2$ and Ni(DMGH)$_2$, and found an activity order, Et$_3$Al $\geqslant$ ^{i}Bu$_3$Al $>$ Et$_2$AlCl $>$ EtAlCl$_2$; also, for iron(III) acetylacetonate, ^{i}Bu$_3$Al $\geqslant$ Et$_2$Al(OEt). The reducing capacity of these alkyls follows the same trends (see Section B).

Sokol'skii and coworkers (Table 78; 166, 1173) have considered in particular the stability of the Cr(acac)$_3$ and Cp$_2$TiCl$_2$ systems with triisobutylaluminum, and conclude that the former is of practical use at 40° in heptane, while the latter deteriorates rapidly (see below for further details of the titanium system). Activation energies of 10, 7, and 5 kcal mole^{-1} were reported for cyclohexene reduction when using the chromium catalyst in heptane, toluene, and chlorobenzene, respectively (1173). The same group (Table 78; 1427, 1429) have also reported studies in these solvents using acetylacetonates with Et$_3$Al; the nickel system is more active and stable than the chromium or iron systems. The optimum alkyl: nickel ratio (2 to 5) depends on the solvent used.

Tikhomirov and coworkers (1178) have concluded from studies on monoene and diene reductions (Tables 78 and 79) when using a Cr(acac)$_3$-^{i}Bu$_3$Al (1:10) catalyst under mild conditions, that hydrogenation rates, which were first-order in olefin, decrease with increasing substitution at an olefin bond.

Ciardelli and coworkers (1353) have used triisobutylaluminum in conjunction with the optically active alkoxide complex, titanium tetra-($-$)-menthoxide, for the hydrogenation of racemic terminal olefins in benzene at 25° and 1 atm. However, the reduction of 3,4-dimethylpent-1-ene and 3,7-dimethyloct-1-ene was essentially nonasymmetric.

Takegami and coworkers have used catalyst systems composed of chlorides of iron(III), cobalt(II), and nickel(II) plus (*a*) Grignard reagents such as PhMgBr, PhCH$_2$MgCl, C$_6$H$_{11}$MgBr, and Ph$_2$Mg in the presence of hydrogen (58b, 1148–1152) and (*b*) lithium aluminum hydride (58b, 1152–1156). Hydrogenations were effectively carried out usually in ethers, using mild conditions under hydrogen or nitrogen, particularly with styrenes and cyclohexene substrates and also with isoprene, indene, ethyl acrylate, *d*-limonene, α-pinene, ethyl vinyl ether, squalene, safrole, and isosafrole. The reactions appear to be heterogeneously catalyzed involving a metal hydride species, but they are retarded by addition of donors such as triphenylphosphine and show the activity sequence Ni > Co > Fe usually observed in the homogeneous Ziegler catalysts. They are clearly closely related to these, and the possible participation of some homogeneous activity cannot be completely ruled out. Indeed, this group have concluded that the same chloride systems with triethylaluminum (Table 78; 173) that gave rise to a black precipitate and a black solution have an activity associated with both phases. The reduction of the substrates was studied in cyclohexane and THF at 0 to 35° with 1 atm hydrogen. As with Grignard and aluminum hydride systems, some reduction of styrene was observed on treatment with the catalyst system under nitrogen; the extent of reduction depended on the catalyst history, particularly its pretreatment with hydrogen. This pretreatment resulted in no gas uptake, although ethane was formed and the transition metal hydride sites were thought to be formed via eq. 469.

The same group (1172) has reported that the reaction of Grignard reagents with Cp$_2$TiCl$_2$ (Table 78, footnote *e*) gives homogeneous catalysts in hydrocarbon solution for cyclohexene reduction under mild conditions. Interaction of a Grignard with the dichloride is believed to give the hydride intermediate [Cp$_2$TiH]$_2$ (1335).

Closely related to some of Takegami's studies and the Ziegler systems is the work of Sajus and coworkers (168, 1169, 1187, 1452) who have reported that reduction of transition metal compounds with alcoxyalanates such as LiAlH(O^{t}Bu)$_3$, LiAlH$_2$(O^{t}Bu)$_2$, LiAlH$_2$(OPh)$_2$, or AlH(O^{t}Bu)$_2$ produces extremely effective, stable, and apparently homogeneous catalysts for the hydrogenation of monoenes and dienes with 100% conversion (Table 82). The reaction rates are little affected by the ligands in the cobalt catalysts, although in the titanium cyclopentadienyl complexes, those containing a phenoxy ligand were most effective, while titanium acetylacetonate with alkoxides were inactive. The aluminum/metal ratio generally seems unimportant, although using LiAlH$_2$(O^{t}Bu)$_2$ with Cp$_2$Ti(OPh)$_2$ an optimum ratio of 6 was noted.

Table 82 Reduction of Cyclopentene[a] (1 M) Using Transition Metal (M)-Alcoxy-alanate Catalysts (168, 1169, 1187)[b]

Catalyst[c]-alcoxyalanate[d]	Al/M	$[M] \times 10^3$ (M)	Rate (M min^{-1})
CoBr$_2$—I	50	0.23	0.36
CoBr$_2$L$_2$—I	30	0.38	0.46
CoBr$_2$L$_2$—II	20	0.57	0.52
Co(acac)$_2$—I	25	0.45	0.21
Co(O$_2$CCH$_3$)$_2$/6L—I	15	0.47	0.35
NiBr$_2$L$_2$—I	40	0.52	0.03
FeCl$_3$—III	3	2.32	0.08[e]
(C$_5$H$_5$)$_2$TiCl$_2$—IV	45	0.92	0.49[e]
(C$_5$H$_5$)$_2$TiCl$_2$—V	5	0.28	0.57[f]

[a] Styrene, methylbutenes, pentenes, methylpentenes, isoprene, hexenes, cyclohexene, cyclooctene, cyclopentadienes, and cycloocta-1,3-diene also reduced.
[b] Heptane solution with about 2% THF, 20°, 1 atm H$_2$.
[c] L = P(OC$_8$H$_{17}$)$_3$; Zn(acac)$_2$, CrCl$_3$/4L, PdCl$_2$/4L, (C$_5$H$_5$)$_2$Ti(OPh)$_2$, (C$_5$H$_5$)$_2$-Ti(SPh)$_2$, and (C$_5$H$_5$)$_2$TiCl(OPh) are also effective.
[d] I = LiAlH(O^{t}Bu)$_3$, II = NaAlH(O^{t}Bu)$_3$, III = AlH(O^{t}Bu)$_2$, IV = 30 parts LiAlH(O^{t}Bu)$_3$ + 15 parts AlH$_3$, V = LiAlH$_2$(O^{t}Bu)$_2$.
[e] At 40°.
[f] 0.9 M Hex-1-ene substrate.

Marko and coworkers (570–572) have reported on homogeneous catalyst systems formed from reaction between hydrocarbon solutions of iron, cobalt and nickel stearates, and ether solutions of Grignard reagents for very effective hydrogenation of monoenes at 20° and 1 atm pressure. (Table 78): for example, 100 ml of the dark brown solution, 0.03 M in Co(stearate)$_2$-^{n}BuMgBr (1:5), hydrogenates 0.4 M hept-1-ene in less than 5 min. Optimum magnesium:metal ratios of 3 to 8 are observed; below 3, brightly colored but inactive solutions are formed, while above 8 the activity falls. The nature of the alkyl seems unimportant. In the absence of substrate, 2 moles of hydrogen are absorbed per cobalt. This has been observed in similar systems by other workers (570) including Takegami's group (1148–1150), and has been attributed to reactions such as 477, but Marko and coworkers (570) sug-

$$CoR_2 + 2H_2 \longrightarrow CoH_2 + 2RH \quad (M = Fe, Co, Ni) \qquad (477)$$

gested that this could be due to catalytic hydrogenation of olefin formed via an intermediate alkyl, as in eq. 478 (cf. eq. 474).

$$Co(stearate)_2 \xrightarrow{\text{EtMgBr}} (L_xCoEt_2) \longrightarrow L_xCoH_2 + 2C_2H_4 \qquad (478)$$

Ethylene and isobutene were indeed found in the required amounts in the distillates from the reaction of stearate with ethylmagnesium bromide and t-butylmagnesium chloride, respectively. Also slow stoichiometric hydrogenations, including that of 1 mole of olefin generated by reaction 478, could be carried out with the stearate-Grignard solutions (Table 78, footnote d), thus giving good evidence for the existence of the dihydride; later studies by the same group has provided further evidence (see Section X-G).

The disparity in the rates of the catalytic and stoichiometric reductions indicates different mechanisms. It is worth noting that the final overall decomposition of the catalyst solution (e.g., eq. 479) corresponds to the familiar disproportionation reaction of unstable transition metal alkyls (570)

$$CoEt_2 \longrightarrow CoH_2 + 2C_2H_4 \longrightarrow Co + C_2H_4 + C_2H_6 \qquad (479)$$

(Section I-A). The solubility of these stearate catalysts was attributed to the ligands L, which were thought to be unsaturated alkoxymagnesium halides, such as $C_{16}H_{33}CH{=}C(OMgX)_2$ or $C_{16}H_{33}C{\equiv}C(OMgX)$, formed from the stearate. These workers have also studied the cobalt stearate-Et_3Al catalysts for reduction of monoenes and aromatics (1174).

Henrici-Olivé and Olivé (1186) have presented e.s.r. evidence for the formation of a cobalt(II) hydride complex, possibly $HCoR(diphos)_2$, when $CoCl_2(diphos)_2$ reacts with magnesium Grignard compounds in 1,2-dichloroethane at Mg:Co $\leqslant$ 8; reduction to cobalt(I) occurs at higher ratios.

Accompanying slower double-bond migration and *cis-trans* isomerization processes have frequently been observed during olefin reduction with Ziegler-type hydrogenation catalysts (e.g., 147, 165, 206, 1141, 1147, 1157, 1176, 1187, 1353). Deuterium exchange with substrate, as observed by Sloan and coworkers (56), is consistent with equilibria such as eq. 470. This also readily accounts for any observed isomerization, although complexing of olefin followed by π-allyl hydride formation and subsequent hydride transfer cannot be precluded. A decrease in isomerization rate with increasing hydrogen pressure has been readily rationalized in terms of eqs. 470 and 471 (147, 1157).

B. THE ROLE OF THE ALKYL COMPONENT

This has been discussed in part in Section A, the studies of Marko being particularly relevant. Alkylation of the transition metal complex (eq. 468) is undoubtedly the prime role of the alkyl component of Ziegler-type catalyst systems. The reducing properties of aluminum alkyls are well known, and there is the possibility of complexing with the transition metal moiety.

It should be noted that the organometallic compounds may behave also as Lewis acids (1143, 1144), which can abstract ligands to form coordinatively unsaturated species of greater activity; activation of the cobalt hydrides $HCo(CO)(PPh_3)_3$ and $HCo(diphos)_2$ by aluminum alkyls has been explained in this way (see Sections X-F-2 and X-G).

The reaction of orange-red hydrocarbon solutions of $(C_5H_5)_2TiCl_2$ with aluminum alkyls has been considered by Long and Breslow (1162) and Shilov and coworkers (1163), especially in regard to the polymerization properties of this catalyst system. The active catalyst is thought to be a titanium(IV) complex **151** (eq. 480) or possibly an ionic form $(C_5H_5)_2TiR^+$,

$$2(C_5H_5)_2TiCl_2 + [R_2AlX]_2 \rightleftharpoons 2(C_5H_5)TiCl_2AlXR_2 \rightleftharpoons$$

$$(C_5H_5)_2Ti \begin{matrix} Cl \\ \diagup \quad \diagdown \\ \quad\quad AlClRX \\ \diagdown \quad \diagup \\ R \end{matrix} \quad (480)$$

151

$$(X = R \text{ or } Cl)$$

both of which can furnish the radical R. Complex **151** is an intermediate in the overall reaction which finally yields blue solutions containing the trivalent species, for example, $(C_5H_5)_2TiCl \cdot AlEt_3$ in the triethylaluminum system. Clauss and Bestian (1164) have shown that under hydrogen these blue solutions readily give titanium(II) species with liberation of alkane via cleavage of a titanium-alkyl bond, for example, as in eq. 481 (see also Section V-A; eq. 76):

$$(C_5H_5)_2TiCl \cdot AlEt_3 \rightleftharpoons (C_5H_5)_2TiEt \cdot AlEt_2Cl \xrightarrow[Et_3Al]{\frac{1}{2}H_2}$$
$$(C_5H_5)_2Ti \cdot AlEt_2Cl + C_2H_6 \quad (481)$$

In the catalytic hydrogenation studies using the $(C_5H_5)_2TiCl_2$-Et_3Al catalyst, Kalechits and coworkers (Table 78; 165) found that the blue solutions absorb some hydrogen to give inactive solutions probably containing titanium(II and III) species. In the presence of up to 0.08 % of oxygen in the hydrogen, reoxidation to an active brownish titanium(IV)-aluminum catalyst occurred. The optimum aluminum/titanium ratio was 6; at higher ratios the lower rates were attributed to the production of some inactive titanium(III) species formed by the reducing triethylaluminum. It should be noted, however, that sodium-reduced solutions of $(C_5H_5)_2TiCl_2$ have been found to be quite active for hydrogenation of monoenes (see Section V-A). Noskova and coworkers (Table 78; 166) also consider that the $(C_5H_5)_2TiCl_2$-$^iBu_3Al$ system involves an active titanium(IV) catalyst.

Sartori and coworkers (1165) think that chromic acetylacetonate reacts with triethylaluminum according to the process outlined in eq. 482.

$$Cr^{III}(acac)_3 \xrightarrow{3Et_3Al} [Cr^0(acac)_3] \xrightarrow{3Et_3Al} [Cr^0(acac\cdot Et_3Al)_3] \xrightarrow{n\cdot Et_3Al} Et_nCr^0 \tag{482}$$

Chromium(III) is first reduced to chromium(0); three further moles of alkyl then coordinate via the carbonyl oxygens of the acetylacetonate before further reaction gives the requisite alkylated chromium catalyst. Tamai and coworkers (1167) have reached similar conclusions for the $Co(acac)_3$-Et_3Al system, and have shown that coordinated triethylaluminum can stabilize lower valence metal states by utilizing the empty $3p$ aluminum orbitals as π-acceptors; the poisoning effect of excess alkyls in these catalysts can be attributed to such coordination (147). Nasirov and coworkers (1171) conclude that homogeneous solutions, containing the vanadium(II) species $VO[acac(Et_2Al)(Et_3Al)]_2$ with aluminum coordination at both carbonyl and enolate oxygens, are formed in the $VO(acac)_2$-Et_3Al system. A number of other investigations of this kind have been reported (e.g., 1168 and references therein).

Kalechits and coworkers (147, 1163) have invoked reactions such as 482 to rationalize the commonly observed optimum aluminum:metal ratios of >6; they have reported that such ratios are to 6 to 8 for acetylacetonates of bivalent metals, and 8 to 12 for the trivalent metals. Such variations have been reported by the same group using the same catalyst with different substrates; for example, ratios of 8 have been noted for cyclohexene (1157) and hept-1-ene (163) reduction, and 12 for hex-1-ene reduction (147) when using the $Fe(acac)_3$-Et_3Al catalyst under the same conditions. A spectrum of a $1:14$ $Cr(acac)_3$-Et_3Al catalyst was unchanged after a catalytic hydrogenation experiment (163).

Sloan and coworkers (56) observed that in the absence of substrate the $1:6$ $Cr(acac)_3$-$^iBu_3Al$ catalyst in heptane at $20°$ absorbed more than 7 moles of hydrogen per mole of chromium, which, after allowing for reduction of the chromium(III), suggests the possible formation of chromium and/or aluminum hydrides; since ketones were not reduced with this catalyst, the acetylacetonate ligands were thought unlikely to be hydrogenated.

Lapporte and Schuett (175, 1147) have noted the evolution of about 3 equivalents of gas [ethane (90%), ethylene, and hydrogen] when the optimal 3 to 4 equivalents of Et_3Al are added to a heptane solution of nickel(II) 2-ethylhexanoate, and that the carboxylate moiety is not reduced. N.m.r. evidence of the catalyst solution prepared at low temperature suggests the presence of π-bonded ethylene and a transition metal hydride; in conjunction with some magnetic susceptibility data, it was suggested that a Ni(0) complex is formed via unstable ethyl and hydride species (cf. eqs. 474 and 478):

$$(RCO_2)_2Ni + Et_3Al \longrightarrow L_3Ni-C_2H_5 \rightleftharpoons HNi(C_2H_4)L_2 \qquad (483)$$

(L = solvent, C_2H_4, RCO_2AlEt_2). The labile ligand groups would be easily displaced by unsaturated substrates.

Sokol'skii and coworkers (177, 1426) have studied conductivity changes during catalyst formation and hydrogenation of some monoenes and phenyl-acetylene when using the $M(acac)_3$-Et_3Al catalysts (M = Fe, Cr) in toluene. Catalyst formation under argon gives conducting solutions and the conductivity increases on admitting hydrogen, which was attributed to a hydride complex; a maximum difference in these two conductivities occurred at an Al:Cr ratio of 7 which was the optimum ratio for hydrogenation. Interaction of the substrate with the iron catalyst, as measured by conductivity increase, decreased in the order: phenylacetylene > styrene > hept-1-ene ~ cyclohexene. The stronger interaction with the acetylene accounts for some trimerization products during the hydrogen (see Section C, Table 83). The terminal olefins were reduced much more rapidly than cyclohexene, however, despite similar "interactions," indicating presumably differences in the ease of the hydrogenation steps.

C.   REDUCTION OF ACETYLENES AND OTHER FUNCTIONAL GROUPS

A number of the reports considered in Sections A and B, and one by Sokol'skii and coworkers (1428), have referred to attempted reduction of acetylenes (Table 83); temperatures from 20 to 60° with a few atmospheres of hydrogen have been employed. The reactions are sometimes complicated by competing cyclization and polymerization reactions (56, 177, 1157, 1428); acetylene itself gives mainly polymers (1428). A reactivity order of Ni > Fe > Cr has been established for the reduction of phenylacetylene and trimethylethynylsilane using the acetylacetonates in the presence of triphenyl-phosphine and triethylphosphite, respectively (147).

Hex-1-yne and hept-1-yne may be reduced selectively to monoenes, which can be rationalized in terms of stronger π-complexes with acetylenes as compared to olefins. This may also hinder hydrogen activation and favor polymerization; acetylenes are reduced more slowly than olefins (177, 1157, 1428). The rate and extent of hydrogenation has been increased by pretreatment of the catalyst with hydrogen in the absence of the acetylene (1157). The $Ni(acac)_2$-Et_3Al-phosphine systems (Table 83, footnote d) showed decreasing activity for hex-1-yne reduction in the order $P(OPh)_3 > P(OEt)_3 > PPh_3$, the reverse of that found for the polymerization.

Selective reduction of phenylacetylene in a mixture with hex-1-ene has been reported (1141, 1145).

Table 83 Reduction of Acetylenes Using Soluble Ziegler Catalysts

Substrate	Catalyst	Optimum Al:M	Product	Reference
Acetylene	Fe(acac)$_3$-Et$_3$Al	3	Ethylene, polymers	1428
Phenylacetylene	Ti(O^{i}Pr)$_4$-^{i}Bu$_3$Al	3.3	Ethylbenzene	56
Phenylacetylene	Cr(acac)$_3$-^{i}Bu$_3$Al	9	Reduction	1173
Phenylacetylene	Fe(acac)$_3$-Et$_3$Al	3–5	Reduction, triphenylbenzene	177, 1428, 1429
Phenylacetylene[a]	Co(acac)$_2$-^{i}Bu$_3$Al[b]	⩾6	Styrene, ethylbenzene	1141, 1145
Phenylacetylene	M(acac)$_n$-Et$_3$Al[b]	—	Styrene	147, 1427
Me$_3$Si-C≡CH	M(acac)$_n$-Et$_3$Al[c]	—	Me$_3$SiCH=CH$_2$	147
Hex-3-yne	Cr(acac)$_3$-^{i}Bu$_3$Al	6.0	Reduction, hexaethylbenzene	56
Hex-3-yne	Ni(Et.Hex)$_2$-Et$_3$Al	3	n-Hexane	1147
Hex-1-yne	Ni(acac)$_2$-Et$_3$Al[d]		Reduction, polymerization	1157
Hept-1-yne	Fe(acac)$_3$-Et$_3$Al	3	Hept-1-ene	1428
Et.Hex = 2-ethylhexanoate				

[a] In the presence of a large excess of hex-1-ene.

[b,c] Cr(acac)$_3$, Fe(acac)$_2$, and Ni(acac)$_2$ in the presence of [b]PPh$_3$ or [c]P(OEt)$_3$; Al:M ratio not specified.

[d] Carried out in presence of 2 moles of PPh$_3$, P(OEt)$_3$, and P(OPh)$_3$ per mole of Ni.

381

Ketones, aldehydes, nitriles, nitro compounds, azo compounds, and esters are not readily reduced by the soluble Ziegler catalysts (56), although some of these have been reduced under more severe conditions. Lapporte (1147) has used the nickel and copper 2-ethylhexanoates-Et$_3$Al (1:3) systems at 100 to 150° and 70 to 100 atm hydrogen for the reduction of cyclohexanone and pentan-3-one to the alcohols; benzaldehyde underwent hydrogenolysis to toluene and disproportionation to benzyl benzoate, as well as reduction to benzyl alcohol. Chlorobenzene formed benzene almost exclusively; iso-phthalonitrile gave mainly *m*-xylylenediamine. Lapporte suggests that free metal catalysts could again be involved at these higher temperature conditions.

Waddan and Williams (1175) have hydrogenated aliphatic dinitriles, such as adiponitrile, to the diamines by using cobaltous acetate, octanoate and acetylacetonate, dicobalt octacarbonyl, nickel acetate, ferric acetylacetonate, and palladium acetate in conjunction with a range of Group I, II, and III alkyls, borohydride, aluminum hydride and alcoxyalanates at 50 to 200° and 1 to 500 atm.

The range of catalysts patented by Kroll (1160; Section A) were said to reduce nitriles, imines and carbonyls, and the cobalt(II) 2-ethylhexanoate catalyst was used by Bressan and Broggi (1176; Section A) to produce propan-2-ol from acetone, and α-methylbenzyl alcohol from acetophenone.

D. POLYMER REDUCTION

Lapporte (1147) has used nickel and copper 2-ethylhexanoate-Et$_3$Al (1:3) catalysts in cyclohexane at 160 to 225° and 110 atm hydrogen for reducing the aromatic rings of polystyrene to give polymers of vinylcyclohexane and styrene-vinylcyclohexane copolymers. Since double bonds are more readily reduced than aromatic nuclei, styrene-isoprene rubbers may also be hydrogenated selectively.

Tikhomirov and coworkers (176, 1178, 1179, 1188) have used a Cr(acac)$_3$-^{i}Bu$_3$Al (1:10) catalyst for the reduction of butadiene polymers and butadiene-styrene copolymers in decalin solution under mild and elevated conditions; hydrogenation proceeds first for vinyl groups of 1,2 units and then 1,4 units. The hydrogenations at 40° and 1 atm reached a limiting rate with increasing polymer concentration and some kinetic data were analyzed in terms of eq. 475. Duck and coworkers (1180) have used nickel salicylate with butyl lithium or lithium aluminum hydride in ether solutions for effective hydro-genation of polybutadiene at 40° and 1 atm, and a number of patents (1181–1184, 1416–1422), particularly by Yoshimoto and coworkers, have been concerned with the hydrogenation of butadiene polymers and copolymers with styrene under similar mild conditions. These have involved nickel, chromium, manganese, molybdenum and cobalt acetylacetonates, 8-hydroxy-quinolates, naphthenates, octanoates, salicylates, salicylaldoximates, and α nitroso-β-naphtholates with a wide range of Group I, II, and III alkyls or BF_3; a third catalyst component in the form of an unsaturated hydrocarbon such as styrene, cyclohexene or octa-1,7-diene (unsaturate/metal = 1) was sometimes added (1182) for increased efficiency.

Kallenbach (1185) has patented the use of nickel(II) diphenylphosphinate-trialkylaluminum (1:4) catalysts for hydrogenation of polyethylene at 150° with up to 20 atm hydrogen.

XVI

Miscellaneous Systems

A. FIRST-ROW TRANSITION METAL STEARATES (TULUPOV'S SYSTEMS)

In 1957 Tulupov (196) briefly reported that a number of salts of nickel(II), iron(III), cobalt(II), and lead(II), particularly stearates, catalyzed the hydrogen reduction of pent-2-ene, oleic acid, cyclohexanone, and benzene at ambient conditions. Since that time his group have reported data on the hydrogenation of cyclopentene catalyzed by manganous stearate in mineral oil solution (193a), and the hydrogenation of cyclohexene in ethanol solution catalyzed by the stearates of chromium(III) (46), nickel(II) (1189), iron(III) (197), copper(II) (146), cobalt(II) (1190), zinc (148), scandium(III) (161), and titanium(IV) (161). The physical properties of these stearates in the solid state were also examined (1191–1195) and the interaction with cylcohexene in cyclohexene solution from a magnetic point of view was also considered (1195, 1196). Tulupov (57a, 57b) has published reviews on these systems, and also some general papers (1197–1999) on the kinetics and mechanisms of the general processes involved.

The stearate-catalyzed cycloolefin reductions are slow at 20 to 60° and ~1 atm hydrogen; gas-uptake has been measured at constant pressure at catalyst concentrations 3.5×10^{-6} to $10^{-3}M$, depending on the metal, and substrate concentrations up to ~1.5 M. Linear rates were measured, and these either were the initial parts of uptakes first-order in substrate or corresponded to a zero-order dependence on substrate; the linear region kinetics showed first-order in hydrogen, and an olefin *and* catalyst dependence going from first- to zero-order with increasing concentration. Thus at higher catalyst and olefin concentrations, the observed rate law was simply

$$\frac{-d[H_2]}{dt} = k[H_2] \tag{484}$$

Complete rate laws, including slow equilibria, were presented (46, 161, 1197) for unsaturate routes exemplified in eq. 485, when it is possible for the

$$\text{Olefin} + M \rightleftharpoons M(\text{olefin}) \xrightarrow{H_2} H_2M(\text{olefin}) \longrightarrow M + \text{products} \tag{485}$$

rate to become independent of substrate, or substrate and hydrogen, concentrations. However, to rationalize the unusual catalyst independent rate law, the following mechanism was postulated for the cyclohexene reductions:

$$H_2 + C \underset{k_{-1}}{\overset{k_1}{\rightleftharpoons}} H_2{}^* \tag{486}$$

$$M + C_6H_{10} \underset{k_{-2}}{\overset{k_2}{\rightleftharpoons}} M(C_6H_{10}) \tag{487}$$

$$M(C_6H_{10}) + H_2{}^* \underset{k_{-3}}{\overset{k_3}{\rightleftharpoons}} H_2{}^*M(C_6H_{10}) \xrightarrow{k_4} M + C_6H_{12} \tag{488}$$

The C may be the wall or any other colliding species and is involved in activation of the hydrogen molecule. In the general case such a scheme accounts for the observed dependences and in the limiting case, when k_4, $k_3 \gg k_{-1}$ and equilibrium 487 lies to the right and is established rapidly (at high olefin), the rate law simply becomes eq. 484, where $k = k_1[C]$, with the forward reaction of 486 rate-determining. A diffusion control reaction or a limiting hydrogen solubility was ruled out by kinetic arguments (1190), but some experimental evidence on these points would be more convincing. Different pretreatments of the glass reaction vessel did not affect the hydrogenation rates.

Formation of complexes between the stearates and cyclohexene was demonstrated principally by spectrophotometric and magnetic susceptibility measurements, and stearates containing $M(C_6H_{10})_n$ moieties have been reported for each of the systems except manganese. Table 84 gives the n

Table 84 **Activation Parameters for Cyclohexene Reduction Catalyzed by First-Row Transition Metal Stearates (57b[a], 1199)**

Stearate	Sc^{III}	Ti^{IV}	Cr^{III}	Mn^{II}[b]	Fe^{III}	Co^{II}	Ni^{II}	Cu^{II}	Zn^{II}
n^c	3	7	9	—	4	1	7	16	20
$\Delta H^{\ddagger}$ (kcal mole^{-1})	6.7	8.7	4.6	5	7.3	8.0	7.0	1.8	5.9
$-\Delta S^{\ddagger}$ (eu)	55	50	58	64	67	57	66	72	61

[a] The $\Delta S^{\ddagger}$ values in Table 2 of Ref. 57b are mostly incorrectly presented.
[b] For cyclopentene reduction.
[c] n = moles of C_6H_{10} per metal in isolated complexes.

values and the thermodynamic parameters, usually determined with eq. 484 for the different systems (57b, 1199). The estimated $\Delta H^{\ddagger}$ values are said to refer to conditions where either eq. 486 or eq. 488 (k_3 step) is rate-determining. The slow reactions are seen to be due to unfavorable entropy changes; Tulupov (57b) has discussed these in terms of complex formation, presumably in the somewhat obscure activation process of eq. 486, which is thought to involve excitation of rotational levels in the hydrogen molecule (57b, 1190). The activation energy for the copper system was noted to correspond to the transition energy from the $J = 4$ to $J = 5$ energy levels. Furthermore, the orientation of the hydrogen molecule in the field of the olefin complex was considered important, suggesting that the olefin complex did play a role in reaction 486. Based on collisions with particles of energy kT, the maximum concentration of activated hydrogen molecules was estimated from some deactivation probabilities given by Brout (1200) to be reached in the region of reaction rates which were invariant with metal concentration. The activation energies in Table 84 were divided into those of the diamagnetic ions, Zn(II), Ti(IV), and Sc(III), and those of the remaining paramagnetic ones. The general trend of decreasing $\Delta H^{\ddagger}$ with increasing n for each series was attributed to a destabilization of the $M(C_6H_{10})_n$ complex, in reaction 488 (1199). This is somewhat unclear, but it seems from the discussions presented (57b, 1190, 1199) that C in process 486 is regarded as the cyclohexene complex. Some of the discussion on the thermodynamic parameters in papers on particular systems also confuses the issue somewhat. For example, the copper system (146) is discussed incorrectly in terms of a positive entropy of activation.

Magnetic susceptibility data on the stearates themselves indicated (*a*) partial conversion of spin-free chromium(III), manganese(II), iron(III), and copper(II) into the spin-paired configuration, and (*b*) partial conversion of spin-paired zinc(II), nickel(II), and cobalt(II) into the spin-free form. E.s.r. signals were also given by the diamagnetic stearates of scandium(III),

152

(489)

titanium(IV), and zinc(II). These changes in electronic configuration were explained in terms of transitions to the carbonyl π^* state from the nd electrons (for ions with d-electrons) or from the np electrons for the p^6 scandium(III) and titanium(IV) systems. Such transitions give rise to a quasiaromatic structure **152** for the dimeric stearates. Equation 489 shows this for the bivalent metals.

Susceptibility and e.s.r. measurements showed that complex formation with cyclohexene can give an increase, decrease or no change in the number of spins, which was explained in terms of the ring system reacting with the olefin to give a compound in which the olefin is bound to two metal atoms (eq. 490), an occurrence commonly postulated for heterogeneous systems. Tulupov (57b) presented eqs. 491 and 492 for possible interaction with the activated hydrogen molecule, while Coffey (57e) has suggested an intermediate involving homolytic cleavage of hydrogen at the two metal centers (eq. 493).

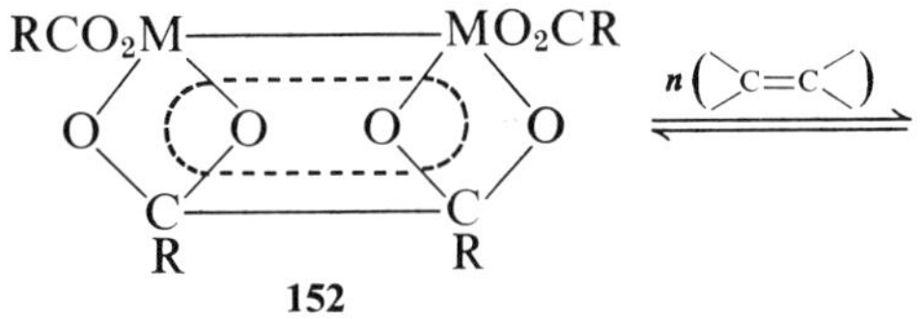

152

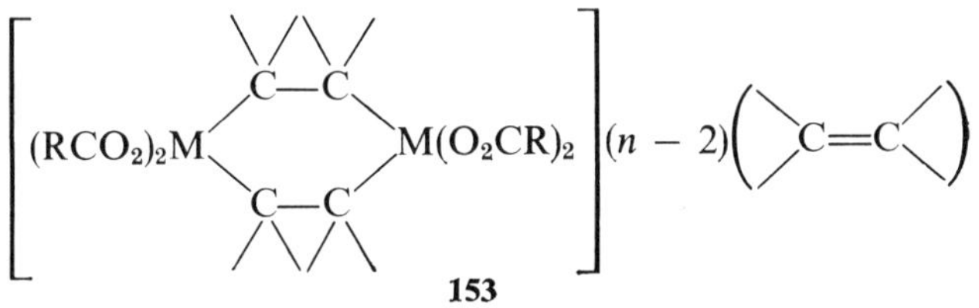

153

$$(490)$$

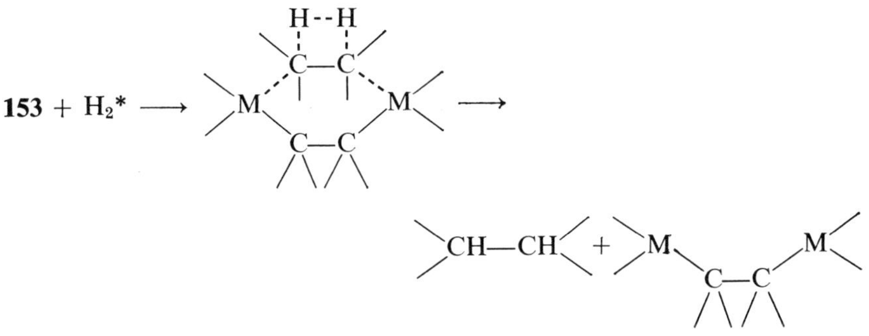

$$(491)$$

$$153 + 4H_2{}^* \longrightarrow \begin{array}{c} \text{H} \quad \text{H} \\ \text{H} \quad C\text{-}C \quad \text{H} \\ M \qquad\qquad M \\ \text{H} \quad C\text{-}C \quad \text{H} \\ \text{H} \quad \text{H} \end{array} \longrightarrow$$

$$2\left(\begin{array}{c}\diagdown \\ \diagup\end{array}CH\text{—}CH\begin{array}{c}\diagup \\ \diagdown\end{array}\right) + 152 + 2H_2 \qquad\qquad (492)$$

$$153 + 2H_2 \longrightarrow \begin{array}{c} C\text{—}C \\ H\text{-}H \\ M \qquad M \\ H \quad H \\ C\text{—}C \end{array} \longrightarrow 2\left(\begin{array}{c}\diagdown \\ \diagup\end{array}CH\text{—}CH\begin{array}{c}\diagup \\ \diagdown\end{array}\right) + 152$$

$$(493)$$

Tulupov refers to the catalytic region obeying rate law 484 as "latent catalysis of the second kind," since the concentrations of both the catalyst and substrate are absent.

The stearate reductions are inhibited by water and oxygen. The copper and cobalt systems were also studied at higher pressures up to 110 atm. The copper system, discussed in Section II-B, becomes complicated because of accompanying reduction to copper(I) and metal; the cobalt system was ineffective for reduction of benzene or the carbonyl group in ethyl acetate.

Tulupov and Shigorin (1201, 1202) have considered in qualitative theoretical terms the orbital interactions involved in metal complex-olefin-hydrogen systems, based mainly on the stearate catalysts.

B. METAL IONS IN AQUEOUS SOLUTION

Earlier sections of this book have included data on the activation of hydrogen by simple inorganic salts of the following metals in aqueous solution: copper(I,II), silver(I), mercury(I,II), manganese(VII), ruthenium-(II,III), rhodium(III), palladium(II), iridium(III), and platinum(IV); the iridium and platinum systems showed relatively little activity. The inactivity of salts of gold(III), magnesium(II), zinc(II), cadmium(II), cerium(IV), vanadium(V), chromium(III,VI), iron(III), cobalt(II), nickel(II), and osmium(IV) was also mentioned (this excludes the cyanides of cobalt and nickel). Halpern's group (25, 26), among others, had investigated these systems and has also mentioned (25, 26) the inactivity of the following ions: calcium(II), aluminum(III), thallium(III), uranium(VI), and lead(II). These tests were usually made at temperatures up to 150° with a few atmospheres of hydrogen.

Fasman and Ikhsanov (1478) have made a cursory study of the activity of ions of 62 metals in aqueous acid solutions at 70° and 1 atm hydrogen. Catalytic reduction of dichromate (cf. Section II-B) was observed for solutions containing initially cupric acetate, permanganate, or anionic chloride complexes of ruthenium(IV), rhodium(III), osmium(IV), iridium(IV), and platinum(IV). The ruthenium(IV) system involves a ruthenium(III) catalyst (see Section IX-B-1); the osmium, iridium, and platinum systems are of low activity and require further study to elucidate the species and mechanisms involved (cf. Section XII-E for iridium and Section XIII-C-1 for platinum). Silver(I) nitrate, tetrachloroaurate(III), and mercury(II) sulfate were hydrogenated to the metal but did not catalyze dichromate reduction. Ceric sulfate was very slowly reduced to cerous. The finding of activity for gold(III) and cerium(IV) contrasts with that of Halpern's group, but in any case few details are available for either study. Fasman and Ikhsanov recorded inactivity for salts of lithium(I), beryllium(II), the alkaline earth metals, aluminum(III), gallium(III), indium(III), thallium(I), germanium(IV), tin(II,IV), lead(II), arsenic(III), antimony(III), bismuth(III), zinc(II), cadmium(II), scandium(III), and lanthanides(III).

C. OXIDATION AND REDUCTION OF METAL IONS BY HYDROGEN ATOMS

Studies on the activation of hydrogen by copper and silver complexes (see Sections II-A to II-D) led Halpern and coworkers (1273) to suggest that oxidation and reduction reactions of metal ions by hydrogen atoms, which were first observed in aqueous solution by Weiss and coworkers (e.g., 1274; eqs. 494 and 495), might involve hydride intermediates.

$$Fe^{2+} + H + H^+ \longrightarrow Fe^{3+} + H_2 \tag{494}$$

$$Fe^{3+} + H \longrightarrow Fe^{2+} + H^+ \tag{495}$$

Reaction 494 was thought to proceed via eqs. 496 and 497,

$$Fe^{2+} + H \underset{k_{-1}}{\overset{k_1}{\rightleftharpoons}} FeH^{2+} \tag{496}$$

$$FeH^{2+} + H^+ \xrightarrow{k_2} Fe^{3+} + H_2 \tag{497}$$

and kinetic studies by Czapski and coworkers (1275, 1276) support such a mechanism ($k_1 = 6 \times 10^4 \, M^{-1} \, s^{-1}$ and $k_{-1}/k_2 = 0.22 \, M$ at $4°$). Joyson and coworkers (938) have since reported spectrophotometric evidence for the formation of FeH^{2+} according to eq. 496 during some pulse radiolysis experiments on acid ferrous perchlorate solutions.

The oxidation of cuprous ions probably occurs via reactions such as 496 and 497, particularly since CuH^+ intermediates seem highly likely in the heterolytic splitting of hydrogen by cupric ions (1273) (cf. eq. 37 in Section II-B, the reverse of eq. 496). Similarly, the AgH^+ species formed from the silver(I) homolytic splitting of hydrogen (eq. 48d, Section II-D) was suggested (1273) as the likely intermediate in the reduction of silver ions by hydrogen atoms (eqs. 498 and 499), and Eachus and Symons (131a) have later obtained

$$Ag^+ + H \longrightarrow AgH^+ \tag{498}$$

$$AgH^+ \longrightarrow Ag + H^+ \tag{499}$$

experimental e.s.r. evidence for reaction 498 (Section II-D).

The reduction of some cobalt(III) complexes by hydrogen atoms was mentioned in Section X-B-1.

D. BORANES

Koster and coworkers (e.g., 31, 1203, 1204) made extensive studies on the hydrogenolysis of the boron-carbon bond at 160 to 220° and 200 to 300 atm hydrogen, and a simple example included the reaction with trialkylborane to give a dialkylborane and a saturated hydrocarbon:

$$BR_3 + H_2 \rightleftharpoons R_2BH + RH \tag{500}$$

About the same time, Brown (33, 1205) had developed the hydroboration reaction, showing the facile addition of boron hydrides to olefins; for example,

$$R_2BH + R'CH{=}CH_2 \rightleftharpoons R'CH_2CH_2BR_2 \tag{501}$$

Clearly, reaction 501 followed by a further hydrogenolysis reaction (eq. 502) would lead overall to olefin hydrogenation catalyzed by the borane BR_3.

$$R'CH_2CH_2BR_2 + H_2 \longrightarrow R'CH_2CH_3 + R_2BH \tag{502}$$

Koster and coworkers (31) first discussed this possibility and mentioned the hydrogenation of a terminal olefin by using a N-trialkylborazane catalyst $H_3B.NR_3$ prepared by reaction 500 in the presence of the amine, but no details of the hydrogenation were given. DeWitt and coworkers (32, 1206) then used tri-n-butyl- and triisobutylborane for reduction of cyclohexene, octenes, and olefinic polymers at 190 to 240° and 35 to 180 atm hydrogen.

Both components of a hex-1-yne/cyclohexene mixture were reduced to saturated products. The reaction rates are highly temperature dependent, but are influenced little by hydrogen pressures. Diglyme and benzene could be used as inert solvents if necessary. Lewis acids (metal chlorides), diethylamine, esters, ketones, alcohols, and carbon tetrachloride completely inhibited the reductions; some inhibition was noted in the presence of butyl chloride, dimethylsulfide, furan, 2,4-dimethylsulfolane, and diethyl acetal. Lewis bases (e.g., diglyme) catalyze the hydroboration reaction but did not enhance the hydrogenation rates, suggesting that reaction 502 is probably rate-determining. Since volatile boranes polymerize at high temperatures to give boron hydrides, which can themselves undergo thermal decomposition, the actual catalyst is uncertain. Hydrogenolysis of boranes can also give rise to a range of alkyl diboranes, $B_2H_{6-x}R_x$, where $x = 0$ to 4 (31). The aliphatic double bonds in 1,2 and 1,4-polybutadienes, cis-1,4-polyisoprene, poly-1,3-pentadiene, and a styrene-butadiene rubber were completely reduced; neoprene was not hydrogenated. Extensive deuterium exchange was observed in the borane-catalyzed hydrogenation of cis-1,4-polyisoprene-3-d, which is again consistent with the addition-elimination reaction (eq. 501) being more rapid than the hydrogenolysis.

E.   LITHIUM ALUMINUM HYDRIDE

The use of lithium aluminum hydride with transition metal salts to give hydrogenation catalysts was considered in Section XV-A. $LiAlH_4$ itself has been widely used for the reduction of organic compounds, (e.g., 57g, 1207), the reactions likely involving addition of the hydride to the unsaturated system followed by hydrolysis of the resulting Al—X bond, where X may be carbon, nitrogen, or oxygen; for example,

$$C{=}X + LiAlH_4 \longrightarrow -\underset{\underset{LiAlH_3}{|}}{C}H-X \xrightarrow{H_2O} -CH-XH \qquad (503)$$

Since Podall and coworkers (1208) had demonstrated the possible hydrogenolysis of aluminum carbon bonds (eq. 504), Slaugh (1207) successfully

$$Na(^iBu_3AlH) + H_2(220 \text{ atm}) \xrightarrow[Pt_2O]{150°} Na(^iBu_2AlH_2) + \text{isobutane} \qquad (504)$$

reasoned that $LiAlH_4$ might function as a homogeneous hydrogenation catalyst via hydride addition and subsequent hydrogenolysis (cf. eqs. 501 and 502 in Section D). Conjugated dienes (penta-1,3-diene, cycloocta-1,3-diene), and pent-2-yne, were reduced to monoenes (an isomeric mixture) in THF solution at 190° with 60 to 100 atm hydrogen when using quite high ratios (1:4) of catalyst to substrate; pentadiene also gave rise to some C_{10} products formed via a Diels-Alder reaction. *Cis*-pent-2-ene and cycloocta-1,5-diene were not reduced. Equations 505 and 506 summarize the proposed mechanism.

$$CH_2{=}CHCH{=}CHCH_3 + LiAlH_4 \rightleftharpoons (C_5H_9)LiAlH_3 \overset{C_5H_8}{\rightleftharpoons}$$
$$(C_5H_9)_2LiAlH_2, \cdots \quad (505)$$

$$C_5H_9LiAlH_3 + H_2 \longrightarrow C_5H_{10} + LiAlH_4 \quad (506)$$

The product isomer distribution (mainly pent-2-enes) will depend on the detailed mechanism of the hydrometalation step (cf. Section X-B-2a). Use of $LiAlD_4$ gave largely monodeuterated pentenes, which is consistent with addition of one hydrogen from the hydride and one from the gaseous phase. Reaction 506 has to be faster than the reverse of reaction 505 to explain the absence of polydeuterated pentenes (cf. Section D). However, reaction 506 could still be rate-determining. Tracer studies involving hydride or deuteride addition followed by solvolysis with D_2O or H_2O, respectively, indicated that $LiAlH_4$ adds in a *trans* manner to pent-2-yne which contrasts to the usual *cis*-addition of alkylaluminum hydrides (R_2AlH) to triple bonds (1209, 1210).

F. ORGANOTIN HYDRIDES

Tin hydrides do not appear to have been used in conjunction with molecular hydrogen for homogeneous hydrogenation as have boron and aluminum hydrides (see Sections D and E). However, the reduction of organic compounds by organotin hydrides alone is well known; Kuivila (1478) reviewed this topic. Corresponding to the reactions in 503, one mechanism involves hydrostannation, followed by replacement of the organotin group by hydrogen, the replacement involving a protonic acid (HY) or a further tin hydride:

$$C{=}X + HSnR_3 \longrightarrow -CH-X-SnR_3 \overset{H-Y}{\longrightarrow} -CH-XH + R_3SnY$$
$$\downarrow \small{HSnR_3} \qquad\qquad\qquad (507)$$
(R = alkyl, aryl) $\qquad -CH-XH + (SnR_3)_2$

Reduction of saturated compounds can be written as an overall displacement reaction (eq. 508), but free radical mechanisms are almost certainly involved.

$$R{-}Y + HSnR_3 \longrightarrow R{-}H + R_3SnY \quad (508)$$

G. ACID-BASE SYSTEMS (NO METALS)

The activation of molecular hydrogen by basic aqueous systems in the absence of metal complexes is evidenced by some reported isotope exchange with the deuterium in heavy water (H_2-D_2O or D_2-H_2O) and the conversion of parahydrogen.

In 1936 Wirtz and Bonhoeffer (1211) established the exchange reaction with 1 M alkali at 100°. The exchange has been studied subsequently by various other workers (217, 218, 1212–1216). Rate law 509 has been established from 80 to 190° at hydroxide concentrations of 10^{-5} to 1 M and hydrogen pressures up to 100 atm (217, 218, 1213–1216):

$$\frac{d[\text{HD}]}{dt} = k[\text{H}_2][\text{OH}^-] \tag{509}$$

Activation energies of 24 to 28 kcal mole^{-1} and activation entropies of about +5 eu have been estimated. The presence of added buffers did not affect the exchange rates at a given pH, thus showing that anions were not involved (1216). A reversible heterolytic splitting of hydrogen readily explains the rate law (1211, 1213):

$$\text{H}_2 + \text{OD}^- \xrightarrow{k} \text{H}^- + \text{HOD}; \text{DOD} + \text{H}^- \xrightarrow{\text{fast}} \text{OD}^- + \text{HD} \tag{510}$$

There was considerable discussion on the postulation of a free H^- ion as an intermediate, and this has been summarized by Halpern (25) and Weller and Mills (23). A later study (1214) in concentrated alkali revealed a close thermodynamic similarity between the transition state and hydroxides suggesting that heterolytic splitting is nearly complete in the transition state which closely resembles a solvated H^- ion. An alternative mechanism involving concerted attack on H_2 by OD$^-$ and D_2O with synchronous proton transfers (eq. 511) differs only in the lifetime of the intermediate H^- ion (25, 1213), and has been favored by Schindewolf (217, 218).

$$\text{DO}^- \boxed{\text{D}^+ + \text{H}^-} \boxed{\text{H}^+ + \text{OD}^-} \rightleftharpoons \text{DO}^- + \text{DH} + \text{HOD} \tag{511}$$

Mechanism 511 possibly predicts (1213) an analogous acid-catalyzed exchange as shown in eq. 512, but this was not observed when using dilute and concentrated acids (1213, 1216) or a BF_3-HF system that was studied by Mills and coworkers (1218).

$$\text{D}_2\text{O} \boxed{\text{D}^+ + \text{H}^-} \boxed{\text{H}^+ + \text{OD}_2} \rightleftharpoons \text{D}_2\text{O} + \text{DH} + \text{HOD}_2 \tag{512}$$

The rates of the hydroxide-catalyzed parahydrogen conversion were up to twice those of the D_2-H_2O exchange (1213), showing the same small isotope effect observed with corresponding metal complex catalyzed reactions (Sections II-A, IX-B-1, and X-A-2).

Bar-Eli and Klein (1217) and Schindewolf (217, 218) have shown that D_2-CH_3OH exchange is catalyzed by methoxide and hydroxide in an analogous manner; an activation energy of 24 kcal mole^{-1} and activation entropy of $+4$ eu were reported.

Analogous exchange and conversion reactions of hydrogen catalyzed by NH_2^- in liquid ammonia were first studied by Wilmarth and coworkers (1215, 1222, 1223), and later by Bar-Eli and Klein (1217, 1219), Bourke and Lee (1220), Dirian and coworkers (1221), and Bigeleisen (1224). A rate law corresponding to 509 with analogous interpretations was reported by Wilmarth's group (1223); the NH_2^--catalyzed reaction at $-50°$ is some 10^4 times more rapid than the OH$^-$-catalyzed reaction at $100°$, and by assuming equal frequency factors for the two reactions an activation energy of about 10 kcal mole^{-1} was estimated for the former. This was attributed to much greater base strength of NH_2^- relative to OH$^-$, and gives some support for the hydride mechanism. Exchange in the very weakly basic acetic acid/acetate system is reported to have an activation of at least 35 kcal mole^{-1}. (217, 218).

The conditions in some of the studies (1220, 1221, 1224) led to diffusion control rates. Bar-Eli and Klein (1217, 1219) observed an exchange rate independent of deuterium pressure, but interpreted the data via a mechanism similar to that of eq. 511; they reported $\Delta H^{\ddagger}$ and $\Delta S^{\ddagger}$ values of 7.5 kcal mole^{-1} and -9.2 eu, respectively. Amide-catalyzed exchange studies with a number of other liquid aliphatic amines yielded activation energies in the range 3.5 to 6.7 kcal mole^{-1} (1219).

The dependence on amide concentration in a number of these studies (1217, 1219, 1221) was not strictly first-order, and slight ordinate intercepts on the rate: concentration plots were attributed to a contribution to the exchange catalyzed by the neutral amide (e.g., KNH_2).

These base-catalyzed systems do not in general catalyze the hydrogen reduction of dissolved substrates (25), but Walling and Bollyky (1225, 1226) have reported hydrogenation of benzophenone to benzhydrol at 130 to 230° and 75 to 180 atm when using t-butoxide as catalyst in t-butyl alcohol, benzene, or preferentially diglyme as solvent, presumably via a hydride mechanism:

$$RO^- + H_2 \rightleftharpoons ROH + H^-; H^- + R_2CO \rightleftharpoons R_2CHO^-;$$

$$R_2CHO^- + ROH \rightleftharpoons R_2CHOH + RO^- \quad (513)$$

Acetone and cyclohexene were not reduced, but nitrobenzene was partially converted into aniline. The reductions were more rapid at higher base: substrate ratios. A very slow reduction of benzophenone was observed using an aqueous solution of concentrated potassium hydroxide. These workers also suggested the possibility of acid-catalyzed reductions via a carbonium ion intermediate; for example,

$$ROH + H^+ \rightleftharpoons R^+ + H_2O \tag{514}$$

$$R^+ + H_2 \rightleftharpoons RH + H^+ \tag{515}$$

However, triphenylcarbinol and triphenylchloromethane in acid solvents did not absorb hydrogen under severe conditions, although benzene, cyclohexene, and isobutene in chloroform-containing Lewis acids, such as aluminum tribromide, to obtain a carbonium ion, were said to show some reactivity. These authors have suggested that the hydrogenolysis of organometallic compounds may involve attack of hydrogen on a strongly basic carbanion or carbanionoid moiety (cf. eq. 515).

Of interest here is the report by Brongersma and coworkers (1227) that the pentamethylbenzyl cation (A^+), dissolved in a strong proton acid, catalyzes the hydrogenation of quinones (and aromatic radical ions) by alkanes at low temperatures.

$$+ \; C_nH_{2n+2} + H_2SO_4 \longrightarrow \tag{516}$$

$$+ \; C_nH_{2n+1}SO_4H$$

(AH_2^+)

Hydride transfer from the alkane to A^+ followed by addition of a proton from the acid yields protonated hexamethylbenzene, which readily transfers hydrogen to the substrate with regeneration of A^+. Reaction 516, which may be compared to 515 appears to be an organic reaction approaching those being sought for activation of saturated hydrocarbons by transition metal systems (cf. Sections X-G and XIII-C-1).

H. PHOTOCHEMICAL HYDROGENATION IN THE ABSENCE OF METAL COMPLEXES

Photochemically initiated catalyzed hydrogenations have occasionally been reported for transition metal complex systems (e.g., see Sections IX-B-2c and XII-B-1).

Elad and Rosenthal (1277) have reported that irradiation of uracil and 1,3-dimethyluracil in propan-2-ol, or in an aqueous solution containing (±)-methionine or EDTA, leads to hydrogenation of the 5,6-double bond:

$$\xrightarrow{hv} \tag{517}$$

R = H or Me

The hydrogen source is thought to be the alcohol or water. Photochemically initiated reductions of uracil in the presence of compounds containing SH groups (e.g., 1278) and in the presence of borohydride (e.g., 1279) have also been reported.

I. HYDROGENASE

A variety of microorganisms have the ability to activate hydrogen for exchange with water, para-ortho conversion, and reduction reactions. This ability was ascribed to an enzyme or system of enzymes named *hydrogenase*, which can exist in soluble of particulate forms: even in the latter form its behavior resembles that of homogeneous catalysts and is thus pertinent to the present theme.

This section will briefly survey the interesting but very complex area of hydrogen activation. In this section alone, the author has not always referred to the original papers; the references listed are not exhaustive and are presented mainly to introduce readers to this biochemical area. These enzyme-catalyzed reactions have been reviewed in English up to the 1960 literature from a chemical standpoint by Weller and Mills (23), Halpern (25), and Volpin and Kolomnikov (57d). A Japanese review by Yagi and Ino-kuchi (1228) appeared later. The biological role of hydrogenase was reviewed in 1954 by Gest (1229).

Strickland and coworkers (1230–1233) first recognized hydrogen activation by microorganisms. Studies by various workers (136, 1234–1238, 1240–1244) suggest that most enzyme preparations contain an active iron site, almost certainly in the ferrous state and possibly with a coordinated thiol group. X-Ray fluorescence spectroscopy (1242) has revealed iron as the only metal component in one extract. Molybdenum may sometimes be an active component (1235, 1236), although the need for it has rarely been demonstrated (1244). Kleiner and Burris (1455) and Peive and coworkers (1245), however, have reported that molybdenum, copper, and especially cobalt increase the activity of some systems.

Deuterium exchange and parahydrogen conversion have been studied by many groups including Rittenberg and coworkers (1238, 1240, 1241, 1246–1248), Farkas and coworkers (1249–1251a), Gest and coworkers (1235, 1252), Cavanagh and coworkers (1253), Hyndman and coworkers (1254), Pethica and coworkers (1255), Couper and coworkers (1256, 1257), Kleiner and Burris (1455), and Tamuja and Miller (1258). Hydrogen reduction of ferricyanide and dyes such as methylene blue and viologens, and the catalyzed evolution of hydrogen from reduced viologen, have been studied by many of these groups (136, 1235–1238, 1240–1248, 1259–1262, 1455), Back and coworkers (1263), Joklik (1264), Shturm (1265), King and Winfield (354), Curtis and Ordal (1266), and Chang and Wolin (1267). Hydrogenation of fumarate, and reduction of oxygen and simple inorganic anions such as nitrate and sulfate have also been studied (e.g., 1251, 1263, 1265).

Inhibitors of hydrogenase include particularly carbon monoxide, cyanide, nitric oxide and various pentacyano compounds; such studies have been reported on by Kempner and Kubowitz (1268), Bone (1269), and many of the workers mentioned above (e.g., 1238, 1240, 1243, 1254, 1260, 1262, 1270–1272).

The observed catalyzed reactions probably involve a similar mechanism, but their rates for any given hydrogenase preparation vary considerably, and the ratio of the reduction to the exchange rates varies for different preparations. The conversion and exchange rates, which are retarded by the same inhibitors, are generally similar although one conversion rate has been reported as less than one-tenth the exchange rate (1258). The rates have been found to be proportional to the extract concentration at low concentration, where activation energies of 7 to 10 kcal mole^{-1} have been estimated; at higher concentration, the rates level off and the activation energy decreases, perhaps because of some diffusion control mechanism. Water is essential for the reactions although the exchange and conversion reactions are strongly inhibited in D_2O, suggesting that exchange with hydrogen is important.

In D_2-H_2O exchange experiments in the presence of substrate, the exchange proceeds at the normal rate but only after the reduction reaction is completed, suggesting a common intermediate for the two processes (1246, 1247). On the basis of mainly HD production in H_2-D_2O exchange, the now familiar heterolytic splitting of H_2 by the enzyme (E) was postulated (1240, 1247):

$$E + H_2 \xrightarrow{k_1} EH^- + H^+ \tag{518}$$

$$EH^- + D^+ \xrightarrow{k_2} E + HD \tag{519}$$

In the presence of a substrate (S), hydride transfer and subsequent protonation yields the reduction product:

$$EH^- + S + H^+ \xrightarrow{k_3} E + SH_2 \tag{520}$$

Later workers (1258) studied the D_2/HD product ratio as a function of enzyme concentration and pH, and modified the mechanism in that EH^- may also exchange with the solvent (cf. Section IX-B-1).

Rittenberg (136) has suggested that a basic site in the enzyme stabilizes the released proton of reaction 518 as shown in eq. 521, and such a model

$$Fe^{2+} \quad :B + H_2 \rightleftharpoons FeH^+ \quad H:B^+ \tag{521}$$

readily explains an observed maximum activity with varying pH, (e.g., 136, 1248, 1258, 1263), since at low pH the basic site may be neutralized, while at high pH the ferrous may be hydrolyzed (see also Section II-B; Fig. 1).

Gest and coworkers (1235, 1259) have proposed a homolytic splitting of hydrogen at two iron centers with subsequent transfer of the hydrogen atoms to a flavin moiety with the isoalloxazine nucleus, which functions as the oxidation-reduction center for two-electron acceptors such as methylene blue. One-electron processes (exchange, viologen reduction) are thought to involve a second metal, possibly molybdenum. Such mechanisms can explain the fact that some hydrogenase preparations give very low exchange rates compared with hydrogenation of methylene blue (1252, 1254, 1255), which is not readily explicable in terms of eqs. 518 to 520.

Systems in which the substrate reduction is slower than exchange or conversion can be rationalized in terms of (a) poisoning of the enzyme by the substrate, particularly oxidants (see below) or (b) k_3 being $<k_2$ in eqs. 518 to 520.

The two mechanisms of hydrogen activation have been discussed further more recently (1455).

Fumarate reduction may involve other enzyme systems that activate the substrate (25).

King and Winfield (354) compared the mode of hydrogen activation by a soluble hydrogenase to that of pentacyanocobaltate(II) which involves two cobalt centers (Section X-A-2). The uptake of hydrogen by $Co(CN)_5^{3-}$ is inhibited by dyes particularly methylene blue, due to irreversible oxidation to cobalt(III), although some catalyzed reduction may still occur depending on concentrations (see Section X-B-2m). Similar deactivation was noted for the hydrogenase system. Dyes had no inactivating effect on particulate hydrogenase, and it was suggested that these were screened from attack by larger molecules (but not hydrogen) by protein or lipid.

The studies of Khidekel and coworkers illustrating the similarity of cobaloxime complexes and some rhodium analogs to hydrogenase, in regard to properties of hydrogen transfer, were discussed in Sections X-H, XI-G, and XI-H.

J. NITROGENASE AND NITROGEN FIXATION

The intriguing problem of nitrogen fixation, the role played by nitrogenase, and the chemistry of the simpler model, molecular nitrogen complexes have evoked a great deal of interest. A significant number of reviews have appeared; for example, those of Shilov (1294), Fergusson and Love (1016), Martino and Sajus (1017), Schrauzer (1068), Allen (1469), Van Tamelen (1470), and Hardy and coworkers (1471). Papers by Shilov and coworkers (1467) and Schrauzer and Doemeny (1468) seem particularly significant.

Nitrogen and other substrates reducible by the nitrogenase system have been reduced by inorganic model systems, including the conversion of nitrogen to ammonia. To date, the source of hydrogen in the ammonia production has been shown to arise from a protonic solvent and not molecular hydrogen.

K. RELATION BETWEEN HOMOGENEOUS AND HETEROGENEOUS CATALYSIS

The topic of relating homogeneous and heterogeneous catalytic systems is an extremely popular one, and indeed always has been. Such a comparison even for hydrogenation would constitute a further review in itself and will not be dealt with further. Some of the studies presented here on hydrogenation have drawn analogies to heterogeneous systems: 57b (Section XVI-A), 128 (Section II-D), 322 (Section X-B-2b), 637 (Section XI-B-3), 644 (Section IX-B-2b), 802, 803 (Section XI-H), and 860 (Section XII-A-1); in addition, some of the more recent and significant references in this area are 13a, 40, 58d, 68a, 79c, 158–160, 240a, 807, 869, 1030, 1103, and 1474.

XVII

Summary

As regards hydrogenation of unsaturated organics by using molecular hydrogen, both "unsaturate" and "hydride" routes have been postulated. The former, involving initial complexation of the unsaturate at the metal center, is exemplified by systems such as those involving ruthenium(II) chlorides (Sections IX-B-2a and IX-B-2c), some $RhCl(CO)L_2$ systems, where L is an unsaturated tertiary phosphine (Section XI-E), some rhodium(I) chloride and sulfide complexes (Section XI-H), $IrCl(CO)(PPh_3)_2$ and analogs (Section XII-A-2), platinum(II) chlorides (Section XIII-C-1), and some metal stearates (Section XVI-A). The latter route involving prior formation of either mono- or dihydrides has been invoked to a much greater extent and is exemplified by systems such as those involving $HRuCl(PPh_3)_3$ (Section IX-B-2b); ruthenium(I) chlorides (Section IX-B-2c); cobalt(II) cyanides (Sections X-A to X-D); cobalt carbonyls and phosphine complexes (Sections X-E to X-G, XIV-C); rhodium(III) ammines (Section XI-A-1); $HRh(CO)$-$(PPh_3)_3$ (Section XI-D); rhodium carbonyls and some complexes containing phosphine type ligands (Sections XI-F and XI-G); rhodium(I,II) carboxylates (Sections XI-H and XI-I); some iridium(I) carbonyl phosphine complexes (Sections XII-A-2 and XII-B-2); some palladium(II) and platinum(II) phosphine complexes (Sections XIII-B, XIII-C-3b, and XIV-F-2); metal acetylacetonates (Section XIV-E), and Ziegler-type catalysts (Sections XIV-E and Chapter XV). It is difficult to show that a particular route occurs exclusively. Both paths have been demonstrated for the $RhCl(PPh_3)_3$ complex (Section XI-B), and both have been invoked for iron and chromium carbonyl catalysts (Sections VII-A, IX-A, XIV-B, and XIV-D), and platinum(II)-trichlorostannate(II) systems (Section XIII-C-2).

Scheme 19 summarizes the various reaction schemes that have commonly been presented for unsaturate routes (steps 4 or 5) and hydride routes (steps 6 or 7) for monoolefin reduction; the scheme is readily modified for hydrogenation of acetylenes and polyenes when vinyl and σ- or π-allyl intermediates, respectively, may be involved:

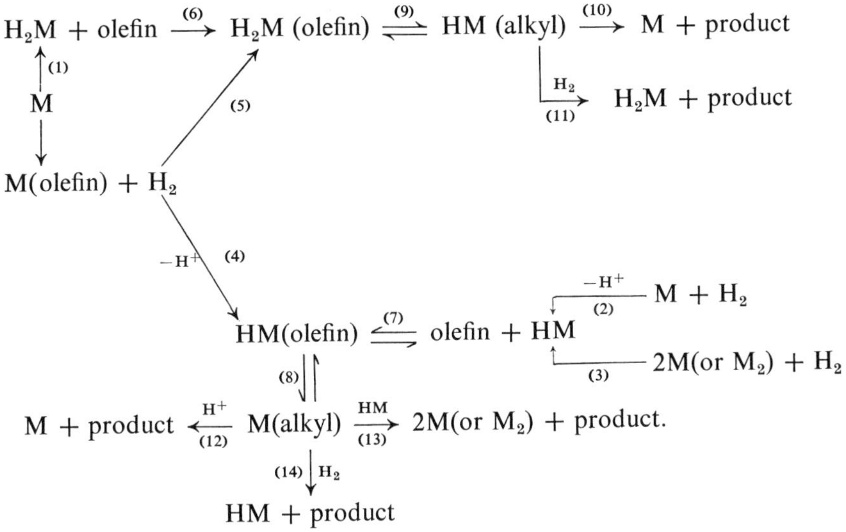

$$H_2M + olefin \xrightarrow{(6)} H_2M \text{ (olefin)} \underset{(9)}{\rightleftharpoons} HM \text{ (alkyl)} \xrightarrow{(10)} M + product$$

$$\uparrow (1)$$

$$M \qquad\qquad (5) \qquad\qquad \downarrow H_2 \atop (11) \quad H_2M + product$$

$$\downarrow$$

$$M(olefin) + H_2$$

$$-H^+ \searrow (4) \qquad\qquad\qquad -H^+ \atop (2) \quad M + H_2$$

$$HM(olefin) \underset{(7)}{\rightleftharpoons} olefin + HM \qquad (3) \quad 2M(or\ M_2) + H_2$$

$$(8)\uparrow\downarrow$$

$$M + product \xleftarrow[(12)]{H^+} M(alkyl) \xrightarrow[(13)]{HM} 2M(or\ M_2) + product.$$

$$(14)\downarrow H_2$$

$$HM + product$$

Scheme 19

The classification of metal hydride formation via molecular hydrogen was outlined in the introduction (Section I-B). Dihydride formation via step 1 is well documented, particularly for d^8 metal systems involving tertiary phosphine ligands (e.g., see Sections X-G, XI-B-1, XI-G, XII-A-1, and XII-B). Step 5 has been invoked for example, in hydrogenations using RhCl(PPh$_3$)$_3$ (Section XI-B-1); rhodium(I) chlorides and sulfides (Section XI-H); and IrCl(CO)(PPh$_3$)$_2$, (Section XII-A-2). Monohydride formation via an overall homolytic splitting of hydrogen (step 3) is exemplified by certain copper(I) and silver(I) systems (Sections II-A, II-C, and II-D); cyanocobaltate(II) complexes (Sections X-A-2, X-C, and X-D); Co$_2$(CO)$_8$ and some phosphine derivatives (Sections X-E-1 and X-F); some cobaloximes (X-H); and [(PPh$_3$)Rh(DMGH)$_2$]$_2$ (Section XI-G). An overall heterolytic splitting of hydrogen to give monohydride (step 2) has been demonstrated or invoked for a large number of systems which include some copper and silver complexes (Sections II-A to II-D); ruthenium(II,III) chlorides (Section IX-B-1); RuCl$_2$(PPh$_3$)$_3$ (Sections IX-B-2b and IX-B-2e); some cobalt tertiary-phosphine and phosphite complexes (Section X-G), rhodium(III) complexes (Sections XI-A-1, XI-B-9, and XI-H); and palladium(II) and platinum(II) complexes (Sections XIII-B and XIII-C-3). As discussed in the appropriate sections, steps 2 and 3 for certain systems at least, almost certainly proceed via dihydride intermediates (e.g., Sections II-A to II-D, IX-B-2e, XI-E, XIII-B, and XIII-C). Heterolytic activation of hydrogen by a metal olefin complex (step 4) is best exemplified by the chlororuthenate(II) system (Section IX-B-2a).

Radical mechanisms for activation of molecular hydrogen involving transfer of hydrogen atoms from the catalyst to the substrate appear to have been considered only for certain of the pentacyanocobaltate(II) systems (Section X-B).

In Scheme 19, steps 4 to 6 are frequently rate-determining and steps 7 to 9 are almost invariably fast; hence hydridoolefin complexes are difficult to detect in catalytic systems, although some such monohydride species are known: $HRuCl(C_7H_8)(PPh_3)_2$ (Section IX-B-2b); $[HIrCl_2(C_8H_{12})]_2$ (Section XII-C); $HPtCN(C_6N_4)(PEt_3)_2$ and $HPt(C_2H_4)(PEt_3)_2^+$ (Section XIII-C-3a); and some complexes of platinum(II)-trichlorostannate(II) systems (Section XIII-C-3b). Some of these complexes have been formed by reaction of olefin with metal hydride (step 7). The production of metal alkyls via such a reaction is well documented and is generally thought to proceed via the hydridoolefin complex (step 8), although some such reactions for $HCo(CN)_5^{3-}$ appear not to require prior coordination of olefin (Section X-B-4).

Intermediate $H_2M(olefin)$ species, formed via steps 5 or 6, generally appear to transfer both coordinated hydrides to the olefinic substrate (steps 9 and 10), as exemplified by systems involving some chromium carbonyl complexes (Sections VII-A and XIV-D); $RhCl(PPh_3)_3$ (Section XI-B); $RhCl(CO)-[Ph_2P(CH_2)_2CH = CH_2]_2$ (Section XI-E); some cationic rhodium and iridium complexes (Sections XI-G and XII-B-1); and $IrCl(CO)(PPh_3)_2$ (Section XII-A-2). Rather than transfer of the second hydrogen, a final step involving hydrogenolysis to product and the dihydride catalyst (step 11) has been invoked for an iron carbonyl system (Section XIV-B).

Formation of the saturated product from a M(alkyl) species can involve protonolysis (step 12), reaction with metal hydride (step 13), or hydrogen-olysis (step 14). Step 12 is exemplified by systems involving platinum(II) alkyls (Sections XIII-C-2 and XIII-C-3a); the $HIrCl_2(DMSO)_3$ catalyst (Section XII-C); $HCo(CN)_5^{3-}$ (Section X-B-2a); chlororuthenate(II) complexes (Section IX-B-2a); and the $MoCl_5-SnCl_2$ catalyst (Section VII-B). Step 13 has been frequently postulated, for example, for some ruthenium systems (Section IX-B-2d); $HCo(CN)_5^{3-}$ (Section X-B-2a); some $Co_2(CO)_8$ systems (Sections X-E-2 and X-E-3); some cobalt carbonylphosphine complexes (Section X-F-2); metal acetylacetonates (Section XIV-E); and for some Ziegler systems (Chapter XV). Diene substrates give rise to σ and/or π-allyl intermediates, and a number of the catalyst systems refer to reaction of these with HM. Step 14, which may be considered to be a specific example of step 2, is exemplified by systems involving tetrabenzylzirconium (Section V-B); $HRuCl(PPh_3)_3$ (Section IX-B-2b); some $Co_2(CO)_8$ systems (Section X-E-3); some cobalt phosphine and carbonylphosphine complexes (Sections X-F-2 and X-G); $HRh(CO)(PPh_3)_3$ (Section XI-D-1); some rhodium(I) carboxylates (Section XI-H); $HIr(CO)(PPh_3)_3$ (Section XII-B-2); some palladium(II) and platinum(II) complexes (Sections XIII-B, XIII-C-3b, and XIV-F-2); Ziegler catalysts (Section XIV-E and Chapter XV); boranes (Section XVI-D); and aluminum hydrides (Section XVI-E). Hydrogenolysis of π-allyl complexes has similarly been invoked, for example, for some cobalt and platinum allyls (Sections X-F-2 and XIII-C-3b) and iron allyls Section XIV-B). Product formation with regeneration of the catalyst M has lso been postulated for hydrogenolysis of metal alkyls, for example, with itanium alkyls (Sections V-A and XV-B), and hydrogenolysis of π-allyls, or example, with some cobalt allyl complexes (Section X-F-2).

Hydrogenolysis of acyl complexes to give aldehyde product and the metal hydride is important for reduction under hydroformylation conditions (Sections X-E, X-F-1, XI-F-1, XI-F-2, and XII-D), and hydrogenolysis of alkoxide complexes may be involved in catalytic reduction of aldehydes to alcohols (e.g., see Sections X-E-2 and XI-F-1).

As discussed under the separate systems, hydrogenolysis reactions are commonly thought to occur via oxidative addition of the H_2 molecule.

The importance of ligand dissociation from a complex to create sites for hydrogen and/or substrate coordination is well established. Innumerable examples include systems involving carbonyls of Group VI, iron, and cobalt (Sections VII, IX-A, X-E, X-F, and XIV-B), and systems involving platinum metal complexes, particularly phosphine complexes of ruthenium, rhodium, and iridium (Sections IX-B-2b, XI-B to XI-H, XII-A to XII-D). Closely related is the importance of solvent effects, which have been studied in reasonable detail for some systems involving copper, silver, and mercury salts (Sections II-A to II-D and III-B); $Co(CN)_5^{3-}$ (Section X-C); a number of rhodium complexes (Sections XI-B-11, XI-B-12, XI-G, and XI-H); $IrCl(CO)(PPh_3)_2$ (Section XII-A); and some platinum(II)-trichlorostannate-(II) complexes (Sections XIII-C-2 and XIII-C-3b); not surprisingly, in view of the multistep nature of the hydrogenation process, they generally appear complex.

Electronic ligand effects are not easy to predict. The stability and lability, for example, of intermediate hydride, olefin, hydridoolefin, alkyl complexes, and so on, will be important, and as with solvent effects, ligand effects can manifest themselves in the relevant reaction steps shown in Scheme 19, including any necessary dissociation reactions. The intermediates must be sufficiently stable to be formed, but not so stable as to prevent further reactivity. Simple examples of ligand effects on dissociation reactions are the observed rate increases for arene-chromium carbonyl complexes and butadiene-iron carbonyl complexes with introduction of electron-withdrawing substituents on the unsaturated moiety, which facilitates dissociation to active metal tricarbonyl intermediates (Sections VII-A and IX-A).

Hydrides of transition metals in low valency states are known to be stabilized by π-acceptors such as carbon monoxide and phosphines and formation of dihydrides at least, which are stabilized by more basic phosphine ligands (e.g., Sections XI-B-7, XII-A-1 and XII-B-3), may be pictured as interaction of a Lewis base metal center with a Lewis acid hydrogen molecule (Section XII-A-1); increasing π-acceptor property in ligands decreases the ease of dihydride formation (e.g., Sections X-G, XI-E, and XII-A-1). Reactivity trends for hydrogenation of organic substrates with dihydride systems have demonstrated increasing activity with both increasing and decreasing basicity of phosphines (e.g., Sections XI-B-7, XI-G, and XII-A-2); arsine and stibine systems are generally, but not always, of lower reactivity than the corresponding phosphine ones (e.g., Sections XI-B-7, XI-E, and XII-B-1); phosphite systems, stronger π-acids than corresponding phosphines, are generally poorer catalysts than the phosphine ones (e.g., Sections XI-B-7b and XI-E), although the reverse has been noted (Section XII-A-2). In monohydride complexes, introduction of stronger σ-donors and weaker π-acceptors reduces the acidity of the metal hydride; this has given rise to both increased hydrogenating activity, as noted for some carbonyl phosphine systems (e.g., Sections X-F and XI-F-2), and decreased activity for some platinum(II)-chlorostannate-phosphine catalysts (Section XIII-C-3b); arsine systems have been both more and less active than corresponding phosphine ones (e.g., Sections XIII-C-3b and X-F).

Replacing a chloride ligand by the more polarizable and basic bromide and iodide ligands has again given rise to both increasing and decreasing activity trends in platinum metal systems (see, e.g., Sections XI-A, XI-B-6, XII-A, XIII-B, and XIII-C-3b).

More predictable ligand effects have sometimes been realized by grosser modification of the coordinated ligands as, for example, in the production of less stable and more reactive hydrides on the replacement of cyanide ligands in $Co(CN)_5^{3-}$ by nitrogen donor ligands (Section X-D).

Some earlier work on formation of intermediate monohydrides by complexes of copper, silver and mercury indicated increasing reactivity with increasing ligand basicity and was rationalized in terms of opposing effects of proton affinity and complexation tendency of the ligand (Sections II-A to II-D and III-B).

The most significant steric ligand effects have been reported for tertiary phosphine type complexes of Group VIII metals (e.g., see Sections IX-B-2b,

X-F-2, XI-B-7, XI-D-1, XI-F-2, and XII-A-1); selectivity control by inhibiting hydride transfer to coordinated substrate seems particularly important.

Early views on hydrogen activation (Section XII-A-1) had suggested that bonding electrons of hydrogen could attack vacant metal orbitals, since the active species discovered to that time (1960) were d^8 to d^{10} metal systems—copper(I,II), silver(I), mercury(II), palladium(II)—which either had a vacant d orbital or made one available by low energy electronic transitions. Halpern's arguments (25, 26) had suggested a correlation between activity of divalent metal ions and their "electron affinities" (as measured by the sum of the first two ionization potentials of the atoms), but such speculations have met with limited success in view of later discoveries on the activity of earlier transition metal species, for example, the d^6 ruthenium(II) species. More useful reactivity correlations generally are likely to be found by comparison with organic species (saturated species, free radicals, carbenes, carbonium ions, and carbanions) and their tendency to achieve closed-shell configurations (37, 62).

The use of multicenter catalysts (dimers, polymers), particularly in respect of more meaningful models for comparison with heterogeneous systems, has been limited thus far. Of those studies in some detail, most appear to require dissociation to an active monomeric form, for example, $[C_5H_5Cr(CO)_3]_2$ (Section VIII-A); Ru_2^I and $[C_6H_6RuCl_2]_n$ (Section IX-B-2c); $Co_2(CO)_8$ and polynuclear cobalt carbonyl phosphines (Sections X-E and X-F); $[(Ph_3P)\text{-}Rh(DMGH)_2]_2$ (Section XI-G); and $[MCl(C_8H_{14})_2]_2$, M=Rh, Ir (Sections XI-H and XII-C). Activity has been attributed to dimeric species in some systems involving, for example, some dimeric chromium tricarbonyl complexes (Section VII-A); rhodium(I) chlorides (Section XI-G); $Rh_2(CO_2Me)_4$ (Section XI-I); and $PtCl_2(PPh_3)_2\text{-}SnCl_2$ (Section XIII-C-3b). A number of binuclear compounds containing two different metals have been reported to be active (e.g., see Chapter VII and Sections IX-A and XIII-B), but the systems have not been studied in detail.

Electronic and steric effects of substituents on the substrates are generally discussed to varying extents for most catalytic systems reported, and because of the variety of mechanisms considered it is very difficult to draw general conclusions. The most extensive studies of this kind have been carried out for systems involving $HRuCl(PPh_3)_3$ (Section IX-B-2b); $Co(CN)_5^{3-}$ (Sections X-B-2 and X-B-4); $Co_2(CO)_8$ (Section X-E); $RhCl(PPh_3)_3$ (Sections XI-B-3 and XI-B-5); $HRh(CO)(PPh_3)_3$ (Section XI-D); $IrCl(CO)(PPh_3)_2$ (Section XII-A-2); $PtCl_2(PPh_3)_2\text{-}SnCl_2$ (Section XIII-C-3b); and some Ziegler catalysts (Chapter XV).

A considerable number of metal-catalyzed homogeneous hydrogenation systems not involving molecular hydrogen have been reported. The solvent is commonly the source of hydride (or proton); borohydride has been used either as a hydride source or for reduction of the valency state of a metal, and hydrogenation can then occur via electron transfer/protonation processes. The hydridic mechanisms are exemplified by some systems involving cyano-cobaltate(II) complexes (Sections X-A to X-D); some rhodium(III) complexes with nitrogen donors or cyanide (Section XI-A-1); some rhodium-(I) carbonyl phosphine complexes (Section XI-D-1); some iridium phosphine and phosphite complexes (Section XII-B-1); iridium(III)-DMSO complexes (Section XII-C), some metal acetylacetonates (Section XIV-E); and some phosphine complexes of nickel(II), palladium(II), and platinum(II) in the absence and presence of stannous chloride (Section XIV-F-2). Protonation-redox mechanisms are exemplified by some systems involving $TiCl_3$, (Section V-A); vanadium(II) salts (Chapter VI); chromium carbonyls (Section VII-A); chromium(II) salts (Section VII-A); and $Co(CN)_5^{3-}$ (Sections X-B-2e and X-B-5). Distinction between the two basic mechanisms is not always easy as can be seen by consideration of the cobalt cyanide systems and also some systems involving iron carbonyls (Section IX-A), cobaloximes (Sections X-B-5 and X-H), and cyanonickelate(I) species (Section XIII-A). Ortho positions of phenyl groups of coordinated tertiary phosphine and phosphite ligands also provide a possible source of hydrogen for appropriate metal systems (e.g., see Scheme 3 in Section IX-B-2b and Sections XI-G and XIV-F-2), although for a *catalytic* process, the hydrogen would have to be refurnished, probably by the use of molecular hydrogen.

In addition to carbon-carbon unsaturated bonds in aliphatic, aromatic, and heterocyclic compounds, common heteronuclear unsaturated bonds such as $>C=O$, $-C\equiv N$, $>C=N-$, $-NO$ and $-NO_2$, and $-N=N-$ have all been effectively catalytically reduced, although severe conditions of temperature and hydrogen pressure are sometimes necessary. The trend, of course, is to develop catalysts active under mild conditions (~ 1 atm hydrogen) and with greater selectivity, and considerable progress has been made in the case of olefinic substrates. The number of catalysts effective *under mild conditions* for hydrogenation (with or without hydrogen) of the other substrates is more limited. For example, acetylenes have been reduced in this way using chromium(II) species (Section VII-A); $HRuCl(PPh_3)_3$ (Section IX-B-2b); $RhCl-(PPh_3)_3$ (Section XI-B-10); $HM(CO)(PPh_3)_3$, M=Rh, Ir (Sections XI-D-1 and XII-B-2); $MCl(CO)(PPh_3)_2$, M=Rh, Ir (Sections XI-E and XII-A-2); $py_2(DMF)RhCl_2(BH_4)$ (Section XI-H); rhodium(I) carboxylate complexes (Section XI-H); and some Ziegler catalysts (Section XV-C). Similarly, carbonyl groups have been reduced using chromium(II) species (Section VII-A); $Co(CN)_5^{3-}$ (Sections X-B-2c and X-B-2e); cationic rhodium phosphine complexes (Section XI-G); iridium phosphine and phosphite complexes (Section XII-B-1); and iridium-DMSO complexes (Section XII-C).

Mild reduction of aromatics and heterocyclics has been effected using some rhodium complexes (Sections XI-B-8 and XI-H); the most effective catalysts for reduction of unsaturated nitrogen moieties appear to be $Co(CN)_5^{3-}$ (Sections X-B-2f and X-B-5) and $py_2(DMF)RhCl_2(BH_4)$ (Section XI-H).

In nearly all cases studied, the overall addition of hydrogen to carbon-carbon double and triple bands appears to be *cis*; this has been demonstrated for systems involving arene-chromium tricarbonyls (Sections VII-A and XIV-D); ruthenium(I,II) chlorides (Sections IX-B-2a and IX-B-2e); HRuCl-$(PPh_3)_3$ (Section IX-B-2b); $Co_2(CO)_8$ (Sections X-E-2 and X-E-3); RhCl-$(PPh_3)_3$ and analogs (Sections XI-B-3, XI-B-4, XI-B-5c, XI-B-10, and XI-C); $py_2(DMF)RhCl_2(BH_4)$ (XI-H); rhodium(I) carboxylates (Section XI-H); iridium-DMSO complexes (Section XII-C); and some Ziegler catalysts (Section XV-A). *Trans* addition products have been observed for some systems involving $Co(CN)_5^{3-}$ (Section X-B-2b); platinum(II)-tin(II) halides (Section XIII-C-2); and $py_2(DMF)RhCl_2(BH_4)$ (XI-H). In the platinum system *cis* addition products were also formed, and in the rhodium system the product resulted from an intermediate formed by an initial *cis* addition process.

A recent interesting development is that of asymmetric hydrogenation, which has been effected by using metal complexes containing optically active ligands (Section XI-C).

Although kinetic isotope effects, when using gaseous deuterium in place of hydrogen, have been studied for a wide range of metal complex systems, they appear thus far to be invariably small ($k_H/k_D = 0.9$ to 1.5), and contribute little to the distinction between the different mechanisms for hydrogen activation. They do suggest, however, that metal-hydride bond formation and hydrogen-hydrogen bond breaking occur to some extent via a concerted process. Such isotope effects have been measured for systems involving copper(I) (Section II-A); copper(II) (Section II-B); methylbenzoate chromium tricarbonyl (Section VII-A); chlororuthenate(II,III) complexes (Sections IX-B-1 and IX-B-2a); chlorocarbonylruthenate(II) (Section IX-B-1); HRuCl(PPh_3)_3 (Section IX-B-2b); ruthenium(I) chlorides (Section IX-B-2c); $Co(CN)_5^{3-}$ (Section X-A-2); RhCl(PPh_3)_3 (Sections XI-B-1 and XI-B-4); HRh(CO)(PPh_3)_3 (Section XI-D-1); $H_2Rh(PPhMe_2)_2$ (solvent)$_2^+$ (Section XI-G); rhodium(I) chlorides and sulfides (Section XI-H); and IrCl(CO)-$(PPh_3)_2$ (Section XII-A-1). A somewhat larger isotope effect of 2.1 has been measured for loss of molecular hydrogen from a cationic $H_2Ir(CO)_2(PPh_2-Me)^+$ complex (Section XII-B-3). Rates of reaction of hydrogen with species in H_2O and D_2O have also been measured (1479); values of k_{H_2O}/k_{D_2O} for reaction with $Cu(OCOCH_3)_2$, MnO_4^-, chlororhodate(III) complexes, and $PdCl_4^{2-}$ were 0.9 to 1.0, while for reaction with the aquo cations Cu^{2+}, Hg^{2+}, Hg_2^{2+} and Ag^+, the value was 1.25 ± 0.1.

In conclusion, this review was intended to be as exhaustive as possible, but a noteworthy fact is that, of the almost 1600 papers and patents referred to (up to the end of 1970), about 45% have appeared since the beginning of 1968. This not only reflects the tremendous increase in studies of homogeneous hydrogenation, but also the upsurge of homogeneous catalysis generally and the simultaneous advances in organometallic chemistry. A fairly detailed picture of the catalytic hydrogenation process has been obtained for some systems. The continuing trend will almost certainly be the development of new catalysts incorporating perhaps novel ligand species with an aim for selective hydrogenation under mild conditions.

The relatively new n.m.r. technique of heteronuclear double resonance, which can furnish the effects of ligand configuration upon metal chemical shifts (e.g., 1480, 1481), could prove a sensitive probe generally for the study of homogeneous systems in which metal complexes function as stereospecific catalysts. The potential of more theoretical studies, such as those based on the Woodward-Hoffmann orbital symmetry conservation rules (e.g., 1482–1484), toward the understanding of metal-catalyzed hydrogenation reactions remains to be fully realized.

Appendix

1971 and Part 1972 Literature

This appendix serves to cover the most recent literature—1971 and part of 1972—on homogeneous hydrogenation. A few earlier studies, which were missed during the attempted coverage to the end of 1970 but have now come to the author's attention, are also mentioned.

The material is arranged by the chapter and section in which the topic is discussed. For example, studies dealing with rhodium systems can be found under the headings XI-A, XI-B, and so on.

I

A review of homogeneous hydrogenation and dehydrogenation has been published by Kwiatek (1485), and a text on transition metal hydrides includes articles by Schunn (1486) and Tolman (1487), dealing in part with hydrogen activation and hydrogenation. Reviews stressing the importance of oxidative addition reactions and electronic structure of the metal ion in catalysis generally, including hydrogenation, have also appeared (1488, 1489). Other general reviews with some dicussion of homogeneous hydrogenation (1490–1492), and an article on the industrial future of hydrogenation (1493), have been published. The third edition of Cotton and Wilkinson's *Advanced Inorganic Chemistry* contains a new chapter on homogeneous catalysis by organometallic compounds (1494).

Vaska (1494a) has presented a review-type paper on the activation of hydrogen, oxygen, and sulfur dioxide by platinum metal complexes, stressing particularly the role of the metal. Iridium(I) and rhodium(I) systems were compared, as were osmium(O), iridium(I) and platinum(II) systems.

The hydrogenation (1495) and deuteration (1496) of steroids using rhodium complexes, particularly $RhCl(PPh_3)_3$, have been reviewed; and more general articles on homogeneous hydrogenation, aimed at the synthetic organic chemist, are now in press (1497). Hanzlik (1497a) has reviewed reduction of organic compounds catalyzed by cobalt cyanide complexes. Hydrogenation of polyenes to monoenes has been reviewed by Ogata (1497b).

Monographs describing organopalladium chemistry (1497c) and metal olefin chemistry (1497d), and a text on the practical aspects of heterogeneous hydrogenation (1498) have been published. Nelson and Jonassen (1498a) have reviewed olefin and acetylene complexes of the nickel subgroup.

In view of the increasing interest in asymmetric hydrogenation and asymmetric synthesis generally, mention is made of reviews dealing with optical activity in transition metal complexes: one is concerned with asymmetry in coordinated olefins (1498b) and the other with asymmetry at the metal center (1498c).

The discovery of catalysts for hydroformylation under mild conditions has restimulated interest in the Oxo reaction, and reviews of a general nature (1499–1503) and some specifically on cobalt systems (1504–1507) have appeared. The article by Usami (1503) focuses on alcohol production.

Transition-metal alkyls are being studied more widely from both synthetic and theoretical points of view (1490c, 1507d–1507f); stability of the metal-carbon bond must play a key role in many homogeneous catalytic systems.

II-A, II-B

Hahn and Peters (1508) have studied the kinetics of the copper(II)- and copper(I)-catalyzed deuterium exchange (D_2—H_2O and H_2—D_2O systems) in sulfuric and perchloric acid solutions at 160° and 5 to 15 atm pressure. The studies support the validity of the copper reduction mechanism, and substantiate the rate law given in eq. 25 (Section II-A).

Haynes and coworkers (1507a) have used $CuCl(PPh_3)_3$ and some Group VIII metal complexes as effective catalysts for the production of formamides from carbon dioxide, amines and hydrogen; hydride and carboxyl intermediates were suggested. A number of copper(I) complexes containing Group V tertiary donor ligands are reported unreactive toward olefins and acetylenes (1507c).

Churchill and coworkers (1508a) have described the first stoichiometric hydrido copper complex, which is a triphenylphosphine hydride cluster, $H_6Cu_6(PPh_3)_6 \cdot DMF$.

Wide ranging patents (1507b, 1588) have described the use of cations of formula M_2^{n+}, $M_2(OCOR)_{4-n}^{n+}$, $M_2(OCSR)_{4-n}^{n+}$, and $M_2(SCSR)_{4-n}^{n+}$, stabilized by donor ligands, for a variety of catalytic reactions including hydrogenation. The M may be Cu, Cr, Re, Mo, Ru, Rh, and Ir, and the catalysts may be used in solution or adsorbed on supports.

II-E

Complexes of the type LAuX (L = PPh$_3$, AsPh$_3$; X = halide, NCO, SCN, SeCN) have been synthesized (1508d).

III

Ashby and coworkers (1508b) have described the hydrogenolysis of dialkyl- or diarylaminomagnesium alkyls using H$_2$ at ~200 atm; for example,

$$sec\text{-}C_4H_9Mg(N^iPr_2) + H_2 \longrightarrow HMg(N^iPr_2)_2 + C_4H_{10} \qquad (522)$$

Some zinc alkyl bonds also undergo hydrogenolysis under similar conditions (1508c).

IV

Il'chenko and coworkers (1509) have described the conversion of hydrogen–oxygen mixtures to water using Y$_2$O$_5$—MoO$_3$ catalysts at elevated temperatures and pressures. The results were discussed in terms of a heterogeneous-homogeneous mechanism.

V-A

Britzinger's group (1510–1512) have isolated a polymeric [(C$_5$H$_5$)$_2$TiH]$_n$ complex which slowly loses hydrogen in ether solution to give the metastable titanocene [(C$_5$H$_5$)$_2$Ti]$_2$. This reaction is reversible at room temperature; exposure to deuterium finally results in exchange with all the hydrogens.

$$[(C_5H_5)_2Ti]_2 + H_2 \rightleftharpoons \frac{2}{n} [(C_5H_5)_2TiH]_n \qquad (523)$$

The titanocene, which is readily converted to the more stable hydride isomer [(C$_5$H$_5$)(C$_5$H$_4$)TiH]$_2$, also forms a triphenylphosphine derivative which again reacts reversibly with hydrogen:

$$[(C_5H_5)_2Ti(PPh_3)]_2 + H_2 \rightleftharpoons 2(C_5H_5)_2TiH(PPh_3) \qquad (524)$$

A monomeric decamethyltitanocene complex [C$_5$(CH$_3$)$_5$]$_2$Ti, prepared by a hydrogenolysis reaction (eq. 525), was also found to react reversibly with hydrogen under mild conditions (eq. 526).

$$[C_5(CH_3)_5][C_5(CH_3)_4CH_2]TiCH_3 + H_2 \longrightarrow [C_5(CH_3)_5]_2Ti + CH_4 \quad (525)$$

$$[C_5(CH_3)_5]_2Ti + H_2 \rightleftharpoons [C_5(CH_3)_5]_2TiH_2 \quad (526)$$

Both the titanocene and decamethyltitanocene complexes reversibly form molecular nitrogen complexes, and evidence was presented for the occurrence of titanocene and its hydride derivatives in certain nitrogen-fixation systems as well as in catalytic hydrogenations. These workers (1512) also suggested that the titanocene intermediate is a likely intermediate for hydrogenations catalyzed by $(C_5H_5)_2Ti(CO)_2$.

Van Tamelen's group (1513) and Shilov's group (1514) have also reported on titanocene and related species, and a number of the hydrides appear efficient catalysts for homogeneous hydrogenation of monoenes at ambient conditions (1513).

Martin and coworkers (1515, 1516) have presented further details of their studies on π-allyl-dicyclopentadienyltitanium(III) complexes for hydrogenation and isomerization of monoenes and dienes. Deuterium exchange with the ring hydrogen atoms was studied for the 1-methylallyl complex and the allylbis(methylcyclopentadienyl)titanium(III) complex; the former exchanges all the ring hydrogen atoms, while the latter exchanges only those β to the methyl substituent, and intermediates involving a bridged π- as well as a σ-bonded C_5H_4 or $C_5H_3CH_3$ group were proposed.

Volpin and coworkers (1517) have shown that in the Grignard reduction of $(\pi\text{-}C_5H_5)_2TiCl_2$ to hydrides, the cyclopentadienyl ligand acts as the hydride source. Hydrogen abstraction from this ligand also occurs during the thermal decomposition of $(C_5H_5)_2Ti(CH_2Ph)_2$ to $[(C_5H_4)_2Ti]_n$ and toluene (1518). Kenworthy and coworkers (1519) have discussed the structures of biscyclopentadienyltitanium(III) hydrides from e.s.r. studies.

V-B

Wailes and coworkers (1520) have studied the reactions of $(C_5H_5)_2ZrH_2$ and $(C_5H_5)_2Zr(H)Cl$ with acetylenes to give *trans*-alkenyl derivatives.

VI

Tebbe and Parshall (1521) have reported on $(\pi\text{-}C_5H_5)_2NbH_3$, which behaves similarly to the tantalum analog (eq. 78, Chapter VI), but is more reactive. Reversible loss of hydrogen at 80° leads to $[(C_5H_5)(C_5H_4)NbH]_2$ via a likely $(C_5H_5)_2NbH$ intermediate; this also leads to formation of

$(C_5H_5)_2Nb(C_2H_4)(C_2H_5)$ by reaction of the trihydride with ethylene. The hydrido-ethylene complex $(C_5H_5)_2NbH(C_2H_4)$ was also isolated; this readily undergoes hydrogenolysis to ethane and "$(C_5H_5)_2NbH$," and accounts for a catalytic hydrogenation of ethylene. The tantalum complex $(C_5H_5)_2TaH_3$ also effected this reduction. Reaction of the niobium trihydride with dienes gives saturated hydrocarbon and the complexes $(C_5H_5)_2Nb(\pi\text{-allyl})$. Other studies investigating reduction products of $(C_5H_5)_2NbCl_n$ ($n = 2$ or 3) have been noted (1521a); hydride products may be anticipated.

VII-A

Cais and coworkers (1522) have published in more detail the use of arene-chromium tricarbonyl complexes for the hydrogenation of dienes to mono-enes. The use of these catalysts for partial reduction of unsaturated fats (Section XIV-D) has been patented (1523); bi- and tricyclic olefins have also been hydrogenated (1523a).

Nasielski and coworkers (1524) have reported the photoinduced hydro-genation of 1,3-cyclohexadiene to cyclohexene, and 2,3-dimethyl-1,3-buta-diene to tetramethylethylene, using decalin solutions of $Cr(CO)_6$. Wilputte-Steinhert and Kirsch (1524a) have also used the photoactivated hexacarbonyl for diene reduction.

Tucci (1524b) has patented the use of $M(CO)_5PBu_3$ ($M = Cr, W, Mo$) complexes, with $HCo(CO)_4$ as cocatalyst, for hydroformylation of hex-1-ene; the major products are alcohols via reduction of the saturated C_7 aldehydes.

Cationic chromium carboxylates have been used as hydrogenation catalysts (1507b, 1588, Appendix II-A, II-B).

Sneeden and Zeiss (1525) have shown by using deuterium tracer studies that intermolecular hydrogen transfer from chromium and iron alkyls to olefinic substrates (1,7-octadiene, 3-phenylpropene) occurs from the β-position of the coordinated alkyl.

The reduction of organic compounds using solutions of chromous com-plexes in the absence of hydrogen, which was reviewed in 1968 (1526), still attracts interest (e.g. Ref. 1527).

VII-B

Green and coworkers (1528, 1529) have reported further on the interaction of the $(C_5H_5)_2MH_2$ complexes ($M = Mo, W$) with monoenes to yield alkyls, and with aromatic compounds such as benzene in the presence of a con-jugated diene to yield $(C_5H_5)_2MH(R)$ complexes ($R = aryl$). Nakamura and

Otsuka (1530, 1530a) have similarly studied the interaction of $(C_5H_5)_2MoH_2$ with activated olefins and acetylenes, the latter giving hydrido-alkenyl complexes. Reaction of some σ-alkenyl complexes with hydrogen chloride can yield the chloroalkyl complex:

$$(C_5H_5)_2MoH(CR\!\!=\!\!CHR) \xrightarrow{\text{HCl}} (C_5H_5)_2MoCl(CHRCH_2R) \quad (527)$$

$$(R = CO_2CH_3)$$

A related hydrogen transfer process was noted for reaction of some hydridoalkyls in the presence of excess olefin $(R = CO_2CH_3)$:

$$(C_5H_5)_2MoH(CHRCH_2R) + RCH\!\!=\!\!CHR \longrightarrow$$

$$(C_5H_5)_2Mo(RCH\!\!=\!\!CHR) + RCH_2CH_2R \quad (527a)$$

Thomas and Brintzinger (1531, 1532) have suggested that the formation of $(C_5H_5)_2MoH_2$ by amalgam reduction of $(C_5H_5)_2MoCl_2$ under H_2 pressure occurs by way of the active transient molybdenocene $(C_5H_5)_2Mo$. An isolated polymeric compound $[(C_5H_5)_2Mo]_n$ was unreactive toward H_2. The dimeric decamethyl derivative $[(C_5Me_5)_2Mo]_2$ was also made, and it reacted with H_2 under pressure to yield $(C_5Me_5)_2MoH_2$. Wolframocene $[(C_5H_5)_2W]$ was also said to occur as an unstable reaction intermediate and behaves similarly to molybdenocene as a coordinately unsaturated species.

These studies parallel those reported for the titanium analogs (Appendix V-A).

Insertion of dichlorocarbene into a tungsten-hydrogen bond of $(C_5H_5)_2WH_2$ has been reported (1533).

The existence of anionic species such as $(C_5H_5)_2MH^-$ seems likely (1529). Hydrides of the type $C_6H_6MoH_2(PR_3)_2$ and $C_6H_6MoH(PR_3)_3{}^+$ (1534, 1535), and the tetrahydrides $MH_4(PR_3)_4$, where PR_3 = tertiary phosphine or $P(OEt)_2Ph$ (1536–1538), have been synthesized.

Reaction of H_2 with $trans$-$Mo(N_2)_2(diphos)_2$ under mild conditions has yielded $trans$-$H_2Mo(diphos)_2$ and $[trans$-$H_2Mo(diphos)]_2$-μ-diphos (1539).

Shortland and Wilkinson (1540) have synthesized hexamethyltungsten, and report that it reacts rapidly with hydrogen at low temperatures to give blue solutions; these decompose rapidly to metal with liberation of methane. The hydrogen reaction is inhibited by phosphines, and the existence of hydride complexes of high coordination number (>6) in the blue solutions seems likely (1541). Reaction of $W(CH_3)_6$ with strong acid also liberates methane.

The use of cationic molybdenum species for hydrogenation has been patented (1507b, 1588, Appendix II-A, II-B). The use of $M(CO)_5PBu_3$ complexes (M = Mo, W) for hydroformylation was mentioned in Appendix VII-A.

VIII

Bennett and Watt (1542) have reported that the hydrides $HM(CO)_4L$, where M is Mn or Re, and L is o-styryldiphenylphosphine, undergo intramolecular hydride addition to the free vinyl group of the phosphine ligand to give chelate M—C σ-bonded complexes. The direction of addition was predominantly Markownikoff for the manganese system, and anti-Markownikoff for the rhenium. The same workers (1543) have also isolated the first π-oxapropenyl complex from the reaction of the same phosphine with $MeMn(CO)_5$. Such a complex was a suggested intermediate in the hydrogenation of α,β-unsaturated aldehydes and ketones by $HCo(CO)_4$ (eq. 282 in Section X-E-3).

Kaesz and coworkers (1544) have synthesized $H_3Re_3(CO)_{12}$ and $H_4Re_4(CO)_{12}$ by treatment of $Re_2(CO)_{10}$ with 1 atm hydrogen.

A patent has mentioned the use of cationic rhenium complexes for hydrogenation (1588, Appendix II-A, II-B).

IX-A

A detailed paper by Cais and Maoz (1545) has described the use of the diene-$Fe(CO)_3$ and monoene-$Fe(CO)_4$ complexes for hydrogenation of methyl sorbate.

Arresta and coworkers (1546, 1547) have reported on some iron(IV) tetrahydride complexes which lose hydrogen reversibly at ambient conditions in organic solvents:

$$H_4FeL_3 \rightleftharpoons H_2FeL_3 + H_2; \quad L = \text{tertiary phosphine} \quad (528)$$

Under N_2 or CO, the 6-coordinate complexes, such as $H_2FeL_3(N_2)$, are formed irreversibly. These complexes were said to become active hydrogenation catalysts only when mixed with triethylaluminum, which effectively removes a phosphine ligand by adduct formation. However, Newton and coworkers (1548) have used $H_2Fe(Ph_2PEt)_3$ and the nitrogen complex as potential models for nitrogenase, and report that both complexes slowly reduce ~ 0.4 mol of acetylene per mol of complex in benzene or ethanol to give ethylene (0.35 mol) and ethane (0.04 mol). Shilov and coworkers (1549) have reported that the bridged nitrogen complex $HFePr^i(PPh_3)_2N_2FePr^i(PPh_3)_2$ reacts with HCl to give hydrazine; a suggested mechanism involved transfer of the coordinated hydride.

Gerlach and coworkers (1550) have also studied the $H_2FeL_3(N_2)$ and $H_2FeL_3(CO)$ complexes, as well as the dihydrides H_2FeL_4, where L is a variety of trivalent phosphorus ligands. Deuterium exchange with the hydride ligands and the phenyl hydrogens of some aryl phosphine ligands was observed for a number of the complexes. Cis-H_2FeL_4, $trans$-$HFeXL_4$ (X = Cl, SCN), $trans$-$HFe(CH_3CN)L_4^+$, and $HFeL_5^+$, where L is diphenylphosphine, have also been made (1551).

A patent has described the synthesis of hydridoiron complexes containing chelating phosphorus or arsenic ligands for use as hydrogenation catalysts (1551a).

Nishiguchi and Fukuzumi (1552) have used $FeX_2(PPh_3)_2$ complexes (X = halide) in alcohols and dihydroxybenzenes at 240° under vacuum for the reduction of cyclooctadienes; the hydroxybenzenes were the most efficient hydrogen donors.

IX-B-2a

Mercer and Dumas (1553) have isolated the mixed valence (II, III) cations $Ru_2Cl_{3+n}^{(2-n)+}$, where $n = 0$, 1, 2, from the so-called blue ruthenium(II) solutions.

IX-B-2b

Vaska and coworkers (1554, 1555) have used toluene solutions of $RuCl_2(PPh_3)_3$ for the catalytic conversion of hydrogen-oxygen mixtures to water at ambient conditions, and have studied also the catalyzed hydrogen-deuterium exchange between D_2 and acetic acid, alcohols, and morpholine, using the same complex and the osmium analog $OsBr_2(PPh_3)_3$. $RuCl_2(AsPh_3)_3$ has been used by Khan and coworkers (1556) for the hydrogenation of maleic acid in benzene solution.

Blum and coworkers (1557, 1558) have hydrogenated α,β-unsaturated carbonyl compounds, especially ketones, using $RuCl_2(PPh_3)_3$ in hydrogen-donor solvents in the absence of molecular hydrogen. Solvents used included benzyl alcohol, formic acid, N-methylformamide, p-tolualdehyde, and α-naphthaldehyde. The complex also catalyzes the dehydrogenation of aromatics such as 9,10-dihydroanthracene (1558a) and has been used for the catalytic formation of amides from amines, hydrogen, and carbon dioxide (1507a, Appendix II-A, II-B).

Observations on the reaction of $RuCl_2(PPh_3)_3$ with olefins have been noted (1559, 1560). Schunn (1561) has reported on the ligand–hydrogen transfer observed in D_2 and C_2D_4 exchange studies of $HRuCl(CO)(PPh_3)_3$ and $HRuCl(PPh_3)_3$; he suggests that stoichiometric hydrogenation of olefins using such catalysts in the absence of molecular hydrogen might involve hydrogen abstraction from the ortho positions of the phenyl groups (cf. Scheme 3 in Section IX-B-2b). James and Markham (1560) have reported the ready protonolysis of solutions believed to contain a ruthenium ethyl derivative:

$$C_2H_5RuCl(PPh_3)_3 + HCl \longrightarrow C_2H_6 + RuCl_2(PPh_3)_3 \qquad (529)$$

Wells and coworkers (1562, 1563) have reported further on the isomerization of pent-1-ene, and on deuterium distribution in *trans*-$C_2H_2D_2$ catalyzed by $HRuCl(PPh_3)_3$ and $RuCl_2(PPh_3)_3$ in benzene. Lyons (1564, 1565) has studied the interaction of $RuCl_2(PPh_3)_3$ with hydroperoxides, including those present in unpurified olefins, and has isolated complexes such as $RuCl_2(CO)(PPh_3)_2$ (olefin). Metal peroxocomplexes were detected in the presence of small amounts of oxygen.

The use of $RuCl_3$–organophosphine–triethylamine mixtures in methanol for hydrogenation has been patented (1566).

The catalyst "ruthenium(II) acetate," $Ru_2(CO_2Me)_4$, and a triphenylphosphine adduct have been shown to contain an oxygen-centered triangle Ru_3O, and have thus been reformulated as $[Ru_3O(CO_2Me)_6(H_2O)_3][O_2CMe]$ and $Ru_3O(CO_2Me)_6(PPh_3)_3$, respectively (1567). The use of these complexes for hydrogenation of terminal alkenes has been patented (1568).

The complete crystal structures of the naphthyl hydrides, *cis*-$HM(C_{10}H_7)$-$(Me_2PCH_2CH_2PMe_2)_2$, M = Ru, Os (see eq. 140b in Section IX-B-2b), have been reported (1569).

Gol'dschleger and coworkers (1569a) have used ruthenium phosphine complexes to catalyze exchange between alkanes and D_2O.

IX-B-2c

James and coworkers (1570) have isolated dimethylacetamide complexes of ruthenium(II) and ruthenium(I) by hydrogenation of $RuCl_3$, $3H_2O$ at 1 atm pressure. The $RuCl_3(HDMA)$ and $Ru_2Cl_3(HDMA)$ complexes, where HDMA is protonated dimethylacetamide, are efficient catalysts for olefin hydrogenation, and the former complex catalytically converts hydrogen-oxygen mixtures to water. Hydrogen reduction of ruthenium halides in DMSO yields $RuX_2(DMSO)_4$ complexes (1571). Ruthenium tribromide catalyzes the hydrogen reduction of DMSO to dimethyl sulfide and water (cf. eq. 384 in Section XI-H). Sulphide complexes such as $RuCl_3(PhS^nPr)_3$ do not catalyze the hydrogenation of olefinic and acetylenic substrates at ambient conditions in a variety of solvents (1572).

Litvin and coworkers (1573) have reported the use of a ruthenium complex with *N*-phenylalanine in DMF under mild conditions for the hydrogenation of monoenes and dienes. The complexes appear closely related to some corresponding rhodium catalysts (Section XI-H). The use of ruthenium(III) complexes of poly(L-methylethylenimine) and similar ligands in aqueous acetate solutions for the asymmetric hydrogenation (80 atm H_2) of the ketonic group in methyl acetoacetate, and the olefinic group in mesityl oxide, has been described by Hirai and coworkers (1574–1576).

Wilson and Osborne (1577) have reported that some $HRu(NO)L_3$ complexes (L = tertiary phosphine) are effective hydrogenation and isomerization catalysts for monoenes under mild conditions in benzene; the hydride ligand readily exchanges with deuterium. The structures of the triphenylphosphine catalyst (1578), and a $Ru(NO)(diphos)_2^+$ complex (1579) have been reported. Details of the $RuCl(NO)_2(PPh_3)_2^+$ structure have appeared (1580). The 6-coordinate complexes such as $trans$-$RuCl(NO)(diars)_2^{2+}$ (1581), $RuCl_2(NO)$-$(PPh_3)_3^+$ (1582), and $MCl_3(NO)(PPh_3)_2$ (M = Ru, Os) (1583, 1584) are unlikely to be very reactive toward hydrogen. The 5-coordinate $RuCl_2(NO)$ $(PPh_3)_2$ probably exists (1584).

The cluster hydride $H_4Ru_4(CO)_{12}$ has been synthesized by reaction of hydrogen with $Ru_3(CO)_{12}$ (1544, 1585). The same tetrahydride is formed from $H_2Ru_4(CO)_{13}$ or $Ru_2Fe(CO)_{12}$ using 1 atm H_2; $H_2FeRu_3(CO)_{13}$ similarly affords $H_4FeRu_3(CO)_{12}$; corresponding reactions using D_2 give complexes such as $H_2D_2Ru_4(CO)_{12}$ (1544). $Ru(CO)_3(PMe_2Ph)_2$ has been used for hydroformylation of terminal olefins to aldehydes (1586).

A number of homogeneous ruthenium catalysts have been made insoluble by binding to ion-exchange resins, polymeric ligands, or alumina, and have subsequently been used for a variety of catalytic reactions including hydrogenation (1507b, 1587–1589).

IX-B-2e

A more detailed paper by Knoth (1590) has reported further on hydridotriphenylphosphine complexes. $H_4Ru(PPh_3)_3$ can lose hydrogen reversibly to give $H_2Ru(PPh_3)_3$ (cf. Section IX-B-2e). Pennella and coworkers (1591) have studied the hydrogenation of alkenes and alkynes at a few atmospheres pressure using $H_4Ru(PPh_3)_4$ and $H_2Ru(N_2)(PPh_3)_3$ in toluene. Further H_2RuL_4 complexes, where L is a tertiary phosphine or phosphite ligand, have been made; $H_2Ru[P(OEt)_3]_4$ does not undergo exchange with deuterium under mild conditions (1548). The following related complexes have also been synthesized: cis-H_2RuL_4, $trans$-$HRuXL_4$, $HRuL_5^+$ and $trans$-$HRuL_4Y^+$ (L = diphenylphosphine; X = halogen, SCN, and $SnCl_3$; Y = CO and N_2) (1551, 1592); the cationic trialkylphosphite complexes RuL_6^{2+}, $RuXL_5^+$, $HRuL_5^+$, [L = $P(OMe)_3$, $P(OEt)_3$, and X = Cl, Br], and the cyclooctadiene complexes $HRu(C_8H_{12})L_3^+$ (L = hydrazines, pyridines) (1593, 1594).

Roper and coworkers (1595) have noted that molecules such as H_2, O_2, C_2H_4, and $PhC{\equiv}CPh$, add rapidly to solutions of $Ru(CO)_2(PPh_3)_3$ probably via predissociation of a phosphine. The 5-coordinate complexes, $Ru(CO)_3$-$(PR_3)_2$, where R is p-C_6H_4Me and p-C_6H_4OMe (1596), and cationic complexes such as $HM(CO)_2(PPh_3)_3^+$ and $HM(CO)(CH_3CN)_2(PPh_3)_2^+$ (M = Ru, Os) (1595) have been made.

IX-C

A number of studies involving osmium complexes were mentioned in the preceding Appendix sections—IX-B-2b, IX-B-2c, and IX-B-2e.

$HOs(NO)(PPh_2R)_3$ complexes (R = Ph, Me) have been made, but whether they are hydrogenation catalysts, like the ruthenium analogs, was not reported (1577). Other 5-coordinate nitrosyl complexes such as $HOs(CO)(NO)(PPh_3)_2$, and cationic derivatives such as $Os(CO)_2(NO)(PPh_3)_2^+$, $Os(CO)(NO)L_3^+$, and $Os(CO)(NO)(RNC)(PPh_3)_2^+$ (where L = tertiary phosphine and R = p-tolyl) have also been made (1597), as well as $Os_3(CO)_{12}(NO)_2$ (1598).

Reaction of $Os_3(CO)_{12}$ with 1 atm H_2 at 110° yields initially $H_2Os_3(CO)_{10}$, and then $H_4Os_4(CO)_{12}$ (1544). Irradiation of $Os_3(CO)_{12}$ in cyclohexane leads to a mixture of polynuclear hydridocarbonyls as a result of hydrogen abstraction from the hydrocarbon (1599). The complex $Os(CO)_2(PPh_3)_3$ is reported to dissociate slowly in solution (1595), and is thus likely to form $H_2Os(CO)_2(PPh_3)_2$ on reaction with hydrogen. L'Eplattenier and coworkers (1599a–1599c) have reacted hydrogen with $Os(CO)_5$ or $Os_3(CO)_{12}$ to form $H_2Os(CO)_4$; complexes such as $H_2Os(CO)_3(PPh_3)$, and the alkyls $Os(CO)_4R_2$ (R = Me, Et) were synthesized, and the carbonyl insertion reactions of the latter studied.

The synthesis of $H_2OsCl_2(PR_3)_3$ complexes (R_3 = $MePh_2$, Et_2Ph, $EtPh_2$, Ph^nPr_2) described by Chatt and coworkers (1600) involves amalgam reduction of mer-$OsCl_3(PR_3)_3$ complexes under 1 atm H_2, and likely involves dihydride formation via an osmium(II) intermediate (cf. eq. 149 in Section IX-C). The 5-coordinate $HOsCl(CO)(PR_3)_2$ (R = cyclohexyl) has been reported by Moers (1601).

Malin and Taube (1602) have synthesized $H_2Os(en)_2^{2+}$ and report that the hydrides readily exchange with deuterium in acidic D_2O. Cis-$H_2Os(HPPh_2)_4$ (1592) and $OsX(PR_3)_5^+$, where X is Cl, Br, and R is OMe, OEt (1593), have been made.

X-A-2

Deuterium exchange between D_2 and H_2O catalyzed by $Co(CN)_5^{3-}$ has been reinvestigated by Strathdee and Quinn (1603).

X-B-1

Alkaline solutions of $HCo(CN)_5^{3-}$ have been used to reduce ferricyanide (1604) and mercuric cyanide (1605), and the kinetic data obtained have been analyzed in terms of the cobalt(I) intermediate $Co(CN)_5^{4-}$, produced via eq. 213 (Section X-B-2e). Carbonylation of the alkaline hydride solutions gives rise to the cobalt(I) cyanocarbonyl anion $Co(CN)_3(CO)_2^{2-}$ in a similar manner (1606). Halpern's group (1607) have presented further pulse radiolysis data giving evidence for the transient $Co(CN)_5^{4-}$ intermediate and its reaction with water to give $HCo(CN)_5^{3-}$. A number of the earlier reported $Co(CN)_5^{3-}$-catalyzed hydrogen reductions of inorganic substrates may occur via cobalt(I) intermediates.

X-B-2a

Funabiki and Tarama (1608) have studied the reaction of $HCo(CN)_5^{3-}$ with butadiene by n.m.r. and i.r. methods, and obtained evidence supporting the hydrogenation mechanism proposed by Burnett and coworkers. The Japanese workers (1609–1610a) have also studied further the effects of solvent (water, aqueous methanol, ethylene glycol-methanol, glycerol-methanol-water) on the butadiene hydrogenation.

X-B-2b

Basters and coworkers (1611) have studied the catalyzed hydrogenation and deuteration of a series of substituted *trans*-cinnamic acids using $HCo(CN)_5^{3-}$, and have discussed the kinetic data in terms of the free radical hydrogen atom transfer mechanism. Simandi and coworkers (1612) have reported further on the catalyzed hydrogenation of cinnamic, sorbic, and muconic acids.

X-B-5

Fleischer and Krishnamurthy (1613) have reported that *meso*-tetra(*p*-sulfonatophenyl)porphinatocobaltate(III) catalyzes the borohydride reduction of acetylene to ethylene in aqueous solution at pH 10 and ambient conditions, by way of cobalt(I) intermediates (see Appendix X-H for other borohydride reductions).

X-C

Some studies on $Co(CN)_5^{3-}$-catalyzed hydrogenation in nonaqueous solvents were mentioned in Appendix X-B-2a. Zakhariev and coworkers (1614) have studied the kinetics of styrene hydrogenation in aqueous ethanol.

X-D

Ohgo and coworkers (1615, 1616) have used aqueous solutions of cyano-amine-cobaltate(II) complexes for the catalytic hydrogenation of α,β-unsaturated acids. Use of optically active amines (1,2-propanediamine and N,N'-dimethyl-1,2-propanediamine) resulted in asymmetric hydrogenation of atropate to hydratropic acid in 7% optical yield.

X-E-1

Hydroformylation of olefins to aldehydes using $Co_2(CO)_8$ still demands attention from both synthetic (1617–1620) and mechanistic (1621–1626) view-points (see also Appendix X-E-2, X-E-3). The use of the $Co_2(CO)_8$ catalyst supported on divinylbenzenevinylpyridine copolymers has been patented (1627, 1628). Pino and coworkers (1619) have studied the hydroformylation of S-(+)-3-methyl-1-pentene to 2,3-dimethylpentanals; the *sec*-Bu group exerted a weak asymmetric induction, and the ratio of *erythro* to *threo* products differed from the diastereoisomeric equilibrium. The same group (1619a) have also used optically active bis(N-α-methylbenzylsalicylaldiminato)cobalt(II), and $Co_2(CO)_8$ with the salicylaldimine, to promote asymmetric hydroformyl-ation of styrene and α-methylstyrene.

Ungvary (1629) has studied the rates of reaction between $Co_2(CO)_8$ and H_2 in heptane to give $HCo(CO)_4$. The reaction is first-order in both octa-carbonyl and H_2, and at low carbon monoxide partial pressure is inversely dependent on carbon monoxide; the data were analyzed in terms of a direct reaction of $Co_2(CO)_8$ and H_2, and a competing mechanism involving the following steps:

$$Co_2(CO)_8 \underset{}{\overset{fast}{\rightleftharpoons}} Co_2(CO)_7 + CO \qquad (530)$$

$$Co_2(CO)_7 + H_2 \overset{k}{\longrightarrow} H_2Co_2(CO)_7 \longrightarrow HCo(CO)_4 + HCo(CO)_3 \qquad (531)$$

$$HCo(CO)_3 + CO \longrightarrow HCo(CO)_4 \qquad (532)$$

Ellgen (1630) has concluded that reaction of $Co_2(CO)_8$ with alkynes also pro-ceeds via $Co_2(CO)_7$.

The decomposition of $C_3H_7COCo(CO)_4$ under nitrogen is reported by Rupilius and Orchin (1624) to yield n- and isobutyraldehyde and propylene. The data were discussed in terms of the reverse of the reactions shown in eqs. 247 to 249 (Section X-E); the olefin metal hydride intermediate $HCo(CO)_3C_3H_6$ effects hydrogenolysis of the acyl to produce aldehyde and olefin.

X-E-2, X-E-3

An article by Marko and Bathory (1631) and patents (1632–1635) have described the production of alcohols from terminal olefins using $Co_2(CO)_8$ or cobalt salts under hydroformylation conditions, sometimes in the presence of amines. Ethylene has been converted into diethyl ketone using $Co_2(CO)_8$-amine catalysts (1635a). Matsubara (1636) has hydroformylated cyclopentadiene and dicyclopentadiene to saturated cyclic alcohols (via the monoformylation products); the second olefinic bond is hydrogenated. Mistrik and Mateides (1637) have hydroformylated α-substituted furans to give aldehyde and alcohol products, as well as products of just hydrogenation of the heterocyclic ring. Derbesy and coworkers (1638) have converted methyl 3-methyl-2-butenoate to the δ-lactone, 3-methyl-5-pentanolide (eq. 252b in Section X-E-2). $HCo(CO)_4$ has been used as a cocatalyst with some Group VI $M(CO)_5PBu_3$ complexes for hydroformylation (1524a, Appendix VII-A).

Taylor and Orchin (1622) have studied the reaction of dimethyl maleate with $DCo(CO)_4$. Deuterium is incorporated into the hydrogenation product (succinate) and the isomerization product (fumarate), but is not present in recovered maleate; hydrogenation occurs via the fumarate and was thought to involve 1,4-addition of $DCo(CO)_4$ to carbon and carbonyl oxygen. The hydridocarbonyl reduces 1,2-diphenylcyclobutene to the cyclobutane by a mechanism which must involve 1,2-addition (1639). The addition of $HCo(CO)_4$ to propylene in the gas phase at ambient temperatures proceeds mainly by Markownikov addition (1640). Orchin and Rupilius (1623) have also isolated 1-methyl-π-allylcobalt tricarbonyl complexes from reaction of $HCo(CO)_4$ with butadiene; these complexes and an intermediate σ-allyl complex undergo hydrogenolysis by further $HCo(CO)_4$ to yield the butene products (eq. 280a in Section X-E-3).

X-E-6

Sato and coworkers (1641, 1642) have converted acrylonitrile to γ-hydroxylbutyronitrile using $Co_2(CO)_8$ under Oxo conditions; the reaction proceeds via hydrogenation of the hydroformylation product, β-formyl-propionitrile. Under similar conditions, acid anhydrides $(RCO)_2O$ undergo hydrogenolysis to RCO_2H and RCHO, and cyclic anhydrides give the ω-formyl acids (1643). The anhydride reactions reported by Wakamatsu and co-workers appear to be the first of their type.

X-F-1, X-F-2

Tucci (1644) and Orchin and coworkers (1645) have discussed the influence of the phosphine ligand in complexes such as $HCo(CO)_3(PR_3)$ (R = alkyl or aryl) on the distribution of reaction products (aldehydes, alcohols, and paraffins) from hydroformylation of terminal olefins. Ogata and Asakawa (1646) have reported similarly on α,β-unsaturated ester and styrene substrates. A number of patents further indicate that, relative to $Co_2(CO)_8$, the phosphine-substituted catalysts generally increase the amount of hydrogenation products from alkene hydroformylation (1647–1653a), although such systems have been modified to give high aldehyde yields (1654–1656).

Ogata and Kubota (1657) have used $[Co(CO)_3PR_3]_2$ (R = Bu, cyclohexyl, Ph) complexes for selective hydrogenation of polyunsaturated fatty acid esters to monoenes under $\sim$50 atm H_2, with $\sim$2 to 5 atm CO. Some alcohol products from the Oxo reaction were observed. Kogami and coworkers (1658) have used $Co_2(CO)_8$ in the presence of tertiary phosphine ligands for the selective hydrogenation of p-isopropyl-α-methylcinnamaldehyde to cyclamen aldehyde under Oxo conditions. Lai and Ucciani (1659) have similarly obtained mainly propylbenzene from $trans$-1-phenyl-1-propene using $Co_2(CO)_8$ or $[Co(CO)_3(PPh_3)]_2$.

Further patents using cobalt carbonyl–tertiary phosphine catalysts have appeared describing the hydrogenation of 1,5-cyclooctadiene to cyclooctene (1660), and 1,5,9-cyclododecatriene to cis- and $trans$-cyclododecene (1661).

X-G

A full paper has appeared by Yamamoto and coworkers (1662) on the properties of $HCo(N_2)(PPh_3)_3$ and related complexes such as $[C_2H_4Co(PPh_3)_3]_2$ and $CH_3Co(PPh_3)_3$. The hydride catalyzes ethylene hydrogenation at ambient conditions via an ethyl intermediate which undergoes hydrogenolysis to regenerate a hydride catalyst (cf. eqs. 306 to 308 in Section X-G). Redistribution of hydrogen isotopes in the reaction of $HCo(N_2)(PPh_3)_3$ with $trans$-$C_2H_2D_2$ has been studied (1563). Interaction of $HCo(N_2)(PPh_3)_3$, and other $HCoL_4$ and $HCo(CO)L_3$ complexes (L = tertiary phosphine or phosphite), with D_2, C_2D_4, and but-1-ene has also been studied (1561); exchange with ligand hydrogen atoms as well as with the metal hydride ligand was observed (see also Appendix IX-B-2b). The complex $H_3Co(PPh_3)_3$ is said to catalyze exchange between methane and deuterium (1569a).

Vaska and coworkers (1663) have studied the addition of hydrogen and oxygen to the square planar $M(PP)_2^+$ cations (M = Co, Rh, Ir; PP = cis-$Ph_2PCH{=}CHPPh_2$); the affinity for hydrogen decreased in the order Co > Ir ≫ Rh. The tetrahedral complex $CoCl(PPh_3)_3$ was unreactive toward H_2 at 65° and 1 atm. The addition of hydrogen to $Co(diphos)^+$ has been reported on further (1664). $HCo(diphos)$ effectively catalyzes the formation of DMF from dimethylamine, hydrogen and carbon dioxide (1507a, Appendix II-A, II-B). A patent (1551a) has described the synthesis of such chelated complexes for use as hydrogenation catalysts.

The structure of $HCo[PhP(OEt)_2]_4$ is approximately trigonal bipyramidal with the hydride occupying a specific coordination site (1665); an X-ray study of $HCo(PF_3)_4$ failed to give a hydrogen atom position although slight distortion of the $Co(PF_3)_4$ structure from tetrahedral was observed (1666).

Porri and coworkers (1667) have isolated anti-π-but-2-enyl butadiene-(triphenylphosphine)cobalt, $(C_4H_7)(C_4H_6)CoPPh_3$, from reaction of butadiene with $CoCl_2(PPh_3)_2$ in reducing conditions; hydride intermediates seem likely. Butadiene and butenes are formed on treatment of the butenyl complex with aqueous hydrochloric acid. $CoX_2(PPh_3)_2$ complexes (X = halide) catalyze hydrogen transfer to cyclooctadienes from alcohol and dihydroxybenzene solvents at 240° (1552).

X-H

The kinetic study of the reaction between hydrogen and $Co(DMGH)_2$ in aqueous methanol, in the absence and presence of added pyridine, has been reported in more detail (1612). Ohgo and coworkers (1668) have reported that $Co(DMGH)_2$ catalyzes at ambient conditions the hydrogenation of activated olefins, α-diketones, α-oxo acid esters, nitrobenzene, azobenzene, and azoxybenzene. The same group (1669) have reported the asymmetric hydrogenation of benzil to $S(+)$-benzoin by a $Co(DMGH)_2$-quinine catalyst.

Green and coworkers (1670, 1671) have reported on the borohydride reduction of nitro- and nitroso-arenes catalyzed by methanolic solutions of trans-$Co(DMGH)_2Br(H_2O)$. Hill and coworkers (1672) have studied similar systems catalyzed by hydroxocobalamin; that the reductions occur via cobalt(I) intermediates was confirmed by reaction of the substrates and the initial reduction products with vitamin B_{12s}, a cobalt(I) cobalamin. Schrauzer's group (1673, 1674) have synthesized hydridocobaloximes, such as $HCo(DMGH)_2P^nBu_3$, and hydridocobalamin solutions. The hydrides in protic media react with olefins to yield alkyl or substituted-alkyl derivatives.

XI-A-1

The crystal structure of the $HRh(NH_3)_5^+$ cation has been reported (1675). Jewsbury and Maher (1676) consider that the hydride previously formulated as $HRh(CN)_4(H_2O)^{2-}$ is in fact $HRh(CN)_5^{3-}$.

XI-A-2

Gillard and coworkers (1677) have reported further on substitution reactions of rhodium(III), which are catalyzed by rhodium(I) species formed via rhodium(III) hydride intermediates.

XI-B-1, XI-B-2

N.m.r. and spectrophotometric studies by Meakin and coworkers (1678) have confirmed that $RhCl(PPh_3)_3$ is little dissociated in dichloromethane; furthermore, the degree of dissociation of phosphine in the dihydride $H_2RhCl(PPh_3)_3$, is also small, although a phosphorus *trans* to hydrogen is reasonably labile (at 30°, the dissociation rate constant is $\sim$400 s^{-1}). An equilibrium constant of $\sim$5 × 10^3 M^{-1} was estimated for the reaction of $RhCl(PPh_3)_3$ with hydrogen at 25°. A subsequent step in the hydride route to hydrogenation must involve coordination of olefin with loss of the labile phosphine. This mechanism is consistent with a postulate that, in general, only 16- and 18-electron complexes are present in homogeneous reactions catalyzed by Group VIII metal complexes (1489).

Arai and Halpern (1679) have estimated spectrophotometrically an equilibrium constant of $\sim$1.4 × 10^{-4} M for the loss of PPh_3 from $RhCl(PPh_3)_3$ in benzene at 25°, but Meakin and coworkers (1678) have suggested that contributions from equilibrium 533 could be important under the conditions of measurement.

$$2RhCl(PPh_3)_3 \rightleftharpoons [RhCl(PPh_3)_2]_2 + 2PPh_3 \qquad (533)$$

Strohmeier and Endres (1680) have measured the rates of reaction of H_2 with $RhX(PPh_3)_3$ complexes (X = Cl, Br) in toluene.

Siegel and Ohrt (1681) have studied kinetic hydrogen isotope effects for the hydrogenation of cyclohexene catalyzed by $RhCl(PPh_3)_3$, and they suggest that k_H/k_D should decrease by a factor of 1.38 as the order of the reaction changes from first to zero with respect to hydrogen [cf. eq. 329 for the hydride route ($k'' = 0$), Section XI-B-1], that is, depending on whether oxidative addition of hydrogen or hydrogen transfer becomes rate-determining. The same workers (1682) reinvestigated the kinetics of the cyclohexene reduction, and analyzed the data in terms of a rate-determining conversion of the $H_2RhCl(PPh_3)_2$(alkene) complex to alkane and $RhCl(PPh_3)_2$; nine other reactions were written as various equilibria involving H_2, alkene, and PPh_3, with $RhCl(PPh_3)_2$, $H_2RhCl(PPh_3)_2$, $RhCl(PPh_3)_2$(alkene), and $RhCl(PPh_3)_3$ species. A dependence of the rate on catalyst concentration similar to that shown in Fig. 5 (Section XI-B-7a) was analyzed in terms of phosphine dissociation from the $H_2RhCl(PPh_3)_3$ complex.

Since the discovery of the importance of trace oxygen on the dissociation of $RhCl(PPh_3)_3$ to the bisphosphine complex, Wilkinson and coworkers (1683) have reported further on the catalyzed hydrogenation of hex-1-ene and cyclohexene; using more stringent experimental precautions, essentially the same kinetic rates were obtained.

XI-B-3

Initial rate data for hydrogenation of monoolefins and monoacetylenes (cf. Table 63, Section XI-E) catalyzed by $RhCl(PPh_3)_3$ and $IrCl(PPh_3)_3$ indicate that the rhodium complex is up to 100 times more active (1684).

XI-B-4

Shuford (1685) has compared homogeneous and heterogeneous hydrogenation and exchange of substituted olefinic bonds using $RhCl(PPh_3)_3$ and rhodium metal catalysts. Wahren and Bayerl (1686) have noted that, in the presence of small amounts of oxygen, the $RhCl(PPh_3)_3$ catalyzed deuteration of cyclohexene is accompanied by a heterogeneously induced exchange reaction involving the cyclohexene hydrogens.

XI-B-5

Birch and Walker (1687) have used $RhCl(PPh_3)_3$ solutions for the synthesis of 1,2,3,4,5,6,7,8-octahydronaphthalene from 1,4,5,8-tetrahydronaphthalene,

1,2,3,4,5,6,7,8,9,10-decahydroanthracene from 1,4,5,8,9,10-hexahydro-anthracene, 4,5,6,7-tetrahydroindane from 4,7-dihydroindane, substituted cyclohex-1-enes from the 1,4-dienes, 3β-acetoxypregn-5-en-20-one-16α,17α-d_2 from the 5,16-dien-20-one, 4-methyl-4-trichloromethyl(cyclohex-2-ene-1-one) from the cyclohexa-2,5-diene-1-one, and griseofulvin from ($-$)-dehydro-griseofulvin; butadiene rubbers were also hydrogenated. Senda and coworkers (1688) have reduced substituted 2-cyclohexenols to the cyclohexanols, and Khidekel's group (1689) have reduced an olefinic bond in some spirocyclic compounds to give saturated products. Catalytic cis-deuteration of various unsaturated 5α-spirostane derivatives has yielded steroids labeled in the side chains (1690), and cis-tritiation of 17-β-hydroxyandrosta-1,4-dien-3-one gives mainly testosterone-1α,2α-t_2 (1691). Biellman and coworkers (1692) have noted that the catalytic acetalization of steroid ketones (eq. 338, Section XI-B-5c) is promoted by traces of oxygen in hydrogenation procedures using RhCl(PPh$_3$)$_3$. Gagnaire and Vottero (1693) have specifically cis-deuterated 2,5-dihydrofurans to the tetrahydrofurans. Long-chain olefins (C$_9$, C$_{10}$, C$_{17}$, C$_{19}$) and cyclohexene are similarly reduced (1694–1696), but cycloheptene, -octene, and -dodecene underwent H—D scrambling during deuteration (1695). A paper (1697) has now described the hydrogenation of the mentha-2,8-diene derivative, **100** (Section XI-B-5a). The acetylenic compound 1,7,13-tridehydro[18]annulene is not hydrogenated using RhCl(PPh$_3$)$_3$ (1698).

Chloroform solutions containing H$_2$RhCl(PPh$_3$)$_3$ react with diazonium cations to yield complexes in which the diazonium group has been reduced to a hydrazine derivative (1699).

The RhCl(PPh$_3$)$_3$-catalyzed hydrogen-deuterium exchange between D$_2$ and acetic acid, alcohols, and morpholine has been studied (1555). RhCl(PPh$_3$)$_3$ has been used to catalyze hydrogen transfer from formyl groups of solvents to α,β-unsaturated ketones (1558, Appendix IX-B-2b). The complex, and RhCl$_3$(AsPh$_3$)$_3$, also catalyze the dehydrogenation of compounds such as 9,10-dihydroanthracene (1558a).

Also reported are the catalytic production of DMF from amines, hydrogen and carbon dioxide (1507a, Appendix II-A, II-B), and the catalyzed hydrogenolysis of oxygen to water (1554) using the RhCl(PPh$_3$)$_3$ complex.

XI-B-6

Solutions containing Rh$_2^{4+}$ yield RhX(PPh$_3$)$_3$ (X = Cl, OCOR) complexes on treatment with triphenylphosphine in the presence of excess X; initial hydrogenation rates of hex-1-ene using the carboxylates in benzene were reported (1683). These catalysts as well as rhodium trichloride, have been heterogenized on inert supports (1587). The use of a wide range of complexes prepared *in situ* from the cationic species has been patented (1507b, 1588, Appendix II-A, II-B).

The complex $(Me_3SiCH_2)Rh(PPh_3)_3$ has been used for hydrogenation (1700).

XI-B-7, XI-B-9

Horner and Siegel (1701) have reported further on the rate of hex-1-ene hydrogenation using various tertiary phosphine-rhodium(I) systems. The di-*t*-butylphosphine complex $RhCl(PH^tBu_2)_3$ is inactive for hex-1-ene hydrogenation at ambient conditions (1702). Shaw and coworkers (1702, 1703) have isolated the 5-coordinate complexes $H_2RhCl(P^tBu_3)_2$ and $HRhCl_2L_2$ (L = tertiary *t*-butylphosphines), which are highly effective catalysts for hydrogenation of olefins and alk-1-ynes, although the monohydrides require the presence of a base. $HRhCl_2(P^tBu_2Me)_2$ reacts with hydrogen in the presence of base to give $H_2RhCl(P^tBu_2Me)_2$.

Grubbs and Kroll (1704) have used $RhCl(PPh_3)_3$ absorbed on polystyrene-divinylbenzene-diphenylphosphine beads for hydrogenation of monoenes; they observed a decrease in reduction rate with molecular size, which was attributed to restriction of the size of the solvent channels by the polymer crosslinks. In a related study, Collman and coworkers (1705) have synthesized the resin-bound complexes $MClL_3$ and $MCl(CO)L_2$, where M = Rh, Ir, and L is a resin-substituted triphenylphosphine:

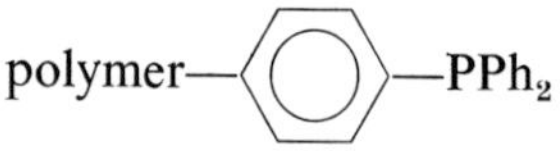

The complexes exhibited reactions analogous to the homogeneous complexes with L = PPh_3 but were superior for hydrogenation in terms of catalyst lifetime and ease of product separation; resin chelation unfortunately prevents production of high concentrations of coordinatively unsaturated complexes.

XI-B-10

The $RhCl(PPh_3)_3$-catalyzed hydrogenations of dipentyne, dihexyne, *o*-nitro-phenylacetylene, and 2,7-dimethylocta-2,6-dien-4-ynedial ceased with precipitation of rhodium complexes (1687). Some hydrido-butylphosphine complexes effectively catalyze reduction of alk-1-ynes (1703, Appendix XI-B-7, XI-B-9).

XI-C

A number of reports have been concerned with asymmetric hydrogenation. Morrison and coworkers (1706) used tris(neomenthyldiphenylphosphine)-rhodium(I) chloride solutions at elevated conditions for reduction of α- and β-methylcinnamic acids and atropic acid. Knowles and coworkers (1707, 1708) have reported the use of rhodium complexes with LMePPh phosphine ligands (L = Pr, o-MeOC$_6$H$_4$, cyclohexyl, Me$_2$CH, PhCH$_2$, m-MeOC$_6$H$_4$, 3-cholesteryl), and with methylcyclohexyl-o-anisylphosphine, for hydrogenation of β-substituted α-acylaminoacrylic acids. Dang and Kagan (1709, 1710) have used *in situ* RhCl(P-P)(solvent) complexes, where P-P is the chelating chiral diphosphine $(-)$-2,3-0-isopropylidene-2,3-dihydroxy-1,4-bis-(diphenylphosphino)butane, for hydrogenation of various precursors of alanine, phenylalanine, tyrosine, dopa, and leucine; optical purities of the products were as high as 80%, and the catalyst was more efficient than RhCl(PPh$_3$)$_3$ for hydrogenation of α-acetamidocinnamic acid. Horner and Siegel (1711) have used the R-$(-)$-PrMePPh phosphine complex to reduce α-substituted styrenes; for example, PhC(^{i}Pr)=CH$_2$ gave $(-)$-PhCH(^{i}Pr)CH$_3$ through *cis* hydrogen-addition. Such asymmetric syntheses also lead to a new method for determination of the absolute configuration of molecules.

XI-D

Patents (1712–1714) have described the use of HRh(CO)(PPh$_3$)$_3$ and its analogs on inert supports for hydrogenation as well as hydroformylation.

Ueda (1715) has reported on the mechanism of propylene deuteration using HRh(CO)(PPh$_3$)$_3$. Reaction of the complex with *trans*-C$_2$H$_2$D$_2$ leads to a redistribution of isotopes in the ethylene (1563). Some kinetic data on the reaction of maleate, and hept-1-ene with HRh(CO)(PPh$_3$)$_3$ solutions have been presented by Strohmeier and Rehder-Stirnweiss (1716).

Wilkinson's group (1717) have studied the reaction of HRh(CO)(PPh$_3$)$_3$ with allenes and conjugated dienes to give the π-allyls Rh(allyl)(CO)(PPh$_3$)$_2$. Decomposition of the π-C$_3$H$_5$ complex yields propene, probably via activation of an ortho hydrogen of a phenyl ring of the phosphine; reaction of the same complex with hydrogen (1 atm, 25°) yields propene and HRh(CO)(PPh$_3$)$_2$, which then catalyzes the hydrogenation of the propene to propane. The same group (1718) have studied the interaction of HRh(CO)(PPh$_3$)$_3$ with alkynes (HC≡CR; R = alkyl) to give the acetylides Rh(C≡CR)(CO)(PPh$_3$)$_3$, which dissociate a phosphine in solution; reaction of the acetylide with hydrogen (50 atm, 70°) yields alkane and regenerates the hydride. Booth and Lloyd

(1719), however, have synthesized $Rh(RCH{=}CR)(CO)(PPh_3)_2$ complexes from $HRh(CO)(PPh_3)_3$ with RC_2R alkynes (R = CO_2Me, CO_2Et, CO_2H, Ph, CF_3); reaction of the CO_2Me or Ph complexes with HCl yielded the *trans* olefins, fumarate and stilbene, respectively.

The crystal structure of the binuclear rhodium(o) hydrogenation and hydroformylation catalyst $[Rh(CO)(PPh_3)_2]_2$, $2CH_2Cl_2$ (**117**) has been reported (1720).

XI-E

Strohmeier and Rehder-Stirnweiss (1721) have reported further hydrogenation rates using $RhX(CO)L_2$ complexes in toluene at 80°. For hydrogenation of heptenes, cyclohexene, β-bromostyrene, phenylacetylene, and hexyne the activity of $RhX(CO)(PPh_3)_2$ complexes decreased in the order X = Cl > Br > I > SCN; for styrene, 1,3-cyclooctadiene, and α,β-unsaturated esters no dependence on halide was observed. In all cases, the rate decreased in the order L = PPh_3 > $P(C_6H_{11})_3$ > $P(OPh)_3$ for $RhCl(CO)L_2$ complexes.

The complexes $RhX(CO)(PPh_3)_2$ (X = Cl, ClO_4) catalyze the hydrogenolysis of oxygen to water (1554). The perchlorate complex reacts with ethylene to give the cationic species $Rh(C_2H_4)(CO)(PPh_3)_2^+$ (1494a). The *trans*-$Rh(OCOR)CO(PPh_3)_2$ carboxylate complexes act as hydrogenation catalysts for alkenes and alkynes (1683).

Phosphine ligand exchange in $MX(CO)L_2$ complexes [M = Rh, Ir; X = halide; L = PPh_3, $P(OPh)_3$] has been studied by Strohmeier and coworkers (1722); the rhodium complexes generally seem more labile. Further studies have appeared on the syntheses of $MX(CO)(PPh_3)_2$ complexes (M = Rh, Ir; X = NCO, NCS, NCSe) (1508d, 1722a). Some analogous compounds with Group VI donors $RhCl(CO)(LEt_2)_2$ (L = S, Se, Te) have been made (1723), as well as the complexes $Rh(GeCl_3)CO(PPh_3)_2$, $RhCl(CO)(GeCl_3)_2^{2-}$, and $RhX_n(CO)(SnX_3)_{3-n}^{2-}$ (X = certain halides; n = 0, 1, 2) (681, 1724–1726).

Collman and coworkers (1727) have made $MCl(CO)L$ complexes, where M is Rh, Ir and L is the tripodal ligand $CH_3C(CH_2PPh_2)_3$; the complexes may be 4- or 5-coordinate in solution, but reactivity toward H_2 was not reported. A study in which L was a resin bound phosphine (1705) was mentioned in Appendix XI-B-7, XI-B-9.

XI-F-1

Heil and coworkers (1728) have reported in more detail on the reaction of $Rh_4(CO)_{12}$ with ethylene under hydroformylation conditions (eq. 363a,

Section XI-F-1); the by-product carboxy derivative $[Rh(CO)_2(O_2CC_2H_5)]_2$ reacts with CO/H_2 mixtures to give propionic acid and $Rh_4(CO)_{12}$. Chini and coworkers (1729) have reported the following stoichiometric hydroformylation at ambient conditions in toluene:

$$3Rh_4(CO)_{12} + 4CH_3CH{=}CH_2 + 4H_2 \longrightarrow$$
$$2CH_3CH_2CH_2CHO + 2(CH_3)_2CHCHO + 2Rh_6(CO)_{16} \quad (534)$$

Some hydrogenations using polymer-supported $Rh_4(CO)_{12}$ and $Rh_6(CO)_{16}$ catalysts have been reported (1705).

Of a number of patents (1730–1735) describing hydroformylation of olefins using rhodium carbonyls, those by Falbe (1734, 1735) are concerned with alcohol production: bis(hydroxymethyl)bicycloheptanes are formed from bicycloheptene monoaldehydes via the bicycloheptane dialdehydes, and tricyclodecanedimethanol is formed by hydrogenation of the dialdehyde hydroformylation product of dicyclopentadiene.

XI-F-2

A number of patents and reports (1659, 1736–1743) have described the use of $RhCl(CO)(PPh_3)_2$, and $HRh(CO)(PPh_3)_3$ or their analogs, including aminophosphine complexes, for the hydroformylation of terminal olefins to aldehydes. Unsaturated fats have been selectively hydroformylated with these catalysts (Appendix XIV). The catalysts have also been used heterogeneously on inert supports (1712–1714).

Some organophosphorus-rhodium carbonyl hydroformylation catalysts have been prepared by adding phosphines or phosphites to $Rh(CO)_2L$ complexes, where L = salicylaldoximate (1744), a diketonate (1745), or a chelating nitrogen donor ligand, such as N,N'-(1,3-dimethyl-1-propen-1-yl-3-ylidene)di-o-toluidinate (1746). A carboxyalkylphosphine ligand system (1653a) and use of a π-$C_5H_5Rh(CO)(PBu_3)$ catalyst (1747) give high alcohol yields directly from terminal olefins.

Booth and coworkers (1748) have studied the reaction of $Rh_4(CO)_{12}$ and $Rh_6(CO)_{16}$ with phosphines and phosphites (L); under carbon monoxide, complexes such as $[Rh(CO)_nL_{4-n}]_2$ ($n = 1, 2$) were isolated, some of which were hydroformylation catalysts under mild conditions. $Rh_4(CO)_{10}(PPh_3)_2$ and related complexes containing a tertiary amine are similarly effective (1729, 1749).

Cis-$RhCl(CO)_2L$ complexes, where L is an amine, a monomeric tertiary phosphine, or similar resin polymeric ligands, have been made by Rollman (1750); the resin amine complexes reacted with 1:1 H_2/CO mixtures to give carbonyl anion aggregates such as $Rh_3(CO)_{10}{}^-$ and $Rh_{12}(CO)_{30}{}^{2-}$.

XI-G

The use of supported rhodium-phosphine catalysts, where the donor phosphorus atoms are built into a divinylbenzene-styrene copolymer, has been described by Capka and coworkers (1751); the olefinic bonds in hept-1-ene, crotonaldehyde, and vinyl acetate were readily reduced. Manassen (1587) has used a similar system to reduce camphene to a mixture of *endo*- and *exo*-isocamphanes.

A detailed paper by Schrock and Osborne (1752) has further discussed cationic species of the type $Rh(diene)L_n^+$ ($n = 2$, 3), $H_2RhL_2S_2^+$, RhL_4^+, and $H_2RhL_4^+$ (L = tertiary phosphine or arsine, and S = solvent); the $H_2RhL_2S_2^+$ complexes lose hydrogen to give $RhL_2S_2^+$ species, and also react with excess dienes to yield $Rh(diene)L_2^+$ complexes together with liberated alkene. The cations $Rh(CO)L_3^+$, $Rh(CO)_3L_2^+$, and $Rh(CO)L_2S^+$ were also made, and the last-mentioned was ineffective for hex-1-ene hydrogenation under mild conditions. $Rh(diene)L_2^+$ complexes catalyze the reduction of internal alkynes specifically to *cis*-alkenes (1753). Johnson and coworkers (1754) have given details of their studies on the $Rh(diene)L_2^+$ complexes.

Couch and Robinson (1593), Green and coworkers (1755) and Haines (1756, 1757) have also further reported on the cationic species RhL_4^+, RhL_5^+, and $Rh(diene)L_n^+$ ($n = 2$, 3; L = tertiary phosphine, arsine, or phosphite), and their addition and oxidative addition reactions, including dihydride formation (1757). Kong and Roundhill (1758) have reported on the protonation of the complex $Rh[Ph_2P(OMe)]_4^+$. Barefield and Parshall (1759) have reported on the activation of ortho carbon-hydrogen bonds in the triphenylphosphite complexes, $RhCl[P(OPh)_3]_3$ and $Rh[P(OPh)_3]_4^+$, which leads to hydrogen-deuterium exchange; dihydride intermediates are involved (cf. Scheme 3 in Section IX-B-2b). The diphenylphosphine complex $Rh(HPPh_2)_4^+$ has been made (1592). The low reactivity of $Rh(Ph_2PCH=CHPPh_2)_2^+$ toward H_2 was mentioned in Appendix X-G (1663); rhodium(I) complexes of this type appear less basic than the iridium analogs (1494a). Exchange of D_2 and C_2D_4 with the hydride ligand of the complex $HRh(diphos)_2$ has been studied; the deuterium exchange is thought to involve a 7-coordinate intermediate (1561).

Intille (1760) has reported on the synthesis of a wide range of rhodium(I) and rhodium(III) tertiary phosphine(L) complexes: $RhClL_3$, $RhCl_3L_3$, $HRhCl_2L_3$, H_2RhClL_3, $HRhCl_2L_2$, $RhCl(CO)L_2$, $RhCl_3(CO)L_2$, and $HRhCl_2(CO)L_2$.

The complexes $L_2RhCl(P_4)$ (L = tertiary phosphine or arsine), containing a tetrahedral P_4 molecule, are unreactive toward both ethylene and hydrogen at $-78°$ (1761).

Oxidative addition reactions of the cationic isonitrile complexes $M(CNR)_4^+$ and $M(PPh_3)_2(CNR)_n^+$ ($M =$ Rh, Ir; $R =$ alkyl or aryl; $n = 2, 3$) have been described but reactivity toward hydrogen was not reported (1762–1765). Rhodium(III) complexes of the type $RhL_2(PPh_3)_2^+$ and $RhL_3(PPh_3)_n$, where L is an anionic sulfur donor ligand, and $n = 1, 2,$ or 3 have been made (1683).

Nappier and Meek (1766) have synthesized the halide complexes RhXL, where L is $PhP[CH_2CH_2CH_2PPh_2]_2$, a tritertiary phosphine with trimethylene linkages; the complex reacts with a variety of small molecules, although H_2 was not mentioned. Cationic species such as $HRhCl^+$, $RhL(solvent)^+$, $HRhL(solvent)^+$, and $RhL(CO)_n^+$ ($n = 1, 2$) were also synthesized.

Complexes such as $[RhClL]_2$ (L = 1,2,5-triphenylphosphole oxide), in which the metal is bonded to the π-system of the phosphole ring, have been synthesized by Holah and coworkers (1767) for potential use as catalysts.

Studies on some rhodium nitrosyls containing Group V donors have been summarized by Baird (1583) and Robinson and coworkers (1596, 1768); included were $Rh(NO)L_3$, $RhCl_2(NO)L_2$ (L = PPh_3, $AsPh_3$, $SbPh_3$), $RhCl(NO)(PPh_3)_2$, $RhCl(NO)_2(PPh_3)_2$, and $RhCl(NO)(NO_2)(PPh_3)_2$. Collman and coworkers (1769–1770) have also synthesized $RhCl_2(CO)(NO)$-$(PPh_3)_2$ and $Rh(NO)_2(PPh_3)_2^+$. Hydrogenation of the complexes $Rh(NO)(PPh_3)_2L$, where L is 1,4-benzoquinone or maleic anhydride, yields hydroquinone or succinic anhydride and a rhodium product containing no nitrosyl (1770a). The formally rhodium(II) nitrosyls are given in Appendix XI-I.

Bennett and Patmore (1771) and Nixon and Swain (1772) have reported further on trifluorophosphine complexes of rhodium and iridium. Under mild conditions, $Rh_2(PF_3)_8$ reacts irreversibly with H_2 to give $HRh(PF_3)_4$, and with Group IV hydrides RH to give $RRh(PF_3)_4$ and $HRh(PF_3)_4$; $Ph_3SiRh(PF_3)_4$ undergoes hydrogenolysis at 25° with 1 atm H_2 (1771). The iridium analogs react similarly but more slowly.

Bresadola and coworkers (1773) have synthesized rhodium and iridium phosphine complexes containing a σ-bonded carborane nucleus. A rhodium(I) complex, $1\text{-}[(PPh_3)_3Rh]\text{-}2\text{-}C_6H_5\text{-}1,2\text{-}(\sigma\text{-}B_{10}C_2H_{10})$, was said to react with both hydrogen and ethylene.

Weber and Schrauzer (1774) have reported on bisdimethylglyoximato-rhodium complexes—or rhodoximes. The chemistry seems similar to that of the corresponding cobaloximes, for example, the rhodoxime(III) hydrides with axial ligands such as pyridine, triphenylphosphine, or water behave as acids and in alkaline solution liberate the anionic rhodoxime(I) nucleophiles; organorhodoximes were sythesized; the dimeric rhodoxime(II) derivatives disproportionate into Rh^I and Rh^{III}; however, the rhodoxime(III) hydrides, unlike the corresponding cobaloximes, do not readily decompose with hydrogen evolution (cf. eqs. 312b to 312f in Section X-H).

XI-H

Full papers have appeared by James and Ng (1775, 1776) on hydrogen activation by the rhodium complexes containing sulfide ligands, and by McQuillin and coworkers (1777, 1778) describing hydrogenation using the $py_2(DMF)RhCl_2(BH_4)$ complex in DMF and in optically active amide solvents.

The cationic $M(diene)S_n{}^+$ complexes (M = Rh, Ir; S = solvent, $n = 2, 3$) have been reported on further (1753), and are useful intermediates for synthesis of a wide range of cationic complexes, for example $RhL_2S_2{}^+$, where L = diene, C_2H_4, $P(OPh)_3$, CO (see also Appendix XI-G). The complex $Rh(norbornadiene)_2{}^+$ in acetone is an effective hydrogenation catalyst for the conversion of dienes to monoenes, but norbornadiene itself undergoes hydrodimerization to a $C_{14}H_{18}$ compound (1753, 1779). Green and coworkers (1755, 1780) have also reported further on $M(diene)_2{}^+$ and $M(diene)L_2{}^+$ complexes (M = Rh, Ir; L = CH_3CN, L_2 = bipyridyl); rhodium complexes with L = PhCN, $PhNH_2$, DMSO, and arenes were also synthesized (1753, 1780).

The $Rh_2A_2Cl^-$ complexes (A = N-phenylanthranilate), and a fluorescein complex, have been used to hydrogenate olefinic bonds in some unsaturated spirocyclic compounds (1689) and some tetrahydrobenzo-1,3-dimethoxy-tetrahydrofurans (1782); patents (1781–1781b) have described the general use of these catalysts. Kinetic data for the hydrogenation of pentenes using the phenylanthranilate and tyrosine complexes have been given together with corresponding data for some heterogeneous rhodium catalysts (1783).

Bruce and coworkers (1784) have noted that $[Rh(CO)_2Cl]_2$ catalyzes the $LiAlH_4$ reduction of azobenzene to hydrazobenzene; azophenetole gives 4-ethoxyaniline, and azoxyanisole gives hydrazoanisole. A number of complexes, including chloride-bridged dimers containing Rh^I and Rh^{III}, were obtained from reaction between azobenzene and $[Rh(CO)_2Cl]_2$ (1784, 1785).

Attridge and Wilkinson (1786) have used ethanolic rhodium trichloride solutions to catalyze the disproportionation of 4-vinylcyclohexene into ethylbenzene and ethylcyclohexene, while disproportionation of cyclohexadienes to cyclohexene and benzene is catalyzed by $(\pi\text{-}C_5Me_5)Rh(C_6H_8)$ (1787). Maitlis and coworkers (1788) have used the hydride-bridged complexes $[M(C_5Me_5)]_2HCl_3$ (M = Rh, Ir) for hydrogenation of monoolefins. A $RhCl_3$-DMSO-isopropanol system becomes active for hydrogenation of styrene, cyclohexene, and phenylacetylene, only in the presence of molecular H_2 (1788a, cf. Section XII-C and Appendix XII-C).

The relative complexing abilities of ethylene, propylene, butenes, and butadiene with rhodium(I), in complexes such as $Rh(CO)_2(acac)$, have been determined (1789).

XI-I

James and Stynes (1790) have made a rhodium(II) tetraphenylporphyrin complex Rh(TPP), which is reduced by hydrogen to H[Rh(TPP)]; the rhodium(II) complex in DMF at ambient temperatures slowly catalyzes acetylene hydrogenation, and the conversion of hydrogen-oxygen mixtures to water.

The rhodium(II) complexes $trans$-RhCl$_2$(P^{t}Bu$_2$R)$_2$ (R = Me, Et, ^{n}Pr) have been isolated (1702, 1703); the dibutylmethylphosphine complex reacts with hydrogen in dichloromethane to give a 1 : 1 mixture of HRhCl$_2$(P^{t}Bu$_2$Me)$_2$ and H$_2$RhCl(P^{t}Bu$_2$Me)$_2$ (see Appendix XI-B-7, XI-B-9). Baird (1583) has synthesized Rh(NO)Cl$_3$(PPh$_3$)$_2$ and a solvated Rh(NO)Cl$_3$ complex.

XII-A

Studies on the synthesis of IrX(CO)(PPh$_3$)$_2$, X = pseudohalide, and of IrCl(CO)L, L = a tridentate tertiary phosphine ligand, were referred to in Appendix XI-E(1508d, 1722a, 1727). Phosphine ligand exchange in Vaska-type complexes has been studied (1722, Appendix XI-E).

The complexes IrX(CO)(P^{t}BuR$_2$)$_2$ (X = Cl, Br; R = Me, Et, ^{n}Pr, ^{n}Bu) form dihydrides slowly in comparison to IrX(CO)(PPh$_3$)$_2$ (1791). $Trans$-IrCl(CS)(PPh$_3$)$_2$ (1792) and [IrCl(CO)(PPh$_3$)S$_2$]$_n$ (1793) are unreactive toward hydrogen.

Chen and Halpern (1794) have shown that IrCl(CO)L$_2$ (L = PMe$_2$Ph) readily adds further phosphine ligands to give 5-coordinate complexes IrCl(CO)L$_3$ and Ir(CO)L$_4^+$. Dihydride formation with IrCl(CO)L$_3$ in chlorobenzene and in DMF was about five times faster than with IrCl(CO)L$_2$, and was attributed to a reactive ion-pair intermediate [Ir(CO)L$_3$]Cl.

Strohmeier and coworkers (1795–1797) have given further details on the IrX(CO)L$_2$ catalysts (L = tertiary phosphine or phosphite); no relationship was observed between the hydrogenation rate of monoolefins or phenyl-acetylene and the π-acceptor strength of L. Two reports (1795, 1797) concluded a reactivity order X = Cl > Br > I; the other report (1796) concluded that hydrogenation rates were not greatly dependent on the halide. A report has appeared on the use of IrCl(CO)(PPh$_3$)$_2$ for hydrogenation of maleic acid in a range of solvents (1798). Osumi and Yamaguchi (1799, 1800) have patented the use of IrCl(CO)(PPh$_3$)$_2$ and H$_a$IrX$_b$(CO)$_c$L$_d$ complexes ($a, b, c =$ 0 or integers, $d =$ an integer; X = halide; L = tertiary Group V ligand) for hydrogenation of olefins usually under H$_2$/CO mixtures.

The complexes IrX(CO)(PPh$_3$)$_2$ (X = Cl, Br), as well as fac-H$_3$Ir(CO)

$(PPh_3)_2$, $HIr(CO)(PPh_3)_3$, and $H_3Ir(PPh_3)_3$, have been used to catalyze the conversion of hydrogen-oxygen mixtures to water (1554). The mechanism of propylene deuteration using $IrCl(CO)(PPh_3)_2$, $HIr(CO)(PPh_3)_3$, and $HIrCl_2$ $(PPh_3)_3$ has been studied (1715).

A study of catalysts corresponding to $IrCl(CO)(PPh_3)_2$ and $IrCl(PPh_3)_3$ on a cross-linked polystyrene-diphenylphosphine resin (1705) was referred to in Appendix XI-B-7, XI-B-9.

$IrCl(CO)(PPh_3)_2$ catalyzes deuterium exchange with acetic acid, methanol, and morpholine (1555), and the complex has been used to reduce α,β-unsaturated ketones using other hydrogen donor solvents in the absence of molecular hydrogen (1558, Appendix IX-B-2b). C_6 to C_{20} paraffins and aromatics such as 9,10-dihydroanthracene have been catalytically dehydrogenated using $IrCl(CO)(PPh_3)_2$ (1558a, 1800a).

Vaska's compound has been used to catalyze the formation of DMF from dimethylamine, carbon dioxide, and hydrogen (1507a).

XII-B-1

Some rate data for hydrogenation of olefins and acetylenes catalyzed by $IrCl(PPh_3)_3$ have been reported (1684, Appendix XI-B-3). The complexes $H_3Ir(PPh_3)_3$ and $H_2IrCl(PPh_3)_3$ catalyze deuterium exchange with alcohols, acetic acid, and morpholine (1555). The trihydride also catalyzes exchange between methane and deuterium (1569a). Reaction of the trihydride with diazonium cations leads to insertion products such as $H_2Ir(NH{=}NAr)(PPh_3)_3{}^+$, where Ar is aryl (1700).

Maspero and Perrotti (1801) have used $H_3Ir(PPh_3)_2$ in isopropanol to reduce methylbutynol to methylbutenol in the absence of hydrogen, and Fanta and Erman (1801a) have used the iridium-trimethylphosphite-isopropanol system to reduce 3-isobornylcyclohexanone specifically to the *trans*-cyclohexanol product.

Dihydride formation by $Ir(Ph_2PCH{=}CHPPh_2)^+$ was considered in Appendix X-G (1663). Doronzo and Bianco (1802) have also reported on the same system, and on the corresponding reaction with the 5-coordinate carbonyl complex, which forms the dihydride with loss of carbon monoxide. In contrast to the corresponding diphos system, the oxygen of the complex $IrO_2(Ph_2PCH{=}CHPPh_2)^+$ is not readily replaced by molecular hydrogen. Some rate data have appeared for the reaction of hydrogen with $Ir(diphos)^+$ in chlorobenzene (1494a).

Further studies on cationic complexes of the type $Ir(diene)L_n{}^+$, $IrL_4{}^+$, and $IrL_5{}^+$ (L = tertiary phosphine or phosphite; $n = 2$, 3) have been reported (1593, 1755, 1803). Cationic isonitrile complexes, with and without coordinated triphenylphosphine, have been reported (1762, Appendix XI-G). The use of cationic catalysts containing Group V or VI donor ligands has been patented (1588, Appendix II-A, II-B).

Robinson and coworkers (1804) and Bennett and Charles (1804a) have reported details of the studies on iridium triarylphosphite complexes and their intramolecular ligand-metal hydrogen transfer and elimination reactions. The complex $IrCl[P(OPh)_2(o\text{-}OC_6H_4)]P(OPh)_3$ has been studied crystallographically (1805).

Haszeldine and coworkers (1806) have investigated compounds of the type $(C_8H_{12})IrClL$ (L = tertiary phosphine or arsine, py, quinoline). The triphenylphosphine complex does not react with hydrogen but catalyzes hydrogenation of hex-1-ene; the pyridine and quinoline complexes are reduced by hydrogen to the metal. The hydride $HIr(C_8H_{12})Cl_2(PPh_3)$ on reaction with ethylene transfers hydrogen chloride to give ethyl chloride. Van der Ent and Cuppers (1807) have patented the use of $[IrCl(C_8H_{14})_2]_2$ in the presence of phosphines and arsines as a hydrogenation catalyst.

Shaw and coworkers (1808) have described the 5-coordinate hydrides $HIrCl_2(P^tBu_2R)_2$ (R = Me, Et, ^nPr) for which the hydride ligands have the largest τ-values ($\sim$60) yet observed.

The chemistry of iridium complexes containing trifluorophosphine, such as $IrCl(PF_3)_4$, $[IrCl(PF_3)_2]_2$, and $Ir_2(PF_3)_8$, has been studied further (1771; see Appendix XI-G).

XII-B-2

Burnett and Morrison (1809) have studied the kinetics and mechanism of ethylene hydrogenation catalyzed by $HIr(CO)(PPh_3)_3$ in DMF. The dihydride adduct was shown to be 6-coordinate $H_3Ir(CO)(PPh_3)_2$, but no adduct was formed with ethylene at ambient conditions. The rate law, of the form $k[Ir]_T[C_2H_4]/(1 + k'[H_2] + k''[PPh_3])$, shows an inverse dependence on hydrogen; the active species is thought to be $HIr(CO)(PPh_3)_2$ which reacts in the rate-determining step with ethylene to form the ethyl complex. A faster subsequent hydrogenolysis completes the catalytic cycle.

Studies on exchange of D_2 and C_2D_4 with the metal hydride in $HIr(CO)(PPh_3)_3$ have appeared (1561).

$HIr(CO)(PPh_3)_3$ or $HIr(CO)_2(PPh_3)_2$ reacts with alk-1-ynes to yield the acetylides $Ir(C{\equiv}CR)(CO)(PPh_3)_n$ (n = 3,2) and hydrogen; under severe conditions (50 atm H_2, 70°) the acetylides undergo hydrogenolysis to $HIr(CO)(PPh_3)_3$ and presumably alkane (1718, 1810).

The crystal structure of the $(C_6N_4H)Ir(CO)(PPh_3)_2(C_6N_4)$ complex has been reported (1811).

XII-B-3

The cationic complex $Ir(CO)(PPh_3)_2(C_2H_4)^+$ has been made (1494a), and such intermediates have been postulated for olefin hydrogenations catalyzed by $IrCl(CO)(PPh_3)_2$ in sulfolane (1798). Complexes such as $Ir(CO)L_2(C_2H_4)_2^+$, $Ir(CO)L_2(diene)_2^+$, L = PMe_2Ph (1812), and the diphenylphosphine complex $Ir(CO)(HPPh_2)_4^+$ (1592) have also been made.

Shaw and Stainbank (1813) have extended their studies on protonation of Vaska-type complexes, by using phosphine ligands containing t-butyl groups.

Mays and Stefanini (1814) have synthesized cationic thiocarbonyls of the type $Ir(CO)_2(CS)L_2^+$ and $Ir(CO)(CS)L_3^+$ (L = tertiary phosphine), and $Ir(CS)(diphos)_2^+$. $Ir(CO)_2(CS)(PPh_3)_2^+$ does not react with hydrogen, but the corresponding tris(cyclohexyl)phosphine complex reversibly forms the dihydride $H_2Ir(CO)(CS)[P(C_6H_{11})_3]_2^+$.

XII-B-4

The crystal structures of the following nitrosyls have been reported: $IrCl_2(NO)(PPh_3)_2$ (1815), $IrI(CH_3)(NO)(PPh_3)_2$ (1816), $Ir(NO)(PPh_3)_3$ (1817), and $HIr(NO)(PPh_3)_3^+$ (1818). New ways of synthesizing $IrX_2(NO)$-$(PR_3)_2$ (X = halide, R = alkyl or aryl) have been reported (1768). $IrCl(NO)$-$(CS)L_2^+$ and $IrCl(CO)(NO)[P(C_6H_{11})_3]_2^+$ have been made (1814).

XII-C

$Ir(C_8H_{12})_2^+$ in acetone effectively catalyzes the reduction of 1,5-cyclo-octadiene to cyclooctene (1753); the iridium cation is synthesized from solvated IrL_2^+ species (L_2 = cyclooctadiene; L = cyclooctene), which are useful sources for a wide range of cationic complexes (1753, 1755, Appendix XI-H), some of which are hydrogenation catalysts (Section XII-B-1). Chan and James (1819) have used the cyclooctene dimer $[IrCl(C_8H_{14})_2]_2$ for selective hydrogenation of cyclooctene in mixtures with hex-1-ene; kinetics and a possible mechanism were presented.

The hydride-bridged complex $[Ir(C_5Me_5)]_2HCl_3$ has been used for hydrogenation (1788, Appendix XI-H).

Gullotti and coworkers (1788a) have used the chloroiridate-DMSO-isopropanol system without hydrogen for reduction of acetophenone and methyl α-naphthyl ketone to the alcohols; some dehydration to styrene and α-vinylnaphthalene occurs with subsequent reduction of the olefinic bonds. Aliphatic aldehydes were reduced more slowly, although crotonaldehyde was readily reduced to 1-butanol; reduction of aromatic aldehydes was difficult. Only olefins and acetylenes activated by a substituent electronegative group were reduced effectively; cyclohexa-1,3-diene gave cyclohexene, phenyl-acetylene gave ethylbenzene. The catalytic system was found to be more efficient under hydrogen. Other sulfoxides with bulkier alkyl groups, gave lower rates.

Some amine-stabilized iridium(III) hydrides, such as $HIr(piperidine)_4I_2$, have been reported (1820).

XII-D

The complex $HIr(CO)_2(PPh_3)_2$ reacts with allene and conjugated dienes to give the π-allyl complexes $Ir(allyl)(CO)(PPh_3)_2$ (1717). The allyls undergo reversible carbon monoxide insertion reactions; hydrogenolysis of the allyls liberates the olefin, and $HIr(CO)(PPh_3)_2$, which is subsequently converted to $H_3Ir(CO)(PPh_3)_2$.

XIII-A

Bingham and Burnett (1821) have made a detailed kinetic study of the stoichiometric reduction of butadiene to butenes using hexacyanodinickel-ate(I) in aqueous solution. Formation of a π-butadiene and a π-1-methylallyl complex are rate-determining; butene production occurs via reduction steps involving $Ni(CN)_3^{3-}$ and/or water (cf. eqs. 413, 413a, and 413g in Section XIII-A). Hashimoto and coworkers (1822) have used aqueous solutions of $Ni_2(CN)_6^{4-}$ for the hydrogenolysis of certain organic halides (cf. Section X-B-3).

Details of the $trans$-$HNiX(PR_3)_2$ complexes (X = halide, NCS, CN, BH_4; R = ^iPr or cyclohexyl) have appeared (1823, 1824). Nickel(II) alkyls have been isolated (e.g., 1824a).

Ando and coworkers (1825) have studied the slow D_2—H_2 exchange using $NiX_2(PPh_3)_2$ (X = halide) in various solvent systems at 35°; an intermediate $HNiX(PPh_3)_2$ was postulated. Exchange between C_2H_4 and C_2D_4, in the presence of hydrogen, occurs via reversible formation of an ethyl intermediate from the hydride. Hydrogenation must be negligible under these conditions. Maruyama and coworkers (1826) have similarly studied isotopic exchange of C_2H_4—C_2D_4 using bis(triphenylphosphine)-σ-1-naphthyl-nickel(II) bromide; the intermediate hydride is formed by reaction of the complex with ethylene. Further patents have described the use of $NiI_2(PPh_3)_2$ for hydrogenation of 1,5,9-cyclododecatriene to cyclododecenes (1827), and the synthesis of complexes containing chelating phosphines or arsines for use as hydrogenation catalysts (1551a). The $NiX_2(PPh_3)_2$ complexes (X = halide) also catalyze the hydrogenation of cyclooctadienes in alcohols and especially dihydroxy-benzenes, at elevated temperatures under vacuum (1552). Of interest is a report by Kanai (1828) on the formation of active isomerization catalysts by adding tin(II) chloride to solutions of reduced $NiX_2(PPh_3)_2$ complexes.

The allyl complexes $(\pi\text{-}C_3H_5)MClL_2$ (M = Ni, Pd; L = tertiary phosphine, arsine or phosphite) catalyze the hydrogenation of conjugated polyenes to monoenes at 100 atm H_2 and 120° (1829).

The isomerization of but-1-ene catalyzed by acid solutions of $Ni[P(OEt)_3]_4$ involves addition-elimination of the hydride $HNi[P(OEt)_3]_3$. Tolman (1830) considers that the small amounts of butane observed arise from protonation of the butyl intermediates.

Coordinated azobenzene in the zerovalent complex $(Me_3P)_2Ni(PhN{=}NPh)$ is reduced to hydrazobenzene by treatment of the complex with aqueous ethanol (1831).

The complexes $Ni(CO)_nL_{n-4}$ (L = tertiary phosphine or phosphite) are said to be effective hydroformylation catalysts (1740). Nickel(II) carbonyl halide species have been identified by a matrix isolation procedure (1832).

XIII-B

Alvarez and Pino (1833) have reported that aqueous sulfuric acid solutions of palladous chloride at 50–90° and 1 atm H_2 catalytically reduce vanadium(V) to vanadium(III).

Freidlin and coworkers (1834) have used palladium(II)-DMSO complexes at ambient conditions for catalytic hydrogenation and isomerization of pentenes. Olefin hydrogenation has also been effected using $PdCl_2(PhCN)_2$ supported on an ion-exchange resin (1587). Benzene solutions of $PdCl_2(PhCN)_2$ in the absence of hydrogen do not catalyze isotope redistribution in *trans*-$C_2H_2D_2$, although isomerization to the *cis* ethylene occurs (1563). Mechanisms via a free radical or a carbonium ion were suggested.

Stern and Maples (1835, 1836) have hydrogenated and deuterated mono-olefins and conjugated dienes using catalysts such as $PdCl_2(diphos)$, $Pd(diphos)_2$, and $Pd_2(Ph_2PCH_2PPh_2)_3$ in toluene at room temperature with a few atmospheres of H_2 or D_2.

Green and coworkers (1823) have reported further on the hydrides *trans*-$HPdX(PR_3)_2$ (X = halide, NCS, BH_4; R = ^iPr or cyclohexyl); the cationic species $HPd(diphos)(PR_3)^+$ have also been made (1837). Formation of species containing metal-carbon bonds from the phosphite complexes $MX_2[P(OR)_3]_2$ (M = Pd, Pt; X = halide; R = Ph, *o*-, *m*- or *p*-tolyl) by hydrogen abstraction from an orthophenyl position has been reported by Ainscough and Robinson

(1838). Complexes such as $PtCl[P(OC_6H_4)(OPh)_2][P(OPh)_3]$ (1838), and the cationic species $ML_4{}^{2+}$, $MXL_3{}^+$, and $PdL_5{}^{2+}$ [$L = P(OMe)_3$ or $P(OEt)_3$] (1593) have been isolated. Solvated species such as $Pd(\pi\text{-allyl})^+$ appear useful for the synthesis of a wide range of cationic palladium(II) complexes (1753).

The allyl complex $(\pi\text{-}C_3H_5)PdCl(PEt_3)_2$ has been used as a hydrogenation catalyst (1829, Appendix XIII-A), and $PdCl_2(PPh_3)_2$ and $Pd(CO)_nL_{4-n}$ complexes ($L = $ tertiary phosphine or phosphite) have been used for hydroformylation of olefins (1740, 1839).

Kingston and Scollary (1840) have reported further on $Pd(CO)Cl(SnCl_3)_2{}^-$ and its reactions with tertiary Group V ligands. Matrix isolation techniques (1841) have yielded evidence for $Pd(CO)_4$.

The carbonate complex $Pd(CO_3)(PPh_3)_2$ catalyzes the formation of amides from amines, carbon dioxide, and hydrogen (1507a).

XIII-C-1

Shilov's group (1842) have reported further on deuterium exchange with methane and ethane. For $PtX_4{}^{2-}$ catalysts, the activity order was $Cl > Br > I \gg CN$; $Pt(PPh_3)_3$ and $H_2Pt(CO)Cl_2$ were ineffective; an activation energy of 18.6 kcal mole^{-1} was measured for the $PtCl_4{}^{2-}$-catalyzed exchange with ethane. Details of the similar studies by Hodges and coworkers (1843) have appeared; a solvated neutral platinum(II) dichloride species was thought to be the active catalyst. Garnett and coworkers (1844) have reported further on hydrogen exchange in aromatic compounds catalyzed by tetrachloroplatinate(II), and have noted similarities in exchange in the side chain of alkylbenzenes to the exchange observed in simple alkanes (1845). Garnett (1847) has reviewed these platinum-catalyzed hydrogen exchange systems. Gill and Shaw (1847a) have reported the formation of platinum complexes involving metallation of alkyl groups attached to a benzene ring.

Attempted hydrogenation of the complex dichloro(pentamethyl-5-vinylcyclopentadiene)platinum led to an isomerization reaction involving a stereochemical change at the 5-position, the *endo* vinyl group becoming *exo* to the metal. A dihydride intermediate seems likely (1849).

XIII-C-2

Yasumori and coworkers (1850, 1851) have carried out kinetic studies on the hydrogenation of ethylene and acetylene, and on but-1-ene isomerization using H_2PtCl_6—$SnCl_2$ (1:10) in methanol at 0 to 20° and <1 atm H_2. The active catalyst was written as $Pt(SnCl_3)_5{}^{3-}$, and the mechanism for ethane production involved protonolysis of an ethyl intermediate formed via a mono-hydride species and ethylene.

The complex $Pt(CO)Cl(SnCl_3)_2^-$ and its reactions with triphenylphosphine have been described more fully (1840).

XIII-C-3a

Clark and Kurosawa (1852, 1853) have studied insertion reactions of olefins and dienes with $trans$-$HPtXL_2$ complexes (X = Cl, Br; L = tertiary phosphine) in the presence of $AgBF_4$ or $AgPF_6$; the olefin reactions occur via the 4-coordinate cationic $trans$-$HPtL_2(olefin)^+$ complexes; thus at $-50°$ ethylene gives $trans$-$HPtL_2(C_2H_4)^+$ while at room temperature the product is $trans$-$Pt(C_2H_5)L_2(C_2H_4)^+$. Interesting preparative solvated intermediates $trans$-$HPt(PPh_2Me)_2(acetone)^+$ (1854) and $Pt(dicyclopentadiene$-$OCH_3)^+$ (1753) have been made.

Cherwinski and Clark (1855) have reported further on the reaction of $trans$-$PtX(CO)L_2^+$ complexes (X = anion, L = tertiary phosphine) with water to give $trans$-$HPtXL_2$, and with methanol to give $trans$-$PtX(CO_2CH_3)L_2$. Of interest is the development by Clark's group of the chemistry of cationic acetylenic platinum complexes, which show reactivity characteristic of carbonium ions (e.g., 1856).

Roundhill and coworkers (1857) have reported details of their study on reaction of protonic acids (HX) with $(PPh_3)_2Pt(acetylene)$ complexes to yield the olefin and $PtX_2(PPh_3)_2$; the reaction proceeds via a platinum(II) hydride to the isolable vinyl complex, and then via a platinum(IV) hydride to products. Since the acetylene complex could readily be prepared from $PtX_2(PPh_3)_2$, the mechanism was described as "pseudocatalytic." Shaw and coworkers (1858) have protonated such acetylene complexes to the alkenyl stage. Earlier reports of these protonation studies were mentioned in Section XIII-C, in relation to catalysis by iridium complexes.

Intramolecular metal-ligand hydrogen transfer occurs in some tertiary aryl phosphite complexes; cationic complexes such as $Pt[P(OMe)_3]_4^{2+}$ and $PtCl[P(OMe)_3]_3^+$ are known (1593, 1838; Appendix XIII-B).

Cationic azo complexes of platinum(II), $(PPh_3)_3Pt(N{=}N{-}C_6H_4R$-$p)^+$ (R = H, F, NO_2, Me, OMe, NMe_2, NEt_2) react with hydrogen at ambient conditions to yield $HPt(PPh_3)_3^+$, C_6H_5R, and nitrogen (1859; cf. Ref. 1699, Appendix XI-B-5). Quinone complexes such as $Pt(PPh_3)_2(quinone)$ are inert toward hydrogen, although reaction with hydrogen chloride yields hydroquinone and cis-$PtCl_2(PPh_3)_2$ (1860).

Since formic acid has been used as a hydrogen donor in catalytic hydrogenations (e.g., Section XII-B-1), it is of interest to note a mechanistic study on the catalyzed decomposition of the acid to hydrogen and carbon dioxide. Dixon and Hawke (1861) used platinum(II) phosphine complexes, and formulated a reaction scheme via a carboxylate intermediate, which lost carbon dioxide to give a hydride; protonation of the hydride liberated hydrogen and regenerated catalyst.

Blake and Nyman (1862) have described some interesting photochemical reactions of the oxalate complex $Pt(PPh_3)_2(C_2O_4)$. In an inert atmosphere, photolysis yields $Pt_2(PPh_3)_4$ via a postulated $Pt(PPh_3)_2$ intermediate. In the presence of hydrogen, the product is $Pt_4(PPh_3)_5(CO)_3$, the oxalate being the source of carbon monoxide; the hydrogen is thought to reduce the cationic platinum(I) intermediate $Pt(PPh_3)_2{}^+$ to platinum(O). Small amounts of hexane and cyclohexane products were detected when hex-1-ene or cyclohexene were present in the photolysis experiments under hydrogen.

An ill-defined compound $PtO(PPh_3)_n$, formed from reaction of hydrogen with oxygen complexes of $Pt(PPh_3)_n$, catalyzes the hydrogenolysis of oxygen to water (1554). The complexes $Pt(PPh_3)_3$, $Pt(CO_3)(PPh_3)_2$ and $H_2PtCl_4 \cdot 6H_2O$ have been studied as catalysts for the conversion of amines to amides under H_2/CO_2 atmospheres (1507a). The complex $Pt(PEt_3)_3$, in contrast to the triphenylphosphine analog, readily forms a 5-coordinate dihydride by reaction with molecular hydrogen (1862a) (cf. Section XIII-A).

XIV

Hydrogenation of unsaturated fats has been reviewed by Ucciani (1863) and Linsen (1864).

Organophosphine-cobalt carbonyl catalysts have been used to hydrogenate unsaturated fats (1657; see Appendix X-F-1, X-F-2).

Frankel and coworkers (1865) have used methyl cis-9, cis-15-octadecadienoate as a model substrate for hydrogenation of methyl linolenate catalyzed by platinum-tin, palladium and nickel complexes. Isomerization to conjugated dienes precedes reduction to a mixture of isomeric monoenes for the diene and linolenate systems. Nickel(II) chloride-borohydride solutions (cf. Section XV-A) have been used by Hinze and Frost (1866) to hydrogenate linoleate, soybean oil, as well as methyl sorbate.

Frankel and Thomas (1867, 1868) have used the rhodium-carbonyl-triphenylphosphine system (Section XI-F-2) to selectively hydroformylate unsaturated fats. For example, oleate gave formyl stearate; safflower methyl esters yielded formyl stearate, diformyl stearate, as well as formyl oleate which could subsequently be hydroformylated and hydrogenated to mono- and diformyl stearate.

Soybean oil fatty acids and methyl linoleate have been hydrogenated using the $[IrCl(C_8H_{14})_2]_2$ complex in the presence of added triphenylphosphine (1807).

XV

Sharifkanova and coworkers (1869, 1870) and Shmidt and coworkers (1871) have reported further on the activity of $M(acac)_n$-Et_3Al catalysts (M = Fe, Cr, Co, Ni; $n = 2$ or 3) for hydrogenation of styrene, cyclohexene, hept-1-ene, acetylene, hept-1-yne, and phenylacetylene. Similar ferric catalysts have been used under hydrogen for isomerization of hept-1-ene (1872).

Falk (1873) has used $M(2$-ethylhexanoate$)_2$-LiR catalysts (M = Co, Ni; R = alkyl or aryl) for reduction of cyclic and acyclic olefins with a few atmospheres of hydrogen. Increasing substitution at an olefinic bond decreases the hydrogenation rate. It was thought that the catalysts might exist as finely suspended particles. Cobalt acetylacetonate-butyllithium catalysts have been used by Butte to effect reduction of C_2-C_8 monoolefins (1874, 1875), and Morris and coworkers (1875a) have reduced adiponitrile to ω-aminocapronitrile and hexamethylenediamine using a cobalt alkanoate-$^iBu_3Al$ catalyst. Faraday and Marko (1876) have reported on the alkyl products formed in reactions between olefins and Grignard reagents in the presence of nickel chloride.

Kroll (1877) has patented the use of π-$C_5H_5M(CO)_3$-R_2Al systems (M = Mo, W; R = alkyl), with and without added nitrogen-, oxygen-, or phosphorus-donors, for catalytic hydrogenation.

Sajus and coworkers have extended their work on the transition metal-alcoxyalanate catalysts. A paper (1878) reports the reduction of cyclododecatriene to cyclododecane, and patents (1879–1883) report on hydrogenation of dienes, cycloolefins, alk-1-ynes, and nitriles.

The considerable interest in partial hydrogenation of polymers using Ziegler catalysts is evidenced by the number of patents that have appeared (1884–1899). Diene (particularly butadiene) -copolymers are the usual substrates, and nickel- or cobalt-aluminum alkyls with and without co-catalysts usually comprise the Ziegler catalyst system. Falk and Schlott (1900–1902) have published papers in the open literature on such systems.

XVI-C

Redox reactions of atomic hydrogen with platinum(II and IV) complexes containing chloride and ethylenediamine ligands have been studied by Shagisultanova and coworkers (1903).

XVI-G

Symons and Buncel (1904) have studied exchange of molecular deuterium in DMSO-water mixtures in the presence of base (tetrabutylammonium hydroxide). The mechanism was written as follows:

$$HO^- + D—D \xrightarrow{\text{slow}} [H—O—D—D]^- \tag{535}$$

$$[H—O—D—D]^- + HOH \longrightarrow HOD + H—D + HO^- \tag{536}$$

Formation of an intermediate hydride (eq. 510, Section XVI-G), which is considered unlikely on theoretical arguments (1905), was ruled out.

Botter and coworkers (1906) have described an interesting process for deuterium enrichment of hydrogen by exchange with D_2O using ammonia as a homogeneous catalyst.

XVI-I

Legall and coworkers (1907) have presented evidence for the involvement of nonheme iron, bonded to a sulfhydryl group, in the active site of a hydrogenase preparation.

XVI-J

Nitrogen fixation remains topical, and a review by Chatt and Leigh (1908) covers the literature to the end of 1971. Recent model systems have incorporated particularly molybdenum and/or iron centers (e.g., 1548, 1909–1911).

XVI-K

Studies on comparisons of homogeneous and heterogeneous catalysis have intensified, particularly since a further number of homogeneous catalysts have been "heterogenized" by adsorption onto inert supports or by building the catalyst into polymeric ligands: 1507b, 1587–1589 (Appendix II-A, II-B, IX-B-2c, XI-B-6, and XIII-B), 1627, 1628 (Appendix X-E-1), 1704, 1705 (Appendix XI-B-7, XI-B-9, XI-F-1, and XII-A), 1712–1714 (Appendix XI-D and XI-F-2), 1750 (Appendix XI-F-2), and 1751 (Appendix XI-G). Further development of this potentially useful modification, in which product and catalyst recovery are very much simplified, can be expected. Other studies reported in this Appendix have drawn analogies to heterogeneous systems (1509, 1685, 1686, 1783, 1793).

In this context, mention should be made of the developments by Tamaru and coworkers (1912–1914) who have used graphite-lamellar compounds for a range of catalytic reactions, including hydrogen-deuterium exchange, hydrogenation of carbon monoxide, olefins and acetylenes, methane-deuterium exchange, and conversion of nitrogen-hydrogen mixtures to ammonia.

XVII

Undoubtedly the most promising and interesting areas that have been investigated during 1971 and 1972 are those concerning asymmetric hydrogenation, and those involving the heterogenizing of homogeneous catalysts (see Appendix XVI-K).

Reports on asymmetric synthesis are found in Section XI-C, and Appendix IX-B-2c (1574–1576), Appendix X-D (1615, 1616), Appendix X-E-1 (1619, 1619a), Appendix X-H (1669), Appendix XI-C (1706–1711), and Appendix XI-H (1778). The study of asymmetry in organometallic chemistry is not new, but since, for example, L-dopa has now been produced in 80% optical yields by catalytic hydrogenation of β-substituted α-acylaminoacrylic acids, the efforts to produce high stereospecificity in reactions such as hydrogenation and some hydroformylation processes will almost certainly intensify.

Many biological systems operate essentially by the use of heterogenized homogeneous catalysts, the catalyst support being a large, semiordered, insoluble polymer. The high stereospecificity of enzyme systems is well known, and the advances being made in supported homogeneous catalysts and asymmetric synthesis suggest that studies toward improved model systems for metalloenzymes will also be intensified.

References

1. K. B. Harvey and G. B. Porter, *Introduction to Physical Inorganic Chemistry*, Addison-Wesley, Reading, Mass., 1963, p. 326.
2. W. M. Latimer, *Oxidation Potentials*, 2nd ed., Prentice-Hall, Englewood Cliffs, N.J., 1952, p. 35.
3. National Bureau of Standards, Technical note 270-3, *Selected Values of Chemical Thermodynamic Properties*, U.S. Government Printing Office, Washington, D.C., 1968, p. 12.
4. J. Jortner and R. M. Noyes, *J. Phys. Chem.*, **70**, 770 (1966).
5. Report of the Organometallic Chemistry Panel, The Science Research Council, H.M. Stationery Office, November 1968.
5a. C. N. Satterfield, *Ind. Eng. Chem.*, **61** (6), 4 (1969).
5b. *Chem. Eng. News*, May 4, **1970**, 36.
6. D. O. Mayward and B. M. W. Trapnell, *Chemisorption*, Butterworths, London, 1964, p. 226.
7. G. C. Bond, *Catalysis by Metals*, Academic Press, New York-London, 1962.
8. G. C. Bond and P. B. Wells, *Adv. Catal.*, **15**, 91 (1964).
9. R. L. Augustine, *Catalytic Hydrogenation*, Marcel Dekker, New York, 1965.
10. E. K. Rideal, *Concepts in Catalysis*, Academic Press, New York-London, 1968.
11. S. J. Thomson and G. Webb, *Heterogeneous Catalysis*, Oliver and Boyd, Edinburgh, 1969.
12. G. C. Bond, *Ann. Rep. Chem. Soc.*, *A*, **65**, 121 (1968).
13. J. E. Germain, *Catalytic Conversion of Hydrocarbons*, Academic Press, New York-London, 1969.
13a. G. C. Bond, *Advances in Chemistry*, Vol. 70, American Chemical Society, Washington, D.C., 1968, p. 25.
13b. J. M. Sinfelt, *Ind. Eng. Chem.*, **62**, (10), 66 (1970).
14. M. Calvin, *Trans. Faraday Soc.*, **34**, 1181 (1938).
15. M. Calvin, *J. Am. Chem. Soc.*, **61**, 2230 (1939).

16. V. Ipatieff and W. Werchowsky, *Chem. Ber.*, **42**, 2078 (1909).

17. R. G. Dakers and J. Halpern, *Can. J. Chem.*, **32**, 969 (1954).

18. J. Halpern and R. G. Dakers, *J. Chem. Phys.*, **22**, 1272 (1954).

19. M. Iguchi, *J. Chem. Soc. Jap.*, **60**, 1287 (1939).

20. M. Iguchi, *J. Chem. Soc. Jap.*, **63**, 634, 1752 (1942).

21. J. Kwiatek and J. K. Seyler, *Advances in Chemistry*, Vol. 70, American Chemical Society, Washington, D.C., 1968, p. 207.

22. J. Halpern, *Quart. Rev.*, **10**, 463 (1956).

23. S. W. Weller and G. A. Mills, *Adv. Catal*, **8**, 163 (1956).

24. J. Halpern, *Adv. Catal*, **9**, 302 (1957).

25. J. Halpern, *Adv. Catal*, **11**, 301 (1959).

26. J. Halpern, *J. Phys. Chem.*, **63**, 398 (1959).

26a. V. K. Bykhovskii and O. N. Temkin, *Kinet. Catal. (USSR)*, **1**, 347 (1960).

27. J. Halpern, J. F. Harrod, and B. R. James, *J. Am. Chem. Soc.*, **83**, 753 (1961).

28. J. Halpern, Abstracts, 141st Meeting of the American Chemical Society, Washington, D.C., 1962, p. 10Q.

29. J. Halpern and B. R. James, Abstracts, 142nd Meeting of the American Chemical Society, Atlantic City, 1962, p. 23N.

30. R. D. Cramer, E. L. Jenner, R. V. Lindsey Jr., and U. G. Stolberg, *J. Am. Chem. Soc.*, **85**, 1691 (1963).

31. R. Köster, G. Bruno, and P. Binger, *Ann. Chem.*, **644**, 1 (1961).

32. E. J. DeWitt, F. L. Ramp, and L. E. Trapasso, *J. Am. Chem. Soc.*, **83**, 4672 (1961).

33. H. C. Brown, *Tetrahedron*, **12**, 117 (1961).

34. L. Vaska and J. W. Diluzio, *J. Am. Chem. Soc.*, **84**, 679 (1962).

35. L. Vaska and J. W. Diluzio, *J. Am. Chem. Soc.*, **83**, 2784 (1961).

36. L. Vaska, *Acc. Chem. Res.*, **1**, 335 (1968).

37. J. Halpern, *Advances in Chemistry*, Vol. 70, American Chemical Society, Washington, D.C., 1968, p. 1.

37a. J. Halpern, *Discuss. Faraday Soc.*, **46**, 7 (1968).

37b. J. Halpern, *Pure Appl. Chem.*, **20**, 59 (1969).

38. J. P. Collman, *Acc. Chem. Res.*, **1**, 136 (1968).

39. J. P. Collman and W. R. Roper, *Adv. Organomet. Chem.*, **7**, 54 (1968).

39a. J. P. Collman, *Trans. N.Y. Acad. Sci.*, **30**, 479 (1968).

40. S. Carrà and R. Ugo, *Inorg. Chim. Acta Rev.*, **1**, 49 (1967).

41. R. Cramer, *Acc. Chem. Res.*, **1**, 186 (1968).

41a. W. Strohmeier, "Problem und Modell der homogenen Katalyse," in C. K. Jorgensen, J. B. Neilands, R. S. Nyholm, D. Reinen, and R. J. P. Williams, Eds., *Structure and Bonding*, Vol. 5, Springer-Verlag, New York, 1968, p. 96.

42. G. Wilkinson, *Bull. Soc. Chim. Fr.*, **1968**, 5055.

43. F. H. Jardine, J. A. Osborn, G. Wilkinson, and J. F. Young, *Chem. Ind. (Lond.)*, **1965**, 560.

43a. J. F. Young, J. A. Osborn, F. H. Jardine, and G. Wilkinson, *Chem. Commun. (Lond.)*, **1965**, 131.

44. D. Evans, J. A. Osborn, F. H. Jardine, and G. Wilkinson, *Nature*, **208**, 1203 (1965).

45. R. S. Coffey, Imperial Chemical Industries, Brit. Pat. 1,121,642 (filed 1965).

46. V. A. Tulupov, *Russ. J. Phys. Chem.*, **36**, 873 (1962).

47. E. N. Frankel, E. A. Emken, and V. L. Davison, *J. Org. Chem.*, **30**, 2739 (1965).

48. I. Ogata and A. Misono, *J. Jap. Oil Chem. Soc.*, **13**, 644 (1964); through *Chem. Abstr.*, **63**, 17828 (1965).

49. L. Marco, *Proc. Chem. Soc. (Lond.)*, **1962**, 67.

50. H. Greenfield, S. Metlin, M. Orchin, and I. Wender, *J. Org. Chem.*, **23**, 1054 (1958).

51. J. Falbe, *Synthesen mit Kohlenmonoxyd*, Springer-Verlag, Berlin-Heidelberg, New York, 1967; English translation (C. R. Adams), Springer-Verlag, 1970.

52. C. W. Bird, *Chem. Rev.*, **62**, 283 (1962).

53. C. W. Bird, *Transition Metal Intermediates in Organic Synthesis*, Logos Press, London 1967, p. 117.

54. A. J. Chalk and J. F. Harrod, *Adv. Organomet. Chem.*, **6**, 119 (1968).

55. D. Evans, J. A. Osborn, and G. Wilkinson, *J. Chem. Soc.*, A, **1968**, 3133.

56. M. F. Sloan, A. S. Matlock, and D. S. Breslow, *J. Am. Chem. Soc.*, **85**, 4014 (1963).

57. Ref. 53, p. 248.

57a. V. A. Tulupov, *Proc. Symp. Coord. Chem.*, Tihany, Hungary, 1964, Akademiai Kiado, Budapest, 1965, p. 57.

57b. V. A. Tulupov, *Russ. J. Phys. Chem.*, **39**, 1251 (1965).

57c. J. E. Lyons, L. E. Rennick, and J. L. Burmeister, *Ind. Eng. Chem. Prod. Res. Develop.*, **9**, 2 (1970).

57d. M. E. Volpin and I. S. Kolomnikov, *Russ. Chem. Rev.*, **38**, 273 (1969).

57e. R. S. Coffey, "Recent Advances in Homogeneous Hydrogenation of Carbon-Carbon Multiple Bonds," in R. Ugo, Ed., *Aspects of Homogeneous Catalysis*, Vol. 1, Carlo Manfredi, Milan, 1970, p. 3.

57f. A. Andreetta, F. Conti and G. F. Ferrari, "Selective Homogeneous Hydrogenation of Dienes and Polyenes to Monoenes," in R. Ugo, Ed., *Aspects of Homogeneous Catalysis*, Vol. 1, Carlo Manfredi, Milan, 1970, p. 203.

57g. J. P. Candlin and R. A. C. Rennie, "Reduction Methods," in K. W. Bentley and G. W. Kirby, Eds., *Technique of Organic Chemistry*, Vol. XI, 2nd ed., Interscience, in press.

57h. J. Halpern, *Proc. Symp. Coord. Chem.*, Tihany, Hungary, 1964, Akademiai Kiado, Budapest, 1965, p. 45.

58. A. Misono and I. Ogata, *Kagaku To Kogyo*, **20**, 1083 (1968).

58a. J. C. Lauer, *Ann. Chim. (Fr.)*, **10**, 301 (1965).

58b. T. Ueno and Y. Takegami, *J. Chem. Soc. Jap., Ind. Chem. Sect.*, **72**, 1621 (1969).

58c. D. V. Sokol'skii, *Progr. Elektrokhim, Org. Soedin.*, **1**, 328 (1969); through *Chem. Abstr.*, **72**, 85509 (1970).

58d. N. Hagihara, *Kagaku Kogyo*, **21**, 490 (1970); through *Chem. Abstr.*, **72**, 131651 (1970).

58e. L. Sajus, *Rev. Inst. Fr. Petrole*, **24**, 1477 (1969).

58f. H. Itatani, *Yuki Gosei Kagaku Kyokai Shi*, **27**, 632 (1969); through *Chem. Abstr.*, **71**, 90693 (1969).

450 References

58g. M. Shiota and S. Nishimura, *Yuki Gosei Kagaku Kyokai Shi*, **27**, 418 (1969); through *Chem. Abstr.*, **71**, 102098 (1969).

58h. K. Sonogashira and N. Hagihara, *Kagaku No Ryoiki Zokan*, **1970**, No. 89, 5; through *Chem. Abstr.*, **73**, 81030 (1970).

59. J. Halpern, *Ann. Rev. Phys. Chem.*, **16**, 103 (1965).

60. J. Halpern, *Proc. 3rd. Intern. Congr. Catal.*, North-Holland, Amsterdam, 1965, p. 146.

61. R. S. Nyholm, *Proc. 3rd Intern. Congr. Catal.*, North-Holland, Amsterdam, 1965, p. 25.

62. J. Halpern, *Chem. Eng. News*, October 31, **1966**, 68.

62a. H. Schmidt, *Chem.-Ztg.*, **90**, 351 (1966).

63. B. R. James, *Coord. Chem. Rev.*, **1**, 505 (1966).

64. G. C. Bond, *Ann. Rep. Chem. Soc.*, **63**, 27 (1966).

65. J. P. Collman, "Reactions of Ligands Coordinated with Transition Metals," in R. L. Carlin, Ed., *Transition Metal Chemistry*, Vol. 2, Marcel Dekker, New York, 1966, p. 1.

66. S. Ikeda and A. Yamamoto, *Kagaku To Kogyo*, **20**, 1070 (1968).

66a. Y. Trambouze, *Catal. Lab. Ind., Rec. Trav. Sess.*, **1966**, 19; through *Chem. Abstr.*, **70**, 118510 (1969).

66b. S. Otsuka, *Yuki Gosei Kagaku Kyokai Shi*, **25**, 946 (1967); through *Chem. Abstr.*, **68**, 81669 (1968).

66c. K. Tarama, *Kagaku Kogyo*, **19**, 995 (1968); through *Chem. Abstr.*, **70**, 50952 (1969).

67. E. Ochiai, *Coord. Chem. Rev.*, **3**, 49 (1968).

68. R. F. Heck, *Acc. Chem. Res.*, **2**, 10 (1969).

68a. R. Ugo, *Chim. Ind. (Milan)*, **51**, 1319 (1969).

68b. J. P. Candlin, K. A. Taylor, and D. T. Thompson, *Ind. Chim. Belge*, **35**, 1085 (1970).

68c. J. Chatt, *Science*, **160**, 723 (1968).

68d. J. F. Biellmann, H. Hemmer, and J. Levisalles, *Chem. Alkenes* **2**, 215 (1970).

68e. R. S. Coffey, *Ann. Rep. Chem. Soc., B*, **66**, 321 (1969).

69. F. A. Cotton and G. Wilkinson, *Advanced Inorganic Chemistry*, Interscience, 2nd ed., New York-London, 1966, p. 791.

70. F. Basolo and R. G. Pearson, *Mechanisms of Inorganic Reactions*, 2nd ed., Wiley, New York, 1967, p. 609.

71. J. P. Candlin, K. A. Taylor, and D. T. Thompson, *Reactions of Transition Metal Complexes*, Elsevier, Amsterdam, 1968, p. 144.

72. G. E. Coates, M. L. H. Green, and K. Wade, *Organometallic Compounds*, Vol. 2, 3rd ed., Methuen, London, 1968, p. 321.

73. P. L. Pauson, *Organometallic Chemistry*, Arnold, London, 1967, p. 168.

74. G. E. Coates, M. L. H. Green, P. Powell, and K. Wade, *Principles of Organometallic Chemistry*, Methuen, London, 1968, p. 236.

74a. M. M. Jones, *Ligand Reactivity and Catalysis*, Academic Press, New York, 1968, pp. 152, 192.

74b. M. Tsutsui, M. N. Levy, A. Nakamura, M. Ichikawa, and K. Mori, *Introduction to Metal π-Complex Chemistry*, Plenum, New York-London, 1970, p. 171.

75. W. P. Griffith, *The Chemistry of the Rarer Platinum Metals*, Interscience, New York-London, 1967, pp. 95, 101, 141, 144, 178, 194, 279, 284, 326, 347, 354, 367, 389, 403.

76. G. C. Bond, *Plat. Met. Rev.*, **8**, 92 (1964).

77. B. W. Malerbi, *Plat. Met. Rev.*, **9**, 47 (1965).

78. G. Wilkinson, *Plat. Met. Rev.*, **12**, 50 (1968).

79. J. A. Osborn, *Endeavour*, **26**, 144 (1967).

79a. J. Tsuji, *Adv. Org. Chem.*, **6**, 109 (1969).

79b. J. P. Candlin, *Chem. Ind. (Lond.)*, **1970**, 76.

79c. P. N. Rylander and L. Hasbrouck, *Englehard Ind. Tech. Bull.*, **10**, 85 (1969).

79d. S. Takahashi and N. Hagihara, *J. Chem. Soc. Jap. (Ind. Chem. Sect.)*, **72**, 1637 (1969).

79e. B. R. James, *Inorg. Chim. Acta Rev.*, **4**, 73 (1970).

79f. E. Schleitzer, *Chem.-Ztg.*, **94**, 189 (1970).

79g. J. F. Biellmann, *Bull. Soc. Chim. Fr.*, **1968**, 3055.

79h. T. D. Inch, *Synthesis*, **1970**, 466.

80. "Homogeneous Catalysis with Special Reference to Hydrogenation and Oxidation," *Discuss. Faraday Soc.*, **46**, 1969.

80a. "Second Conference on Catalytic Hydrogenation and Analogous Pressure Reactions," *Ann. N.Y. Acad. Sci.*, **158** (2), 1969.

80b. "Third Conference on Catalytic Hydrogenation and Analogous Pressure Reactions," *Ann. N.Y. Acad. Sci.*, **172** (9), 1970.

81. E. N. Frankel and H. J. Dutton, "Hydrogenation with Homogeneous and Heterogeneous Catalysts," in F. D. Gunstone, Ed., *Topics in Lipid Chemistry*, Vol. 1, Logos Press, London, 1970, p. 161.

82. J. Chatt, *Proc. Chem. Soc., (Lond.)*, **1962**, 318.

83. M. L. H. Green and D. J. Jones, *Adv. Inorg. Chem. Radiochem.*, **7**, 115 (1965).

84. A. P. Ginsberg, "Hydride Complexes of the Transition Metals," in R. L. Carlin, Ed., *Transition Metal Chemistry*, Vol. 1, Marcel Dekker, New York, 1965, p. 111.

85. W. K. Wilmarth, M. K. Barsh, and S. S. Dharmatti, *J. Am. Chem. Soc.*, **74**, 5035 (1952).

86. W. K. Wilmarth and M. K. Barsh, *J. Am. Chem. Soc.*, **75**, 2237 (1953).

87. S. Weller and G. A. Mills, *J. Am. Chem. Soc.*, **75**, 769 (1953).

88. M. Calvin and W. K. Wilmarth, *J. Am. Chem. Soc.*, **78**, 1301 (1956).

89. W. K. Wilmarth and M. K. Barsh, *J. Am. Chem. Soc.*, **78**, 1305 (1956).

90. L. W. Wright and S. Weller, *J. Am. Chem. Soc.*, **76**, 3345 (1954).

91. L. W. Wright, S. Weller, and G. A. Mills, *J. Phys. Chem.*, **59**, 1060 (1955).

92. A. J. Chalk and J. Halpern, *J. Am. Chem. Soc.*, **81**, 5846 (1959).

93. A. J. Chalk and J. Halpern, *J. Am. Chem. Soc.*, **81**, 5852 (1959).

94. M. Parris and R. J. P. Williams, *Discuss. Faraday Soc.*, **29**, 240 (1960).

95. W. J. Dunning and P. E. Potter, *Proc. Chem. Soc.*, **1960**, 244; P. E. Potter, Ph.D. Thesis, University of Bristol, 1958.

96. E. A. Hahn and E. Peters, *J. Phys. Chem.*, **69**, 547 (1965).

97. H. J. Emeleus and J. S. Anderson, *Modern Aspects of Inorganic Chemistry*, 3rd ed., Routledge and Kegan Paul, London, 1960, p. 415.

98. Ref. 2, p. 184.

99. F. D. Rossini, D. D. Wagman, W. H. Evans, S. Levine, and I. Jaffe, National Bureau of Standards, Circular 500, U.S. Government Printing Office, Washington, D.C. 1952, p. 208.

452 References

100. E. A. Hahn, Ph. D. Thesis, University of British Columbia, 1963.

101. H. W. van der Linden, B. Stouthamer, and J. C. Vlugter, *Chem. Weekblad*, **60,** 254 (1964).

102. B. Stouthamer and J. C. Vlugter, *J. Am. Oil Chem. Soc.*, **42,** 646 (1965).

103. E. Peters and J. Halpern, *Can. J. Chem.*, **33,** 356 (1955).

104. E. Peters and J. Halpern, *J. Phys. Chem.*, **59,** 793 (1955).

105. J. Halpern, E. R. Macgregor, and E. Peters, *J. Phys. Chem.*, **60,** 1455 (1956).

106. E. A. Hahn and E. Peters, *Can. J. Chem.*, **39,** 162 (1961).

107. E. R. Macgregor and J. Halpern, *Trans. Met. Soc.*, *A.I.M.E.*, **212,** 244 (1958).

108. E. Peters and E. A. Hahn, in *Unit Processes in Hydrometallurgy, Part II: Pressure Leaching and Reduction*, M. E. Wadsworth and F. T. Davis, Eds., Gordon and Breach, New York, 1966.

109. E. Peters and J. Halpern, *Can. J. Chem.*, **34,** 554 (1956).

109a. M. T. Beck, *Chemistry of Complex Equilibria*, Van Nostrand Reinhold, London, 1970, p. 110.

110. W. P. Jencks, *Catalysis in Chemistry and Enzymology*, McGraw-Hill, New York, 1969, p. 84.

111. H. F. McDuffie, E. L. Compere, H. H. Stone, L. F. Woo, and C. H. Secoy, *J. Phys. Chem.*, **62,** 1030 (1958).

112. A. A. Frost and R. G. Pearson, *Kinetics and Mechanism*, Wiley, New York, 2nd ed., 1965, p. 147.

113. W. K. Wilmarth, private communication to J. Halpern, quoted in ref. 25, p. 317.

114. M. N. Beketoff, *Compt. Rend.*, **48,** 442 (1859).

115. F. Hein and W. Daniel, *Z. Anorg. Allg. Chem.*, **181,** 78 (1929).

116. F. Hein, W. Daniel, and H. Schwedler, *Z. Anorg. Allg. Chem.*, **233,** 161 (1937).

117. W. K. Wilmarth, Ph.D. Thesis, University of California, 1942.

118. W. K. Wilmarth and A. F. Kapauan, *J. Am. Chem. Soc.*, **78,** 1308 (1956).

119. A. H. Webster and J. Halpern, *J. Phys. Chem.*, **60,** 280 (1956).

120. A. H. Webster and J. Halpern, *J. Phys. Chem.*, **61,** 1239 (1957).

121. E. T. Blues and D. Bryce-Smith, *Discuss. Faraday Soc.*, **47,** 190 (1969).

122. A. H. Webster and J. Halpern, *J. Phys. Chem.*, **61,** 1245 (1957).

123. A. J. Chalk, J. Halpern, and A. C. Harkness, *J. Am. Chem. Soc.*, **81,** 5854 (1959).

124. A. H. Webster and J. Halpern, *Trans. Faraday Soc.*, **53,** 51 (1957).

125. J. Halpern and J. B. Milne, *Proc. 2nd Intern. Congr. Catal.*, Ed. Technip, Paris, **1961,** p. 445.

126. A. Lowenstein and S. Meiboom, *J. Chem. Phys.*, **27,** 1067 (1957).

127. S. Meiboom, A. Lowenstein, and S. Alexander, *J. Chem. Phys.*, **29,** 968 (1958).

128. F. Nagy and L. I. Simandi, *Acta Chim. Acad. Sci. Hung.*, **38,** 213 (1963).

129. F. Nagy and L. I. Simandi, *Acta Chim. Acad. Sci. Hung.*, **38,** 373 (1963).

130. L. I. Simandi, *Discuss. Faraday. Soc.*, **46,** 93 (1968).

131. F. Nagy and L. Simandi, *Kinet. Catal. (USSR)*, **7,** 7 (1966).

131a. R. S. Eachus and M. C. R. Symons, *Chem. Commun. (Lond.)*, **1969,** 285.

132. M. T. Beck, I. Gimesi, and J. Farkas, *Nature*, **197,** 73 (1963).

133. M. T. Beck and I. Gimesi, *Acta Chim. Acad. Sci. Hung.*, **42,** 343 (1964).

134. *Stability Constants*, Chemical Society Special Publication No. 17, Eds., L. G. Sillen and A. E. Martell, London, 1964.

135. J. Halpern and J. F. Harrod, unpublished; quoted in ref. 25.

135a. A. B. Fasman and Zh. A. Ikhsanov, *Kinet. Catal. (USSR)*, **8**, 53 (1967).

136. D. Rittenberg, *Proc. Intern. Symp. Enzyme Chem.*, *Tokyo and Kyoto*, **2**, 256 (1957).

137. J. Halpern, G. J. Korinek, and E. Peters, *Research (Lond.)* **7**, S61 (1954).

138. G. J. Korinek and J. Halpern, *J. Phys. Chem.*, **60**, 285 (1956).

139. J. Halpern and A. C. Harkness, unpublished; quoted in ref. 59.

140. A. R. Topham and A. G. White, *J. Chem. Soc.*, **1952**, 105.

141. J. Halpern and E. Peters, *J. Chem. Phys.*, **23**, 605 (1955).

142. G. J. Korinek and J. Halpern, *Can. J. Chem.*, **34**, 1372 (1956).

143. J. D. Richter and P. J. Van den Berg, *J. Am. Oil Chem. Soc.*, **46**, 155 (1969).

144. E. A. Emken, E. N. Frankel, and R. O. Butterfield, *J. Am. Oil Chem. Soc.*, **43**, 14 (1966).

145. S. Nan'ya, M. Hanai, and K. Fukuzumi, *J. Chem. Soc. Jap.*, *Ind. Chem. Sect.*, **72**, 2005 (1969).

146. V. A. Tulupov and M. I. Gagarina, *Russ. J. Phys. Chem.*, **38**, 926 (1964).

147. I. V. Kalechits, V. G. Lipovich, and F. K. Shmidt, *Preprint of paper 25, 4th Intern. Congr. Catal.*, Moscow, 1968.

148. V. A. Tulupov, *Russ. J. Phys. Chem.*, **39**, 397 (1965).

149. F. R. Hartley, *Chem. Rev.*, **69**, 799 (1969).

150. H. W. Quinn and J. H. Tsai, *Adv. Inorg. Radiochem.*, **12**, 217 (1969).

151. U. Belluco, B. Crociani, R. Pietropaolo, and P. Uguagliati, *Inorg. Chim. Acta Rev.*, **3**, 19 (1969).

152. M. Orchin, *Adv. Catal.*, **16**, 1 (1966).

153. N. R. Davies, *Rev. Pure Appl. Chem.*, **17**, 83 (1967).

154. A. J. Hubert and H. Reimlinger, *Synthesis*, **1970**, 405.

155. R. P. A. Sneedon and H. H. Zeiss, *J. Organomet. Chem.*, **22**, 713 (1970).

156. G. Yagupsky, W. Mowat, A. Shortland, and G. Wilkinson, *Chem. Commun. (Lond.)*, **1970**, 1369.

157. E. Ucciani, *Jour. Etud. Hydrogenation Corps Gras*, **1968**, 11; through *Chem. Abstr.*, **72**, 80628 (1970).

158. "The Role of the Adsorbed State in Heterogeneous Catalysis," *Discuss. Faraday Soc.*, **41**, 1966.

159. S. Carrà, R. Ugo, and L. Zanderighi, *Inorg. Chim. Acta Rev.* **3**, 55 (1969).

160. R. L. Augustine and J. F. Van Peppen, *Ann. N.Y. Acad. Sci.*, **172**, (9), 244 (1970).

161. V. A. Tulupov, *Russ. J. Phys. Chem.*, **41**, 456 (1967).

162. D. S. Breslow and N. R. Newburg, *J. Am. Chem. Soc.*, **81**, 81 (1959).

163. I. V. Kalechits and F. K. Shmidt, *Kinet. Catal. (USSR)*, **7**, 541 (1966).

164. V. G. Lipovich, F. K. Shmidt, and I. V. Kalechits, *Kinet. Catal. (USSR)*, **8**, 812 (1967).

165. I. V. Kalechits, V. G. Lipovich, and F. K. Shmidt, *Kinet. Catal. (USSR)*, **9**, 16 (1968).

454   References

166.   N. F. Noskova, M. I. Popandopulo, and D. V. Sokol'skii, *Izv. Akad. Nauk Kaz. SSR, Ser. Khim.*, **19**, 22 (1969); through *Chem. Abstr.*, **71** 64686 (1969).

167.   Y. Tajima and E. Kunioka, *J. Org. Chem.*, **33**, 1689 (1968).

168.   R. Stern and L. Sajus, *Tetrahedron Lett.*, **1968**, 6313.

169.   K. Sonogashira and N. Hagihara, *Bull. Chem. Soc. Jap.*, **39**, 1178 (1966).

170.   R. Tsumura and N. Hagihara, *Bull. Chem. Soc. Jap.*, **38**, 861 (1965).

171.   K. Shikata, K. Nishino, and K. Azuma, *J. Chem. Soc., Jap., Ind. Chem. Sect.*, **68**, 490 (1965).

172.   L. M. Slaugh, *J. Org. Chem.*, **32**, 108 (1967).

172a.   A. Kanai and Miyake, Jap. Pat. 69/12,125; through *Chem. Abstr.*, **71**, 102460 (1969).

173.   Y. Takegami, T. Ueno, and T. Fujii, *Bull. Chem. Soc. Jap.*, **42**, 1663 (1969).

173a.   P. Karrer, Y. Yen, and I. Reichstein, *Helv. Chem. Acta*, **13**, 1308 (1930).

174.   V. A. Tulupov, *Russ. J. Phys. Chem.*, **36**, 873 (1962).

175.   S. J. Lapporte and W. R. Schuett, *J. Org. Chem.*, **28**, 1947 (1963).

176.   B. I. Tikhomirov, I. A. Klopotova, and A. I. Yakubchik, *Vysokomol. Soedin.*, **B9**, 427 (1967); through *Chem. Abstr.*, **67**, 82418 (1967).

177.   D. V. Sokol'skii, N. F Noskova, and G. N. Sharifkanova, *Russ. J. Phys. Chem.*, **185**, 263 (1969).

178.   M. Cais, E. N. Frankel, and A. Rejoan, *Tetrahedron Lett.*, **1968**, 1919.

179.   A. Rejoan and M. Cais, in M. Cais Ed., *Progress in Coordination Chemistry*, Elsevier, Amsterdam, 1968, p. 32.

180.   M. Cais, N. Maoz, and A. Rejoan, *Proc. 2nd Intern. Symp. Adv. Chem. Met-Carbon, Met.-Hydrogen and Met.-Olefin Complexes*, Venice, 1969, p. 25.

181.   E. N. Frankel, E. Selke, and C. A. Glass, *J. Am. Chem. Soc.*, **90**, 2446 (1968).

182.   E. N. Frankel, *Proc. 4th Intern. Conf. Organomet. Chem.*, Bristol 1969, L8.

183.   E. N. Frankel and F. L. Little, *J. Am. Oil Chem. Soc.*, **46**, 256 (1969).

183a.   E. N. Frankel, *J. Am. Oil Chem. Soc.*, **47**, 11 (1970).

184.   A. Miyake and H. Kondo, *Angew. Chem. Intern. Edit.*, **7**, 631 (1968).

185.   A. Miyake and H. Kondo, *Angew. Chem. Intern. Edit.*, **7**, 880 (1968).

185a.   H. Kondo and A. Miyake, Jap. Pat. 69 12,126; through *Chem. Abstr.*, **71**, 90774 (1969).

186.   E. O. Fisher, *Inorg. Synth.*, **7**, 136 (1963).

187.   D. A. Brown, J. P. Hargaden, C. M. McMullin, N. Gogan, and H. Sloan, *J. Chem. Soc.*, **1963**, 4914.

187a.   P. Legzdins, G. L. Rempel, and G. Wilkinson, *Chem. Commun. (Lond.)*, **1969**, 825.

188.   O. N. Eremenko, O. N. Efimov, A. G. Ovcharenko, M. L. Khidekel, and P. S. Chekrii, Paper presented at 2nd All-Union Conference on Catalytic Reactions in the Liquid Phase, *Izd. Akad. Nauk. Kaz. SSR, Alma-Ata*, **1966**; through ref. 57d.

188a.   N. T. Denisov, V. F. Shuvalov, N. I. Shuvalova, and E. V. Artamonova, *Kinetika i Kataliz*, in press; through ref. 1294.

189.   G. Just and Y. Kauko, *Z. Physik. Chem.*, **76**, 601 (1911).

190.   E. Wilke and H. Kuhn, *Z. Physik. Chem.*, **113**, 313 (1924).

191.   M. L. Delwaulle, *Compt. Rend.*, **192**, 1736 (1931).

192.   R. Stewart, *Oxidation Mechanisms*, Benjamin, New York, 1964, p. 58.

193. U. Giannini and U. Zucchini, *Chem. Commun. (Lond.)*, **1968**, 940.

193a. V. A. Tulupov, *Zhur. Fiz. Khim.*, **32**, 727 (1958).

194. U. Zucchini, U. Giannini, E. Albizzati, and R. D'Angelo, *Chem. Commun. (Lond.)*, **1969**, 1174.

195. I. V. Kalechits, V. G. Lipovich, and F. K. Schmidt, *Katal. Reakts. Zhidk. Faze, Tr. Vses, Konf., 2nd Alma-Ata, Kaz. SSR*, **1966**, 425; through *Chem. Abstr.*, **69**, 76204 (1968).

196. V. A. Tulupov, *Zhur. Fiz. Khim.*, **31**, 519 (1957).

197. A. I. Tulupov and V. A. Tulupov, *Russ. J. Phys. Chem.*, **37**, 1449 (1963).

198. E. N. Frankel, H. M. Peters, E. P. Jones, and H. J. Dutton, *J. Am. Oil Chem. Soc.* **41**, 186 (1964).

199. E. N. Frankel, E. A. Emken, H. M. Peters, V. L. Davison, and R. O. Butterfield, *J. Org. Chem.*, **29**, 3292 (1964).

200. E. N. Frankel, T. L. Mounts, R. O. Butterfield, and H. J. Dutton, *Advances in Chemistry*, Vol. 70, American Chemical Society, Washington, D.C. 1968, p. 177.

201. A. Misono, I. Ogata, and F. Funami, *J. Jap. Oil Chem. Soc.*, **13**, 21 (1964); through ref. 202.

202. I. Ogata and A. Misono, *Bull. Chem. Soc. Jap.*, **37**, 439 (1964).

203. I. Ogata and A. Misono, *Bull. Chem. Soc. Jap.*, **37**, 900 (1964).

204. I. Ogata and A. Misono, *J. Chem. Soc. Jap.*, **85**, 748 (1964).

205. I. Ogata and A. Misono, *J. Chem. Soc. Jap.*, **85**, 753 (1964).

206. Y. Tajima and E. Kunioka, *J. Am. Oil Chem. Soc.*, **45**, 478 (1968).

207. A. Misono and Y. Uchida, Jap. Pat., 7021,082, through *Chem. Abstr.*, **73**, 87459 (1970).

208. H. W. Sternberg, unpublished work quoted in ref. 198.

209. E. N. Frankel, N. Maoz, A. Rejoan, and M. Cais, *Proc. 3rd Intern. Conf. Organomet. Chem.*, Munich, 1967, Abstr., p. 210.

210. N. Maoz and M. Cais, *Isr. J. Chem.*, **6**, 32P (1968).

211. D. R. Levering, U.S. Pat. 3,152,184; through *Chem. Abstr.*, **62**, 427 (1965).

212. A. Misono, Y. Uchida, K. Tamai, and M. Hidai, *Bull. Chem. Soc. Jap.*, **40**, 931 (1967).

213. B. M. Trost and G. M. Bright, *J. Am. Chem. Soc.*, **91**, 3689 (1969).

214. J. F. Harrod, S. Ciccone, and J. Halpern, *Can. J. Chem.*, **39**, 1372 (1961).

215. D. A. Fine, Ph.D. Dissertation, University of California, Berkeley, California, 1960.

216. J. Halpern and B. R. James, *Can. J. Chem.*, **44**, 671 (1966).

217. U. Schindewolf, *Ber. Busenges. Physik. Chem.*, **67**, 219 (1963).

218. U. Schindewolf, *J. Chim. Phys.*, **60**, 124 (1963).

219. J. Halpern, J. F. Harrod, and B. R. James, *J. Am. Chem. Soc.*, **88**, 5150 (1966).

220. B. Hui and B. R. James, *Chem. Commun. (Lond.)*, **1969**, 198.

221. B. Hui, Ph.D. Dissertation, University of British Columbia, Vancouver, British Columbia, 1969.

222. B. Hui and B. R. James, *Proc. 4th Intern. Conf. on Organomet. Chem.*, Bristol, 1969, L6.

222a. B. R. James and L. D. Markham, unpublished results.

223. B. R. James and E. Ochiai, to be published.
224. G. Charlot, *Selected Constants, Oxidation Reduction Potentials*, Pergamon, Paris, 1958.
225. J. Halpern, B. R. James, and A. L. W. Kemp, *J. Am. Chem. Soc.*, **88**, 5142 (1966).
226. R. E. Dessy and F. Paulik, *Bull. Soc. Chim. Fr.*, **1963**, 1373.
226a. D. S. Matteson, *Organomet. Chem. Rev.*, *A*, **4**, 263 (1969).
227. B. R. James and J. Louie, *Inorg. Chim. Acta*, **3**, 568 (1969).
228. J. Louie, M.Sc. Thesis, University of British Columbia, Vancouver, British Columbia, 1968.
229. B. R. James and R. J. King, to be published.
229a. J. F. Brennan, U.S. Pat. 3,478,123; through *Chem. Abstr.*, **72**, 21275 (1970).
230. J. Halpern and B. R. James, *Can. J. Chem.*, **44**, 495 (1966).
230a. B. C. Hui and B. R. James, *Inorg. Nucl. Chem. Lett.*, **6**, 367 (1970).
231. J. Chatt and R. G. Hayter, *J. Chem. Soc.*, **1961**, 2605.
232. S. J. LaPlaca and J. A. Ibers, *Inorg. Chem.*, **4**, 778 (1965).
233. T. A. Stephenson and G. Wilkinson, *J. Inorg. Nucl. Chem.*, **28**, 945 (1966).
234. P. S. Hallman, D. Evans, J. A. Osborn, and G. Wilkinson, *Chem. Commun. (Lond.)*, **1967**, 305.
235. P. S. Hallman, B. R. McGarvey, and G. Wilkinson, *J. Chem. Soc.*, *A*, **1968**, 3143.
236. A. C. Skapski and P. G. H. Troughton, *Chem. Commun. (Lond.)*, **1968**, 1230.
237. K. W. Muir and J. A. Ibers, *Inorg. Chem.*, **9**, 440 (1970).
238. D. Rose, J. D. Gilbert, R. P. Richardson, and G. Wilkinson, *J. Chem. Soc.*, *A*, **1969**, 2610.
239. A. C. Skapski and F. A. Stephens, *Chem. Commun. (Lond.)*, **1969**, 1008.
240. I. Jardine and F. J. McQuillin, *Tetrahedron Lett.*, **1966**, 4871.
240a. I. Jardine, R. W. Howsam, and F. J. McQuillin, *J. Chem. Soc. C*, **1969**, 260.
241. B. Hudson, P. C. Taylor, D. E. Webster, and P. B. Wells, *Discuss. Faraday Soc.*, **46**, 37 (1968).
242. G. W. Parshall, W. H. Knoth, and R. A. Schunn, *J. Am. Chem. Soc.*, **91**, 4990 (1969).
242a. G. W. Parshall, *Acc. Chem. Res.*, **3**, 139 (1970).
243. P. Abley and F. J. McQuillin, *Discuss. Faraday Soc.*, **46**, 31 (1968).
244. Imperial Chemical Industries Ltd., Fr. Pat. 1,538,700; *Chem. Abstr.*, **71**, 83317 (1969).
244a. K. C. Dewhirst, U.S. Pat. 3,454,644; through *Chem. Abstr.*, **71**, 91647 (1969).
245. J. P. Candlin, R. S. Coffey, and A. R. Oldham, private communication.
245a. S. Nishimura and K. Tsuneda, *Bull. Chem. Soc. Jap.*, **42**, 852 (1969).
246. J. Chatt and J. M. Davidson, *J. Chem. Soc.*, **1965**, 843.
246a. S. D. Ibekwe, B. T. Kilbourn, U. Raeburn, and D. R. Russell, *Chem. Commun. (Lond.)*, **1969**, 433.
247. P. B. Chock and J. Halpern, *J. Am. Chem. Soc.*, **88**, 3511 (1966).
248. P. N. Rylander, N. Himelstein, D. Steele, and J. Kreidl, *Englehard Ind. Tech. Bull.*, **3**, 61 (1962).
249. G. A. Rechnitz and H. A. Catherino, *Inorg. Chem.*, **4**, 112 (1965).

250.   L. Vaska, *Proc. 8th Intern. Conf. Coord. Chem.*, V. Gutmann, Ed., Springer-Verlag, Wein-New York, 1964, p. 99.

250a. L. Vaska, *Proc. 1st Intern. Conf. Organomet. Chem.*, Madison, 1965, p. 79.

251.   K. C. Dewhirst, W. Keim, and C. A. Reilly, *Inorg. Chem.*, **7**, 546 (1968).

252.   A. Yamamoto, S. Kitazume, and S. Ikeda, *J. Am. Chem. Soc.*, **90**, 1089 (1968).

253.   W. H. Knoth, *J. Am. Chem. Soc.*, **90**, 7172 (1968).

254.   J. J. Levinson and S. D. Robinson, *Chem. Ind. (Lond.)*, **1969**, 1514.

254a. T. Anderson and R. V. Lindsey, Jr., U.S. Pat. 3,081,357; through *Chem. Abstr.*, **59**, 8594 (1963).

255.   J. F. Harrod, D. F. R. Gilson, and R. Charles, *Can. J. Chem.*, **47**, 1431 (1969).

255a. B. R. James and L. D. Markham, *Inorg. Nucl. Chem. Lett.*, **7**, 373 (1971).

256.   S. D. Robinson, *Chem. Commun. (Lond.)*, **1968**, 521.

256a. P. Smith and H. H. Jaeger, Ger. Pat. 1,159,926; through *Chem. Abstr.*, **60**, 14389 (1964).

257.   W. H. Knoth and R. A. Schunn, *J. Am. Chem. Soc.*, **91**, 2400 (1969).

257a. F. L'Eplattenier, P. Matthys, and F. Calderazzo, *Inorg. Chem.*, **9**, 342 (1970).

258.   J. D. Gilbert and G. Wilkinson, *J. Chem. Soc.*, A, **1969**, 1749.

258a. J. Norton, D. Valentine, and J. P. Collman, *J. Am. Chem. Soc.*, **91**, 7537 (1969).

259.   T. A. Stephenson, *Inorg. Nucl. Chem. Lett.*, **4**, 687 (1968).

259a. M. H. B. Stiddard and R. E. Townsend; *Chem. Commun. (Lond.)*, **1969**, 1372.

260.   D. Valentine and J. P. Collman, unpublished results quoted in ref. 38.

260a. F. L'Eplattenier and F. Calderazzo, *Inorg. Chem.*, **7**, 1290 (1968).

261.   L. Vaska, *Inorg. Nucl. Chem. Lett.*, **1**, 89 (1965).

262.   J. P. Collman and W. R. Roper, *J. Am. Chem. Soc.*, **88**, 3504 (1966).

263.   F. L'Eplattenier and F. Calderazzo, *Inorg. Chem.*, **6**, 2092 (1967).

264.   L. Vaska, *J. Am. Chem. Soc.*, **88**, 4100 (1966).

265.   K. R. Laing and W. R. Roper, *J. Chem. Soc.*, A, **1969**, 1889.

265a. J. R. Moss and W. A. G. Graham, *Chem. Commun. (Lond.)*, **1969**, 800.

265b. J. Knight and M. J. Mays, *Chem. Commun. (Lond.)*, **1969**, 384.

265c. P. G. Douglas and B. L. Shaw, *Chem. Commun. (Lond.)*, **1969**, 624.

265d. P. G. Douglas and B. L. Shaw, *J. Chem. Soc.*, A, **1970**, 334.

265e. G. J. Leigh, J. J. Levinson, and S. D. Robinson, *Chem. Commun. (Lond.)*, **1969**, 705.

266.   K. R. Laing and W. R. Roper, *Chem. Commun. (Lond.)*, **1968**, 1568.

267.   D. N. Hume and I. M. Kolthoff, *J. Am. Chem. Soc.*, **71**, 867 (1949).

268.   A. W. Adamson, *J. Am. Chem. Soc.*, **73**, 5710 (1951).

268a. T. Suzuki and T. Kwan, *J. Chem. Soc. Jap.*, **87**, 395 (1966).

268b. B. M. Chadwick and A. G. Sharpe, *Adv. Inorg. Chem. Radiochem.*, **8**, 83 (1966) and ref. therein.

268c. J. Kwiatek, *Catal. Rev.*, **1**, 37 (1967).

269.   M. E. Winfield, *Revs. Pure Appl. Chem. (Aust.)*, **5**, 217 (1955).

270.   A. Haim and W. K. Wilmarth, *J. Am. Chem. Soc.*, **83**, 509 (1961).

271.   W. P. Griffith and G. Wilkinson, *J. Chem. Soc.*, **1959**, 2757.

272.   G. A. Mills, S. Weller and A. Wheeler, *J. Phys. Chem.*, **63**, 403 (1959).

273. N. K. King and M. E. Winfield, *J. Am. Chem. Soc.*, **80**, 2060 (1958).

274. J. H. Bayston, R. N. Beale, N. K. King, and M. E. Winfield, *Aust. J. Chem.*, **16**, 954 (1963).

275. J. M. Pratt and R. J. P. Williams, *J. Chem. Soc.*, *A*, **1967**, 1291.

276. J. Pratt, unpublished results, quoted in ref. 275.

277. B. DeVries, *J. Catal.*, **1**, 489 (1962).

278. B. DeVries, *Konikl. Ned. Akad. Wetenschap.*, *Proc. Ser. B*, **63**, 443 (1960); through *Chem. Abstr.*, **55**, 9142 (1961).

279. N. K. King and M. E. Winfield, *J. Am. Chem. Soc.*, **83**, 3366 (1961).

280. J. M. Pratt and P. R. Silverman, *J. Chem. Soc.*, *A*, **1967**, 1286.

281. J. P. Candlin, J. Halpern, and S. Nakamura, *J. Am. Chem. Soc.*, **85**, 2517 (1963).

282. J. Bayston, N. K. King, and M. E. Winfield, *Adv. Catal.*, **9**, 312 (1957).

283. R. G. S. Banks and J. M. Pratt, *J. Chem. Soc.*, *A*, **1968**, 854.

284. M. G. Burnett, P. J. Connolly, and C. Kemball, *J. Chem. Soc.*, *A*, **1967**, 800.

284a. T. Mizuta and T. Kwan, *J. Chem. Soc. Jap.*, **86**, 1010 (1965).

285. R. G. S. Banks and J. M. Pratt, *Chem. Commun. (Lond.)*, **1967**, 776.

286. O. Piringer and A. Farcas, *Z. Physik. Chem. (Frankf.)*, **46**, 190 (1965).

287. O. Piringer, *Abh. Deut. Acad. Wiss. Berl., Kl. Chem. Geol. Biol.*, **1964**, 821; through *Chem. Abstr.*, **67**, 6097 (1967).

288. L. Simandi, *Magy. Kem. Foly.*, **71**, 141 (1965); through *Chem. Abstr.*, **63**, 6613 (1965).

289. G. A. Mills, *Adv. Catal.*, **9**, 376 (1957).

290. R. A. Ogg, Jr., *Abstr.*, *123rd Meet. Am. Chem. Soc.*, **1953**, 24P.

291. R. G. S. Banks, P. K. Das, H. A. O. Hill, J. M. Pratt, and R. J. P. Williams, *Discuss. Faraday Soc.*, **46**, 80 (1968).

292. L. Simandi and F. Nagy, *Acta Chim. Akad. Hung.*, **46**, 101 (1965).

292a. L. Simandi and F. Nagy, *Proc. Symp. Coord. Chem.*, Tihany, Hungary, 1964, Akademiai Kiado, Budapest, 1965, p. 83.

293. A. A. Vlcek, *Proc. 7th Intern. Conf. Coord. Chem.*, Stockholm, 1962, p. 286.

294. G. Pregaglia, D. Morelli, F. Conti, and G. Gregorio, *Discuss. Faraday. Soc.*, **46**, 110 (1968).

295. J. Kwiatek, I. L. Mador, and J. Y. Seyler, *J. Am. Chem. Soc.*, **84**, 304 (1962).

296. M. S. Spencer and D. A. Dowden, Imperial Chemical Industries Ltd., U.S. Pat. 3,009,969; German Pat. 1,114,183; through *Chem. Abstr.*, **56**, 8558 (1962).

297. J. Kwiatek, I. L. Mador, and J. Y. Seyler, *Advances in Chemistry*, Vol. 37, American Chemical Society, Washington, D.C., 1963, p. 201.

298. A. Haim and W. K. Wilmarth, *Inorg. Chem.*, **1**, 573 (1962).

298a. J. P. Birk and J. Halpern, *J. Am. Chem. Soc.*, **90**, 305 (1968).

299. J. Halpern and J. Rabani, *J. Am. Chem. Soc.*, **88**, 699 (1966).

300. J. H. Bayston and M. E. Winfield, *J. Catal.*, **3**, 123 (1964).

300a. H. L. Roberts and W. R. Symes, *J. Chem. Soc.*, *A*, **1968**, 1450.

301. A. W. Adamson, *J. Am. Chem. Soc.*, **78**, 4260 (1956).

302. A. Haim, R. J. Grassi, and W. K. Wilmarth, *Advances in Chemistry*, Vol. 49, American Chemical Society, Washington, D.C., 1965, p. 31.

303. J. Kwiatek and J. K. Seyler, unpublished results, quoted in ref. 268c.

304. T. Suzuki and T. Kwan, *J. Chem. Soc. Jap.*, **86**, 713 (1965).

305. M. G. Burnett, P. J. Conolly, and C. Kemball, *J. Chem. Soc.*, *A*, **1968**, 991.

306. J. Kwiatek and J. K. Seyler, *Proc. 8th Intern. Conf. Coord. Chem.*, Vienna, 1964, p. 308.

307. J. Kwiatek and J. K. Seyler, *J. Organomet. Chem.*, **3**, 421 (1965).

308. J. Kwiatek and J. K. Seyler, *J. Organomet. Chem.*, **3**, 433 (1965).

309. M. L. H. Green and P. L. I. Nagy, *J. Chem. Soc.*, **1963**, 189.

310. P. D. Sleezer, S. Winstein, and W. G. Young, *J. Am. Chem. Soc.*, **85**, 1890 (1963).

311. T. Suzuki and T. Kwan, *J. Chem. Soc. Jap.*, **86**, 1198 (1965).

312. T. Suzuki and T. Kwan, *J. Chem. Soc. Jap.*, **88**, 440 (1967).

313. T. Suzuki and T. Kwan, *J. Chem. Soc. Jap.*, **86**, 1341 (1965).

313a. W. Strohmeier and N. Iglauer, *Z. Physik. Chem. (Frankf.)*, **61**, 29 (1968).

314. A. De Jonge, unpublished results, quoted in ref. 278.

315. A. F. Mabrouk, H. J. Dutton, and J. C. Cowan, *J. Am. Oil Chem. Soc.*, **41**, 153 (1964).

316. A. F. Mabrouk, E. Selke, W. K. Rohwedder, and H. J. Dutton, *J. Am. Oil Chem. Soc.*, **42**, 432 (1965).

317. M. Murakami and J. Kang, *Bull. Chem. Soc. Jap.*, **36**, 763 (1963).

318. M. Murakami, K. Suzuki, H. Itatani, M. Kyo, and S. Senoh, Jap. Pat. 21,974 (1965); through *Chem. Abstr.*, **64**, 4945 (1966).

319. A. F. Mabrouk, unpublished results, quoted in ref. 81.

320. W. K. Rohwedder, A. F. Mabrouk, and E. Selke, *J. Phys. Chem.*, **69**, 1711 (1965).

321. L. Simandi and F. Nagy, *Magy. Kem. Foly.*, **71**, 6 (1965); through *Chem. Abstr.*, **62**, 13006 (1965).

322. L. Simandi and F. Nagy, *Acta Chim. Akad. Sci. Hung.*, **46**, 137 (1965).

322a. M. Takahashi, Y. Hisamatsu, and M. Iguchi, *J. Chem. Soc. Jap.*, **91**, 46 (1970).

323. L. M. Jackman, J. A. Hamilton, and J. M. Lawlor, *J. Am. Chem. Soc.*, **90**, 1914 (1968).

323a. P. B. Chock, R. B. K. Dewar, J. Halpern, and L. Y. Wong, *J. Am. Chem. Soc.*, **91**, 82 (1969).

324. M. Takahashi, *Bull. Chem. Soc. Jap.*, **36**, 622 (1963).

325. M. Murakami, R. Kawai, and K. Suzuki, *J. Chem. Soc. Jap.*, **84**, 669 (1963).

326. M. Murakami, K. Suzuki, and J. Kang, *J. Chem. Soc. Jap.*, **83**, 1226 (1962).

327. J. Kwiatek and I. L. Mador, U.S. Pat. 3,185,727; through *Chem. Abstr.*, **63**, 2900 (1965).

328. J. Halpern and L. Y. Wong, *J. Am. Chem. Soc.*, **90**, 6665 (1968).

329. L. Simandi, F. Nagy, and E. Budo, *Acta Chim. Akad. Sci. Hung.*, **58**, 39 (1968).

330. L. Simandi, F. Nagy, and E. Budo, *Magy. Kem. Foly.*, **74**, 441 (1968); through *Chem. Abstr.*, **70**, 10771 (1969).

331. L. Simandi, F. Nagy, and E. Budo, *Magy. Kem. Foly.*, **74**, 451 (1968); through *Chem. Abstr.*, **70**, 10794 (1969).

331a. L. Simandi and E. Budo, *Magy. Kem. Foly.*, **76**, 53 (1970); through *Chem. Abstr.*, **72**, 120820 (1970).

331b. L. Simandi and E. Budo, *Acta Chim. Akad. Sci. Hung.*, **64**, 125 (1970).

332. J. Kwiatek and I. L. Mador, U.S. Pat. 3,205,217; through *Chem. Abstr.*, **63**, 17978 (1965).

333. W. Strohmeier and N. Iglauer, *Z. Physik. Chem. (Frankf.)*, **51**, 50 (1966).

334. J. Kwiatek and I. L. Mador, U.S. Pat. 3,205,266; through *Chem. Abstr.*, **63**, 17979 (1965).

335. A. A. Vlcek and J. Hanzlik, *Inorg. Chem.*, **6**, 2053 (1967).

336. J. Hanzlik and A. A. Vlcek, *Inorg. Chem.*, **8**, 669 (1969).

337. J. Hanzlik and A. A. Vlcek, *Chem. Commun. (Lond.)*, **1969**, 47.

338. K. Isogai and Y. Hazeyama, *J. Chem. Soc. Jap.*, **86**, 869 (1965).

339. M. Murakami, K. Suzuki, H. Itatani, M. Kyo, and S. Senoh, Jap. Pat. 21,058 (1963); through *Chem. Abstr.*, **60**, 3093 (1964).

340. D. J. Cram and G. S. Hammond, *Organic Chemistry*, McGraw-Hill, New York–San Francisco, 1964, p. 417.

340a. M. Murakami, K. Suzuki, M. Fujishige, and J. Kang, *J. Chem. Soc. Jap.*, **85**, 235 (1964).

340b. M. Murakami, S. Senoh, T. Matsusato, H. Itaya, and M. Kyo, Jap. Pat. 18,711 (1962); through *Chem. Abstr.*, **59**, 11660 (1963).

341. M. Murakami and J. Kang, *Bull. Chem. Soc. Jap.*, **35**, 1243 (1962).

342. M. Murakami, K. Suzuki, M. Itaya, M. Kyo, and S. Senoh, Jap. Pat. 17,306 (1963); through *Chem. Abstr.*, **60**, 3092 (1964).

343. M. Murakami, K. Suzuki, H. Itatani, M. Kyo, and S. Senoh, Jap. Pat. 21,059, (1963); through *Chem. Abstr.*, **60**, 3093 (1964).

344. M. Murakami, J. Kang, H. Itatani, S. Senoh, and N. Matsusato, *J. Chem. Soc. Jap.*, **84**, 48 (1963).

345. M. Murakami, J. Kang, H. Itatani, S. Senoh, and N. Matsusato, *J. Chem. Soc. Jap.*, **84**, 51 (1963).

345a. M. Murakami, S. Senoh, T. Matsusato, H. Itatani, and M. Kyo, Jap. Pat. 10,765 (1962); through *Chem. Abstr.*, **59**, 10237 (1963).

346. M. Murakami, J. Kang, H. Itatani, S. Senoh, and N. Matsusato, *J. Chem. Soc. Jap.*, **84**, 53 (1963).

347. J. Kang, *J. Chem. Soc. Jap.*, **84**, 56 (1963).

348. M. Murakami, *Proc. 7th Intern. Conf. Coord. Chem.*, Stockholm and Uppsala, 1962, p. 268.

349. M. Murakami, S. Senoh, N. Matsusato, H. Itatani, and J. Kang, *J. Chem. Soc. Jap.*, **83**, 747 (1962).

350. W. P. Griffith and G. Wilkinson, *J. Chem. Soc.*, **1959**, 1629.

351. M. J. Mays and G. Wilkinson, *J. Chem. Soc.*, **1965**, 6629.

352. M. J. Mays and G. Wilkinson, *Nature*, **203**, 1167 (1964).

353. R. Mason and D. R. Russell, *Chem. Commun. (Lond.)*, **1965**, 182.

354. N. K. King and M. E. Winfield, *Biochim. Biophys. Acta*, **18**, 431 (1955).

355. T. Mizuta and T. Kwan, *J. Chem. Soc. Jap.*, **88**, 471 (1967).

356. J. Halpern and J. P. Maher, *J. Am. Chem. Soc.*, **86**, 2311 (1964).

357. J. Halpern and J. P. Maher, *J. Am. Chem. Soc.*, **87**, 5361 (1965).

358. P. B. Chock and J. Halpern, *J. Am. Chem. Soc.*, **91**, 582 (1969).

359. R. F. Heck, *Advances in Chemistry*, Vol. 49, American Chemical Society, Washington D.C., 1965, p. 181.

360. A. Kasahara and T. Hongu, *Yamagata Daigaku Kujo, Shizen Kagaku*, **6**, 263 (1965); through *Chem. Abstr.*, **65**, 2120 (1966).

361. A. Kasahara and T. Hongu, *J. Chem. Soc. Jap.*, **86**, 1343 (1965).

362. D. Goerrig, Ger. Pat. 1,099,506 (1961); through *Chem. Abstr.*, **56**, 13806 (1962).

363. Ciba Ltd., Neth. Pat. 6,400,332; through *Chem. Abstr.*, **62**, 9292 (1965).

364. A. A. Vlcek and A. Rusina, *Proc. Chem. Soc. (Lond.)*, **1961**, 161.

365. A. Rusina, H. P. Schroer, and A. A. Vlcek, *Z. Anorg. Allg. Chem.*, **351**, 275 (1967).

366. K. Tamara and T. Funabiki, *Bull. Chem. Soc. Jap.*, **41**, 1744 (1968).

367. T. Suzuki and T. Kwan, *J. Chem. Soc. Jap.*, **87**, 342 (1966).

368. T. Suzuki and T. Kwan, *J. Chem. Soc. Jap.*, **87**, 926 (1966).

369. R. Ripan, A. Farcas, and O. Piringer, *Z. Anorg. Allg. Chem.*, **346**, 211 (1966).

370. O. Piringer and A. Farcas, *Nature*, **206**, 1040 (1965).

371. O. Piringer, A. Farcas, and U. Luca, *Proc. 9th Intern. Conf. Coord. Chem.*, St. Moritz, 1966, p. 202.

372. O. Piringer and A. Farcas, *Z. Physik. Chem. (Frankf.)*, **49**, 321 (1966).

373. A. Farcas, U. Luca, and O. Piringer, in M. Cais, Ed., *Progress in Coordination Chemistry*, Elsevier Amsterdam, 1968, p. 29.

374. A. Farcas, U. Luca, N. Morar, and O. Piringer, *Z. Physik. Chem. (Frankf.)*, **58**, 87 (1968).

375. G. Mandre, in M. Cais., Ed., *Progress in Coordination Chemistry*, Elsevier Amsterdam, 1968, p. 35.

376. G. M. Schwab, and G. Mandre, *J. Catal.*, **12**, 103 (1968).

377. M. Orchin, L. Kirch, and I. Goldfarb, *J. Am. Chem. Soc.*, **78**, 5450 (1956).

377a. I. Wender, H. W. Sternberg, and M. Orchin, *J. Am. Chem. Soc.*, **75**, 3041 (1953).

378. M. Orchin, in *The Chemistry of Petroleum Hydrocarbons*, B. T. Brooks, C. E. Boord, S. S. Kuntz, and L. Schmerling, Eds., Vol. 3, Reinhold, New York, 1955, p. 343.

378a. G. Natta, R. Ercoli, and S. Castellano, *Chim. Ind. (Milan)*, **37**, 6 (1955).

379. O. Roelen, U.S. Pat. 2,327,066; through *Chem. Abstr.*, **38**, 550 (1944).

380. W. Hieber and H. Schulten, *Z. Anorg. Allg. Chem.*, **232**, 29 (1937).

381. H. W. Sternberg, I. Wender, R. A. Friedel, and M. Orchin, *J. Am. Chem. Soc.*, **75**, 2717 (1953).

382. I. Wender, H. W. Sternberg, R. A. Friedel, S. J. Metlin, and R. E. Markby, *U.S. Bur. Mines, Bull.* **600** (1962).

383. P. Pino, R. Ercoli, and F. Calderazzo, *Chim. Ind. (Milan)*, **37**, 782 (1955).

384. S. Metlin, I. Wender, and H. W. Sternberg, *Nature*, **183**, 457 (1959).

385. C. L. Aldridge and H. B. Jonassen, *Nature*, **188**, 404 (1960).

386. M. Almasi and L. Szabó, *Acad. Rep. Pop. Rom., Studii Ceretari Chim.*, **8**, 531 (1960); through *Chem. Abstr.*, **55**, 12268 (1961).

387. C. L. Aldridge and H. B. Jonassen, *J. Am. Chem. Soc.*, **85**, 886 (1963).

388. H. W. Sternberg, I. Wender, and M. Orchin, *Anal. Chem.*, **24**, 174 (1952).

389. R. Iwanaga, *Bull. Chem. Soc. Jap.*, **35**, 774 (1962).

462 References

390. R. Iwanaga, T. Fujii, H. Wakamatsu, T. Yoshida, and J. Kato, *J. Chem. Soc. Jap.*, *Ind. Chem. Sect.*, **63**, 960 (1960).
391. L. Kirch and M. Orchin, *J. Am. Chem. Soc.*, **80**, 4428 (1958).
392. L. Kirch and M. Orchin, *J. Am. Chem. Soc.*, **81**, 3597 (1959).
393. G. L. Karopinka and M. Orchin, *J. Org. Chem.*, **26**, 4187 (1961).
394. D. S. Breslow and R. F. Heck, *Chem. Ind. (Lond.)*, **1960**, 467.
395. R. F. Heck and D. S. Breslow, *J. Am. Chem. Soc.*, **83**, 4023 (1961).
396. R. F. Heck and D. S. Breslow, *Actes du Deuxième Congrès International de Catalyse*, Vol. I, Editions Technip, Paris, 1960, p. 671.
397. R. F. Heck and D. S. Breslow, *Advances in the Chemistry of the Coordination Compounds*, S. Kirschner, Ed., Macmillan Co., New York, 1961, p. 281.
398. R. F. Heck and D. S. Breslow, *J. Am. Chem. Soc.*, **83**, 1097 (1961).
399. L. Marko, G. Bor, G. Almasy, and P. Szabo, *Brennst.-Chem.*, **44**, 184 (1963).
400. R. W. Goetz and M. Orchin, *J. Org. Chem.*, **27**, 3698 (1962).
401. R. W. Goetz and M. Orchin, *J. Am. Chem. Soc.*, **85**, 2782 (1963).
402. Y. Takegami, C. Yokokawa, Y. Watanabe, H. Masada, and Y. Okuda, *Bull. Chem. Soc. Jap.*, **37**, 1190 (1964).
403. I. Wender, M. Orchin, and H. H. Storch, *J. Am. Chem. Soc.*, **72**, 4842 (1950).
404. L. Marko, *Khim. i Tekhnol. Topliv i Masel*, **5**, 19 (1960); through *Chem. Abstr.*, **55**, 2075 (1961).
405. C. L. Aldridge, U.S. Pat. 3,091,644; through *Chem. Abstr.*, **59**, 11260 (1963).
406. Esso, Brit. Pat. 864,142; through *Chem. Abstr.*, **55**, 18597, (1961).
407. V. Macho, *Chem. Zvesti*, **17**, 525 (1963); through *Chem. Abstr.*, **60**, 10534 (1964).
408. V. Macho, *Ropa Uhlie*, **6**, 297 (1964); through *Chem. Abstr.*, **62**, 8996 (1965).
409. V. Macho, E. J. Mistrik, and M. Ciha, *Coll. Czech. Chem. Comm.*, **29**, 826 (1964).
410. Shell International Research Maatschappij, N.V., Belg. Pat. 616,141; through *Chem. Abstr.*, **59**, 1495 (1963).
411. J. Falbe and F. Korte, *Angew. Chem. Intern. Edit.*, **1**, 657 (1962).
412. J. Falbe, N. Huppes, and F. Korte, *Chem. Ber.*, **97**, 863 (1964).
413. H. W. Sternberg and I. Wender, *Chem. Soc. Spec. Publ. 13 (Lond.)*, 1959, p. 35.
414. L. Marko and P. Szabo, *Chem. Tech. (Berl.)*, **13**, 482 (1961).
415. I. Wender, R. Levine, and M. Orchin, *J. Am. Chem. Soc.*, **72**, 4375 (1950).
416. I. Wender and M. Orchin, U.S. Pat. 2,614,107; through *Chem. Abstr.*, **47**, 5422 (1953).
417. V. W. Dawydoff, *Chem. Tech. (Berl.)*, **11**, 431 (1959).
418. I. Wender, H. Greenfield, and M. Orchin, *J. Am. Chem. Soc.*, **73**, 2656 (1951).
419. H. Adkins and G. Krsek, *J. Am. Chem. Soc.*, **71**, 3051 (1949).
420. J. Falbe and F. Korte, *Chem. Ber.*, **97**, 1104 (1964).
421. J. Berty, E. Altay, and L. Marko, *Chem. Tech. (Berl.)*, **9**, 283 (1957).
422. L. Marko and O. Budavari, *Magyar Tud. Akad. Kém. Tud. Oszt.Közl*, **13**, 153 (1960); through *Chem. Abstr.*, **55**, 206 (1961).
423. A. I. M. Keulemans, R. Kwantes, and Th. van Bavel, *Rec. Trav. Chim.*, **67**, 298 (1948).
424. R. C. Schreyer, U.S. Pat. 2,564,130, through *Chem. Abstr.*, **47**, 142 (1953).

425. V. L. Hughes and L. Kirshenbaum, *Ind. Eng. Chem.*, **49**, 1999 (1957).

426. N. Kutepow and H. Kindler, *Angew. Chem.*, **72**, 802 (1960).

427. L. Marko, *Chem. Ind. (Lond.)*, **1962**, 260.

428. C. R. Greene and R. E. Meeker, Belg. Pat. 621,833; through *Chem. Abstr.*, **59**, 11259 (1963).

429. L. G. Cannell, L. H. Slaugh, and R. D. Mullineaux, Ger. Pat. 1,186,455; through *Chem. Abstr.*, **62**, 16054 (1965).

430. G. F. Cox and G. H. Whitfield, Brit. Pat. 999,461; through *Chem. Abstr.*, **63**, 9811 (1965).

431. E. I. du Pont de Nemours and Co., Brit. Pat. 614,010; through *Chem. Abstr.*, **43**, 4685 (1949).

432. P. L. Barrick, U.S. Pat. 2,542,747; through *Chem. Abstr.*, **46**, 7584 (1951).

433. I. Wender, J. Feldman, S. Metlin, B. H. Gwynn, and M. Orchin, *J. Am. Chem. Soc.*, **77**, 5760 (1955).

434. N. S. Imyanitov and D. M. Rudkovskii, U.S.S.R. Pat. 149,422; through *Chem. Abstr.*, **58**, 6633 (1963).

435. A. W. C. Taylor, Brit. Pat. 798,541; through *Chem. Abstr.*, **53**, 2089 (1959).

436. J. Habeshaw and L. S. Thornes, Brit. Pat. 702,195; through *Chem. Abstr.*, **49**, 5513 (1955).

437. A. W. C. Taylor and S. A. Lamb, Brit. Pat. 684,673; through *Chem. Abstr.*, **48**, 1421 (1954).

438. Inventa A-G. für Forschung und Patentsverwertung, Fr. Pat. 1,404,182; through *Chem. Abstr.*, **63**, 16230 (1965),

439. Esso, Brit. Pat, 728,913; through *Chem. Abstr.*, **50**, 7852 (1956).

440. C. Bordenca and W. A. Lazier, U.S. Pat. 2,584,539; through *Chem. Abstr.*, **46**, 8676 (1952).

441. W. H. Clement and M. Orchin, *Ind. Eng. Chem. Prod. Res. Dev.*, **4**, 283 (1965).

442. Inventa A-G. für Forschung und Patentsverwertung, Brit. Pat. 1,007,627 through *Chem. Abstr.*, **64**, 4968 (1966).

443. W. Reppe, O. Schlichting, K. Klager, and T. Toepel, *Ann. Chem.*, **560**, 1 (1948).

444. Chemische Werke Huels A.G., Fr. Pat. 1,411,448; through *Chem. Abstr.*, **64**, 3380 (1966).

445. Ruhr-Chemie A.G., Brit. Pat. 779,241; through *Chem. Abstr.*, **52**, 1224 (1958).

446. A. L. Nussbaum, T. L. Popper, E. P. Oliveto, S. Friedman, and I. Wender, *J. Am. Chem. Soc.*, **81**, 1228 (1959).

447. P. F. Beal and M. A. Rebenstorf, Ger. Pat. 1,124,942; through *Chem. Abstr.*, **58**, 8021 (1963).

448. P. F. Beal, M. A. Rebenstorf, and J. E. Pike, *J. Am. Chem. Soc.*, **81**, 1231 (1959).

449. Esso, Neth. Pat. 6,400,701; through *Chem. Abstr.*, **62**, 5194 (1965).

450. D. M. Rudkovskii and N. S. Imyanitov, *J. Appl. Chem. (USSR)*, **35**, 2608 (1962).

451. F. Gaslini and L. Z. Nahum, *J. Org. Chem.*, **29**, 1177 (1964).

452. J. Falbe and F. Korte, U.S. Pat. 3,159,653; through *Chem. Abstr.*, **62**, 9112 (1965).

453. A. Rosenthal and D. Abson, *Can. J. Chem.*, **42**, 1811 (1964).

454. A. Rosenthal and D. Abson, *Can. J. Chem.*, **43**, 1318 (1965).

455. A. Rosenthal and H. J. Koch, *Can. J. Chem.*, **43**, 1375 (1965).

456. A. Rosenthal and D. Read, *Methods Carbohydr. Chem.*, **2**, 450 (1963).

457. A. Rosenthal and D. Abson, *Can. J. Chem.*, **43**, 1985 (1965).

458. A. Rosenthal and D. Read, *Can. J. Chem.*, **35**, 788 (1957).

459. A. Rosenthal, D. Abson, T. D. Field, H. J. Koch, and R. E. J. Mitchell, *Can. J. Chem.*, **45**, 1525 (1967).

460. R. F. Heck, *J. Am. Chem. Soc.*, **85**, 1460 (1963).

461. Y. Takegami, C. Yokokawa, and Y. Watanabe, *Bull. Chem. Soc. Jap.*, **37**, 935 (1964).

462. Y. Takegami, C. Yokokawa, Y. Watanabe, and H. Masada, *Bull. Chem. Soc. Jap.*, **37**, 672 (1964).

463. Y. Takegami, C. Yokokawa, Y. Watanabe, and H. Masada, *Bull. Chem. Soc. Jap.*, **38**, 1649 (1965).

464. C. Yokokawa, Y. Watanabe, and Y. Takegami, *Bull. Chem. Soc. Jap.*, **37**, 677 (1964).

465. J. L. Eisenmann and R. L. Yamartino, Ger. Pat. 1,146,485; through *Chem. Abstr.*, **59**, 11291 (1963).

465a. J. L. Eisenmann, U.S. Pat. 3,290,379; through *Chem. Abstr.*, **66**, 46100 (1967).

466. H. Greenfield, J. H. Wotiz, and I. Wender, *J. Org. Chem.*, **22**, 542 (1957).

467. H. Adkins and J. L. Williams, *J. Org. Chem.*, **17**, 980 (1952).

468. M. Morikawa, *Bull. Chem. Soc. Jap.*, **37**, 379 (1964).

469. C. Bordenca, U.S. Pat. 2,790,006; through *Chem. Abstr.*, **51**, 14786 (1957).

470. G. Gut, M. H. El-Markhzangi, and A. Guyer, *Helv. Chim. Acta*, **48**, 1151 (1965).

471. I. Wender, R. Markby, and H. Greenfield, unpublished work quoted in ref. 466.

472. S. Friedman, S. Metlin, A. Svedi, and I. Wender, *J. Org. Chem.*, **24**, 1287 (1959).

473. I. Wender, R. Levine, and M. Orchin, *J. Am. Chem. Soc.*, **71**, 4160 (1949).

474. I. Wender, H. Greenfield, S. Metlin, and M. Orchin, *J. Am. Chem. Soc.*, **74**, 4079 (1952).

475. I. Wender, R. A. Friedel, and M. Orchin, *Science*, **113**, 206 (1957).

476. W. Reppe and H. Friederich, Ger. Pat. 894,403; through *Chem. Abstr.*, **50**, 16830 (1956).

477. K. H. Ziesecke, *Brennst.-Chem.*, **33**, 385 (1952).

478. G. R. Burns, *J. Am. Chem. Soc.*, **77**, 6615 (1955).

479. I. Wender, S. Metlin, and M. Orchin, *J. Am. Chem. Soc.*, **73**, 5704 (1951).

480. R. A. Schunn, *Abstr. M52, Am. Chem. Soc., 155th Meet.*, San Francisco, California, 1968; through ref. 242a.

481. S. Murahashi and S. Horiie, *Bull. Chem. Soc. Jap.*, **33**, 78 (1960).

482. R. C. Schreyer, U. S. Pat. 3,206,498; through *Chem. Abstr.*, **64**, 605 (1966).

483. A. Nakamura and N. Hagihara, *Mem. Inst., Sci. Ind. Res., Osaka Univ.*, **15**, 195 (1958).

484. S. Horiie and S. Murahashi, *Bull. Chem. Soc. Jap.*, **33**, 247 (1960).

485. S. Murahashi, S. Horiie, and T. Jo, *J. Chem. Soc. Jap.*, **79**, 68 (1958).

486. S. Murahashi and S. Horiie, *Ann. Rep. Sci. Works, Fac. Sci., Osaka Univ.*, **7**, 89 (1959).

487. J. T. Shaw and F. T. Ryson, *J. Am. Chem. Soc.*, **78**, 2538 (1956).

488. P. Pino and R. Ercoli, *Ric. Sci.*, **23**, 1231 (1953).

489. M. Freund, L. Marko, and J. Laki, *Acta Chim. Acad. Sci. Hung.*, **31**, 77 (1962).

490. M. Orchin, *Adv. Catal.*, **5**, 385 (1953).

491. I. Wender, H. W. Sternberg, and M. Orchin, *Catalysis*, Vol. V, Reinhold, New York, 1957, p. 73.

492. R. F. Heck, *J. Am. Chem. Soc.*, **85**, 651 (1963).

493. P. Thuring and A. Perrett, *Helv. Chim. Acta*, **36**, 13 (1953).

494. J. Kato, T. Ito, Y. Yabe, R. Iwanaga, and T. Yashida, *J. Chem. Soc., Jap., Ind. Chem. Sect.*, **65**, 184 (1962).

495. D. M. Rudkovskii, N. S. Imyanitov, and V. Y. Gankin, *Tr. Vses. Nauchn. Issled. Inst. Neftekhim. Protsessov*, **2**, 121 (1960); through *Chem. Abstr.*, **57**, 10989 (1962).

496. F. Calderazzo and F. A. Cotton, *Inorg. Chem.*, **1**, 30 (1962).

497. F. Calderazzo and F. A. Cotton, *Proc. 7th Intern. Conf. Coord. Chem.*, Sweden, 1962, Abstr., p. 296.

498. R. F. Heck and D. S. Breslow, *J. Am. Chem. Soc.*, **84**, 2499 (1962).

499. I. Wender and M. Orchin, unpublished work quoted in ref. 490.

500. R. E. Brooks, U.S. Pat. 2,517,383; through *Chem. Abstr.*, **45**, 1158 (1951).

501. H. B. Jonassen, R. J. Stearns, J. Kenttamaa, D. W. Moore, and A. G. Whittaker, *J. Am. Chem. Soc.*, **80**, 2586 (1958).

502. D. W. Moore, H. B. Jonassen, T. B. Joyner, and A. J. Bertrand, *Chem. Ind. (Lond.)*, **1960**, 1304.

503. S. Friedman, H. F. Kauffman, and I. Wender, unpublished work quoted in ref. 472.

504. L. E. Craig, R. M. Elofson, and I. J. Ressa, *J. Am. Chem. Soc.*, **72**, 3277 (1950).

505. I. Wender, H. Greenfield, S. Metlin, R. Markby, and M. Orchin, unpublished work quoted in ref. 490.

506. S. Horiie and S. Murahashi, *Bull. Chem. Soc. Jap.*, **33**, 88 (1960).

507. K. Hamada, K. Baba, and N. Hagihara, *Mem. Inst. Sci. Ind. Res., Osaka Univ.*, **14**, 207 (1957).

508. M. Hondo, T. Koga, and G. Noyori, *J. Chem. Soc. Jap., Ind. Chem. Sect.*, **70**, 1346 (1967).

509. P. R. Rony, *J. Catal.*, **14**, 142 (1969).

510. D. Commereuc, I. Douek, and G. Wilkinson, *J. Chem. Soc., A*, **1970**, 1771.

511. G. M. Bancroft, M. H. Mays, B. E. Prater, and F. P. Stefanini, *J. Chem. Soc., A*, **1970**, 2146.

512. A. J. Deeming and B. L. Shaw, *J. Chem. Soc., A*, **1970**, 2705.

513. M. J. Cleare and W. P. Griffith, *J. Chem. Soc., A*, **1970**, 2788.

514. S. B. Laing and P. J. Sykes, *J. Chem. Soc., C*, **1968**, 421.

515. P. Wieland and G. Anner, *Helv. Chim. Acta*, **51**, 1698 (1968).

516. H. J. Brodie, K. J. Kripalani, and G. Possanza, *J. Am. Chem. Soc.*, **91**, 1241 (1969).

517. A.-B. Hörnfeldt, J. S. Gronowitz, and S. Gronowitz, *Acta Chem. Scand.*, **22**, 2725 (1968).

518. W. H. Dennis, Jr., D. H. Rosenblatt, R. R. Richmond, G. A. Finseth, and G. T. Davis, *Tetrahedron Lett.*, **1968**, 1821.

519. K. Jonas and G. Wilke, *Angew. Chem. Intern. Edit.*, **8**, 519 (1969).

520. R. Nast and T. von Krakkay, *Z. Naturforsch.*, **9b**, 798 (1954).

521. L. Roos and M. Orchin, *J. Org. Chem.*, **31**, 3015 (1966).

522. F. Ungvary and L. Marko, *New Aspects Chem. Met. Carbonyls Deriv., 1st Intern. Symp.*, Venice, 1968, Abstr., E5.

523. F. Ungvary and L. Marko, *Acta Chim. Akad. Sci. Hung.*, **62**, 425 (1969).

524. L. H. Slaugh, Belg. Pat. 621,662; through *Chem. Abstr.*, **59**, 11268 (1963).

525. L. H. Slaugh and R. D. Mullineaux, U.S. Pat. 3,239,566; through *Chem. Abstr.*, **64**, 15745 (1966).

526. L. H. Slaugh and R. D. Mullineaux; U.S. Pat. 3,239,569; through *Chem. Abstr.*, **64**, 15745 (1966).

527. L. H. Slaugh and R. D. Mullineaux; U.S. Pat. 3,239,570; through *Chem. Abstr.*, **64**, 19420 (1966).

528. L. H. Slaugh and R. D. Mullineaux, *J. Organomet. Chem.*, **13**, 469 (1968).

529. E. R. Tucci, *Ind. Eng. Chem., Prod. Res. Dev.*, **7**, 32 (1968).

530. E. R. Tucci, *Ind. Eng. Chem., Prod. Res. Dev.*, **7**, 125 (1968).

531. E. R. Tucci, *Ind. Eng. Chem., Prod. Res. Dev.*, **8**, 286 (1969).

532. E. R. Tucci, *Ind. Eng. Chem., Prod. Res. Dev.*, **7**, 227 (1968).

533. A. Hershman and J. H. Craddock, *Ind. Eng. Chem. Prod. Res. Dev.*, **7**, 226 (1968).

534. W. Kniese, H. J. Nienburg, and R. Fischer, *J. Organomet. Chem.*, **17**, 133 (1969).

535. F. Piacenti, M. Bianchi, and E. Benedetti, *Chim. Ind. (Milan)*, **49**, 245 (1967).

536. B. Fell, W. Rupilius, and F. Asinger, *Tetrahedron Lett.*, **1968**, 3261.

537. R. Cramer, *J. Am. Chem. Soc.*, **89**, 4621 (1967).

538. I. J. Goldfarb and M. Orchin, *Adv. Catal.*, **9**, 609 (1957).

539. E. R. Tucci, *Ind. Eng. Chem. Prod. Res. Dev.*, **8**, 215 (1969).

540. F. Asinger, B. Fell, and W. Rupilius, *Ind. Eng. Chem. Prod. Res. Dev.*, **8**, 214 (1969).

541. C. Palm, German Pat. 1,230,010; through *Chem. Abstr.*, **66**, 104696 (1967).

542. J. F. Deffner, E. R. Tucci, J. V. Ward, H. I. Thayer, and H. J. Elder, Fr. Pat. 1,534,510; through *Chem. Abstr.*, **71**, 101314 (1969).

543. A. Misono and I. Ogata, *Bull. Chem. Soc. Jap.*, **40**, 2718 (1967).

544. A. Misono and I. Ogata, Ger. Pat., 1,807,827; through *Chem. Abstr.*, **71**, 38442 (1969).

545. I. Ogata and A. Misono, *Discuss. Faraday Soc.*, **46**, 72 (1968).

546. I. Ogata, *J. Chem. Soc. Jap.*, **72**, 1710 (1969).

547. W. Strohmeier, J. F. Guttenberger, and H. Hellmann, *Z. Naturforsch.*, **19b**, 353 (1964).

548. K. E. Atkins, U.S. Pat. 3,308,177; through *Chem. Abstr.*, **67**, 21504 (1967).

549. A. Andreetta, G. F. Ferrari, and G. F. Pregaglia, *Proc. 4th Intern. Conf. Organomet. Chem.*, Bristol, 1969, Abstr., L3.

550. G. F. Pregaglia, A. Andreetta, and R. Ugo, *Chim. Ind. (Milan)*, **50**, 1332 (1968).

551. G. F. Pregaglia, R. Castelli, and A. Andreetta, Fr. Pat. 1,514,495; through *Chem. Abstr.*, **70**, 79633 (1969).

552. R. L. Pruett and J. A. Smith, U.S. Pat. 3,505,408; through *Chem. Abstr.*, **73**, 3476 (1970).

553. M. Hidai, T. Kuse, T. Hikita, Y. Uchida, and A. Misono, *Tetrahedron Lett.*, **1970**, 1715.

554. A. Misono, Y. Uchida, T. Saito, and K. M. Song, *Chem. Commun. (Lond.)*, **1967**, 419.

555. A. Misono, Y. Uchida, M. Hidai, and T. Kuse, *Chem. Commun. (Lond.)*, **1968**, 981.

556. G. Speier and L. Marko, *Inorg. Chim. Acta*, **3**, 126 (1969).

557. A. Yamamoto, L. S. Pu, S. Kitazume, and S. Ikeda, *J. Am. Chem. Soc.*, **89**, 3071 (1967).

558. A. Sacco and M. Rossi, *Chem. Commun. (Lond.)*, **1967**, 316.

559. A. Sacco and M. Rossi, *Inorg. Chim. Acta*, **2**, 127 (1968).

560. A. Yamamoto, S. Kitazume, L. S. Pu, and S. Ikeda, *Chem. Commun. (Lond.)*, **1967**, 79.

561. P. H. Enemark, B. R. Davis, J. A. McGinety, and J. A. Ibers, *Chem. Commun. (Lond.)*, **1968**, 96.

562. A. Misono, Y. Uchida, M. Hidai, and M. Araki, *Chem. Commun. (Lond.)*, **1968**, 1044.

563. A. Misono, Y. Uchida, and T. Saito, *Bull. Chem. Soc. Jap.*, **40**, 700 (1967).

564. L. S. Pu, A. Yamamoto, and S. Ikeda, *J. Am. Chem. Soc.*, **90**, 3896 (1968).

565. S. Otsuka and M. Rossi, *J. Chem. Soc.*, *A*, **1969**, 497 and refs. therein.

566. A. Sacco and R. Ugo, *J. Chem. Soc.*, **1964**, 3274.

567. M. Rossi and A. Sacco, *Chem. Commun. (Lond.)*, **1969**, 471.

567a. G. Speier and L. Marko, *Proc. 13th Intern. Conf. Coord. Chem.*, Cracow-Zakopane, Poland, 1970, Abstr., 217.

568. A. Simon, G. Speier, and L. Marko, *Proc. 13th Intern. Conf. Coord. Chem.*, Cracow-Zakopane, Poland, 1970, Abstr., 84.

569. A. Simon, Z. Nagy-Magos, J. Palagyi, G. Palyi, G. Bor, and L. Marko, *J. Organomet. Chem.*, **11**, 634 (1968).

570. F. Ungvary, B. Babos, and L. Marko, *J. Organomet. Chem.*, **8**, 329 (1967).

571. B. Babos, F. Ungvary, L. Papp, and L. Marko, *Magy. Kem. Foly.*, **75**, 126 (1969); through *Chem. Abstr.*, **70**, 114484 (1969).

572. G. Speier, B. Babos, and L. Marko, *Magy. Kem. Foly.*, **76**, 59 (1970); through *Chem. Abstr.*, **72**, 100791 (1970).

573. I. Loberth, H. Nöth, and P. V. Rinze, *J. Organomet. Chem.*, **16**, P1 (1969).

574. S. Tyrlik and H. Stepowska, *Tetrahedron Lett.*, **1969**, 3593.

575. R. F. Heck, U.S. Pat. 3,270,087; through *Chem. Abstr.*, **65**, 16857 (1966).

576. N. S. Imyanitov and D. M. Rudkovskii, *J. Prakt. Chem.*, **311**, 712 (1969).

577. N. F. Gol'dshleger, M. B. Tyabin, A. E. Shilov, and A. A. Shteinman, *Russ. J. Phys. Chem.*, **43**, 1222 (1969).

578. A. Sacco and M. Rossi, *Chem. Commun. (Lond.)*, **1965**, 602.

579. A. Sacco, M. Rossi, and C. F. Nobile, *Chem. Commun. (Lond.)*, **1966**, 589.

580. H. Wakamatsu and K. Sakamaki, *Chem. Commun. (Lond.)*, **1967**, 1140.

581. M. Iwamoto, *J. Chem. Soc. Jap., Ind. Chem. Sect.*, **71**, 1510 (1968); through *Chem. Abstr.*, **70**, 28310 (1969).

582. R. D. Mullineaux, U.S. Pat. 3,270,348; through *Chem. Abstr.*, **66**, 65,080 (1967).

583. L. S. Pu, A. Yamamoto, and S. Ikeda, *Chem. Commun. (Lond.)*, **1969**, 189.

584. G. Pregaglia, A. Andreetta, G. Ferrari, and R. Ugo, *Chem. Commun. (Lond.)*, **1969**, 590.

585. A. Misono, Y. Uchida, M. Hidai, and T. Kuse, *Chem. Commun. (Lond.)*, **1969**, 208.

586. B. M. Chadwick and L. Shields, *Chem. Commun. (Lond.)*, **1969**, 650.

587. J. P. Maher, *J. Chem. Soc., A*, **1968**, 2918.

588. M. G. Swanwick and W. A. Waters, *Chem. Commun. (Lond.)*, **1970**, 930.

589. L. Marko, *Acta Chim. Akad. Sci. Hung.*, **59**, 378 (1969); through *Chem. Abstr.*, **70**, 118516 (1969).

590. Ref. 53, Chapters 7 and 8.

591. R. F. Heck, "Organic Syntheses via Alkyl and Acylcobalt Tetracarbonyls," in I. Wender and P. Pino, Eds., *Organic Syntheses via Metal Carbonyls*, Vol. 1, Interscience, New York, 1968, p. 373.

592. M. E. Volpin and I. S. Kolomnikov, *Proc. Acad. Sci. USSR*, **170**, 997 (1966).

593. M. E. Volpin and I. S. Kolomnikov, *Bull. Acad. Sci., Chem. Div.*, **1966**, 1980.

594. M. E. Volpin and I. S. Kolomnikov, "*Soveshchanie po Gomogennomu Katalizu (Tezisy Dokladov)*," summaries of papers presented at the Conference on Homogeneous Catalysis, Kiev, 1966; through ref. 57d.

595. M. E. Volpin and I. S. Kolomnikov, *Katal. Reakts. Zhidk, Faze, Tr. Vses. Konf., 2nd, Alma-Ata, Kez. SSR*, **1966**, 429; through *Chem. Abstr.*, **69**, 46340 (1968).

596. B. R. James and G. L. Rempel, unpublished results.

597. G. Wilkinson, *Proc. Chem. Soc. (Lond.)*, **1961**, 72.

598. R. D. Gillard and G. Wilkinson, *J. Chem. Soc.*, **1963**, 3594.

599. J. A. Osborn, R. D. Gillard, and G. Wilkinson, *J. Chem. Soc.*, **1964**, 3168.

600. J. A. Osborn, A. R. Powell, and G. Wilkinson, *Chem. Commun. (Lond.)*, **1966**, 461.

601. A. R. Powell, *Plat. Met. Rev.*, **11**, 58 (1967).

602. K. Thomas, J. A. Osborn, A. R. Powell, and G. Wilkinson, *J. Chem. Soc., A*, **1968**, 1801.

603. H. C. Brown and C. A. Brown, *J. Am. Chem. Soc.*, **84**, 1495 (1962).

604. A. A. Vlcek, *Progr. Inorg. Chem.*, **5**, 371 (1963).

605. J. F. Harrod and J. Halpern, *Can. J. Chem.*, **37**, 1933 (1959).

606. B. R. James and G. L. Rempel, *Can. J. Chem.*, **44**, 233 (1966).

607. W. C. Wolsey, C. A. Reynolds, and J. Kleinberg, *Inorg. Chem.*, **2**, 463 (1963).

608. B. R. James and M. L. Kastner, unpublished results.

609. D. N. Lawson, M. J. Mays, and G. Wilkinson, *J. Chem. Soc., A*, **1966**, 52.

610. B. N. Figgis, R. D. Gillard, R. S. Nyholm, and G. Wilkinson, *J. Chem. Soc.*, **1964**, 5189.

611. B. Martin and G. M. Waind, *J. Chem. Soc.*, **1958**, 4284.

612. J. A. Osborn, F. H. Jardine, J. F. Young, and G. Wilkinson, *J. Chem. Soc., A*, **1964**, 1711.

613. B. R. James and G. Rosenberg, unpublished results.

614. H. B. Charman, *Nature*, **212**, 278 (1966).

615. H. B. Charman, *J. Chem. Soc., B*, **1967**, 629.

616. A. G. Davies, G. Wilkinson, and J. F. Young, *J. Am. Chem. Soc.*, **85**, 1692, (1963).

617. R. D. Gillard, J. A. Osborn, and G. Wilkinson, *J. Chem. Soc.*, **1965**, 4107.

618. R. D. Gillard, J. A. Osborn, P. B. Stockwell, and G. Wilkinson, *Proc. Chem. Soc. (Lond.)*, **1964**, 284.

619. M. Delepine, *Bull. Soc. Chim. Fr.*, **45**, 235 (1929).

620. M. Delepine, *Compt. Rend.*, **236**, 559 (1953).

621. R. D. Gillard, J. A. Osborn, and G. Wilkinson, *J. Chem. Soc.*, **1965**, 1951.

622. F. Basolo, E. J. Bounsall, and A. J. Poë, *Proc. Chem. Soc. (Lond.)*, **1963**, 366.

623. G. C. Kulasingham and W. R. McWhinnie, *J. Chem. Soc.*, **1965**, 7145.

624. B. Martin, W. R. McWhinnie, and G. M. Waind, *J. Inorg. Nucl. Chem.*, **23**, 207 (1961).

625. J. V. Rund, F. Basolo, and R. G. Pearson, *Inorg. Chem.*, **3**, 658 (1964).

625a. J. V. Rund, *Inorg. Chem.*, **7**, 24 (1968).

626. B. R. James, M. Kastner, and G. L. Rempel, *Can. J. Chem.*, **47**, 349 (1969).

627. B. R. James and G. L. Rempel, *J. Chem. Soc.*, *A*, **1969**, 78.

628. R. D. Gillard, *Discuss. Faraday. Soc.*, **46**, 90 (1968).

629. D. J. Baker and R. D. Gillard, *Chem. Commun. (Lond.)*, **1967**, 520.

630. R. D. Gillard, B. T. Heaton, and D. H. Vaughan, *Chem. Commun. (Lond.)*, **1969**, 974.

631. J. P. Maher, *Chem. Commun. (Lond.)*, **1966**, 785.

632. L. E. Johnston and J. A. Page, *Can. J. Chem.*, **47**, 2123 (1969).

633. L. E. Johnston and J. A. Page, *Can. J. Chem.*, **47**, 4241 (1969).

634. R. D. Gillard and G. Wilkinson, *J. Chem. Soc.*, **1964**, 1224.

635. M. A. Bennett and P. A. Longstaff, *Chem. Ind. (Lond.)*, **1965**, 846.

636. J. P. Candlin and A. R. Oldham, *Discuss. Faraday. Soc.*, **46**, 60 (1968).

637. F. M. Jardine, J. A. Osborn, and G. Wilkinson, *J. Chem. Soc.*, *A*, **1967**, 1574.

638. S. Montelatici, A. van der Ent, J. A. Osborn, and G. Wilkinson, *J. Chem. Soc.*, *A*, **1968**, 1054.

639. R. J. Cvetanovic, F. J. Duncan, W. E. Falconer, and R. S. Irwin, *J. Am. Chem. Soc.*, **87**, 1827 (1965).

640. Ref. 7, Chapter 11.

641. S. Takahashi, H. Yamazaki and N. Hagihara, *Bull. Chem. Soc. Jap.*, **41**, 254 (1968).

642. J. F. Biellmann and M. J. Jung, *J. Am. Chem. Soc.*, **90**, 1673 (1968).

643. G. C. Bond and R. A. Hillyard, *Discuss. Faraday Soc.*, **46**, 20 (1968).

644. R. L. Augustine and J. Van Peppen, *Ann. N.Y. Acad. Sci.*, **158** (2), 482 (1969).

645. J. F. Harrod and A. J. Chalk, *J. Am. Chem. Soc.*, **86**, 1776 (1964).

646. R. Ugo, *Discuss. Faraday Soc.*, **46**, 89 (1968).

647. A. J. Birch and K. A. M. Walker, *J. Chem. Soc.*, *C*, **1966**, 1894.

648. A. J. Birch and K. A. M. Walker, *Tetrahedron Lett.*, **1966**, 4939.

649. J. F. Biellman and H. Liesenfelt, *Bull. Soc. Chim. Fr.*, **1966**, 4029.

650. J. F. Biellman and H. Liesenfelt, *Compt. Rend.*, **263**, 251 (1966).

651. C. Djerassi and J. Gutzwiller, *J. Am. Chem. Soc.*, **88**, 4537 (1966).

652. A. Tanaka, H. Uda, and A. Yoshikoshi, *Chem. Commun. (Lond.)*, **1969**, 308.

653. R. W. Britton, Ph.D. Dissertation, University of British Columbia, Vancouver, British Columbia, 1970.

654. E. Piers and K. F. Cheng, *Can. J. Chem.*, **46**, 377 (1968).

655. M. Brown and L. W. Piszkiewicz, *J. Org. Chem.*, **32**, 2013 (1967).

656. J. J. Sims, V. K. Honwad, and L. H. Selman, *Tetrahedron Lett.*, **1969**, 87.

657. A. J. Birch and K. A. M. Walker, *Tetrahedron Lett.*, **1967**, 1935.

658. E. Piers and K. F. Cheng, *Chem. Commun.* (*Lond.*), **1969**, 562.

659. R. E. Harmon, J. L. Parsons, D. W. Cooke, S. K. Gupta, and J. Schoolenberg, *J. Org. Chem.*, **34**, 3684 (1969).

660. F. H. Jardine and G. Wilkinson, *J. Chem. Soc.*, *C*, **1967**, 270.

661. K. Ohno and J. Tsuji, *J. Am. Chem. Soc.*, **90**, 99 (1968).

662. M. C. Baird, C. J. Nyman and G. Wilkinson, *J. Chem. Soc.*, *A*, **1968**, 348.

663. R. H. Prince and K. A. Raspin, *J. Chem. Soc.*, *A*, **1969**, 612.

664. W. Voelter and C. Djerassi, *Chem. Ber.*, **101**, 58 (1968).

665. W. Voelter and C. Djerassi, *Chem. Ber.*, **101**, 1154 (1968).

666. A. J. Birch and K. A. M. Walker, *Tetrahedron Lett.*, **1967**, 3457.

667. A. S. Hussey and Y. Takeuchi, *J. Am. Chem. Soc.*, **91**, 672 (1969).

668. A. S. Hussey and Y. Takeuchi, *J. Org. Chem.*, **35**, 643 (1970).

669. C. H. Heathcock and S. R. Poulter, *Tetrahedron Lett.*, **1969**, 2755.

670. A. J. Birch and G. S. R. Subba Rao, *Tetrahedron Lett.*, **1968**, 3797.

671. G. V. Smith and R. J. Shuford, *Tetrahedron Lett.*, **1970**, 525.

672. H. van Bekkum, F. van Rantwijk, and T. van de Putte, *Tetrahedron Lett.*, **1969**, 1.

673. M. C. Baird, D. N. Lawson, J. T. Mague, J. A. Osborn, and G. Wilkinson, *Chem. Commun.* (*Lond.*), **1966**, 129.

674. J. Blum, J. Y. Becker, H. Rosenman, and E. D. Bergmann, *J. Chem. Soc.*, *B*, **1969**, 1000.

675. S. Otsuka, A. Nakamura, Y. Tatsuno, and Y. Yamamoto, *Proc. 4th Intern. Conf. Organomet. Chem.*, Bristol, 1969, Abstr., 03.

676. M. Takesada, H. Tamazaki, and N. Hagihara, *Bull. Chem. Soc. Jap.*, **41**, 270 (1968)

677. J. Blum, H. Rosenmann, and E. D. Bergmann, *Tetrahedron Lett.*, **1967**, 3665.

678. A. L. Odell, J. B. Richardson, and W. R. Roper, *J. Catal.*, **8**, 393 (1967).

679. H. Simon and O. Berngruber, *Tetrahedron*, **26**, 161 (1970).

680. H. Simon and O. Berngruber, *Tetrahedron*, **26**, 1401 (1970).

681. J. F. Young, R. D. Gillard, and G. Wilkinson, *J. Chem. Soc.*, **1964**, 5176.

682. J. T. Mague and G. Wilkinson, *J. Chem. Soc.*, *A*, **1966**, 1736.

683. M. C. Baird, J. T. Mague, J. A. Osborn, and G. Wilkinson, *J, Chem. Soc.*, *A*, **1967**, 1347.

684. R. Stern, Y. Chevallier, and L. Sajus, *Compt. Rend.*, **264**, 1740 (1967).

685. Y. Chevallier, R. Stern and L. Sajus, *Tetrahedron Lett.*, **1969**, 1197.

686. L. Horner, H. Büthe, and H. Siegel, *Tetrahedron Lett.*, **1968**, 4023.

687. C. O'Connor and G. Wilkinson, *Tetrahedron Lett.*, **1969**, 1375.

688. W. S. Knowles and M. J. Sabacky, *Chem. Commun.* (*Lond.*), **1968**, 1445.

689. W. S. Knowles, M. J. Sabacky, and B. D. Vineyard, *Ann. N. Y. Acad. Sci.*, **172** (9), 232 (1970).

690. L. Horner, H. Siegel, and H. Büthe, *Angew. Chem. Intern. Edit.*, **7**, 942 (1968).

691. I. Jardine and F. J. McQuillin, *Chem. Commun. (Lond.)*, **1969**, 477.

692. P. Abley and F. J. McQuillin, *Chem. Commun. (Lond.)*, **1969**, 477.

693. G. Wilkinson, Ger. Pat. 1,816,063; through *Chem. Abstr.*, **71**, 90766 (1969).

694. B. R. James, F. T. T. Ng, and G. L. Rempel, *Inorg. Nucl. Chem. Lett.*, **4**, 197 (1968).

695. K. C. Dewhirst, U.S. Pat. 3,489,786; through *Chem. Abstr.*, **72**, 89833 (1970).

696. R. E. Harmon, J. L. Parsons, and S. K. Gupta, *Chem. Commun. (Lond.)*, **1969**, 1365.

697. R. E. Harmon, J. L. Parsons, and S. K. Gupta, *Org. Prep. Proc.*, **2**, 25 (1970).

698. F. N. Jones, Fr. Pat. 1,545,065; through *Chem. Abstr.*, **72**, 3056 (1970).

699. G. Wilkinson, Fr. Pat. 1,459,643; through *Chem. Abstr.*, **67**, 53652 (1967).

700. J. A. Osborn, G. Wilkinson, and J. F. Young, *Chem. Commun. (Lond.)*, **1965**, 17.

701. B. I. Tikhomirov, I. A. Kloptova, and A. I. Yakubchik, U.S.S.R. Pat. 265,432; through *Chem. Abstr.*, **73**, 26397 (1970).

702. P. R. Rony, *Ann. N. Y. Acad. Sci.*, **172** (9), 238 (1970).

703. M. A. Bennett, R. Bramley, and P. A. Longstaff, *Chem. Commun. (Lond.)*, **1966**, 806.

704. M. A. Bennett and P. A. Longstaff, *J. Am. Chem. Soc.*, **91**, 6266 (1969).

705. F. de Charentenay, J. A. Osborn, and G. Wilkinson, *J. Chem. Soc.*, A, **1968**, 787.

706. H. Singer and G. Wilkinson, *J. Chem. Soc.*, A, **1968**, 2516.

707. H. Singer and G. Wilkinson, *J. Chem. Soc. A.*, **1968**, 849.

708. J. P. Collman and J. W. Kang, *J. Am. Chem. Soc.*, **89**, 844 (1967).

709. R. L. Augustine and J. F. Van Peppen, *Chem. Commun. (Lond.)*, **1970**, 495.

710. R. L. Augustine and J. F. Van Peppen, *Chem. Commun. (Lond.)*, **1970**, 497 (#266).

711. R. L. Augustine and J. F. Van Peppen, *Chem. Commun. (Lond.)*, **1970**, 497 (#267).

712. R. L. Augustine and J. F. Van Peppen, *Chem. Commun. (Lond.)*, **1970**, 571.

713. J. C. Orr, M. Mersereau, and A. Sanford, *Chem. Commun. (Lond.)*, **1970**, 162.

714. J. R. Morandi and H. B. Jensen, *J. Org. Chem.*, **34**, 1889 (1969).

715. B. Zeeh, G. Jones, and C. Djerassi, *Chem. Ber.*, **100**, 3204 (1967).

716. D. R. Eaton and S. R. Suart, *J. Am. Chem. Soc.*, **90**, 4170 (1968).

717. S. S. Bath and L. Vaska, *J. Am. Chem. Soc.*, **85**, 3500 (1963).

718. C. O'Connor, G. Yagupsky, D. Evans, and G. Wilkinson, *Chem. Commun. (Lond.)*, **1968**, 420.

719. C. O'Connor and G. Wilkinson, *J. Chem. Soc.*, A, **1968**, 2665.

720. D. Evans, G. Yagupsky, and G. Wilkinson, *J. Chem. Soc.*, A, **1968**, 2660.

721. Monsanto Co., Brit. Pat. 1,185,453; through *Chem. Abstr.*, **72**, 132059 (1970).

722. C. K. Brown and G. Wilkinson, *Tetrahedron Lett.*, **1969**, 1725.

723. M. Yagupsky and G. Wilkinson, *J. Chem. Soc.*, A, **1970**, 941.

724. M. Yagupsky, C. K. Brown, G. Yagupsky, and G. Wilkinson, *J. Chem. Soc.*, A, **1970**, 937.

725. G. Yagupsky, C. K. Brown, and G. Wilkinson, *Chem. Commun. (Lond.)*, **1969**, 1244.

726. G. Yagupsky, C. K. Brown, and G. Wilkinson, *J. Chem. Soc.*, A, **1970**, 1392.

472   References

727. S. J. LaPlaca and J. A. Ibers, *Acta Cryst.*, **18**, 511 (1965).
728. W. Keim, *J. Organomet. Chem.*, **14**, 179 (1968).
729. W. Keim, *J. Organomet. Chem.*, **19**, 161 (1969).
730. L. Vaska and R. E. Rhodes, *J. Amer. Chem. Soc.*, **87**, 4970 (1965).
731. K. C. Dewhirst, unpublished results quoted in ref. 728.
732. L. Vallarino, *J. Chem. Soc.*, **1957**, 2287.
733. D. Evans, J. A. Osborn, and G. Wilkinson, *Inorg. Synth.*, **11**, 99 (1968) and refs. therein.
734. W. Strohmeier and W. Rehder-Stirnweiss, *J. Organomet. Chem.*, **18**, P28 (1969).
735. W. Strohmeier and W. Rehder-Stirnweiss, *J. Organomet. Chem.*, **19**, 417 (1969).
736. W. Strohmeier and W. Rehder-Stirnweiss, *Z. Naturforsch.*, **24b**, 1219 (1969).
737. L. Vaska, *Science*, **143**, 769 (1966).
738. H. B. Gray and A. Wojcicki, *Proc. Chem. Soc. (Lond.)*, **1960**, 358.
739. A. Wojcicki and F. Basolo, *J. Am. Chem. Soc.*, **83**, 525 (1961).
740. J. Chatt and B. L. Shaw, *Chem. Ind. (Lond.)*, **1961**, 290.
741. J. A. McCleverty and G. Wilkinson, *Inorg. Synth.*, **8**, 214 (1966).
742. G. Schiller, Ger. Pat. 953,605; through *Chem. Abstr.*, **53**, 11226 (1959).
743. Esso Research and Eng. Co., Brit. Pat. 801,734 (1959); through *Chem. Abstr.*, **53**, 7014 (1959).
744. N. S. Imyanitov and D. M. Rudkovskii, *Neftekhimia* **3**, 198 (1963); through *Chem. Abstr.*, **59**, 7369 (1963).
745. N. S. Imyanitov and D. M. Rudkovskii, *Oksosint., Poluch. Metodom Oksosint. Al'degidov, Spirtov i Vtorichnykh Prod. na ikh Osnove, Vses. Nauchn. Issled. Inst. Neftekhim. Profsessov*, **1963**, 30; through *Chem. Abstr.*, **60**, 9072 (1964).
746. V. Yu Gankin, L. S. Genender, and D. M. Rudkovskii, in D. M. Rudkovskii, Ed., *Karbonilirovanie Nenasyshchennykh Uglevodorodov*, Khimiya, Leningrad, 1968, p. 57; through *Chem. Abstr.*, **71**, 2753 (1969).
747. B. Heil and L. Marko, *Chem. Ber.*, **99**, 1086 (1966).
748. B. Heil and L. Marko, *Acta Chim. Acad. Sci. Hung.*, **55**, 107 (1968).
749. B. Heil and L. Marko, *Magy. Asvanyolaj, Foldgaz Kiserl. Intez. Kozlem*, **9**, 121 (1968); through *Chem. Abstr.*, **71**, 12705 (1969).
750. B. Heil and L. Marko, *Magy. Kem. Lapja*, **23**, 669 (1968); through *Chem. Abstr.*, **70**, 56861 (1969).
751. B. Heil and L. Marko, *Chem. Ber.*, **101**, 2209 (1968).
752. B. Heil and L. Marko, *Chem. Ber.*, **102**, 2238 (1969).
753. M. Yamaguchi and T. Onoda, Jap. Pat. 69 02,445; through *Chem. Abstr.*, **70**, 105950 (1969).
754. M. Yamaguchi, *J. Chem. Soc. Jap., Ind. Chem. Sect.*, **72**, 671 (1969).
755. H. Wakamatsu, *J. Chem. Soc. Jap.*, **85**, 227 (1964).
756. T. Alderson, U.S. Pat. 3,020,314; through *Chem. Abstr.*, **56**, 9969 (1962).
757. J. H. Bartlett and V. L. Hughes, U.S. Pat. 2,894,038; through *Chem. Abstr.*, **54**, 2216 (1960).
758. W. Hieber and H. Lagally, *Z. Anorg. Allg. Chem.*, **251**, 96 (1943).
759. B. R. James, G. L. Rempel, and F. T. T. Ng, *J. Chem. Soc.*, *A*, **1969**, 2454.

760.   P. Chini and S. Martinengo, *Inorg. Chim. Acta*, **3**, 315 (1969).

761.   P. Chini and S. Martinengo, *Inorg. Chim. Acta*, **3**, 21 (1969).

762.   J. Falbe, N. Huppes, and F. Korte, *Brennst.-Chem.*, **47**, 207 (1966).

763.   J. Falbe and N. Huppes, *Brennst.-Chem.*, **47**, 314 (1966).

764.   J. Falbe and N. Huppes, *Brennst.-Chem.*, **48**, 46 (1967).

765.   R. L. Pruett and J. A. Smith, *J. Org. Chem.*, **34**, 327 (1969).

766.   J. H. Craddock, A. Hershman, F. E. Paulik, and J. F. Roth, *Ind. Eng. Chem. Prod. Res. Develop.*, **8**, 291 (1969).

767.   C. K. Brown and G. Wilkinson, *J. Chem. Soc.*, *A*, **1970**, 2753.

768.   D. R. Eaton, *Can. J. Chem.*, **47**, 2645 (1969).

769.   B. Heil, L. Marko, and G. Bor, *Proc. 13th Intern. Conf. Coord. Chem.*, Cracow-Zakopane, Poland, 1970, Abstr., 84.

770.   M. Yamaguchi and T. Onoda, Jap. Pat. 69 10,765; through *Chem. Abstr.*, **71**, 30064 (1969).

771.   Y. Takegami, Y. Watanbe, and H. Masada, *Bull. Chem. Soc. Jap.*, **40**, 1459 (1967).

772.   A. Sacco, R. Ugo, and A. Moles, *J. Chem. Soc.*, *A*, **1966**, 1670.

773.   A. Hershman, K. K. Robinson, J. H. Craddock, and J. F. Roth, *Ind. Eng. Chem. Prod. Res. Dev.*, **8**, 372 (1969).

774.   B. R. James and F. T. T. Ng, *Chem. Commun. (Lond.)*, **1970**, 908.

775.   F. T. T. Ng, Ph.D. Dissertation, University of British Columbia, Vancouver, British Columbia, 1970.

776.   B. R. James and G. L. Rempel, *Discuss. Faraday Soc.*, **46**, 48 (1968).

777.   L. Porri and A. Lionetti, *J. Organomet. Chem.*, **6**, 422 (1966).

778.   B. R. James and G. L. Rempel, *Can. J. Chem.*, **46**, 571 (1968).

779.   B. R. James and D. L. Pavlis, unpublished results.

780.   V. Gutmann, *Coord. Chem. Rev.*, **2**, 239 (1967).

781.   J. Chatt and S. A. Butter, *Chem. Commun. (Lond.)*, **1967**, 501.

782.   J. L. Mague and J. P. Mitchener, *Chem. Commun. (Lond.)*, **1968**, 911.

783.   W. R. Cullen and J. A. J. Thompson, *Can. J. Chem.*, **48**, 1730 (1970).

784.   J. J. Levison and S. D. Robinson, *Chem. Commun. (Lond.)*, **1968**, 1405.

785.   J. J. Levison and S. D. Robinson, *J. Chem. Soc.*, *A*, **1970**, 96.

786.   J. R. Shapley, R. R. Schrock, and J. A. Osborn, *J. Am. Chem. Soc.*, **91**, 2816 (1969).

787.   R. R. Schrock and J. A. Osborn, *Chem. Commun. (Lond.)*, **1970**, 567.

788.   L. M. Haines, *Inorg. Nucl. Chem. Lett.*, **5**, 399 (1969).

789.   M. Takesada, H. Yamazaki, and N. Hagihara, *Bull. Chem. Soc. Jap.*, **41**, 270 (1968).

790.   K. C. Dewhirst, U.S. Pat. 3,480,659; through *Chem. Abstr.*, **72**, 78105 (1970).

791.   B. Ilmaier and R. S. Nyholm, *Naturwissenschaften*, **56**, 415, 636 (1969).

792.   W. Keim, *J. Organomet. Chem.*, **8**, P25 (1967).

793.   N. Hagihara, H. Yamazaki, and M. Takesada, *Proc. 3rd Intern. Conf. Organomet. Chem.*, Munich, 1967, p. 330.

794.   D. A. Clement, J. F. Nixon, and M. D. Sexton, *Chem. Commun. (Lond.)*, **1969**, 1509 and refs. therein.

795.   M. A. Bennett and D. J. Patmore, *Chem. Commun. (Lond.)*, **1969**, 1510.

796. J. Blum, E. Oppenheimer, and E. D. Bergmann, *J. Am. Chem. Soc.*, **89**, 2338 (1967).

797. T. Kruck, *Angew. Chem. Intern. Edit.*, **6**, 53 (1967).

798. M. C. Baird, G. Hartwell, Jr., and G. Wilkinson, *J. Chem. Soc., A*, **1967**, 2037.

799. J. P. Collman, N. W. Hoffman, and D. E. Morris, *J. Am. Chem. Soc.*, **91**, 5659 (1969).

800. B. W. Graham, K. R. Laing, C. O'Connor, and W. R. Roper, *Chem. Commun. (Lond.)*, **1970**, 1272.

801. K. C. Dewhirst, U.S. Pat. 366,646; through *Chem. Abstr.*, **68**, 95311 (1968).

802. I. Jardine and F. J. McQuillin, *Chem. Commun. (Lond.)*, **1969**, 502.

803. I. Jardine and F. J. McQuillin, *Chem. Commun. (Lond.)*, **1969**, 503.

804. P. Abley and F. J. McQuillin, *Chem. Commun. (Lond.)*, **1969**, 1503.

805. I. Jardine and F. J. McQuillin, *Chem. Commun. (Lond.)*, **1970**, 626.

806. F. J. McQuillin, private communication.

807. I. Jardine and F. J. McQuillin, *J. Chem. Soc., C*, **1966**, 458.

808. B. R. James, F. T. T. Ng, and G. L. Rempel, *Can. J. Chem.*, **47**, 4521 (1969).

809. J. Trocha-Grimshaw and H. B. Henbest, *Chem. Commun. (Lond.)*, **1968**, 1035.

810. R. Stern, and L. Sajus, Fr. Pat. 1,568,741; through *Chem. Abstr.*, **72**, 104451 (1970).

811. M. Takesada, H. Yamazaki, and N. Hagihara, *J. Chem. Soc. Jap.*, **89**, 1126 (1968).

812. S. A. Shchepinov, E. N. Sal'nikova, and M. L. Khidekel, *Bull. Acad. Sci. USSR, Chem. Sect.*, **1967**, 2057.

813. B. G. Rogachev and M. L. Khidekel, *Bull. Acad. Sci. USSR, Chem. Sect.*, **1969**, 127.

814. V. A. Avilov, O. N. Eremenko, and M. L. Khidekel, *Bull. Acad. Sci. USSR, Chem. Sect.*, **1967**, 2655.

815. M. L. Khidekel, O. N. Efimov, O. N. Eremenko, and A. G. Ovcharenko, U.S.S.R. Pat. 259,836; through *Chem. Abstr.*, **73**, 19009 (1970).

816. G. N. Schrauzer and K. C. Dewhirst, *J. Am. Chem. Soc.*, **86**, 3265 (1964).

817. S. McVey and P. M. Maitlis, *Can. J. Chem.*, **44**, 2429 (1966).

818. G. Winkhaus and H. Singer, *J. Organomet. Chem.*, **7**, 487 (1967).

819. V. A. Avilov, Yu. G. Borod'ko, V. B. Panov, M. L. Khidekel, and P. S. Chekrii, *Kinet. Catal. (USSR)*, **9**, 582 (1968).

820. O. N. Efimov, M. L. Khidekel, V. A. Avilov, P. S. Chekrii, O. N. Eremenko, and A. G. Ovcharenko, *J. Gen. Chem. (USSR)*, **38**, 2581 (1968).

821. N. V. Borunova, L. Kh. Friedlin, M. L. Khidekel, S. S. Danielova, V. A. Avilov, and P. S. Chekrii, *Bull. Acad. Sci. USSR, Chem. Sect.*, **1968**, 432.

822. O. N. Efimov, O. N. Eremenko, A. G. Ovcharenko, M. L. Khidekel, and P. S. Chekrii, *Bull. Acad. Sci. USSR, Chem. Sect.*, **1969**, 778.

823. L. Dubicki and R. L. Martin, *Inorg. Chem.*, **9**, 673 (1970).

824. B. C. Hui and G. L. Rempel, *Chem. Commun. (Lond.)*, **1970**, 1195.

825. P. Legzdins, R. W. Mitchell, G. L. Rempel, J. D. Ruddick, and G. Wilkinson, *J. Chem. Soc., A*, **1970**, 3322.

826. R. Whyman, *Chem. Commun. (Lond.)*, **1970**, 1194.

827. S. A. Johnson, H. R. Hunt, and H. M. Neumann, *Inorg. Chem.*, **2**, 960 (1963).

828. T. A. Stephenson, S. M. Morehouse, A. R. Powell, J. P. Heffner, and G. Wilkinson, *J. Chem. Soc.*, **1965**, 3632.

829. L. A. Nazarova, I. I. Cherniaev, and A. S. Morozova, *Russ. J. Inorg. Chem.*, **10**, 291 (1965).

830. G. N. Schrauzer, *Acc. Chem. Res.*, **1**, 97 (1968).

831. G. N. Schrauzer, *Ann. N.Y. Acad. Sci.*, **158** (2), 526 (1969).

832. G. N. Schrauzer and R. J. Windgassen, *Chem. Ber.*, **99**, 602 (1966).

833. G. N. Schrauzer, R. J. Windgassen, and J. Kohnle, *Chem. Ber.*, **98**, 3324 (1965).

834. H. A. O. Hill, J. M. Pratt, and R. J. P. Williams, *Chem. Brit.*, **5**, 156 (1969).

835. G. N. Schrauzer, J. Sibert, and R. J. Windgassen, *J. Am. Chem. Soc.*, **90**, 6681 (1968)

836. G. N. Schrauzer and R. J. Windgassen, *J. Am. Chem. Soc.*, **88**, 3738 (1966).

837. G. N. Schrauzer and R. J. Windgassen, *J. Am. Chem. Soc.*, **89**, 3607 (1967).

838. G. N. Schrauzer, E. J. Deutsch, and R. J. Windgassen, *J. Am. Chem. Soc.*, **90**, 2441 (1968).

839. V. V. Abalyaeva, A. S. Astakhova, R. B. Ivanova, E. N. Sal'nikova, and M. L. Khidekel, *Katal, Reakts. Zhidk. Faze, Tr. Vses, Konf. 2nd, Alma-Ata, Kaz. SSR*, **1966**, 415; through *Chem. Abstr.*, **69**, 26937 (1968).

840. E. N. Sal'nikova and M. L. Khidekel, *Bull. Acad. Sci. USSR, Chem. Sect.*, **1967**, 223.

841. J. A. Osborn, *Chem. Commun. (Lond.)*, **1968**, 1231.

842. A. C. Skapski and P. G. H. Troughton, *Chem. Commun. (Lond.)*, **1969**, 666.

843. P. Uguagliati, G. Deganello, L. Busetto, and U. Belluco, *Inorg. Chem.*, **8**, 1625 (1969).

844. G. Deganello, P. Uguagliati, B. Crociani, and U. Belluco, *J. Chem. Soc., A*, **1969**, 2726.

845. R. D. W. Kemmitt, D. I. Nichols, and R. D. Peacock, *J. Chem. Soc., A*, **1968**, 1898.

846. R. D. W. Kemmitt, and D. I. Nichols, *J. Chem. Soc., A*, **1969**, 1577.

847. G. Dolcetti, M. Nicolini, M. Giustiniani, and U. Belluco, *J. Chem. Soc., A*, **1969**, 1387.

848. J. Lewis, R. S. Nyholm, and G. K. N. Reddy, *Chem. Ind. (Lond.)*, **1960**, 1386.

849. G. K. N. Reddy and E. G. Leelamani, *Z. Anorg. Allg. Chem.*, **319**, 362 (1969).

850. G. K. N. Reddy and E. G. Leelamani, *Ind. J. Chem.*, **7**, 929 (1969).

851. W. H. Baddley, *J. Am. Chem. Soc.*, **88**, 4545 (1966).

852. W. Voelter, *Proc. 13th Intern. Conf. Coord. Chem.*, Cracow-Zakopane, Poland, 1970, Abstr., 225.

853. W. H. Baddley and M. S. Fraser. *J. Am. Chem. Soc.*, **91**, 3661 (1969).

854. G. Gregorio, G. Pregaglia, and R. Ugo, *Inorg. Chim. Acta*, **3**, 89 (1969).

855. B. F. G. Johnson, J. Lewis and D. A. White, *J. Am. Chem. Soc.*, **91**, 5186 (1969).

856. M. Green, T. A. Kuc, and S. H. Taylor, *Chem. Commun. (Lond.)*, **1970**, 1553.

857. R. W. Baker, B. Ilmaier, P. J. Pauling, and R. S. Nyholm, *Chem. Commun. (Lond.)*, **1970**, 1077.

858. G. E. Hartwell and P. W. Clark, *Chem. Commun. (Lond.)*, **1970**, 1115.

859. D. D. Lehman, D. F. Shriver, and I. Wharf, *Chem. Commun. (Lond.)*, **1970**, 1486.

860. L. Vaska and M. F. Werneke, *Trans. N. Y. Acad. Sci.*, **31**, 70 (1971).

861. W. Strohmeier and T. Onoda, *Z. Naturforsch.*, **23b**, 1377 (1968).

862. W. Strohmeier and T. Onoda, *Z. Naturforsch.*, **23b**, 1527 (1968).

863. W. Strohmeier and T. Onoda, *Z. Naturforsch.*, **24b**, 515 (1969).

864. W. Strohmeier and F. J. Muller, *Z. Naturforsch.*, **24b**, 770 (1969).

865. W. Strohmeier and F. J. Muller, *Z. Naturforsch.*, **24b**, 931 (1969).

866. W. Strohmeier and T. Onoda, *Z. Naturforsch*, **24b** 1185 (1969).

867. B. R. James and N. A. Memon, *Can. J. Chem.*, **46**, 217 (1968).

868. C. Y. Chan and B. R. James, to be published.

869. G. G. Eberhardt and L. Vaska, *J. Catal.*, **8**, 183 (1967).

870. L. Vaska and M. E. Tadros, unpublished results, quoted in ref. 860.

871. L. Vaska, *Chem. Commun. (Lond.)*, **1966**, 614.

872. W. Strohmeier and T. Onoda, *Z. Naturforsch.*, **24b**, 461 (1969).

873. W. Strohmeier and R. Fleischmann, *Z. Naturforsch.*, **24b**, 1217 (1969).

874. W. Strohmeier and T. Onoda, *Z. Naturforsch.*, **24b**, 1493 (1969).

875. D. M. Blake and M. Kubota, *Inorg. Chem.*, **9**, 989 (1970) and references therein.

876. K. Takao, Y. Fujiwara, T. Imanaka, and S. Teranishi, *Bull. Chem. Soc., Jap.*, **43**, 1153 (1970).

877. G. Reischig, Diplomarbeit, University of Wurzburg, 1969; through ref. 874.

878. R. Cramer and G. W. Parshall, *J. Am. Chem. Soc.*, **87**, 1392 (1965).

879. G. W. Parshall and F. N. Jones, *J. Am. Chem. Soc.*, **87**, 5356 (1965).

880. W. H. Baddley, *J. Am. Chem. Soc.*, **90**, 3705 (1968).

881. W. H. Baddley, *Inorg. Chim. Acta Rev.*, **2**, 7 (1968).

882. M. Cooke, M. Green, and D. C. Wood, *Chem. Commun. (Lond.)*, **1968**, 733.

883. M. Camia, M. P. Lachi, L. Benzoni, C. Zanzottera, and M. Tacchi Venturi, *Inorg. Chem.*, **9**, 251 (1970).

884. M. Angoletta, *Gazz. Chim. Ital.*, **89**, 2359 (1959).

885. L. Manojlovic-Muir, K. W. Muir, and J. A. Ibers, *Discuss. Faraday Soc.*, **47**, 84 (1969).

886. B. R. James and N. A. Memon, unpublished results.

887. R. S. Nyholm and K. Vrieze, *J. Chem. Soc.*, **1965**, 5331.

888. R. S. Nyholm and K. Vrieze, *J. Chem. Soc.*, **1965**, 5337.

889. W. B. Hughes, *Chem. Commun. (Lond.)*, **1969**, 1126.

890. L. Malatesta, G. Caglio, and M. Angoletta, *J. Chem. Soc.*, **1965**, 6974.

891. F. Glockling and M. D. Wilbey, *Chem. Commun. (Lond.)*, **1969**, 286.

892. L. Malatesta, Symposium, *Current Trends in Organometallic Chemistry*, Cincinnati, Ohio, 1963; through ref. 717.

893. J. A. McGinnety and J. A. Ibers, *Chem. Commun. (Lond.)*, **1968**, 235.

894. D. M. P. Mingos and J. A. Ibers, *Inorg. Chem.*, **9**, 1105 (1970).

895. J. S. Ricci, J. A. Ibers, M. S. Fraser, and W. H. Baddley, *J. Am. Chem., Soc.*, **92**, 3489 (1970).

896. W. Hieber and V. Frey, *Chem. Ber.*, **99**, 2607 (1966).

897. K. Taylor, *Advances in Chemistry*, Vol. 70, American Chemical Society, Washington, D.C., 1968, p. 195.

898. L. Vaska and D. L. Catone, *J. Am. Chem. Soc.*, **88**, 5324 (1966).

899. J. A. J. Jarvis, R. H. B. Mais, P. G. Owston, and K. A. Taylor, *Chem. Commun.* (*Lond.*), **1966**, 906.

900. H. C. Volger, K. Vrieze, and A. P. Praat, *J. Organomet Chem.*, **14**, 429 (1968).

901. M. Angoletta, *Gazz. Chim. Ital.*, **92**, 811 (1962).

902. M. Yamaguchi, *J. Chem. Soc. Jap.*, *Ind. Chem. Sect.*, **70**, 675 (1967).

903. D. Giusto, *Inorg. Nucl. Chem. Lett.*, **5**, 767 (1969).

904. J. Chatt, R. S. Coffey, and B. L. Shaw, *J. Chem. Soc.*, **1965**, 7391, and references therein.

905. R. S. Coffey, *Chem. Commun.* (*Lond.*), **1967**, 923 (#779).

906. M. Angoletta and A. Araneo, *Gazz. Chim. Ital.*, **93**, 1343 (1963).

907. R. S. Coffey, Brit. Pat. 1,135,979; through *Chem. Abstr.*, **70**, 67840 (1969).

908. R. S. Coffey, *Chem. Commun.* (*Lond.*), **1967**, 923 (#780).

909. M. Giustiniani, G. Dolcetti, M. Nicolini, and U. Belluco, *J. Chem. Soc.*, *A*, **1969**, 1961.

910. F. Canziani, U. Sartorelli, and F. Zingales, *Rend. Ist. Lombardo Sci. Lett. A*, **96**, 21 (1965); through *Chem. Abstr.*, **65**, 3297 (1966).

911. M. A. Bennett and D. L. Milner, *Chem. Commun.* (*Lond.*), **1967**, 581.

912. M. A. Bennett and D. L. Milner, *J. Am. Chem. Soc.*, **91**, 6983 (1969).

913. J. P. Collman, M. Kubota, F. D. Vastine, J. Y. Sun, and J. W. Kang, *J. Am. Chem. Soc.*, **90**, 5430 (1968).

914. J. P. Collman and J. W. Kang, *J. Am. Chem. Soc.*, **88**, 3459 (1966).

915. D. J. Cardin and M. F. Lappert, *Chem. Commun.* (*Lond.*), **1967**, 1034.

916. D. J. Cardin, G. Chandra, and M. F. Lappert, *Proc. 3rd Intern. Conf. Organomet. Chem.*, Munich, 1967, p. 190.

917. D. J. Cardin, M. F. Lappert, and N. F. Travers, in M. Cais, Ed., *Progress in Co-ordination Chemistry*, Elsevier, Amsterdam, 1968, p. 821.

918. J. E. Lyons, *Chem. Commun.* (*Lond.*), **1969**, 564.

919. Y. M. Y. Haddad, H. B. Henbest, J. Husbands, and T. R. B. Mitchell, *Proc. Chem. Soc.* (*Lond.*), **1964**, 361.

920. P. A. Browne and D. N. Kirk, *J. Chem. Soc.*, *C*, **1969**, 1653.

921. J. Trocha-Grimshaw and H. B. Henbest, *Chem. Commun.* (*Lond.*), **1967**, 544.

922. M. McPartlin and R. Mason, *Chem. Commun.* (*Lond.*), **1967**, 545.

923. J. Trocha-Grimshaw and H. B. Henbest, *Chem. Commun.* (*Lond.*), **1967**, 757.

924. H. Van Gaal, H. G. A. M. Cuppers, and A. van der Ent, *Chem. Commun.* (*Lond.*), **1970**, 1694.

925. B. L. Shaw and E. Singleton, *J. Chem. Soc.*, *A*, **1967**, 1683.

926. A. van der Ent and T. C. van Soest, *Chem. Commun.* (*Lond.*), **1970**, 225.

927. G. Winkhaus and H. Singer, *Chem. Ber.*, **99**, 3610 (1966).

928. M. A. Bennett, R. J. H. Clark, and D. L. Milner, *Inorg. Chem.*, **6**, 1647 (1967).

929. B. E. Mann, C. Masters and B. L. Shaw, *Chem. Commun.* (*Lond.*), **1970**, 703.

930. G. Yagupsky and G. Wilkinson, *J. Chem. Soc. A*, **1969**, 725.

931. R. C. Taylor, J. F. Young, and G. Wilkinson, *Inorg. Chem.*, **5**, 20 (1966).

932. A. J. Deeming and B. L. Shaw, *J. Chem. Soc.*, *A*, **1969**, 1128.

933. A. J. Deeming and B. L. Shaw, *J. Chem. Soc., A*, **1969**, 1802.
934. A. J. Deeming and B. L. Shaw, *J. Chem. Soc., A*, **1968**, 1887.
935. J. F. Harrod, D. F. R. Gilson, and R. Charles, *Can. J. Chem.*, **47**, 2205 (1969).
936. J. F. Harrod and C. A. Smith, *Can. J. Chem.*, **48**, 870 (1970).
937. A. J. Chalk, *Chem. Commun. (Lond.)*, **1969**, 1207.
938. G. G. Joyson, D. A. Stirling, and A. J. Swallow, *Chem. Commun. (Lond.)*, **1967**, 931.
939. M. F. Lappert and N. F. Travers, *Chem. Commun. (Lond.)*, **1968**, 1569.
940. J. K. Nicholson and B. L Shaw, *Tetrahedron Lett.*, **1965**, 3533.
941. R. S. Coffey, *Tetrahedron Lett.*, **1965**, 3809.
942. S. D. Robinson and B. L. Shaw, *Tetrahedron Lett.*, **1964**, 1301.
943. L. Malatesta, M. Angoletta, and G. Caglio, *Angew. Chem. Intern. Edit.*, **2**, 739 (1963).
944. C. A. Reed and W. R. Roper, *Chem. Commun. (Lond.)*, **1969**, 155.
945. C. A. Reed and W. R. Roper, *Chem. Commun. (Lond.)*, **1969**, 1459.
946. D. J. Hodgson, N. C. Payne, J. A. McGinnety, R. G. Pearson, and J. A. Ibers, *J. Am. Chem. Soc.*, **90**, 4486 (1968).
947. D. J. Hodgson and J. A. Ibers, *Inorg. Chem.*, **8**, 1282 (1969).
948. P. Carty, A. Walker, M. Mathew, and G. J. Palenik, *Chem. Commun. (Lond.)*, **1969**, 1374.
949. T. Hashimoto and H. Shiina, *J. Jap. Oil Chem. Soc.*, **8**, 259 (1959); through *Chem. Abstr.*, **54**, 25898 (1960).
950. R. D. Gillard and B. T. Heaton, *Chem. Commun. (Lond.)*, **1968**, 75.
951. R. A. Bauer and F. Basolo, *Chem. Commun. (Lond.)*, **1968**, 458.
952. D. C. Olson and W. Keim, *Inorg. Chem.*, **8**, 2028 (1969).
953. H. C. Clark and P. K. Mittal, *Can. J. Chem.*, **48**, 119 (1970).
954. M. J. Church and M. J. Mays, *Chem. Commun. (Lond.)*, **1968**, 435.
955. S. D. Robinson and B. L. Shaw, *J. Chem. Soc.*, **1965**, 4997.
956. B. L. Shaw, *Chem. Commun. (Lond.)*, **1968**, 464.
957. J. W. Dawson, D. G. E. Kerfoot, C. Preti, and L. M. Venanzi, *Chem. Commun. (Lond.)*, **1968**, 1687.
958. B. E. Mann, C. Masters, and B. L. Shaw, *Chem. Commun. (Lond.)*, **1970**, 846.
959. E. W. Ainscough and S. D. Robinson, *Chem. Commun. (Lond.)*, **1970**, 863.
960. J. J. Levinson and S. D. Robinson, *J. Chem. Soc., A*, **1970**, 639.
961. J. Powell and B. L. Shaw, *J. Chem. Soc., A*, **1968**, 617.
962. V. Yu. Gankin and D. M. Rudkovskii, *Kinet. Catal. (USSR)*, **8**, 778 (1967).
963. N. S. Imyanitov and D. M. Rudkovskii, *Kinet. Catal. (USSR)*, **8**, 1051 (1967).
964. N. S. Imyanitov and D. M. Rudkovskii, *Zh. Prikl. Khim*, **40**, 2020 (1967).
965. R. Whyman, *Chem. Commun. (Lond.)*, **1969**, 1381.
966. L. Malatesta, G. Caglio, and M. Angoletta, *Chem. Commun. (Lond.)*, **1970**, 532.
967. L. Benzoni, A. Andreeta, C. Zanzoterra, and M. Camia, *Chim. Ind. (Milan)*, **48**, 1076 (1966).
968. L. H. Slaugh and R. D. Mullineaux, U.S. Pat. 3,239,571; through *Chem. Abstr.*, **65**, 618 (1966).

969. J. P. Collman, F. D. Vastine, and W. R. Roper, *J. Am. Chem. Soc.*, **90**, 2282 (1968).

970. P. Fotis, Jr. and J. D. McCollum, U.S. Pat. 3,324,018; through *Chem. Abstr.*, **67**, 53616 (1967).

971. M. S. Spencer and D. A. Dowden, U.S. Pat. 2,966,534; through *Chem. Abstr.*, **55**, 8288 (1961).

972. M. G. Burnett, *Chem. Commun. (Lond.)*, **1965**, 507.

973. A. A. Vlcek, *Coll. Czech. Chem. Comm.*, **22**, 948 (1957).

974. T. Mizuta, H. Samejima, and T. Kwan, *Bull. Chem. Soc. Jap.*, **41**, 727 (1968).

975. H. Samejima, T. Mizuta, and T. Kwan, *J. Chem. Soc. Jap.*, **89**, 1028 (1968).

976. H. Samejima, T. Mizuta, H. Yamamoto, and T. Kwan, *Bull. Chem. Soc. Jap.*, **42**, 2722 (1969).

977. L. W. Gosser, Ger. Pat. 1,940,303; through *Chem. Abstr.*, **72**, 100161 (1970).

978. J. C. Bailar, Jr. and H. Itatani, *J. Am. Chem. Soc.*, **89**, 1592 (1967).

979. H. Itatani and J. C. Bailar, Jr., *J. Am. Chem. Soc.*, **89**, 1600 (1967).

980. J. C. Bailar, Jr., H. Itatani, M. J. Crespi, and J. Geldard, *Advances in Chemistry*, Vol. 62, American Chemical Society, Washington, D.C., 1967, p. 103.

981. M. L. H. Green and T. Saito, *Chem. Commun. (Lond.)*, **1969**, 208.

982. M. L. H. Green, H. Munakata, and T. Saito, *Chem. Commun. (Lond.)*, **1969**, 1287.

983. K. W. Barnett, F. D. Mango, and C. A. Reilly, *J. Am. Chem. Soc.*, **91**, 3387 (1969).

984. D. W. Walker, U.S. Pat. 3,415,898; through *Chem. Abstr.*, **70**, 38279 (1969).

985. W. Keim, *Angew. Chem., Intern. Edit.*, **7**, 879 (1968).

986. R. Ugo, *Coord. Chem. Rev.*, **3**, 319 (1968).

987. B. Corain, M. Bressan, P. Rigo, and A. Turco, *Chem. Commun. (Lond.)*, **1968**, 509.

988. L. Porri, M. C. Gallazzi, and G. Vitulli, *Chem. Commun. (Lond.)*, **1967**, 228.

989. P. Heimback, *Angew. Chem., Intern. Edit.*, **3**, 648 (1964).

990. M. Kumada, Y. Kiso, and M. Umeno, *Chem. Commun. (Lond.)*, **1970**, 611.

991. W. Kruse and R. H. Atalla, *Chem. Commun. (Lond.)*, **1968**, 921.

992. H. C. Clark, K. R. Dixon, and W. J. Jacobs, *Chem. Commun. (Lond.)*, **1968**, 93.

992a. E. S. Brown and E. A. Rick, *Chem. Commun. (Lond.)*, **1969**, 112.

993. V. V. Ipatieff, Jr. and V. G. Tronev, *Compt. Rend. Acad. Sci., USSR*, **1**, 629 (1935).

994. V. V. Ipatieff, Jr. and V. G. Tronev, *Compt. Rend. Acad. Sci., USSR*, **1**, 624 (1935).

995. J. Halpern, J. F. Harrod, and P. E. Potter, *Can. J. Chem.*, **37**, 1446 (1959).

996. J. Chatt, L. A. Duncanson, and B. L. Shaw, *Chem. Ind. (Lond.)*, **1958**, 859.

997. A. B. Fasman, V. A. Golodov, L. M. Pustyl'nikov, A. T. Luk'yanov, B. P. Baranovskii, and Yu. V. Darinskii, *Izv. Sib. Otd. Akad. Nauk. SSSR, Ser. Khim. Nauk.*, **1968**, 144; through *Chem. Abstr.*, **70**, 31962 (1969).

998. E. B. Maxted and S. M. Ismail, *J. Chem. Soc.*, **1964**, 1750.

999. H. A. Tayim and J. C. Bailar, Jr., *J. Am. Chem. Soc.*, **89**, 4330 (1967).

1000. H. Itatani and J. C. Bailar, Jr., *J. Am. Oil Chem. Soc.*, **44**, 147 (1967).

1001. J. C. Bailar, Jr., H. Itatani, and H. A. Tahim, *Kagaku No Ryoiki*, **22**, 337 (1968); through *Chem. Abstr.*, **69**, 44717 (1968).

1002. Imperial Chemical Industries Ltd., Neth Pat. 6,611,373; through *Chem. Abstr.*, **67**, 54266 (1967).

1003. J. A. Scheben, I. L. Mador, and M. Orchin, S. African Pat. 68,04,418; through *Chem. Abstr.*, **71**, 90857 (1969).

1004. M. Sakakibara, Y. Takahashi, S. Sakai, and Y. Ishii, *Inorg. Nucl. Chem. Lett.*, **5**, 427 (1969).

1005. E. H. Brooks and F. Glockling, *Chem. Commun.* (*Lond.*), **1965**, 510.

1006. G. E. Batley and J. C. Bailar, Jr., *Inorg. Nucl. Chem. Lett.*, **4**, 577 (1968).

1007. D. T. Thompson, Brit. Pat., 1,182,932; through *Chem. Abstr.*, **73**, 4029 (1970).

1008. E. H. Brooks and F. Glockling, *J. Chem. Soc.*, *A*, **1967**, 1030.

1009. G. A. Pneumaticakis, *Chem. Commun.* (*Lond.*), **1968**, 275.

1010. B. C. Benson, R. Jackson, K. K. Joshi, and D. T. Thompson, *Chem. Commun.* (*Lond.*), **1968**, 1506.

1011. J. V. Kingston and G. R. Scollary, *Chem. Commun.* (*Lond.*), **1969**, 455.

1012. J. V. Kingston and G. R. Scollary, *Chem. Commun.* (*Lond.*), **1970**, 362.

1013. H. Munakata and M. L. H. Green, *Chem. Commun.* (*Lond.*), **1970**, 881.

1014. E. H. Brooks and F. Glockling, *J. Chem. Soc.*, *A*, **1966**, 1241.

1015. G. Carturan, G. Deganello, T. Boschi, and U. Belluco, *J. Chem. Soc.*, *A*, **1969**, 1142.

1016. J. E. Fergusson and J. L. Love, *Rev. Pure Appl. Chem.*, **20**, 33 (1970).

1017. G. Martino and L. Sajus, *Rev. Inst. Fr. Pét.*, **25**, 36 (1970).

1018. J. H. Flynn and H. M. Hulbert, *J. Am. Chem. Soc.*, **76**, 3393 (1954).

1019. J. Chatt and R. G. Wilkins, *J. Chem. Soc.*, **1952**, 2622.

1020. K. E. Hayes, *Nature*, **210**, 412 (1966).

1021. A. S. Gow, Jr. and H. Heinemann, *J. Phys. Chem.*, **64**, 1574 (1960).

1022. R. Cramer, *Inorg. Chem.* **4**, 445 (1965).

1023. R. Cramer, R. V. Lindsey, Jr., C. T. Prewitt, and U. G. Stolberg, *J. Am. Chem. Soc.*, **87**, 658 (1965).

1024. R. V. Lindsey, Jr., G. W. Parshall, and U. G. Stolberg, *J. Am. Chem. Soc.*, **87**, 658 (1965).

1025. R. Cramer and R. V. Lindsey, *J. Am. Chem. Soc.*, **88**, 3534 (1966).

1026. R. W. Dunning, K. K. Joshi, A. R. Oldham, and M. C. K. Willott, Brit. Pat. 1,154,937; through *Chem. Abstr.*, **71**, 49213 (1969).

1027. G. C. Bond and M. Hellier, *Chem. Ind.* (*Lond.*), **1965**, 35.

1028. G. C. Bond and M. Hellier, *J. Catal.*, **7**, 217 (1967).

1029. H. van Bekkum, J. van Gogh, and G. van Minnen-Pathuis, *J. Catal.*, **7**, 292 (1967).

1030. H. van Bekkum, F. van Rantwijk, G. van Minnen-Pathuis, J. D. Remijnse, and A. van Veen, *Rec. Trav. Chim.*, **88**, 911 (1969).

1031. L. P. van't Hof and B. G. Linsen, *J. Catal.*, **7**, 295 (1967).

1032. L. P. van't Hof, Fr. Pat. 1,518,179; through *Chem. Abstr.*, **72**, 12049 (1970).

1033. L. P. van't Hof, Fr. Pat. 1,518,180; through *Chem. Abstr.*, **72**, 12042 (1970).

1034. L. P. van't Hof, Fr. Pat. 1,518,181; through *Chem. Abstr.*, **72**, 12048 (1970).

1035. A. P. Khrushch, L. A. Tokina, and A. E. Shilov, *Kinet. Catal.* (*USSR*), **7**, 793 (1966).

1036. A. P. Khrushch, N. F. Shvetsova and A. E. Shilov, *Kinet. Catal.* (*USSR*), **10**, 1011 (1969).

1037. G. V. Novikova, V. A. Trukhtanov, A. P. Khrushch, A. E. Shilov, and V. I. Gol'danskii, *Proc. Acad. Sci. USSR, Phys. Chem.*, **189**, 882 (1969).

1038. R. Pietropaolo, M. Graziani, and U. Belluco, *Inorg. Chem.*, **8**, 1506 (1969).

1039. D. M. Adams and P. J. Chandler, *Chem. Ind. (Lond.)*, **1965**, 269.

1040. J. C. Bailar, Jr. and H. Itatani, *J. Am. Oil Chem. Soc.*, **43**, 337 (1966).

1041. E. N. Frankel, E. A. Emken, H. Itatani, and J. C. Bailar, Jr., *J. Org. Chem.*, **32**, 1447 (1967).

1042. V. I. Baranovskii, V. P. Sergeev, and B. E. Dzevitskii, *Proc. Acad. Sci. USSR, Phys. Chem.*, **184**, 55 (1969).

1043. J. Chatt and B. L. Shaw, *J. Chem. Soc.*, **1962**, 5075.

1044. J. Halpern and C. D. Falk, *J. Am. Chem. Soc.*, **87**, 3523 (1965).

1045. U. Belluco, M. Giustiniani, and M. Graziani, *J. Am. Chem. Soc.*, **89**, 6494 (1967).

1046. J. A. Chopoorian, J. Lewis, and R. S. Nyholm, *Nature*, **190**, 528 (1961).

1047. L. Malatesta and R. Ugo, *J. Chem. Soc.*, **1963**, 2080.

1048. J. C. Bailar, Jr. and H. Itatani, *Inorg. Chem.*, **4**, 1618 (1965).

1049. J. Chatt, R. S. Coffey, A. Gough, and D. T. Thompson, *J. Chem. Soc., A*, **1968**, 190.

1050. J. Chatt and B. L. Shaw, *J. Chem. Soc.*, **1959**, 4020.

1051. P. Uguagliati and W. H. Baddley, *J, Am. Chem. Soc.*, **90**, 5446 (1968).

1052. W. H. Baddley and L. M. Venanzi, *Inorg. Chem.*, **5**, 33 (1966).

1053. M. Giustiniani, G. Dolcetti, R. Pietropaolo, and U. Belluco, *Inorg. Chem.*, **8**, 1048 (1969).

1054. H. C. Volger and K. Vrieze, *J. Organomet. Chem.*, **13**, 495 (1968).

1055. R. V. Lindsey, Jr., G. W. Parshall, and U. G. Stolberg, *Inorg. Chem.*, **5**, 109 (1966).

1056. L. J. Guggenberger, *Chem. Commun. (Lond.)*, **1968**, 512.

1057. M. C. Baird, *J. Inorg. Nucl. Chem.*, **29**, 367 (1967).

1058. R. J. Cross and F. Glockling, *J. Chem. Soc.*, **1965**, 5422.

1059. F. Glockling and K. A. Hooton, *J. Chem. Soc., A*, **1968**, 826.

1060. E. H. Brooks, R. J. Cross, and F. Glockling, *Inorg. Chim. Acta*, **2**, 17 (1968).

1061. G. Deganello, G. Carturan, and U. Belluco, *J. Chem. Soc., A*, **1968**, 2873.

1062. R. W. Adams, G. E. Batley, and J. C. Bailar, Jr., *J. Am. Chem. Soc.*, **90**, 6051 (1968).

1063. J. C. Bailar, Jr. and H. Itatani, *Proc. Symp. Coord. Chem.*, Tihany, Hungary, 1964, Akademiai Kiado, Budapest, 1965, p. 67.

1064. H. A. Tayim and J. C. Bailar, Jr., *J. Am. Chem. Soc.*, **89**, 3420 (1967).

1065. R. W. Adams, G. E. Batley, and J. C. Bailar, Jr., *Inorg. Nucl. Chem. Lett.*, **4**, 455 (1968).

1066. Y. Kanai and A. Miyaka, Jap. Pat. 70 05,255; through *Chem. Abstr.*, **72**, 110898 (1970).

1067. F. Cariati, R. Ugo, and F. Bonati, *Chem. Ind. (Lond.)*, **1964**, 1714.

1068. G. N. Schrauzer, *Advances in Chemistry*, Vol. 100, American Chemical Society, Washington, D.C., 1971, p. 1.

1069. H. C. Clark, J. H. Tsai, and W. S. Tsang, *Chem. Commun. (Lond.)*, **1965**, 171.

1070. H. C. Clark and W. S. Tsang, *Chem. Commun. (Lond.)*, **1966**, 123.

1071. H. C. Clark and W. S. Tsang, *J. Am. Chem. Soc.*, **89**, 529 (1967).

1072. H. C. Clark, P. W. R. Çorfield, K. R. Dixon, and J. A. Ibers, *J. Am. Chem. Soc.*, **89**, 3360 (1967).

1073. H. C. Clark, K. R. Dixon, and W. J. Jacobs, *J. Am. Chem. Soc.*, **90**, 2259 (1968).

1074. H. C. Clark and K. R. Dixon, *J. Am. Chem. Soc.*, **91**, 596 (1969).

1075. W. J. Cherwinski and H. C. Clark, *Can. J. Chem.*, **47**, 2665 (1969).

1076. F. Cariati, R. Ugo, and F. Bonati, *Inorg. Chem.*, **5**, 1128 (1966).

1077. H. C. Clark, K. R. Dixon, and W. J. Jacobs, *Chem. Commun. (Lond.)*, **1968**, 548.

1078. H. C. Clark, K. R. Dixon, and W. J. Jacobs, *J. Amer. Chem. Soc.*, **91**, 1346 (1969).

1079. M. Akhtar and H. C. Clark, *J. Organomet. Chem.*, **22**, 233 (1970).

1080. J. L. Garnett and R. J. Hodges, *J. Am. Chem. Soc.*, **89**, 4546 (1967).

1081. J. L. Garnett and R. J. Hodges, *Chem. Commun. (Lond.)*, **1967**, 1001.

1082. J. L. Garnett and R. J. Hodges, *Chem. Commun. (Lond.)*, **1967**, 1220.

1083. R. J. Hodges and J. L. Garnett, *J. Phys. Chem.* **72**, 1673 (1968).

1084. J. L. Garnett, J. H. O'Keefe, and P. J. Claringbold, *Tetrahedron Lett.*, **1968**, 2687.

1085. R. J. Hodges and G. L. Garnett, *J. Catal.*, **13**, 83 (1969).

1086. R. J. Hodges and G. L. Garnett, *J. Phys. Chem.*, **73**, 1525 (1969).

1087. R. J. Hodges, D. E. Webster, and P. B. Wells, *Chem. Commun. (Lond.)*, **1971**, 462.

1088. Y. Sibata and E. Matsumoto, *J. Chem. Soc. Jap.*, **60**, 1173 (1939); through Chem. Abstr., **34**, 1582 (1940).

1089. J. Chatt, C. Eaborn, S. Ibekwe, and P. N. Kapoor, *Chem. Commun. (Lond.)*, **1967**, 869.

1090. J. E. Bentham, S. Cradock, and E. A. V. Ebsworth, *Chem. Commun. (Lond.)*, **1969**, 528.

1091. Y. Ichinohe, N. Kameda, and M. Kujirai, *Bull. Chem. Soc. Jap.*, **42**, 3614 (1969).

1092. P. D. Kaplan, P. Schmidt, and M. Orchin, *J. Am. Chem. Soc.*, **90**, 4175 (1968).

1093. G. Booth and J. Chatt, *J. Chem. Soc., A*, **1966**, 634.

1094. D. Wright, *Chem. Commun. (Lond.)*, **1966**, 197.

1095. R. Ugo, F. Cariati, and G. La Monica, *Chem. Commun. (Lond.)*, **1966**, 868.

1096. A. C. Cope and R. W. Sickman, *J. Am. Chem. Soc.*, **87**, 3272 (1965).

1097. D. M. Roundhill and H. B. Jonassen, *Chem. Commun. (Lond.)*, **1968**, 1233.

1098. G. E. Calf and J. L. Garnett, *Chem. Commun. (Lond.)*, **1969**, 373.

1099. D. M. Barlex, R. D. W. Kemmitt, and G. W. Littlecott, *Chem. Commun. (Lond.)*, **1969**, 613.

1100. P. R. Brookes and R. S. Nyholm, *Chem. Commun. (Lond.)*, **1970**, 169.

1101. A. F. Clemmit and F. Glockling, *Chem. Commun. (Lond.)*, **1970**, 705.

1102. J. E. Bentham and E. A. V. Ebsworth, *Inorg. Nucl. Chem. Lett.*, **1970**, 145.

1103. J. L. Garnett and R. S. Kenyon, *Chem. Commun. (Lond.)*, **1970**, 698.

1104. M. J. Church and M. J. Mays, *J. Chem. Soc., A*, **1968**, 3074.

1105. A. J. Deeming, B. F. G. Johnson, and J. Lewis, *Chem. Commun. (Lond.)*, **1970**, 598.

1106. M. Giustiniani, G. Dolcetti, and U. Belluco, *J. Chem. Soc., A*, **1969**, 2047.

1107. L. Toniolo, M. Giustiniani, and U. Belluco, *J. Chem. Soc., A*, **1969**, 2666.

1108. G. Booth and J. Chatt, *J. Chem. Soc., A*, **1969**, 2131.

1109. A. F. Clemmit and F. Glockling, *J. Chem. Soc.*, *A*, **1969**, 2163.

1110. S. J. Ashcroft and C. T. Mortimer, *J. Chem. Soc.*, *A*, **1967**, 930.

1111. A. J. Layton, R. S. Nyholm, G. A. Pneumaticakis, and M. L. Tobe, *Chem. Ind. (Lond.)*, **1967**, 465.

1112. A. De Jonge and B. DeVries, unpublished results quoted in ref. 277.

1113. E. Piers, R. W. Britton, and W. de Waal, *Chem. Commun. (Lond.)* **1969**, 1069.

1114. T. Hashimoto and H. Shiina, *J. Jap. Oil Chem. Soc.*, **9**, 79 (1960); through *Chem. Abstr.*, **54**, 25899 (1960).

1115. T. Hashimoto and H. Shiina, *J. Jap. Oil Chem. Soc.*, **9**, 376 (1960); through *Chem. Abstr.*, **54**, 25899 (1960).

1116. H. L. Retcofsky, E. N. Frankel, and H. S. Gutowsky, *J. Am. Chem. Soc.*, **88**, 2710 (1966).

1117. C. E. Holloway, G. Hulley, B. F. G. Johnson, and J. Lewis, *J. Chem. Soc.*, *A*, **1970**, 1653.

1118. R. O. Butterfield, E. D. Bitner, C. R. Schofield, and H. J. Dutton, *J. Am. Oil Chem. Soc.*, **41**, 29 (1964).

1119. I. Wender and H. W. Sternberg, *Adv. Catal.*, **9**, 594 (1957).

1120. W. Hieber and G. Brendel, *Z. Anorg. Allg. Chem.*, **289**, 324 (1957).

1121. H. W. Sternberg, R. Markby, and I. Wender, *J. Am. Chem. Soc.*, **79**, 6116 (1957).

1122. P. Krumholz and H. M. A. Stettiner, *J. Am. Chem. Soc.*, **71**, 3035 (1949).

1123. W. Reppe and H. Vetter, *Ann. Chem.*, **582**, 133 (1953).

1124. W. F. Gresham and R. E. Brooks, U.S. Pat. 2,497,303; through *Chem. Abstr.*, **44**, 4492 (1950).

1125. H. Uchida and K. Bando, *Bull. Chem. Soc. Jap.*, **29**, 953 (1956).

1126. H. Ruesch and T. J. Mabry, *Tetrahedron*, **25**, 805 (1969).

1127. H. W. Sternberg, R. Markby, and I. Wender, *J. Am. Chem. Soc.*, **78**, 5704 (1956).

1128. H. W. B. Reed and P. O. Lenel, Brit. Pat. 794,067; through *Chem. Abstr.*, **53**, 218 (1959).

1129. I. G. Farbenindustrie A-G., Ger. Pat. 441,179 (1925); through ref. 1121.

1130. W. Hieber, F. Leutert, and E. Schmidt, *Z. Anorg. Allg. Chem.*, **204**, 145 (1932).

1131. H. W. Sternberg, R. A. Friedel, R. Markby, and I. Wender, *J. Am. Chem. Soc.*, **78**, 3621 (1956).

1132. E. N. Frankel, E. P. Jones, V. L. Davison, E. A. Emken, and H. J. Dutton, *J. Am. Oil Chem. Soc.*, **42**, 130 (1965).

1133. I. Ogata and A. Misono, *J. Jap. Oil Chem. Soc.*, **14** 16 (1965); through *Chem. Abstr.*, **63**, 17828 (1965).

1134. E. N. Frankel, unpublished work quoted in ref. 81.

1135. E. N. Frankel and R. O. Butterfield, *J. Org. Chem.*, **34**, 3930 (1969).

1136. E. N. Frankel, E. Selke, and C. A. Glass, *J. Org. Chem.*, **34**, 3936 (1969).

1137. F. G. Mann and D. Purdie, *J. Chem. Soc.*, **1940**, 1230.

1138. J. C. Bailar, Jr., *Plat. Met. Rev.*, **15**, 2 (1971).

1139. G. E. Batley, R. W. Adams, and J. C. Bailar Jr., *Proc. 10th Intern. Conf. Coord. Chem.*, Sydney, Australia, 1969, p. 206.

1140. K. Ziegler, H. G. Gelbert, E. Holzkamp, G. Wilke, E. W. Duck, and W. R. Kroll, *Ann. Chem.*, **620**, 172 (1960).

1141. W. R. Kroll, *J. Catal.*, **15**, 281 (1969).

1142. C. S. G. Phillips and R. J. P. Williams, *Inorganic Chemistry*, Vol. 2, Oxford University Press, New York–Oxford, 1966, p. 572.

1143. S. Otsuka, T. Kikuchi, and T. Taketomi, *J. Am. Chem. Soc.*, **85**, 3709 (1963).

1144. J. L. Herisson, Y. Chauvin, N. H. Phung, and G. Lefebvre, *Compt. Rend.*, *C*, **269**, 661 (1969).

1145. W. R. Kroll, U.S. Pat. 3,412,174; through ref. 1141.

1146. Y. Tajima and E. Kunioka, *J. Catal.*, **11**, 83 (1968).

1147. S. J. Lapporte, *Ann. N.Y. Acad. Sci.*, **158** (2), 510 (1969).

1148. Y. Takegami, T. Ueno, and K. Kawajiri, *J. Chem. Soc., Jap., Ind. Chem. Sect.*, **66**, 1068 (1963); through *Chem. Abstr.*, **62**, 7661 (1965).

1149. Y. Takegami, T. Ueno, K. Shinoki, and T. Sakata, *J. Chem. Soc. Jap., Ind. Chem. Sect.*, **67**, 316 (1964); through *Chem. Abstr.*, **61**, 6618 (1964).

1150. Y. Takegami, T. Ueno, and K. Kawajiri, *Bull. Chem. Soc. Jap.*, **39**, 1 (1966).

1151. Y. Takegami, T. Ueno, T. Fujii, and T. Sakata, *Shokubai (Catalyst)*, **8**, 54 (1966); through *Chem. Abstr.*, **68**, 104620 (1968).

1152. Y. Takegami, T. Ueno, and T. Sakata, *J. Chem. Soc. Jap., Ind. Chem. Sect.*, **68**, 2373 (1965); through *Chem. Abstr.*, **65**, 16884 (1966).

1153. Y. Takegami and T. Ueno, *J. Chem. Soc. Jap., Ind. Chem. Sect.*, **67**, 246 (1964); through ref. 173.

1154. Y. Takegami, T. Ueno, and T. Fujii, *J. Chem. Soc. Jap., Ind. Chem. Sec.* **67**, 1009 (1964); through *Chem. Abstr.*, **61**, 13931 (1964).

1155. Y. Takegami, T. Ueno, and T. Fujii, *J. Chem. Soc. Jap., Ind. Chem. Sect.*, **69**, 1467 (1966); through *Chem. Abstr.*, **66**, 22605 (1967).

1156. Y. Takegami, T. Ueno, and T. Fujii, *Bull. Chem. Soc. Jap.*, **38**, 1279 (1965).

1157. F. K. Shmidt, V. G. Lipovich, S. M. Krasnopol'skaya, and I. V. Kalechits, *Kinet. Catal. (USSR)*, **11**, 286 (1970).

1158. W. D. Horrocks and R. C. Taylor, *Inorg. Chem.*, **2**, 723 (1963).

1159. S. J. Lapporte, Fr. Pat. 1,39,0570; through *Chem. Abstr.*, **64**, 600 (1966).

1160. W. R. Kroll, Fr. Pat. 1,453,329; through *Chem. Abstr.*, **67**, 53621 (1967).

1161. W. R. Kroll, Ger. Pat. 1,801,923; through *Chem. Abstr.*, **71**, 49388 (1969).

1162. W. P. Long and D. S. Breslow, *J. Am. Chem. Soc.*, **82**, 1953 (1960).

1163. A. E. Shilov, A. K. Shilova, and B. N. Bobkov, *Polym. Sci., USSR*, **4**, 526 (1963).

1164. K. Clauss and H. Bestian, *Ann. Chem.*, **654**, 8 (1962).

1165. G. Sartori, E. Cervone, A. D. Furlani, and I. Collamati, *Ric. Sci., Rend.*, **A2**, 385 (1962); through *Chem. Abstr.*, **58**, 5248 (1963).

1166. W. R. Kroll, Brit. Pat. 1,117,319; through *Chem. Abstr.*, **69**, 106002 (1968).

1167. K. Tamai, T. Saito, Y. Uchida, and A. Misono, *Bull. Chem. Soc., Jap.*, **38**, 1575 (1965).

1168. E. Angelescu, C. Nicolau, and Z. Simon, *J. Am. Chem. Soc.*, **88**, 3910 (1966).

1169. C. Lassau and L. Sajus, *Proc. 13th Intern. Conf. Coord. Chem.*, Cracow-Zakopane, Poland, 1970, Abstr., 212.

1170. I. V. Kalechits, V. G. Lipovich, and F. K. Schmidt, *Neftekhimiya* **6**, 813 (1966); through *Chem. Abstr.*, **66**, 94632 (1966).

1171. F. N. Nasirov, G. P. Korpacheva, B. E. Davidov, and B. A. Krentsel, *Bull. Acad. Sci. USSR, Chem. Sect.*, **1964**, 1603.

1172. K. Shikata, N. Nishino, K. Azuma, and Y. Takegami, *J. Chem. Soc. Jap.*, *Ind. Chem. Sect.*, **68**, 358 (1965).

1173. D. V. Sokol'skii, N. F. Noskova, and M. I. Popandopulo, *Tr. Inst. Khim. Nauk. Akad. Nauk. Kaz. SSR*, **26**, 106 (1969); through *Chem. Abstr.*, **72**, 78193 (1970).

1174. P. Szabo and L. Marko, *Conf. Chem. Process Pet. Nat. Gas, Plenary Lect.*, Budapest 1965, Ed. M. Freund, *Akad. Kiado*, Budapest, 1968, p. 405; through *Chem. Abstr.*, **69**, 106864 (1968).

1175. D. Y. Waddan and D. F. Williams, Ger. Pat. 1,904,613; through *Chem. Abstr.*, **71**, 123525 (1969).

1176. G. Bressan and R. Broggi, *Chim. Ind. (Milan)*, **50**, 1194 (1968).

1177. R. W. Baker and P. Pauling, *Chem. Commun. (Lond.)*, **1969**, 1495.

1178. B. I. Tikhomirov, I. A. Kloptova, and A. I. Yakubchik, *Vestn. Leningr. Univ.*, **22** (22), *Fiz. Khim.* 4, 147 (1967); through *Chem. Abstr.*, **68**, 59020 (1968).

1179. B. I. Tikhomirov, I. A. Kloptova, and A. I. Yakubchik, *Eur. Polym. J.*, **1969**, 561.

1180. E. W. Duck, J. M. Locke, and C. J. Mallison, *Ann. Chem.*, **719**, 69 (1969).

1181. T. Yoshimoto, S. Kaneko, T. Narumiya, and H. Yoshii, S. African Pat. 67,06,384; through *Chem. Abstr.*, **70**, 58770 (1969).

1182. T. Yoshimoto, T. Narumiya, S. Kaneko, and H. Yoshii, S. African Pat. 68,07,486; through *Chem. Abstr.*, **71**, 92454 (1969).

1183. Bridgestone Tire Co. Ltd., Fr. Pat. 1,572,717; through *Chem. Abstr.*, **72**, 44810 (1970).

1184. Bridgestone Tire Co. Ltd., Fr. Pat. 1,581,146; through *Chem. Abstr.*, **73**, 4738 (1970).

1185. L. R. Kallenbach, U.S. Pat. 3,432,518; through *Chem. Abstr.*, **70**, 71441 (1969).

1186. G. Henrici-Olivé and S. Olivé, *Chem. Commun. (Lond.)*, **1969**, 1482.

1187. R. Stern, G. Hillion, and L. Sajus, *Tetrahedron Lett.*, **1969**, 1561.

1188. B. I. Tikhomirov, I. A. Kloptova, G. I. Khramova, and A. I. Yakubchik, *Polym. Sci., USSR*, **10**, 2760 (1968).

1189. V. A. Tulupov, *Russ. J. Phys. Chem.*, **37**, 365 (1963).

1190. V. A. Tulupov and T. I. Evlasheva, *Russ. J. Phys. Chem.*, **39**, 41 (1965).

1191. V. A. Tulupov, *Russ. J. Phys. Chem.*, **38**, 750 (1964).

1192. V. A. Tulupov, D. A. Kivilis, and A. G. Kapyshev, *Russ. J. Phys. Chem.*, **38**, 1303 (1964).

1193. V. A. Tulupov, A. G. Kapyshev, and A. I. Tulupova, *Russ. J. Phys. Chem.*, **38**, 1492 (1964).

1194. V. A. Tulupov, A. G. Kapyshev, and V. L. Postylyakov, *Russ. J. Phys. Chem.*, **40**, 457 (1966).

1195. V. A. Tulupov, D. N. Shigorin, and N. V. Verein, *Russ. J. Phys. Chem.*, **40**, 549 (1966).

1196. V. A. Tulupov, *Russ. J. Phys. Chem.*, **38**, 585 (1964).

1197. V. A. Tulupov, *Russ. J. Phys. Chem.*, **37**, 394 (1963).

1198. V. A. Tulupov, *Russ. J. Phys. Chem.*, **38**, 1601 (1964).

1199. V. A. Tulupov, *Russ. J. Phys. Chem.*, **40**, 1574 (1966).

1200. R. Brout, *J. Chem. Phys.*, **22**, 934 (1954).

1201. V. A. Tulupov, *Dokl. Akad. Nauk, SSSR.*, **177**, 369 (1967).

1202. V. A. Tulupov and D. N. Shigorin, *Russ. J. Phys. Chem.*, **41**, 314 (1967).

1203. R. Köster, *Angew. Chem.*, **68**, 383 (1956).

1204. R. Köster, *Angew. Chem.*, **69**, 94 (1957).

1205. H. C. Brown, *Hydroboration*, Benjamin, New York, 1962.

1206. F. L. Ramp, E. J. De Witt, and L. E. Trapasso, *J. Org. Chem.*, **27**, 4368 (1962).

1207. L. H. Slaugh, *Tetrahedron*, **22**, 1741 (1966).

1208. H. E. Podall, H. E. Petree, and J. R. Zietz, *J. Org. Chem.*, **24**, 1222 (1959).

1209. G. Wilke and H. Müller, *Ann. Chem.*, **618**, 267 (1958).

1210. P. S. Skell and P. K. Freeman, *J. Org. Chem.*, **29**, 2524 (1958).

1211. K. Wirtz and K. F. Bonhoeffer, *Z. Physik. Chem. (Leipz.)*, **A177**, 1 (1936).

1212. S. Abe, *Sci. Pap. Inst. Phys. Chem. Res. (Tokyo)*, **38**, 287 (1941).

1213. W. K. Wilmarth, J. C. Dayton, and J. M. Flourney, *J. Am. Chem. Soc.*, **75**, 4549 (1953).

1214. J. M. Flourney and W. K. Wilmarth, *J. Am. Chem. Soc.*, **83**, 2257 (1961).

1215. Y. Claeys, J. C. Dayton, and W. K. Wilmarth, *J. Chem. Phys.*, **18**, 759 (1950).

1216. S. L. Miller and D. Rittenberg, *J. Am. Chem. Soc.*, **80**, 64 (1958).

1217. K. Bar-Eli and F. S. Klein, *J. Chem. Soc.*, **1962**, 3083.

1218. G. A. Mills, S. Weller and A. Wheeler, unpublished results quoted in ref. 23.

1219. K. Bar-Eli and F. S. Klein, *J. Chem. Soc.*, **1962**, 1378.

1220. P. J. Bourke and J. C. Lee, *Trans. Inst. Chem. Eng.*, **39**, 280 (1961).

1221. G. Dirian, F. Botter, J. Ravoire, and P. Grandcollot, *J. Chim. Phys.*, **60**, 139 (1963).

1222. Y. Claeys, C. Baes, and W. K. Wilmarth, *J. Chem. Phys.*, **16**, 425 (1948).

1223. W. K. Wilmarth and J. C. Dayton, *J. Am. Chem. Soc.*, **75**, 4553 (1953).

1224. J. Bigeleisen, *Proc. Intern. Symp. Isot. Sep.*, North-Holland, Amsterdam, 1958, p. 133.

1225. C. Walling and L. Bollyky, *J. Am. Chem. Soc.*, **83**, 2968 (1961).

1226. C. Walling and L. Bollyky, *J. Am. Chem. Soc.*, **86**, 3750 (1964).

1227. H. H. Brongersma, H. M. Buck, H. P. J. M. Dekkers, and L. J. Oosterhoff, *J. Catal.*, **10**, 149 (1968).

1228. T. Yagi and Y. Inokuchi, *Kagaku To Seibutsu*, **7**, 579 (1959); through *Chem. Abstr.*, **72**, 62885 (1970).

1229. H. Gest, *Bacteriol. Rev.*, **18**, 43 (1954).

1230. M. Stephenson and L. H. Strickland, *Biochem. J.*, **25**, 205 (1931).

1231. M. Stephenson and L. H. Strickland, *Biochem. J.*, **26**, 712 (1932).

1232. M. Stephenson and L. H. Strickland, *Biochem. J.*, **27**, 1617 (1933).

1233. D. E. Green and L. H. Strickland, *Biochem. J.*, **28**, 898, (1934).

1234. W. S. Waring and C. H. Werkman, *Arch. Biochem.*, **4**, 75 (1944).

1235. H. D. Peck, Jr., A. San Pietro, and H. Gest, *Proc. Natl. Acad. Sci. U.S.*, **42**, 13 (1956).

1236. H. Gest, *J. Bacteriol.*, **63**, 111 (1952).

1237. H. D. Peck and H. Gest, *J. Bacteriol.* **73**, 569 (1959).

1238. H. D. Hoberman and D. Rittenberg, *J. Biol. Chem.*, **147**, 211 (1943).

1239. K. Krogmann and W. Binder, *J. Organomet. Chem.*, **11**, P27 (1968).

1240. A. I. Krasna and D. Rittenberg, *J. Am. Chem. Soc.*, **76**, 3015 (1954).

1241. J. C. Sadana and D. Rittenberg, *Arch. Biochem. Biophys.*, **108**, 255 (1964).

1242. E. Riklis and D. Rittenberg, *J. Biol. Chem.*, **236**, 2526 (1961).

1243. J. C. Sadana and D. Rittenberg, *Proc. Natl. Acad. Sci. U.S.*, **50**, 900 (1963).

1244. J. C. Sadana and A. V. Morey, *Biochim. Biophys. Acta*, **50**, 153 (1961).

1245. Y. V. Peive, B. A. Yagodin, and A. D. Popazova, *Agrokhimiya*, **1967**, 94; through *Chem. Abstr.*, **66**, 104379 (1967).

1246. A. I. Krasna and D. Rittenberg, *Proc. Natl. Acad. Sci. U.S.*, **42**, 180 (1956).

1247. D. Rittenberg and A. I. Krasna, *Discuss. Faraday Soc.*, **20**, 185 (1955).

1248. A. I. Krasna, E. Riklis, and D. Rittenberg, *J. Biol. Chem.*, **235**, 2717 (1960).

1249. A. Farkas, L. Farkas, and J. Yudkin, *Proc. Roy. Soc.*, B, **115**, 373 (1934).

1250. L. Farkas and E. Fischer, *J. Biol. Chem.*, **147**, 211 (1943).

1251. L. Farkas and E. Fischer, *J. Biol. Chem.*, **167**, 787 (1947).

1251a. A. Farkas, *Trans. Faraday Soc.*, **32**, 922 (1936).

1252. H. Gest and M. D. Kamen, *J. Biol. Chem.*, **182**, 153 (1950).

1253. B. Cavanagh, J. Horiuchi, and M. Polanyi, *Nature*, **133**, 797 (1934).

1254. L. A. Hyndman, R. H. Burris, and P. W. Wilson, *J. Bacteriol.*, **65**, 522 (1953).

1255. B. A. Pethica, E. R. Roberts, and E. R. S. Winter, *Research* (*Lond.*), **3**, 382 (1950).

1256. A. Couper, D. D. Eley, and A. Hayward, *Discuss. Faraday Soc.*, **20**, 174 (1955).

1257. J. G. Beetlestone and A. Couper, *Discuss. Faraday Soc.*, **20**, 281 (1955).

1258. N. Tamuja and S. L. Miller, *J. Biol. Chem.*, **238**, 2194 (1963).

1259. H. Gest, Proc. *Intern. Symp. Enzyme Chem.*, *Tokyo Kyoto*, **2**, 250 (1957).

1260. A. J. Krasna and D. Rittenberg, *J. Am. Chem. Soc.*, **77**, 5295 (1955).

1261. L. Farkas and B. Schneidmesser, *J. Biol. Chem.*, **167**, 807 (1947).

1262. L. Purec, A. I. Krasna and D. Rittenberg, *Biochemistry*, **1**, 270 (1962).

1263. K. J. C. Back, J. Lascelles, and J. L. Still, *Aust. J. Sci.*, **9**, 25 (1946).

1264. W. K. Joklik, *Aust. J. Exp. Biol. Med. Sci.*, **28**, 321 (1950).

1265. L. D. Shturm, *Mikrobiologiya*, **26**, 710 (1958); through *Chem. Abstr.*, **52**, 9313 (1958).

1266. W. Curtis and E. J. Ordal, *J. Bacteriol.*, **68**, 351 (1954).

1267. L. M. S. Chang and J. Wolin, *J. Bacteriol.*, **98**, 51 (1969).

1268. W. Kempner and F. Kubowitz, *Biochem. Z*, **265**, 245 (1933).

1269. D. M. Bone, *Biochim. Biophys. Acta*, **67**, 589 (1963).

1270. J. C. Sadana and V. Jagannathan, *Biochim. Biophys. Acta*, **19**, 440 (1956).

1271. A. I. Krasna and D. Rittenberg, *Proc. Natl. Acad. Sci.*, **40**, 225 (1954).

1272. W. K. Joklik, *Aust. J. Exp. Biol. Med. Sci.*, **28**, 331 (1950).

1273. J. Halpern, G. Czapski, J. Jortner, and G. Stein, *Nature*, **186**, 629 (1960).

1274. T. Rigg, G. Stein, and J. Weiss, *Proc. Roy. Soc.*, A, **211**, 375 (1952).

1275. G. Czapski, J. Jortner, and G. Stein, *J. Phys. Chem.*, **65**, 960 (1961).
1276. G. Czapski and G. Stein, *J. Phys. Chem.*, **63**, 850 (1959).
1277. D. Elad and I. Rosenthal, *Chem. Commun. (Lond.)*, **1968**, 879.
1278. K. C. Smith and R. T. Aplin, *Biochemistry*, **5**, 2125 (1966).
1279. P. Cerutti, Y. Kondo, W. R. Landis, and B. Witkop, *J. Am. Chem. Soc.*, **90**, 771 (1968).
1280. M. Berthelot, *Ann. Chim.*, **9**, 401 (1866); through ref. 1299.
1281. W. Traube and W. Passarge, *Berichte*, **49**, 1692 (1916).
1282. J. B. Conant and H. Cutter, *J. Am. Chem. Soc.*, **48**, 1016 (1926).
1283. W. I. Patterson and V. du Vigneaud, *J. Biol. Chem.*, **123**, 327 (1938).
1284. R. S. Bottei and N. H. Furman, *Anal. Chem.*, **27**, 1183 (1955).
1285. R. S. Bottei, *Anal. Chim. Acta*, **30**, 6 (1964).
1286. R. S. Bottei and W. A. Joern, *J. Am. Chem. Soc.*, **90**, 297 (1968).
1287. C. E. Castro and R. D. Stevens, *J. Am. Chem. Soc.*, **86**, 4358 (1964).
1288. C. E. Castro, R. D. Stevens, and S. Mojé, *J. Am. Chem. Soc.*, **88**, 4964 (1966).
1289. A. Zurqiyah and C. E. Castro, *Org. Syntheses*, **49**, 98 (1969).
1290. K. D. Kopple, *J. Am. Chem. Soc.*, **84**, 1586 (1962).
1291. A. Malliaris and D. Katakis, *J. Am. Chem. Soc.*, **87**, 3077 (1965).
1292. N. T. Denisov, O. N. Efimov, A. G. Ovcharenko, and A. E. Shilov, *Teor. i Eksp. Khim., Akad. Nauk. Ukr. SSR*, **1**, 762 (1965); through *Chem. Abstr.*, **64**, 18482 (1966).
1293. M. O. Broitman, N. T. Denisov, N. I. Shuvalova, and A. E. Shilov, *Kinet. i Katal.*, in press; through ref. 1294.
1294. A. E. Shilov, *Kinet. Catal. (USSR)*, **11**, 256 (1970).
1295. C. E. Castro, *J. Am. Chem. Soc.*, **83**, 3262 (1961).
1296. C. E. Castro and W. C. Kray, *J. Am. Chem. Soc.*, **85**, 2768 (1963).
1297. A. Zurqiyah and C. E. Castro, *J. Org. Chem.*, **34**, 1504 (1969).
1298. E. Vrachnou-Astra and D. Katakis, *J. Am. Chem. Soc.*, **89**, 6772 (1967).
1299. E. Vrachnou-Astra, P. Sakellaridis, and D. Katakis, *J. Am. Chem. Soc.*, **92**, 811 (1970).
1300. A. Misono, Y. Uchida, M. Hidai, and H. Kanai, *Chem. Commun. (Lond.)*, **1967**, 357.
1301. A. Misono, Y. Uchida, T. Saito, and M. Hidai, *J. Chem. Soc. Jap., Ind. Chem. Sect.*, **20**, 1890 (1967).
1302. A. Misono, Y. Uchida, M. Hidai, H. Shinohara, and Y. Watanabe, *Bull. Chem. Soc. Jap.*, **41**, 396 (1968).
1303. A. Misono, Y. Uchida, M. Hidai, and I. Inomata, *Chem. Commun. (Lond.)*, **1968**, 704.
1304. J. D. McClure, R. Owyang, and L. H. Slaugh, *J. Organomet. Chem.*, **12**, P8 (1968).
1305. J. E. Lydon and M. Truter, *J. Chem. Soc.*, A, **1968**, 362.
1306. E. Billig, C. B. Strow, and R. L. Pruett, *Chem. Commun. (Lond.)*, **1968**, 1307.
1307. S. D. Ibekwe and U. A. Raeburn, *J. Organomet. Chem.*, **19**, 447 (1969).
1308. J. R. Moss and B. L. Shaw, *Chem. Commun. (Lond.)*, **1968**, 632.
1309. B. Bell, J. Chatt and G. J. Leigh, *Chem. Commun. (Lond.)*, **1970**, 576.

1310. K. R. Laing and W. R. Roper, *Chem. Commun. (Lond.)*, **1968**, 1556.

1311. M. S. Lupin and B. L. Shaw, *J. Chem. Soc., A*, **1968**, 741.

1312. G. Leuteritz, *Fette, Seifen, Anstrichm.*, **71**, 441 (1969); through *Chem. Abstr.*, **71**, 72205 (1969).

1313. V. V. Abalyaeva and M. L. Khidekel, *Bull. Acad. Sci., USSR, Chem. Sect.*, **1969**, 1951.

1314. H. A. Martin and F. Jellinek, *J. Organomet. Chem.*, **8**, 115 (1967).

1315. H. A. Martin and R. O. de Jongh, *Chem. Commun. (Lond.)*, **1969**, 1366.

1316. P. B. Hitchcock, M. McPartlin, and R. Mason, *Chem. Commun. (Lond.)*, **1969**, 1367.

1317. Ethyl Corp., Brit. Pat. 863,277; through *Chem. Abstr.*, **56**, 9969 (1962).

1318. T. G. Selin, U.S. Dept. Comm. Office Tech. Serv. P.B. Rept. 133796 (1960); through *Chem. Abstr.*, **56**, 4142 (1962).

1319. L. Porri, M. C. Gallazzi, A. Colombo, and G. Allegra, *Tetrahedron Lett.*, **1965**, 4187.

1320. W. Heiber and T. Kruck, *Z. Naturforsch*, **16b**, 709 (1961).

1321. W. T. Hendrix, F. G. Cowherd, and J. L. von Rosenberg, *Chem. Commun. (Lond.)*, **1968**, 97.

1322. H. Günther, R. Wenzl, and H. Klose, *Chem. Commun. (Lond.)*, **1970**, 605.

1323. K. Farmery, M. Kilner, R. Greatrex, and N. N. Greenwood, *J. Chem. Soc., A*, **1969**, 2339.

1324. G. Hata and A. Miyake, *Proc. 10th Intern. Conf. Coord. Chem.*, Osaka, 1967, p. 130.

1325. P. Pino, G. Braca, F. Piacenti, G. Sbrana, M. Bianchi, and E. Benedetti, *New Aspects Chem. Met. Carbonyls Deriv., 1st Intern. Symp.*, Venice, 1968, Abstr., E2.

1326. C. E. Wymore, *Chem. Eng. News*, April 8, **1968**, 52.

1327. L. I. Simandi and E. Budo-Zahonyi, *Proc. 13th Intern. Conf. Coord. Chem.*, Cracow-Zakopane, 1970, Abstr., 17.

1328. J. P. Birk, P. B. Chock, and J. Halpern, *J. Am. Chem. Soc.*, **90**, 6959 (1968).

1329. J. F. Nixon and D. A. Clement, private communication.

1330. C. A. Reilly and H. Thyret, *J. Am. Chem. Soc.*, **89**, 5144 (1967).

1331. J. Powell and B. L. Shaw, *J. Chem. Soc., A*, **1968**, 583.

1332. G. D. Venerable II, E. J. Hart, and J. Halpern, *J. Am. Chem. Soc.*, **91**, 7538 (1969).

1333. J. Chatt, G. J. Leigh, and R. J. Paske, *Chem. Commun. (Lond.)*, **1967**, 671.

1334. J. Dvorak, R. J. O'Brien, and W. Santo, *Chem. Commun. (Lond.)*, **1970**, 411.

1335. H. Brintzinger, *J. Am. Chem. Soc.*, **88**, 4305 (1966).

1336. B. D. James, R. K. Nanda, and M. G. H. Wallbridge, *Inorg. Chem.*, **6**, 1979 (1967).

1337. B. Kautzner, P. C. Wailes, and H. Weigold, *Chem. Commun. (Lond.)*, **1969**, 1105.

1338. S. Otsuka, A. Nakamura, and H. Minamida, *Chem. Commun. (Lond.)*, **1969**, 1148.

1339. J. Chatt and R. S. Coffey, *J. Chem. Soc., A*, **1969**, 1963.

1340. A. P. Ginsberg, *Chem. Commun. (Lond.)*, **1968**, 857.

1341. K. I. Matveev, N. K. Eremenko, L. N. Rachkovskaja, L. M. Kefeli, and L. M. Plyasova, *Proc. 13th Intern. Conf. Coord. Chem.*, Cracow-Zakopane, 1970, Abstr., 418.

1342. P. Rigo, M. Bressan, and A. Turco, *Inorg. Chem.*, **7**, 1460 (1968).

1343. B. Longato, P. Rigo, and A. Turco, *Proc. 13th Intern. Conf. Coord. Chem.*, Cracow-Zakopane, 1970, Abstr., 224.

1345. A. A. Orio, U. Mazzi, and H. B. Gray, *Proc. 13th Intern. Conf. Coord. Chem.*, Cracow-Zakopane, 1970, Abstr., 186.

1346. S. Tyrlik and H. Stepowska, *Proc. 13th Intern. Conf. Coord. Chem.*, Cracow-Zakopane, 1970, Abstr., 79.

1347. L. Malatesta, M. Angoletta, and G. Caglio, *Proc. 13th Intern. Conf. Coord. Chem.*, Cracow-Zakopane, 1970, Abstr., 397.

1348. M. J. Mays, R. N. F. Simpson, and F. P. Stefanini, *J. Chem. Soc.*, *A*, **1970**, 3000.

1349. G. Martino, *React. Bonding Transition Organomet. Cpds.*, *3rd Intern. Symp.* Venice, 1970, Abstr., A3.

1350. G. F. Pregaglia, A. Andreetta, G. F. Ferrari, G. Montrasi, and R. Ugo, *React. Bonding Transition Organomet. Cpds. 3rd Intern. Symp.*, Venice, 1970, Abstr., D2.

1351. Y. Takahashi, Ts. Ito, S. Sakai, and Y. Ishii, *Chem. Commun. (Lond.)*, **1970**, 1065.

1352. A. J. Cheney, B. E. Mann, B. L. Shaw, and R. M. Slade, *Chem. Commun. (Lond.)*, **1970**, 1176.

1353. C. Carlini, D. Politi, and F. Ciardelli, *Chem. Commun. (Lond.)*, **1970**, 1260.

1354. R. J. Cozens, K. S. Murray, and B. O. West, *Chem. Commun. (Lond.)*, **1970**, 1262.

1355. B. E. Mann, B. L. Shaw, and N. I. Tucker, *Chem. Commun. (Lond.)*, **1970**, 1333.

1356. P. B. Tripathy and D. M. Roundhill, *J. Am. Chem. Soc.*, **92**, 3825 (1970).

1357. K. R. Grundy, K. R. Laing, and W. R. Roper, *Chem. Commun. (Lond.)*, **1970**, 1500.

1358. K. P. Davis, J. L. Garnett, and J. H. O'Keefe, *Chem. Commun. (Lond.)*, **1970**, 1672.

1359. J. L. Garnett, and A. T. T. Oei, *J. Catal.*, in press.

1360. M. L. H. Green and P. J. Knowles, *Chem. Commun. (Lond.)*, **1970**, 1677.

1361. E. K. Barefield, G. W. Parshall, and F. N. Tebbe, *J. Am. Chem. Soc.*, **92**, 5234 (1970).

1362. K. Kudo, M. Hidai, T. Murayama, and Y. Uchida, *Chem. Commun. (Lond.)*, **1970**, 1701.

1363. A. Misono, Y. Uchida, M. Hidai, and K. Kudo, *J. Organomet. Chem.*, **20**, P7 (1969).

1364. W. C. Drinkard, D. R. Eaton, J. P. Jesson, and R. V. Lindsey, Jr., *Inorg. Chem.*, **9**, 392 (1970).

1365. R. A. Schunn, *Inorg. Chem.*, **9**, 394 (1970).

1366. C. A. Tolman, *J. Am. Chem. Soc.*, **92**, 4217 (1970).

1367. T. I. Eliades, R. O. Harris, and M. C. Zia, *Chem. Commun. (Lond.)*, **1970**, 1709.

1368. T. Ito, S. Kitazume, A. Yakamoto, and S. Ikeada, *J. Am. Chem. Soc.*, **92**, 3011 (1970).

1369. C. A. Tolman, *J. Am. Chem. Soc.*, **92**, 6777 (1970).

1370. C. A. Tolman, *J. Am. Chem. Soc.*, **92**, 6785 (1970).

1371. H. H. Brintzinger and J. E. Bercaw, *J. Am. Chem. Soc.*, **92**, 6182 (1970).

1372. C. G. Pierpont, D. G. Van Derveer, W. Durland, and R. Eisenberg, *J. Am. Chem. Soc.*, **92**, 4760 (1970).

1373. G. M. Whitesides, E. R. Stedronsky, C. P. Casey, and J. S. Filippo, Jr., *J. Am. Chem. Soc.*, **92**, 1426 (1970).

1374. G. M. Whitesides, J. S. Filippo, Jr., E. R. Stedronsky, and C. P. Casey, *J. Am. Chem. Soc.*, **91**, 6542 (1969).

1375. J. A. Dilts and D. F. Shriver, *J. Am. Chem. Soc.*, **91**, 4088 (1969).

1376. R. Pietropaolo, G. Dolcetti, M. Giustiniani, and U. Belluco, *Inorg. Chem.*, **9**, 549 (1970).

1377. G. Dolcetti, R. Pietropaolo, and U. Belluco, *Inorg. Chem.*, **9**, 553 (1970).

1378. M. Green and C. J. Wilson, *React. Bonding Transition Organomet. Cpds.*, *3rd Intern. Symp.* Venice, 1970, Abstr., B2.

1379. H. C. Clark and J. D. Ruddick, *Inorg. Chem.*, **9**, 1226 (1970).

1380. H. C. Clark and W. J. Jacobs, *Inorg. Chem.*, **9**, 1229 (1970).

1381. P. W. Clark and G. E. Hartwell, *Inorg. Chem.*, **9**, 1948 (1970).

1382. D. A. White and G. W. Parshall, *Inorg. Chem.*, **9**, 2358 (1970).

1383. L. W. Gosser and C. A. Tolman, *Inorg. Chem.*, **9**, 2350 (1970).

1384. J. Halpern and M. Pribanic, *Inorg. Chem.*, **9**, 2616 (1970).

1385. A. J. Oliver and W. A. G. Graham, *Inorg. Chem.*, **9**, 2653 (1970).

1386. S. D. Ibekwe and K. A. Taylor, *J. Chem. Soc.*, *A*, **1970**, 1.

1387. K. Thomas and G. Wilkinson, *J. Chem. Soc.*, *A*, **1970**, 356.

1388. C. A. Reed and W. R. Roper, *J. Chem. Soc.*, *A*, **1970**, 506.

1389. J. Knight and M. J. Mays, *J. Chem. Soc.*, *A*, **1970**, 711.

1390. T. A. Stephenson, *J. Chem. Soc.*, *A*, **1970**, 889.

1391. R. S. Eachus and M. C. R. Symons, *J. Chem. Soc.*, *A*, **1970**, 1336.

1392. S. A. Butter and J. Chatt, *J. Chem. Soc.*, *A*, **1970**, 1411.

1393. J. Chatt and P. Chini, *J. Chem. Soc.*, *A*, **1970**, 1538.

1394. P. Chini and G. Longoni, *J. Chem. Soc.*, *A*, **1970**, 1542.

1395. G. R. Crooks and B. F. G. Johnson, *J. Chem. Soc.*, *A*, **1970**, 1662.

1396. F. Glockling and M. D. Wilbey, *J. Chem. Soc.*, *A*, **1970**, 1675.

1397. I. Collamati, A. Furlani, and G. Attioli, *J. Chem. Soc.*, *A*, **1970**, 1694.

1398. D. Rose and G. Wilkinson, *J. Chem. Soc.*, *A*, **1970**, 1791.

1399. B. C. Hui and B. R. James, *Can. J. Chem.*, **48**, 3613 (1970).

1400. L. Malatesta, M. Angoletta, and G. Caglio, *J. Chem. Soc.*, *A*, **1970**, 1836.

1401. J. Ashley-Smith, M. Green, and D. C. Wood, *J. Chem. Soc.*, *A*, **1970**, 1847.

1402. K. A. Hooton, *J. Chem. Soc.*, *A*, **1970**, 1896.

1403. M. J. Church and M. J. Mays, *J. Chem. Soc.*, *A*, **1970**, 1938.

1404. D. Bingham and M. G. Burnett, *J. Chem. Soc.*, *A*, **1970**, 2165.

1405. J. Llopis and A. Sanchez Robles, *An. Quim.*, **51b**, 661 (1955); through *Chem. Abstr.*, **50**, 14300 (1956).

1406. M. McPartlin and R. Mason, *J. Chem. Soc.*, *A*, **1970**, 2206.

1407. M. C. Hall, B. T. Kilbourn, and K. A. Taylor, *J. Chem. Soc.*, *A*, **1970**, 2539.

1408. M. H. B. Stiddard and R. E. Townsend, *J. Chem. Soc.*, *A*, **1970**, 2719.

1409. M. J. Church, M. J. Mays, R. N. F. Simpson, and F. P. Stefanini, *J. Chem. Soc.*, *A*, **1970**, 2909.

1410. J. J. Levison and S. D. Robinson, *J. Chem. Soc.*, *A*, **1970**, 2947.

1411. A. J. Deeming, B. F. G. Johnson, and J. Lewis, *J. Chem. Soc.*, *A*, **1970**, 2967.

1412. C. A. Reed and W. R. Roper, *J. Chem. Soc., A*, **1970**, 3054.

1413. R. D. Gillard, B. T. Heaton, and D. H. Vaughan, *J. Chem. Soc., A*, **1970**, 3126.

1414. M. F. Lappert and N. F. Travers, *J. Chem. Soc., A*, **1970**, 3303.

1415. A. J. Deeming and B. L. Shaw, *J. Chem. Soc., A*, **1970**, 3356.

1416. T. Yoshimoto, H. Koyama, Y. Takeda, T. Hujimori, and Y. Kaneko, Jap. Pat. 70 10,945; through *Chem. Abstr.*, **73**, 36297 (1970).

1417. T. Yoshimoto, T. Narumiya, T. Takamatsu, T. Sasaki, H. Yoshii, and S. Kaneko, Ger. Pat. 2,005,731; through *Chem. Abstr.*, **73**, 121401 (1970).

1418. Bridgestone Tire Co. Ltd., Brit. Pat. 1,198,195; through *Chem. Abstr.*, **73**, 56958 (1970).

1419. T. Yoshimoto, S. Kaneko, T. Narumiya, and H. Yoshii, U.S.Pat. 3,541,064; through *Chem. Abstr.*, **74**, 23500 (1971).

1420. Bridgestone Tire Co. Ltd., Brit. Pat. 1,213,411; through *Chem. Abstr.*, **74**, 43299 (1971).

1421. R. P. Zelinski and H. R. Gaeth, U.S. Pat., 3,560,405; through *Chem. Abstr.*, **74**, 65262 (1971).

1422. J. F. Pendleton, R. J. Schlott, and D. F. Hoeg, Ger. Pat. 2,030,641; through *Chem. Abstr.*, **74**, 65264 (1971).

1423. M. Iwamoto, Jap. Pat. 70 21,487; through *Chem. Abstr.*, **73**, 76632 (1970).

1424. G. L. Rempel, personal communication.

1425. M. Iwamoto, Jap. Pat. 7022,322; through *Chem. Abstr.*, **73**, 120049 (1970).

1426. D. V. Sokol'skii, G. N. Sharifkanova, and N. F. Noskova, *Tr. Inst. Khim. Nauk. Akad. Nauk. Kaz. SSR*, **30**, 10 (1970); through *Chem. Abstr.*, **73**, 124696 (1970).

1427. N. F. Noskova, N. I. Marusich, and D. V. Sokol'skii, *Tr. Inst. Khim. Nauk. Akad. Nauk. Kaz. SSR*, **30**, 3 (1970); through *Chem. Abstr.*, **74**, 6837 (1971).

1428. D. V. Sokol'skii, G. N. Sharifkanova, and N. F. Noskova, *Dokl. Akad. Nauk. SSSR.*, **194**, 599 (1970).

1429. N. F. Noskova, M. I. Popandopulo, D. V. Sokol'skii, and G. N. Sharifkanova, *Izv. Akad. Nauk. Kaz. SSR. Ser. Khim*, **20**, 17 (1970); through *Chem. Abstr.*, **74**, 31592 (1971).

1430. J. F. Falbe and N. Huppes, U.S. Pat. 3,509,221; through *Chem. Abstr.*, **73**, 34908 (1970).

1431. J. L. Van Winkle, Brit. Pat. 1,191,815; through *Chem. Abstr.*, **73**, 66020 (1970).

1432. T. Kondo and A. Miyake, Jap. Pat. 7021,283; through *Chem. Abstr.*, **73**, 66143 (1970).

1433. G. C. Bond, Brit. Pat. 1,197,723; through *Chem. Abstr.*, **73**, 98401 (1970).

1434. A. R. Powell, Brit. Pat. 1,196,583; through *Chem. Abstr.*, **73**, 59620 (1970).

1435. I. Nakamori, *Yukagaku*, **19**, 556 (1970); through *Chem. Abstr.*, **74**, 2715 (1971).

1436. K. T. Achaya and D. S. Raju, *J. Sci. Ind. Res. (India)*, **29**, 68 (1970); through *Chem. Abstr.*, **73**, 65060 (1970).

1437. M. Takesada and H. Wakamatsu, *Bull. Chem. Soc. Jap.*, **43**, 2192 (1970).

1438. T. Kan and T. Suzuki, Jap. Pat. 70 21,084; through *Chem. Abstr.*, **73**, 76629 (1970).

1439. T. Suga and T. Suzuki, Jap. Pat. 70 21,287; through *Chem. Abstr.*, **73**, 76630 (1970).

1440. E. Ucciani, *Rev. Fr. Corps. Gras.*, **17**, 395 (1970); through *Chem. Abstr.*, **73**, 89426 (1970).

1441. G. Foster and M. G. Lawrenson, S. African Pat. 69 05,913; through *Chem. Abstr.*, **73**, 87430 (1970).

1442. A. Kanai and A. Miyake, Jap. Pat. 70 22,321; through *Chem. Abstr.*, **73**, 87397 (1970).

1443. A. Zakhariev, V. Ivanova, and D. Shopov, *Izv. Otd. Khim. Nauki, Bulg. Akad. Nauk.*, **2**, 749 (1969); through *Chem. Abstr.*, **73**, 87235 (1970).

1444. A. Zakhariev, V. Ivanova, and D. Shopov, *Dokl. Bolg. Akad. Nauk.*, **23**, 679 (1970); through *Chem. Abstr.*, **73**, 91901 (1970).

1445. K. Aoki and A. Mayake, Jap. Pat. 70 21,284; through *Chem. Abstr.*, **73**, 98355 (1970).

1446. I. Ogata, *Tokyo Kogyo Shikensho Hokoku*, **65**, 89 (1970); through *Chem. Abstr.*, **73**, 102428 (1970).

1447. M. L. Khidekel, O. N. Efimov, O. N. Eremenko, and A. G. Ovcharenko, U.S.S.R. Pat. 271,501; through *Chem. Abstr.*, **73**, 120411 (1970).

1448. V. A. Avilov, O. N. Eremenko, O. N. Efimov, A. G. Ovcharenko, M. L. Khidekel, and P. S. Chekrii, U.S.S.R. Pat. 207,879; through *Chem. Abstr.*, **74**, 6952 (1971).

1449. V. A. Avilov, O. N. Eremenko, O. N. Efimov, A. G. Ovcharenko, M. L. Khidekel, and P. S. Chekrii, U.S.S.R. Pat. 215,200; through *Chem. Abstr.*, **74**, 6953 (1971).

1450. S. Sato, A. Morishima, and H. Wakamatsu, *J. Chem. Soc. Jap.*, **91**, 557 (1970).

1451. L. Marko, P. Szabo, and J. Laky, Hung. Pat. 157,605; through *Chem. Abstr.*, **73**, 124009 (1970).

1452. C. Lassau, R. Stern, and L. Sajus, Ger. Pat. 1,944,382; through *Chem. Abstr.*, **73**, 124016 (1970).

1453. W. Strohmeier and R. Endres, *Z. Naturforsch.*, **25b**, 1068 (1970).

1454. I. Ogata, R. Iwata, and Y. Ikeda, *Tetrahedron Lett.*, **1970**, 3011.

1455. D. Kleiner and R. H. Burris, *Biochim. Biophys. Acta*, **212**, 417 (1970).

1456. Ruhrchemie A-G, Fr. Pat. 2,015,475; through *Chem. Abstr.*, **74**, 12831 (1971).

1457. R. S. Crichton, Ger. Pat. 2,020,550; through *Chem. Abstr.*, **74**, 22425 (1971).

1458. L. A. Kheifits, A. E. Gol'dovskii, I. S. Kolomnikov, and M. E. Volpin, *Izv. Akad. Nauk. SSSR. Ser. Khim*, **1970**, 2078; through *Chem. Abstr.*, **74**, 41593 (1971).

1459. W. Strohmeier and S. Hohmann, *Z. Naturforsch.*, **25b**, 1309 (1970).

1460. G. K. Koch and J. W. Dalenberg, *J. Label. Compd.*, **6**, 395 (1970).

1461. G. Braca, G. Sbrana, F. Piacenti, and P. Pino, *Chim. Ind. (Milan)*, **52**, 1091 (1970).

1462. L. Marko and J. Bathory, *Chem.-Anlagen Verfahren*, **1970**, 65; through *Chem. Abstr.*, **74**, 63852 (1971).

1463. H. B. Charman, *J. Chem. Soc., B*, **1970**, 584.

1464. W. Strohmeier, W. Rehder-Stirnweiss, and R. Fleischmann, *Z. Naturforsch.*, **25b**, 1480 (1970).

1465. W. Strohmeier, W. Rehder-Stirnweiss, and R. Fleischmann, *Z. Naturforsch.*, **25b**, 1481 (1970).

1466. R. Endres, Diplomarbeit, The Physical Chemistry Institute of the University of Wurzburg, 1970; quoted in ref. 1465.

1467. N. T. Denisov, V. F. Shuvalov, N. I. Shuvalova, A. K. Shilova, and A. E. Shilov, *Kinet. Catal. (USSR)*, **11**, 673 (1970).

1468. G. N. Schrauzer and P. A. Doemeny, *J. Am. Chem. Soc.*, **93**, 1608 (1971).

1469. A. D. Allen, *Advances in Chemistry*, Vol. 100, American Chemical Society, Washington, D.C., 1971, p. 79.

1470. E. E. Van Tamelen, *Advances in Chemistry*, Vol. 100, American Chemical Society, Washington, D.C. 1971, p. 95.

1471. R. W. F. Hardy, R. C. Burns, and G. W. Parshall, *Advances in Chemistry*, Vol. 100, American Chemical Society, Washington, D.C., 1971, p. 219.

1472. S. N. Zelenin and M. L. Khidekel, *Russ. Chem. Rev.*, **39**, 103 (1970).

1473. E. N. Frankel, F. L. Thomas, and J. C. Cowan, *J. Am. Oil Chem. Soc.*, **47**, 497 (1970).

1474. R. L. Augustine and J. F. Van Peppen, *J. Am. Oil Chem. Soc.*, **47**, 478 (1970).

1475. J. C. Bailar Jr., *J. Am. Oil Chem. Soc.*, **47**, 475 (1970).

1476. A. P. Khrushch and A. E. Shilov, *Kinet. Catal. (USSR)*, **10**, 389 (1969).

1477. A. P. Khrushch and A. E. Shilov, *Kinet. Catal. (USSR)*, **11**, 67 (1970).

1478. H. G. Kuivila, *Synthesis*, **1970**, 499.

1479. J. F. Harrod and J. Halpern, *J. Phys. Chem.*, **65**, 563 (1961).

1480. W. McFarlane, *Chem. Commun. (Lond.)*, **1969**, 700.

1481. T. H. Brown and P. J. Green, *J. Am. Chem. Soc.*, **92**, 2359 (1970).

1482. L. Cassar, P. E. Eaton, and J. Halpern, *J. Am. Chem. Soc.*, **92**, 3515 (1970).

1483. R. G. Pearson, *Chem. Eng. News*, September 28, **1970**, 66.

1484. R. G. Pearson, and W. R. Muir, *J. Am. Chem. Soc.*, **92**, 5519 (1970).

1485. J. Kwiatek, "Hydrogenation and Dehydrogenation," in G. N. Schrauzer, Ed., *Transition Metals in Homogeneous Catalysis*, Marcel Dekker, New York, 1971, p. 13.

1486. R. A. Schunn, "Systematics of Transition Metal Hydride Chemistry," in E. L. Muetterties, Ed., *Transition Metal Hydrides*, Marcel Dekker, New York, 1971, p. 203.

1487. C. A. Tolman, "Role of Transition Metal Hydrides in Homogeneous Catalysis," in E. L. Muetterties, Ed., *Transition Metal Hydrides*, Marcel Dekker, New York, 1971, p. 271.

1488. J. Halpern, *Acc. Chem. Res.*, **3**, 386 (1970).

1489. C. A. Tolman, *Chem. Soc. Rev.*, **1**, 337 (1972).

1490. G. C. Bond, *Mem. Soc. Roy. Sci. Liege, Collect. 8°*, **1** (4), 61 (1971).

1490a. R. J. Kokes, *Catal. Rev.*, **6**, 1 (1972).

1490b. C. W. Bird, *Top. Lipid Chem.*, **2**, 247 (1971).

1490c. J. P. Candlin, K. A. Taylor, and A. W. Parkins, *Ann. Rep. Chem. Soc.*, *B*, **68**, 273 (1971).

1490d. F. G. A. Stone, *Nature*, **232**, 534 (1971).

1490e. G. Henrici-Olivé and S. Olivé, *Angew. Chem. Intern. Edit.*, **10**, 105 (1971).

1490f. R. Ugo, *Engelhard Ind. Tech. Bull.*, **11** (2), 45 (1971).

1491. A. Farkas, *Hydrocarbon Process*, **50** (5), 137 (1971).

1492. D. V. Sokol'skii, *Katal. Gidrirovaniya Rastvorakh, Nauka Kaz. SSR, Alma-Ata*, 1971; through *Chem. Abstr.*, **76**, 129608 (1972).

1493. K. Mitsui, *Kagaku Kogaku*, **35**, 965 (1971).

1494. F. A. Cotton and G. Wilkinson, *Advanced Inorganic Chemistry*, 3rd ed., Interscience, New York-London, 1972, p. 770.

1494a. L. Vaska, *Inorg. Chim. Acta*, **5**, 295 (1971).

1495. R. L. Augustine, "Steroid Hydrogenation," in J. Fried and J. A. Edwards, Eds., *Organic Reactions in Steroid Chemistry*, Vol. 1, 1972, Van Nostrand-Reinhold, New York, 1972, p. 111.

1496. L. Tokes and L. J. Throop, "Introduction of Deuterium into the Steroid System," in J. Fried and J. A. Edwards Eds., *Organic Reactions in Steroid Chemistry*, Vol. 1, Van Nostrand-Reinhold, New York, 1972, p. 145.

1497. A. J. Birch and F. J. McQuillin, *Prog. Org. Chem.*, in press.

1497a. J. Hanzlik, *Chem. Listy*, **65**, 454 (1971); through *Chem. Abstr.*, **75**, 53718 (1971).

1497b. I. Ogata, *J. Jap. Oil Chem. Soc.*, **21**, 53 (1972); through *Chem. Abstr.*, **76**, 152732 (1972).

1497c. P. M. Maitlis, *The Organic Chemistry of Palladium*, Vols. I and II, Academic Press, New York, 1971.

1497d. P. Heimbach and R. Traunmüller, *Metal-Olefin Komplexe*, Verlag Chemie, Weinheim, 1970.

1498. M. Freifelder, *Practical Catalytic Hydrogenation: Techniques and Applications*, Interscience, New York, 1971.

1498a. J. H. Nelson and H. B. Jonassen, *Coord. Chem. Rev.*, **6**, 27 (1971).

1498b. G. Paiaro, *Organomet. Rev.*, *A*, **6**, 319 (1970).

1498c. H. Brunner, *Angew. Chem. Intern. Edit.*, **10**, 249 (1971).

1499. F. E. Paulik, *Catal. Rev.*, **6**, 49 (1972).

1500. K. Furumiya, *Koatsu Gasu*, **7**, 889 (1971); through *Chem. Abstr.*, **74**, 140311 (1971).

1501. D. M. Rudkovskii, *Vyssh. Zhirnye Spirty*, **1970**, 110; through *Chem. Abstr.*, **74**, 41496 (1971).

1502. M. Polievka, V. Macho, and L. Komora, *Petrochemia*, **11**, 78 (1971); through *Chem. Abstr.*, **76**, 139869 (1972).

1503. S. Usami, *Sekiyu Gakkai Shi*, **13**, 254 (1970); through *Chem. Abstr.*, **73**, 34693 (1970).

1504. M. Orchin and W. Rupilius, *Catal. Rev.*, **6**, 85 (1972).

1505. V. Y. Gankin, *Zh. Org. Khim.*, **8**, 424 (1972).

1506. M. Pribanic, *Kem. Ind.*, **20**, 151 (1971); through *Chem. Abstr.*, **76**, 131932 (1972).

1507. H. Weber and J. Falbe, *Ingenieursblad*, **40**, 652 (1961); through *Chem. Abstr.*, **76**, 85267 (1972).

1507a. P. Haynes, L. H. Slaugh, and J. F. Kohnle, *Tetrahedron Lett.*, **1970**, 365.

1507b. G. Wilkinson, Ger. Pat. 2,034,909; through *Chem. Abstr.*, **74**, 91744 (1971).

1507c. W. T. Reichle, *Inorg. Chim. Acta*, **5**, 325 (1971).

1507d. P. S. Braterman, *Chem. Commun. (Lond.)*, **1972**, 761.

1507e. W. Mowat, A. Shortland, G. Yagupsky, N. J. Hill, M. Yagupsky, and G. Wilkinson, *J. C. S. Dalton*, **1972**, 533.

1507f. D. M. P. Mingos, *Chem. Commun. (Lond).*, **1972**, 165.

1508. E. A. Hahn and E. Peters, *J. Phys. Chem.*, **75**, 571 (1971).

1508a. S. A. Bezman, M. R. Churchill, J. A. Osborn, and J. Wormald, *J. Am. Chem. Soc.*, **93**, 2063 (1971).

1508b. R. G. Beach and E. C. Ashby, *Inorg. Chem.*, **10**, 906 (1971).

1508c. E. C. Ashby and R. G. Beach, *Inorg. Chem.*, **10**, 2486 (1971).

1508d. N. J. Destafano and J. L. Burmeister, *Inorg. Chem.*, **10**, 998 (1971).

1509. N. I. Il'chenko, G. I. Golodets, and Y. I. Pyatnitskii, *Dokl. Akad. Nauk. SSSR.*, **103**, 112 (1972); through *Chem. Abstr.*, **76**, 131988 (1972).

1510. J. E. Bercaw and H. H. Brintzinger, *J. Am. Chem. Soc.*, **93**, 2045 (1971).

1511. R. H. Marvich and H. H. Brintzinger, *J. Am. Chem. Soc.*, **93**, 2046 (1971).

1512. J. E. Bercaw, R. H. Marvich, L. G. Bell, and H. H. Brintzinger, *J. Am. Chem. Soc.*, **94**, 1219 (1972).

1513. E. E. Van Tamelen, W. Cretney, N. Klaentschi, and J. S. Miller, *Chem. Commun. (Lond.)*, **1972**, 481.

1514. A. E. Shilov, A. K. Shilova, E. F. Kvashina, and T. A. Vorontsova, *Chem. Commun. (Lond.)*, **1971**, 1590.

1515. H. A. Martin and R. O. de Jongh, *Rec. Trav. Chim.*, **90**, 713 (1971).

1516. H. A. Martin, M. Van Gorkom, and R. O. de Jongh, *J. Organomet. Chem.*, **36**, 93 (1972).

1517. M. E. Volpin, A. A. Belyi, V. B. Shur, Y. I. Lyakhovetskii, R. V. Kudryavtsev, and N. N. Bubnov, *J. Organomet. Chem.*, **27**, C5 (1971).

1518. G. Fachinetti and C. Floriani, *Chem. Commun. (Lond.)*, **1972**, 654.

1519. J. G. Kenworthy, J. Myatt, and M. C. R. Symons, *J. Chem. Soc., A*, **1971**, 1020.

1520. P. C. Wailes, H. Weigold, and A. P. Bell, *J. Organomet. Chem.*, **27**, 373 (1971).

1521. F. N. Tebbe and G. W. Parshall, *J. Am. Chem. Soc.*, **93**, 3793 (1971).

1521a. W. D. Bonds, S. Chandrasekaran, N. H. Kilmer, and C. H. Brubaker, Jr., *Proc. 14th Intern. Conf. Coord. Chem.*, Toronto, 1972, p. 705.

1522. M. Cais and A. Rejoan, *Inorg. Chim. Acta*, **4**, 509 (1970).

1523. M. Cais and E. N. Frankel, U.S. Pat. 3,632,614; through *Chem. Abstr.*, **76**, 72051 (1972).

1523a. M. Cais and D. Fraenkel, *Abstracts, 5th Intern. Conf. Organomet. Chem.*, Moscow, 1971, **II**, 620.

1524. J. Nasielski, P. Kirsch, and L. Wilputte-Steinert, *J. Organomet. Chem.*, **27**, C13 (1971).

1524a. L. Wilputte-Steinert and P. Kirsch, *Abstracts, 5th Intern. Conf. Organomet. Chem.*, Moscow, 1971, **I**, 428.

1524b. E. R. Tucci, U.S. Pat. 3,631,111; through *Chem. Abstr.*, **76**, 58983 (1972).

1525. R. P. A. Sneeden and H. H. Zeiss, *J. Organomet. Chem.*, **27**, 89 (1971).

1526. J. R. Hanson and E. Premuzic, *Angew. Chem. Intern. Edit.*, **7**, 247 (1968).

1527. W. Schmidt, J. H. Swinehart, and H. Taube, *J. Am. Chem. Soc.*, **93**, 1117 (1971).

1528. M. L. H. Green and P. J. Knowles, *J. Chem. Soc., A*, **1971**, 1508.

1529. B. R. Francis, M. L. H. Green, and G. G. Roberts, *Chem. Commun. (Lond.)*, **1971**, 1290.

1530. A. Nakamura and S. Otsuka, *J. Am. Chem. Soc.*, **94**, 1886 (1972).

1530a. A. Nakamura, *Abstracts, 5th Intern. Conf. Organomet. Chem.*, Moscow, 1971, I, 128.

1531. J. L. Thomas and H. H. Brintzinger, *J. Am. Chem. Soc.*, **94**, 1386 (1972).

1532. J. L. Thomas, K. L. Tang, and H. H. Brintzinger, *Proc. 14th Intern. Conf. Coord. Chem.*, Toronto, 1972, p. 704.

1533. K. S. Chen, J. Kleinberg, and J. A. Landgrebe, *Chem. Commun. (Lond.)*, **1972**, 295.

1534. M. L. H. Green and W. E. Silverthorn, *Chen. Commun. (Lond.)*, **1971**, 557.

1535. M. L. H. Green, L. C. Mitchard, and W. E. Silverthorn, *J. Chem. Soc.*, *A*, **1971**, 2929.

1536. B. Bell, J. Chatt, G. J. Leigh, and T. Ito, *Chem. Commun. (Lond.)*, **1972**, 34.

1537. F. Pennella, *Chem. Commun. {Lond.)*, **1971**, 158.

1538. J. P. Jesson, E. L. Muetterties, and P. Meakin, *J. Am. Chem. Soc.*, **93**, 5261 (1971).

1539. M. Hidai, K. Tominari, and Y. Uchida, *J. Am. Chem. Soc.*, **94**, 110 (1972).

1540. A. Shortland and G. Wilkinson, *Chem. Commun. (Lond.)*, **1972**, 318.

1541. G. Wilkinson, private communication.

1542. M. A. Bennett and R. Watt, *Chem. Commun. (Lond.)*, **1971**, 94.

1543. M. A. Bennett and R. Watt, *Chem. Commun. (Lond.)*, **1971**, 95.

1544. H. D. Kaesz, S. A. R. Knox, J. W. Koepke, and R. B. Saillant, *Chem. Commun. (Lond.)*, **1971**, 477.

1545. M. Cais and N. Maoz, *J. Chem. Soc.*, *A*, **1971**, 1811.

1546. M. Arresta, P. Giannoccaro, M. Rossi, and A. Sacco, *Inorg. Chim. Acta*, **5**, 115 (1971).

1547. M. Arresta, P. Giannoccaro, M. Rossi, and A. Sacco, *Inorg. Chim. Acta*, **5**, 203 (1971).

1548. W. E. Newton, J. L. Corbin, P. W. Schneider, and W.A.Bulen, *J. Am. Chem. Soc.*, **93**, 268 (1971).

1549. Y. G. Borodko, M. O. Broitman, L. M. Kachapina, A. E. Shilov, and L. Y. Ukhin, *Chem. Commun. (Lond.)*, **1971**, 1185.

1550. D. H. Gerlach, W. G. Peet, and E. L. Muetterties, *J. Am. Chem. Soc.*, **94**, 4545 (1972).

1551. J. R. Sanders, *J. C. S. Dalton*, **1972**, 1333.

1551a. Y. Lagrange and G. Martino, Fr. Pat. 2,034,147; through *Chem. Abstr.*, **75**, 40967 (1971).

1552. T. Nishiguchi and K. Fukuzumi, *Chem. Commun. (Lond.)*, **1971**, 139.

1553. E. E. Mercer and P. E. Dumas, *Inorg. Chem.*, **10**, 2755 (1971).

1554. L. Vaska and M. E. Tadros, *J. Am. Chem. Soc.*, **93**, 7099 (1971).

1555. G. G. Eberhardt, M. E. Tadros, and L. Vaska, *Chem. Commun. (Lond.)*, **1972**, 290.

1556. M. M. T. Khan, R. K. Andal, and P. T. Manoharan, *Chem. Commun. (Lond.)*, **1971**, 561.

1557. Y. Sasson and J. Blum, *Tetrahedron Lett.*, **1971**, 2167.

1558. J. Blum, Y. Sasson, and S. Iflah, *Tetrahedron Lett.*, **1972**, 1015.

1558a. J. Blum and S. Biger, *Tetrahedron Lett.*, **1970**, 1825.

1559. S. Cenini, A. Fusi, and G. Capparella, *Inorg. Nucl. Chem. Lett.*, **8**, 127 (1972).

1560. B. R. James and L. D. Markham, *J. Catal.*, **27**, 442 (1972).

1561. R. A. Schunn, *Inorg. Chem.*, **9**, 2567 (1970).

1562. D. F. Ewing, B. Hudson, D. E. Webster, and P. B. Wells, *J. C. S. Dalton*, **1972**, 1287.

1563. B. Hudson, D. E. Webster, and P. B. Wells, *J. C. S. Dalton*, **1972**, 1204.

1564. J. E. Lyons, *Chem. Commun.* (*Lond.*), **1971**, 562.

1565. J. E. Lyons, *J. Org. Chem.*, **36**, 2497 (1971).

1566. J. P. Candlin, J. R. Jennings, and P. F. Todd, Brit. Pat. 1,246,123; through *Chem. Abstr.*, **75**, 129935 (1971).

1567. F. A. Cotton, J. G. Norman, A. Spencer, and G. Wilkinson, *Chem. Commun.* (*Lond.*), **1971**, 967.

1568. G. Wilkinson, Ger. Pat. 2,034,908; through *Chem. Abstr.*, **74**, 88127 (1971).

1569. U. A. Gregory, S. D. Ibekwe, B. T. Kilbourn, and D. R. Russell, *J. Chem. Soc., A*, **1971**, 1118.

1569a. N. F. Gol'dschleger, M. B. Tjabin, A. E. Shilov, and A. A. Steinmann, *Abstracts, 5th Intern. Conf. Organomet. Chem.*, Moscow, 1971, **I**, 328.

1570. B. R. James, R. S. McMillan, and E. Ochiai, *Inorg. Nucl. Chem. Lett.*, **8**, 239 (1972).

1571. B. R. James, E. Ochiai, and G. L. Rempel, *Inorg. Nucl. Chem. Lett.*, **7**, 781 (1971).

1572. J. Chatt, G. J. Leigh, and A. P. Storace, *J. Chem. Soc., A*, **1971**, 1380.

1573. E. F. Litvin, L. K. Friedlin, K. G. Karimov, M. L. Khidekel, and V. A. Avilov, *Izv. Akad. Nauk SSSR, Ser Khim*, **1971**, 1539.

1574. H. Hirai and T. Furuta, *J. Polym. Sci.*, **9(B)**, 459 (1971).

1575. H. Hirai and T. Furuta, *J. Polym. Sci.*, **9(B)**, 729 (1971).

1576. H. Hirai, T. Furuta, and S. Makishima, Jap. Pat. 71 39,326; through *Chem. Abstr.*, **76**, 45743 (1972).

1577. S. T. Wilson and J. A. Osborne, *J. Am. Chem. Soc.*, **93**, 3068 (1971).

1578. C. G. Pierpoint and R. Eisenberg, *Inorg. Chem.*, **11**, 1094 (1972).

1579. C. G. Pierpoint, A. Pucci, and R. Eisenberg, *J. Am. Chem. Soc.*, **93**, 3050 (1961).

1580. C. G. Pierpoint and R. Eisenberg, *Inorg. Chem.*, **11**, 1088 (1972).

1581. P. G. Douglas, R. D. Feltham, and H. G. Metzger, *J. Am. Chem. Soc.*, **93**, 84 (1971).

1582. R. E. Townsend and K. J. Coskran, *Inorg. Chem.*, **10**, 1661 (1971).

1583. M. C. Baird, *Inorg. Chim. Acta*, **5**, 46 (1971).

1584. S. D. Robinson and M. F. Uttley, *J. C. S. Dalton*, **1972**, 1.

1585. F. Piacenti, M. Bianchi, P. Frediani, and E. Benedetti, *Inorg. Chem.*, **10**, 2759 (1971).

1586. M. J. Lawrenson and M. Green, Ger. Pat. 2,026,926; through *Chem. Abstr.*, **74**, 124822 (1971).

1587. J. Manassen, *Plat. Met. Rev.*, **15**, 142 (1971).

1588. Johnson, Matthey and Co. Ltd., Fr. Pat. 2,055,060; through *Chem. Abstr.*, **76**, 50629 (1972).

1589. K. Shoda and A. Yasui, Jap. Pat. 71 10,537; through *Chem. Abstr.*, **75**, 77504 (1971).

1590. W. H. Knoth, *J. Am. Chem. Soc.*, **94**, 104 (1972).

1591. F. Pennella, R. L. Banks, and M. R. Rycheck, *Proc. 14th Intern. Conf. Coord. Chem.*, Toronto, 1972. p. 78.

1592. J. R. Sanders, *J. Chem. Soc.*, A, **1971**, 2991.

1593. D. A. Couch and S. D. Robinson, *Chem. Commun. (Lond.)*, **1971**, 1508.

1594. J. J. Hough and E. Singleton, *Chem. Commun. (Lond.)*, **1971**, 371.

1595. B. E. Cavit, K. R. Grundy, and W. R. Roper, *Chem. Commun. (Lond.)*, **1972**, 60.

1596. N. Ahmad, S. D. Robinson, and M. F. Uttley, *J. C. S. Dalton*, **1972**, 843.

1597. G. R. Clark, K. R. Grundy, W. R. Roper, J. M. Waters, and K. R. Whittle, *Chem. Commun. (Lond.)*, **1972**, 119.

1598. J. R. Norton, J. P. Collman, G. Dolcetti, and W. R. Robinson, *Inorg. Chem.*, **11**, 382 (1972).

1599. C. H. Bamford and M. U. Mahmud, *Chem. Commun. (Lond.)*, **1972**, 762.

1599a. F. L'Eplattenier, *Chimia*, **24**, 151 (1970).

1599b. F. L'Eplattenier and F. Calderazzo, U.S. Pat. 3,505,034; through *Chem. Abstr.*, **72**, 134788 (1970).

1599c. F. L'Eplattenier and C. Pelichet, *Helv. Chim. Acta*, **53**, 1091 (1970).

1600. J. Chatt, D. P. Melville, and R. L. Richards, *J. Chem. Soc.*, A, **1971**, 895.

1601. F. G. Moers, *Chem. Commun. (Lond.)*, **1971**, 79.

1602. J. Malin and H. Taube, *Inorg. Chem.*, **10**, 2403 (1971).

1603. G. G. Strathdee and M. J. Quinn, *Can. J. Chem.*, **50**, 3144 (1972).

1604. J. Halpern and M. Pribanic, *Inorg. Chem.*, **11**, 658 (1972).

1605. H. S. Lim and F. C. Anson, *Inorg. Chem.*, **10**, 103 (1971).

1606. G. Guastalla, J. Halpern, and M. Pribanic, *J. Am. Chem. Soc.*, **94**, 1575 (1972).

1607. G. D. Venerable, II and J. Halpern, *J. Am. Chem. Soc.*, **93**, 2176 (1971).

1608. T. Funabiki and K. Tarama, *Chem. Commun. (Lond.)*, **1971**, 1177.

1609. T. Funabiki and K. Tarama, *Bull. Chem. Soc. Jap.*, **44**, 945 (1971).

1610. T. Funabiki and K. Tarama, *Tetrahedron Lett.*, **1971**, 1111.

1610a. T. Funabiki and K. Tarama, *Bull. Chem. Soc., Jap.*, **43**, 3965 (1970).

1611. J. Basters, H. van Bekkum, and L. L. van Reijen, *Rec. Trav. Chim.*, **89**, 491 (1970).

1612. L. Simandi, E. Budo, and F. Nagy, *Kem. Kozl.*, **35**, 129 (1971); through *Chem. Abstr.*, **75**, 53742 (1971).

1613. E. B. Fleischer and M. Krishnamurthy, *J. Am. Chem. Soc.*, **94**, 1382 (1972).

1614. A. Zakhariev, V. Ivanova, and D. Shopev, *Dokl. Bolg. Akad. Nauk.*, **24**, 211 (1971); through *Chem. Abstr.*, **74**, 140551 (1971).

1615. Y. Ohgo, S. Takeuchi, and J. Yoshimura, *Bull. Chem. Soc. Jap.*, **43**, 505 (1970).

1616. Y. Ohgo, K. Kobayashi, S. Takeuchi, and J. Yoshimura, *Bull. Chem. Soc. Jap.*, **45**, 933 (1972).

1617. A. Rosenthal and G. Kan, *J. Org. Chem.*, **36**, 592 (1971).

1618. A. Rosenthal and G. Kan, *Carbohydr. Res.*, **19**, 145 (1971).

1619. P. Pino, S. Pucci, F. Piacenti, and G. Dell'Amico, *J. Chem. Soc.*, C, **1971**, 1640.

1619a. C. Botteghi, G. Consiglio, and P. Pino, *Chimia*, **26,** 141 (1972).

1620. H. Wakamatsu, J. Uda, and N. Yamakami, *Chem. Commun.* (*Lond.*), **1971,** 1540.

1621. C. P. Casey and C. R. Cyr, *J. Am. Chem. Soc.*, **93,** 1280 (1971).

1622. P. Taylor and M. Orchin, *J. Organomet. Chem.*, **26,** 389 (1971).

1623. M. Orchin and W. Rupilius, *J. Org. Chem.*, **36,** 3604 (1971).

1624. W. Rupilius and M. Orchin, *J. Org. Chem.*, **37,** 936 (1972).

1625. G. P. Vysokinskii and V. Y. Gankin, *Zh. Obshch. Khim.*, **41,** 1882 (1971).

1626. J. Falbe, H. Feichtinger, and P. Schneller, *Chem.-Ztg.*, **95,** 644 (1971).

1627. G. R. Kahle and J. W. Cleary, U.S. Pat. 3,652,676; through *Chem. Abstr.*, **76,** 139909 (1972).

1628. P. W. Solomon, U.S. Pat. 3,636,159; through *Chem. Abstr.*, **76,** 72027 (1972).

1629. F. Ungvary, *J. Organomet. Chem.*, **36,** 363 (1972).

1630. P. C. Ellgen, *Inorg. Chem.*, **11,** 691 (1972).

1631. L. Marko and J. Bathory, *Chem.-Anlagen Verfahren*, **1970,** 65.

1632. Badische Anilin- und Soda-Fabrik A.-G., Fr. Pat. 2,053,177; through *Chem. Abstr.*, **76,** 85394 (1972).

1633. G. Ferrari and P. L. Griselli, Ger. Pat. 2,119,334; through *Chem. Abstr.*, **76,** 24678 (1972).

1634. D. M. Rudkovskii, V. K. Pazhitov, and A. G. Trifel, U.S.S.R. Pat. 296,406; through *Chem. Abstr.*, **75,** 63128 (1971).

1635. J. B. Wilkes, Ger. Pat. 2,044,987; through *Chem. Abstr.*, **74,** 140963 (1971).

1635a. K. Nozaki, Ger. Pat. 2,061,798; through *Chem. Abstr.*, **75,** 63135 (1971).

1636. M. Matsubara, *Hokkaido-Ritsu Kogyo Shikenjo Hokoku*, **1968,** 74; through *Chem. Abstr.*, **74,** 12667 (1971).

1637. E. J. Mistrik and A. Mateides, *Chem. Zvesti*, **25,** 350 (1971).

1638. M. Derbesy, R. Lai, and M. Naudet, *Compt. Rend.*, **272,** 86 (1971).

1639. W. L. Fichteman and M. Orchin, *J. Org. Chem.*, **33,** 1281 (1968).

1640. P. Taylor and M. Orchin, *J. Am. Chem. Soc.*, **93,** 6504 (1971).

1641. S. Sato, J. Sato, S. Tatsumi, and H. Wakamatsu, *J. Chem. Soc. Jap., Ind. Chem. Sect.*, **74,** 1830 (1971).

1642. S. Sato, Y. Ono, S. Tatsumi, and H. Wakamatsu, *J. Chem. Soc. Jap.*, **92,** 178 (1971); through *Chem. Abstr.*, **76,** 33755 (1972).

1643. H. Wakamatsu, J. Furukawa, and N. Yamakami, *Bull. Chem. Soc. Jap.*, **44,** 288 (1971).

1644. E. R. Tucci, *Ind. Eng. Chem., Prod. Res. Dev.*, **9,** 516 (1970).

1645. W. Rupilius, J. J. McCoy, and M. Orchin, *Ind. Eng. Chem., Prod. Res. Dev.*, **10,** 142 (1971).

1646. I. Ogata and T. Asakawa, *J. Chem. Soc. Jap., Ind. Chem. Sect.*, **74,** 1640 (1971).

1647. W. Kniese and H. J. Nienburg, Ger. Pat. 2,005,654; through *Chem. Abstr.*, **75,** 117967 (1971).

1648. W. L. Senn, U.S. Pat. 3,576,881; through *Chem. Abstr.*, **75,** 19674 (1971).

1649. R. Platz, H. J. Nienburg, W. Kniese, and R. Kummer, Ger. Pat. 2,033,573; through *Chem. Abstr.*, **76,** 72006 (1972).

1650. H. J. Nienburg, R. Kummer, W. Kniese, and P. Tavs, Ger. Pat. 2,037,783; through *Chem. Abstr.*, **76**, 99099 (1972).

1651. Esso Research and Engineering Co., Brit. Pat. 1,049,291; through *Chem. Abstr.*, **76**, 33793 (1972).

1652. W. Kneise, H. J. Nienburg, and R. Kummer, Ger. Pat. 2,026,163; through *Chem. Abstr.*, **76**, 45731 (1972).

1653. E. R. Tucci, H. I. Thayer, and J. V. Ward, U.S. Pat. 3,644,529; through *Chem. Abstr.*, **76**, 99116 (1972).

1653a. H. J. Neinburg, W. Kniese, R. Kummer, and P. Tavs, Ger. Pat. 1,955,828; through *Chem. Abstr.*, **75**, 48437 (1971).

1654. W. Kneise, H. J. Nienburg, and R. Kummer, Ger. Pat. 2,026,164; through *Chem. Abstr.*, **76**, 58986 (1972).

1655. G. Pregaglia, A. Andreeta, and L. Benzoni, U.S. Pat, 3,627,843; through *Chem. Abstr.*, **76**, 58984 (1972).

1656. J. Fable, H. Tummes, J. Weber, and W. Weisheit, *Tetrahedron*, **27**, 3603 (1971).

1657. I. Ogata and Y. Kubota, *Yukagaku*, **21**, 24 (1972); through *Chem. Abstr.*, **76**, 155925 (1972)

1658. K. Kogami, O. Takahashi, and J. Kumanotani, *Bull. Chem. Soc. Jap.*, **45**, 604 (1972).

1659. R. Lai and E. Ucciani, *Compt. Rend.*, **273**, 1368 (1971).

1660. D. R. Fahey, U.S. Pat. 3,592,862; through *Chem. Abstr.*, **75**, 140363 (1971).

1661. M. Morita, Y. Iwai, J. Itakura, and H. Ito, U.S. Pat. 3,567,790; through *Chem. Abstr.*, **74**, 124983 (1971).

1662. A. Yamamoto, S. Kitazume, L. S. Pu, and S. Ikeda, *J. Am. Chem. Soc.*, **93**, 371 (1971).

1663. L. Vaska, L. S. Chen, and W. V. Miller, *J. Am. Chem. Soc.*, **93**, 6671 (1971).

1664. C. F. Nobile, M. Rossi, and A. Sacco, *Inorg. Chim. Acta*, **5**, 698 (1971).

1665. D. D. Titus, A. A. Orio, R. E. Marsh, and H. B. Gray, *Chem. Commun. (Lond.)*, **1971**, 322.

1666. B. A. Frenz and J. A. Ibers, *Inorg. Chem.*, **9**, 2403 (1970).

1667. G. Vitulli, L. Porri, and A. L. Segre, *J. Chem. Soc.*, A, **1971**, 3246.

1668. Y. Ohgo, S. Takeuchi, and J. Yoshimura, *Bull. Chem. Soc. Jap.*, **44**, 283 (1971).

1669. Y. Ohgo, S. Takeuchi, and J. Yoshimura, *Bull. Chem. Soc. Jap.*, **44**, 583 (1971).

1670. M. Green, R. J. Mawby, and G. Swinden, *Inorg. Nucl. Chem. Lett.*, **5**, 73 (1968).

1671. M. Green and G. Swinden, *Inorg. Chim. Acta*, **5**, 49 (1971).

1672. A. E. Brearley, H. Gott, H. A. O. Hill, M. O'Riordan, J. M. Pratt, and R. J. P. Williams, *J. Chem. Soc.*, A, **1971**, 612.

1673. G. N. Schrauzer and R. J. Holland, *J. Am. Chem. Soc.*, **93**, 1505 (1971).

1674. G. N. Schrauzer and R. J. Holland, *J. Am. Chem. Soc.*, **93**, 4060 (1971).

1675. B. C. Coyle and J. A. Ibers, *Inorg. Chem.*, **11**, 1105 (1972).

1676. R. A. Jewsbury and J. P. Maher, *J. Chem. Soc.*, A, **1971**, 2847.

1677. R. D. Gillard, B. T. Heaton, and D. H. Vaughan, *J. Chem. Soc.*, A, **1971**, 1840.

1678. P. Meakin, J. P. Jesson, and C. A. Tolman, *J. Am. Chem. Soc.*, **94**, 3242 (1972).

1679. H. Arai and J. Halpern, *Chem. Commun. (Lond.)*, **1971**, 1571.

1680. W. Strohmeier and R. Endres, *Z. Naturforsch*, **26b**, 362 (1971).

1681. S. Siegel and D. W. Ohrt, *Chem. Commun.* (*Lond.*), **1971**, 1529.

1682. S. Siegel and D. Ohrt, *Inorg. Nucl. Chem. Lett.*, **8**, 15 (1972).

1683. R. W. Mitchell, J. D. Ruddick, and G. Wilkinson, *J. Chem. Soc.*, *A*, **1971**, 3224.

1684. W. Strohmeier and R. Endres, *Z. Naturforsch.*, **26b**, 730 (1971).

1685. R. J. Shuford, *Diss. Abstr. Int. B*, **32**, 4501 (1972).

1686. M. Wahren and B. Bayerl, *Z. Chem.*, **11**, 263 (1971).

1687. A. J. Birch and K. A. M. Walker, *Aust. J. Chem.*, **24**, 513 (1971).

1688. Y. Senda, T. Iwasaki, and S. Mitsui, *Tetrahedron*, **28**, 4059 (1972).

1689. N. V. Shulyakovskaya, L. V. Vlasova, M. L. Khidekel, and I. A. Markushina, *Izv. Akad. Nauk. SSSR, Ser. Khim.*, **1971**, 1799.

1690. W. H. Faul, A. Failli, and C. Djerassi, *J. Org. Chem.*, **35**, 2571 (1970).

1691. Y. Osawa and D. S. Spaeth, *Biochemistry*, **10**, 66 (1971).

1692. J. F. Biellmann, M. J. Jung, and W. R. Pilgrim, *Bull. Soc. Chim. Fr.*, **1971**, 2720.

1693. D. Gagnaire and P. Vottero, *Bull. Soc. Chim. Fr.*, **1970**, 164.

1694. D. G. Earnshaw, F. G. Doolittle, and A. W. Decora, *Org. Mass Spectrom.*, **5**, 801 (1971).

1695. J. G. Atkinson and M. O. Luke, *Can. J. Chem.*, **48**, 3580 (1970).

1696. M. Ballenegger, A. Ruf, and T. Gäumann, *Helv. Chim. Acta*, **54**, 1373 (1971).

1697. E. Piers, R. W. Britton, and W. De Waal, *Can. J. Chem.*, **49**, 12 (1971).

1698. R. Wolovsky, E. P. Woo, and F. Sondheimer, *Tetrahedron*, **26**, 2133 (1970).

1699. L. Tonioli and R. Eisenberg, *Chem. Commun.* (*Lond.*), **1971**, 455.

1700. National Research Development Corp., Brit. Pat. 1,265,564; through *Plat. Met. Rev.*, **16**, 115 (1972).

1701. L. Horner and H. Siegel, *Ann. Chem.*, **751**, 135 (1971).

1702. C. Masters and B. L. Shaw, *J. Chem. Soc.*, *A*, **1971**, 3679.

1703. C. Masters, W. S. McDonald, G. Raper, and B. L. Shaw, *Chem. Commun.* (*Lond.*), **1971**, 210.

1704. R. H. Grubbs and L. C. Kroll, *J. Am. Chem. Soc.*, **93**, 3062 (1971).

1705. J. P. Collman, L. S. Hegedus, M. P. Cooke, J. R. Norton, G. Dolcetti, and D. N Marquardt, *J. Am. Chem. Soc.*, **94**, 1789 (1972).

1706. J. D. Morrison, R. E. Burnett, A. M. Aquiar, C. J. Morrow, and C. Phillips, *J. Am. Chem. Soc.*, **93**, 1301 (1971).

1707. W. Knowles and M. J. Sabacky, Ger. Pat. 2,123,063; through *Chem. Abstr.*, **76**, 60074 (1972).

1708. W. S. Knowles, M. J. Sabacky, and B. D. Vineyard, *Chem. Commun.* (*Lond.*), **1972**, 10.

1709. T. P. Dang and H. B. Kagan, *Chem. Commun.* (*Lond.*), **1971**, 481.

1710. H. B. Kagan and T. P. Dang, *J. Am. Chem. Soc.*, **94**, 6429 (1972).

1711. L. Horner and H. Siegel, *Phosphorus*, **1**, 199 (1972).

1712. G. C. Bond, Ger. Pat. 2,053,218; through *Chem. Abstr.*, **75**, 76127 (1971).

1713. G. C. Bond, Ger. Pat. 2,047,748; through *Chem. Abstr.*, **75**, 25968 (1971).

1714. G. C. Bond, Ger. Pat. 2,055,539; through *Chem. Abstr.*, **75**, 48429 (1971).

1715. T. Ueda, *Proc. 5th Intern. Congr. Catal.*, North-Holland, Amsterdam, 1972, Preprint 27.

1716. W. Strohmeier and W. Rehder-Stirnweiss, *Z. Naturforsch.*, **26b**, 193 (1971).

1717. C. K. Brown, W. Mowat, G. Yagupsky, and G. Wilkinson, *J. Chem. Soc., A,* **1971**, 850.

1718. C. K. Brown, D. Georgiou, and G. Wilkinson, *J. Chem. Soc., A,* **1971**, 3120.

1719. B. L. Booth and A. D. Lloyd, *J. Organomet. Chem.*, **35**, 195 (1972).

1720. C. B. Dammann, P. Singh, and D. J. Hodgson, *Chem. Commun. (Lond.)*, **1972**, 586.

1721. W. Strohmeier and W. Rehder-Stirnweiss, *Z. Naturforsch.*, **26b**, 61 (1971).

1722. W. Strohmeier, W. Rehder-Stirnweiss, and G. Reischig, *J. Organomet. Chem.*, **27**, 393 (1971).

1722a. L. Vaska and J. Peone, Jr., *Chem. Commun. (Lond.)*, **1971**, 418.

1723. F. Faraone, R. Pietropaolo, and S. Sergi, *J. Organomet. Chem.*, **24**, 797 (1970).

1724. J. V. Kingston, *Inorg. Nucl. Chem. Lett.*, **4**, 65 (1968).

1725. F. Glockling and G. F. Hill, *J. Chem. Soc., A,* **1971**, 2137.

1726. J. V. Kingston and G. R. Scollary, *J. Chem. Soc., A,* **1971**, 3399.

1727. W. O. Siegl, S. J. Lapporte, and J. P. Collman, *Inorg. Chem.*, **10**, 2158 (1971).

1728. B. Heil, L. Marko, and G. Bor, *Chem. Ber.*, **104**, 3418 (1971).

1729. P. Chini, S. Martinengo, and G. Garlaschelli, *Chem. Commun. (Lond.)*, **1972**, 709.

1730. W. Himmele, F. J. Mueller, and W. Aquila, Ger. Pat. 2,039,078; through *Chem. Abstr.*, **76**, 112700 (1972).

1731. W. Himmele, W. Hoffman, H. Pasedach, and W. Aquila, Ger. Pat. 1,964,962; through *Chem. Abstr.*, **75**, 63130 (1971).

1732. M. J. Lawrenson, Ger. Pat. 2,031,380; through *Chem. Abstr.*, **74**, 87397 (1971).

1733. W. Himmele and W. Aquila, Ger. Pat. 1,945,479; through *Chem. Abstr.*, **74**, 111589 (1971).

1734. J. Falbe, Ger. Pat. 1,618,384; through *Chem. Abstr.*, **75**, 129416 (1971).

1735. J. Falbe, Ger. Pat. 1,618,396; through *Chem. Abstr.*, **75**, 88209 (1971).

1736. R. L. Pruett and J. A. Smith, Ger. Pat. 2,062,703; through *Chem. Abstr.*, **75**, 109844 (1971).

1737. G. Wilkinson, Ger. Pat. 2,064,471; through *Chem. Abstr.*, **75**, 109848 (1971).

1738. O. R. Hughes and M. E. Douglas, Ger. Pat. 2,125,382; through *Chem. Abstr.*, **76**, 33794 (1972).

1739. K. L. Olivier, F. B. Booth, and D. E. Mears, U.S. Pat. 3,555,098; through *Chem. Abstr.*, **74**, 124823 (1971).

1740. J. Berthoux, J. P. Martinaud, and R. Poilblanc, Ger. Pat. 2,039,938; through *Chem. Abstr.*, **74**, 99461 (1971).

1741. K. L. Olivier and F. B. Booth, *Am. Chem. Soc., Div. Petrol. Chem., Preprints,* **14** (3), A7 (1969).

1742. C. Lassau, R. Stern, and L. Sajus, Fr. Pat. 2,041,776; through *Chem. Abstr.*, **75**, 129945 (1971).

1743. I. Ogata, Y. Ikeda, and T. Asakawa, *J. Chem. Soc. Jap., Ind. Chem. Sect.*, **74**, 1839 (1971).

1744. G. Foster and M. J. Lawrenson, Brit. Pat. 1,263,720; through *Chem. Abstr.*, **76**, 99114 (1972).

1745. M. J. Lawrenson and G. Foster, Brit. Pat. 1,243,190; through *Chem. Abstr.*, **75,** 109847 (1971).
1746. M. J. Lawrenson, Ger. Pat. 2,058,814; through *Chem. Abstr.*, **75,** 109845 (1971).
1747. M. J. Lawrenson, Brit. Pat. 1,254,222; through *Chem. Abstr.*, **76,** 33787 (1972).
1748. B. L. Booth, M. J. Else, R. Fields, and R. N. Haszeldine, *J. Organomet. Chem.*, **27,** 119 (1971).
1749. B. Fell, A. Geurts, and E. Muller, *Angew. Chem. Intern. Edit.*, **10,** 828 (1971).
1750. L. D. Rollman, *Inorg. Chim. Acta*, **6,** 137 (1972).
1751. M. Capka, P. Svoboda, M. Cerny, and J. Hetfleje, *Tetrahedron Lett.*, **1971,** 4787.
1752. R. R. Schrock and J. A. Osborne, *J. Am. Chem. Soc.*, **93,** 2397 (1971).
1753. R. R. Schrock and J. A. Osborne, *J. Am. Chem. Soc.*, **93,** 3089 (1971).
1754. B. F. G. Johnson, J. Lewis, and D. A. White, *J. Chem. Soc.*, *A*, **1971,** 2699.
1755. M. Green, T. A. Kuc, and S. H. Taylor, *J. Chem. Soc.*, *A*, **1971,** 2334.
1756. L. M. Haines, *Inorg. Chem.*, **10,** 1685 (1971).
1757. L. M. Haines, *Inorg. Chem.*, **10,** 1693 (1971).
1758. P. C. Kong and D. M. Roundhill, *Inorg. Chem.*, **11,** 1437 (1972).
1759. E. K. Barefield and G. W. Parshall, *Inorg. Chem.*, **11,** 964 (1972).
1760. G. M. Intitle, *Inorg. Chem.*, **11,** 695 (1972).
1761. A. P. Ginsberg and W. E. Lindsell, *J. Am. Chem. Soc.*, **93,** 2082 (1971).
1762. J. W. Dart, M. K. Lloyd, J. A. McCleverty, and R. Mason, *Chem. Commun. (Lond.)*, **1971,** 1197.
1763. R. V. Parish and P. G. Simms, *J. C. S. Dalton*, **1972,** 809.
1764. P. R. Branson and M. Green, *J. C. S. Dalton*, **1972,** 1303.
1765. A. L. Balch and J. Miller, *J. Organomet. Chem.*, **32,** 263 (1971).
1766. T. E. Nappier, Jr. and D. W. Meek, *J. Am. Chem. Soc.*, **94,** 306 (1972).
1767. D. G. Holah, A. N. Hughes, and B. C. Hui, *Can. J. Chem.*, **50,** 2442 (1972).
1768. S. D. Robinson and M. F. Uttley, *J. Chem. Soc.*, *A*, **1971,** 1254.
1769. J. P. Collman, P. Farnham, and G. Dolcetti, *J. Am. Chem. Soc.*, **93,** 1788 (1971).
1770. J. P. Collman, private communication, quoted in P. Finn and W. Jolly, *Inorg. Chem.*, **11,** 893 (1972).
1770a. G. La Monica, G. Navazio, P. Sandrini, and S. Cenini, *J. Organomet. Chem.*, **31** 89 (1971).
1771. M. A. Bennett and D. J. Patmore, *Inorg. Chem.*, **10,** 2387 (1971).
1772. J. F. Nixon and J. R. Swain, *J. C. S. Dalton*, **1972,** 1044.
1773. S. Bresadola, B. Longato, and F. Morandini, *Proc. 14th Intern. Conf. Coord. Chem.* Toronto, 1972, p. 448.
1774. J. H. Weber and G. N. Schrauzer, *J. Am. Chem. Soc.*, **92,** 726 (1970).
1775. B. R. James and F. T. T. Ng, *J. C. S. Dalton*, **1972,** 355.
1776. B. R. James and F. T. T. Ng, *J. C. S. Dalton*, **1972,** 1321.
1777. P. Abley, I. Jardine, and F. J. McQuillin, *J. Chem. Soc.*, *C*, **1971,** 840.
1778. P. Abley and F. J. McQuillin, *J. Chem. Soc.*, *C*, **1971,** 844.
1779. R. J. Roth and T. J. Katz, *Tetrahedron Lett.*, **1972,** 2503.
1780. M. Green and T. A. Kuc, *J. C. S. Dalton*, **1972,** 832.

1781. M. L. Khidekel and G. I. Karyakina, U.S.S.R. Pat. 319,336; through *Chem. Abstr.*, **76**, 37882 (1972).

1781a. V. A. Avilov and M. L. Khidekel, U.S.S.R. Pat. 285,902; through *Chem. Abstr.*, **75**, 10881 (1971).

1781b. V. A. Avilov, M. L. Khidekel, and E. G. Chepaikin, U.S.S.R. Pat. 287,908; through *Chem. Abstr.*, **74**, 116491 (1971).

1782. G. E. Marinicheva, M. L. Khidekel, and I. A. Markushina, *Izv. Akad. Nauk. SSSR, Ser. Khim.*, **1971**, 1797.

1783. L. K. Freidlin, E. F. Litvin, and L. F. Topuridze, *Izv. Akad. Nauk. SSSR, Ser. Khim.*, **1971**, 404.

1784. M. I. Bruce, B. L. Goodall, M. Z. Iqbal, and F. G. A. Stone, *Chem. Commun. (Lond.)*, **1971**, 661.

1785. A. R. M. Craik, G. R. Knox, P. L. Pauson, R. J. Hoare, and O. S. Mills, *Chem. Commun. (Lond.)*, **1971**, 168.

1786. C. J. Attridge and P. J. Wilkinson, *Chem. Commun. (Lond.)*, **1971**, 620.

1787. K. Moseley and P. M. Maitlis, *Chem. Commun. (Lond.)*, **1969**, 1156.

1788. P. M. Maitlis, C. White, D. S. Gill, J. W. Kang, and H. B. Lee, *Chem. Commun. (Lond.)*, **1971**, 734.

1788a. M. Gullotti, R. Ugo, and S. Colonna, *J. Chem. Soc., C*, **1971**, 2652.

1789. V. Schurig and E. Gil-Av, *Chem. Commun. (Lond.)*, **1971**, 650.

1790. B. R. James and D. V. Stynes, *J. Am. Chem. Soc.*, **94**, 6225 (1972).

1791. B. L. Shaw and R. E. Stainbank, *J. C. S. Dalton*, **1972**, 223.

1792. M. P. Yagupsky and G. Wilkinson, *J. Chem. Soc., A*, **1968**, 2813.

1793. R. Ugo, G. La Monica, S. Cenini, A. Segre, and F. Conti, *J. Chem. Soc., A*, **1971**, 522.

1794. J. Y. Chen and J. Halpern, *J. Am. Chem. Soc.*, **93**, 4939 (1971).

1795. W. Strohmeier, R. Fleischmann, and T. Onoda, *J. Organomet. Chem.*, **28**, 281 (1971).

1796. W. Strohmeier and R. Fleischmann, *J. Organomet. Chem.*, **29**, C39 (1971).

1797. W. Strohmeier, *J. Organomet. Chem.*, **32**, 137 (1971).

1798. C. Y. Chan and B. R. James, *Proc. 14th Intern. Conf. Coord. Chem.*, Toronto, 1972, p. 70.

1799. Y. Osumi and M. Yamaguchi, Jap. Pat. 71 21,604; through *Chem. Abstr.*, **75**, 129312 (1971).

1800. Y. Osumi and M. Yamaguchi, Jap. Pat. 71 21,605; through *Chem. Abstr.*, **75**, 129313 (1971).

1800a. Ethyl Corp., U.S. Pat. 3,524,898; through *Plat. Met. Rev.*, **15**, 78 (1971).

1801. F. Maspero and E. Perrotti, Ger. Pat. 2,124,925; through *Chem. Abstr.*, **76**, 45715 (1972).

1801a. W. I. Fanta and W. F. Erman, *J. Org. Chem.*, **36**, 358 (1971).

1802. S. Doronzo and V. D. Bianco, *Inorg. Chem.*, **11**, 466 (1972).

1803. L. M. Haines and E. Singleton, *J. Organomet. Chem.*, **25**, C83 (1970).

1804. E. W. Ainscough, S. D. Robinson, and J. J. Levison, *J. Chem. Soc., A*, **1971**, 3413.

1804a. M. A. Bennett and R. Charles, *Aust. J. Chem.*, **24**, 427 (1971).

1805. J. M. Guss and R. Mason, *Chem. Commun. (Lond.)*, **1971**, 58.

1806. R. N. Haszeldine, R. J. Lunt, and R. V. Parish, *J. Chem. Soc.*, *A*, **1971**, 3711.

1807. A. van der Ent and H. G. A. M. Cuppers, Neth. Pat. 70 01,018; through *Chem. Abstr.*, **75**, 151336 (1971).

1808. C. Masters, B. L. Shaw, and R. E. Stainbank, *Chem. Commun. (Lond.)*, **1971**, 209.

1809. M. G. Burnett and R. J. Morrison, *J. Chem. Soc.*, *A*, **1971**, 2325.

1810. C. K. Brown and G. Wilkinson, *Chem. Commun. (Lond.)*, **1971**, 70.

1811. J. S. Ricci and J. A. Ibers, *J. Am. Chem. Soc.*, **93**, 2391 (1971).

1812. A. J. Deeming and B. L. Shaw, *J. Chem. Soc.*, *A*, **1971**, 376.

1813. B. L. Shaw and R. E. Stainbank, *J. Chem. Soc.*, *A*, **1971**, 3716.

1814. M. J. Mays and F. P. Stefanini, *J. Chem. Soc.*, *A*, **1971**, 2747.

1815. D. M. P. Mingos and J. A. Ibers, *Inorg. Chem.*, **10**, 1035 (1971).

1816. D. M. P. Mingos, W. T. Robinson, and J. A. Ibers, *Inorg. Chem.*, **10**, 1043 (1971).

1817. V. G. Albano, P. L. Bellon, and M. Sansoni, *J. Chem. Soc.*, *A*, **1971**, 2420.

1818. D. M. P. Mingos and J. A. Ibers, *Inorg. Chem.*, **10**, 1479 (1971).

1819. C. Y. Chan and B. R. James, *Inorg. Nucl. Chem. Lett.*, in press.

1820. E. R. Birnbaum, *J. Inorg. Nucl. Chem.*, **32**, 1046 (1970).

1821. D. Bingham and M. G. Burnett, *J. Chem. Soc.*, *A*, **1971**, 1782.

1822. I. Hashimoto, N. Tsuruta, M. Ryang, and S. Tsutsumi, *J. Org. Chem.*, **35**, 3748 (1970).

1823. M. L. H. Green, H. Munakata, and T. Saito, *J. Chem. Soc.*, *A*, **1971**, 469.

1824. M. L. H. Green, T. Saito, and P. J. Tanfield, *J. Chem. Soc.*, *A*, **1971**, 152.

1824a. M. L. H. Green and M. J. Smith, *J. Chem. Soc.*, *A*, **1971**, 639.

1825. N. Ando, K. Maruya, T. Mizoroki, and A. Ozaki, *J. Catal.*, **20**, 299 (1971).

1826. K. Maruyama, T. Kuroki, T. Mizoroki, and A. Ozaki, *Bull. Chem. Soc. Jap.*, **44**, 2002 (1971).

1827. M. T. Musser, U.S. Pat. 3,631,210; through *Chem. Abstr.*, **76**, 72116 (1972).

1828. H. Kanai, *Chem. Commun. (Lond.)*, **1972**, 203.

1829. M. Kanai and A. Miyake, Jap. Pat. 71 29,127; through *Chem. Abstr.*, **75**, 150891 (1971).

1830. C. A. Tolman, *J. Am. Chem. Soc.*, **94**, 2994 (1972).

1831. H. F. Klein and J. F. Nixon, *Chem. Commun. (Lond.)*, **1971**, 42.

1832. C. W. DeKock and D. A. VanLeirsburg, *J. Am. Chem. Soc.*, **94**, 3235 (1972).

1833. F. Alvarez and F. Pino, *Inf. Quim. Anal. (Madrid)*, **19**, 142 (1965); through *Chem. Abstr.*, **64**, 13435b (1966).

1834. L. K. Freidlin, N. M. Nazarova, and Y. A. Kopyttsev, *Izv. Akad. Nauk. SSSR.*, *Ser Khim.*, **1972**, 201.

1835. E. W. Stern and P. K. Maples, *J. Catal.*, **27**, 120 (1972).

1836. E. W. Stern and P. K. Maples, *J. Catal.*, **27**, 134 (1972).

1837. M. L. H. Green and H. Munakata, *Chem. Commun. (Lond.)*, **1971**, 549.

1838. E. W. Ainscough and S. D. Robinson, *Chem. Commun. (Lond.)*, **1971**, 130.

1839. D. Fenton, U.S. Pat. 3,641,074; through *Chem. Abstr.*, **76**, 99120 (1972).

1840. J. V. Kingston and G. R. Scollary, *J. Chem. Soc.*, **1971**, 3765.

1841. J. H. Darling and J. S. Ogden, *Inorg. Chem.*, **11**, 666 (1972).

1842. M. B. Tyabin, A. E. Shilov, and A. A. Shteinman, *Dokl. Akad. Nauk, SSSR*, **198**, 380 (1971).

1843. R. J. Hodges, D. E. Webster, and P. B. Wells, *J. Chem. Soc.*, *A*, **1971**, 3230.

1844. K. P. Davis and J. L. Garnett, *J. Phys. Chem.*, **75**, 1175 (1971).

1845. J. L. Garnett and R. S. Kenyon, *Chem. Commun. (Lond.)*, **1971**, 1227.

1846. J. L. Garnett and R. J. Hodges, Aust. Pat. 417,394; through *Chem. Abstr.*, **76**, 59182 (1972).

1847. J. L. Garnett, *Catal. Rev.*, **5**, 229 (1971).

1847a. D. F. Gill and B. L. Shaw, *Chem. Commun. (Lond.)*, **1972**, 65.

1848. J. L. Garnett, R. J. Hodges, and W. A. Sollich-Baumgartner, *Osn. Predvid. Katal. Deist., Tr. Mezhd. Kongr. Katal., 4th*, **1968**, 62; through *Chem. Abstr.*, **75**, 101654 (1971).

1849. P. V. Balakrishnan and P. M. Maitlis, *J. Chem. Soc.*, *A*, **1971**, 1715.

1850. I. Yasumori and K. Hirabayashi, *Trans. Faraday Soc.*, **67**, 3283 (1971).

1851. K. Hirabayashi, S. Saito, and I. Yasumori, *J.C.S. Faraday I*, **1972**, 978.

1852. H. C. Clark and H. Kurosawa, *Chem. Commun. (Lond.)*, **1971**, 957.

1853. H. C. Clark and H. Kurosawa, *Inorg. Chem.*, **11**, 1275 (1972).

1854. H. C. Clark and H. Kurosawa, *J. Organomet. Chem.*, **36**, 399 (1972).

1855. W. J. Cherwinski and H. C. Clark, *Inorg. Chem.*, **10**, 2263 (1971).

1856. M. H. Chisholm and H. C. Clark, *J. Am. Chem. Soc.*, **94**, 1532 (1972).

1857. P. B. Tripathy, B. W. Renoe, K. Adzamli, and D. M. Roundhill, *J. Am. Chem. Soc.*, **93**, 4406 (1971).

1858. B. E. Mann, B. L. Shaw, and N. I. Tucker, *J. Chem. Soc.*, *A*, **1971**, 2667.

1859. S. Cenini, R. Ugo, and G. La Monica, *J. Chem. Soc.*, *A*, **1971**, 3441.

1860. S. Cenini, R. Ugo, and G. La Monica, *J. Chem. Soc.*, *A*, **1971**, 416.

1861. K. R. Dixon and D. J. Hawke, *Can. J. Chem.*, **49**, 3252 (1971).

1862. D. M. Blake and C. J. Nyman, *J. Am. Chem. Soc.*, **92**, 5359 (1970).

1862a. D. H. Gerlach, A. R. Kane, G. W. Parshall, J. P. Jesson, and E. L. Muetterties, *J. Am. Chem. Soc.*, **93**, 3543 (1971).

1863. E. Ucciani, *Rev. Fr. Corps Gras*, **18**, 373 (1971); through *Chem. Abstr.*, **76**, 5133 (1972).

1864. B. G. Linsen, *Fette, Seifen, Anstrichm.*, **73**, 753 (1971); through *Chem. Abstr.*, **76**, 155923 (1972).

1865. E. N. Frankel, H. Itatani, and J. C. Bailar, Jr., *J. Am. Oil Chem. Soc.*, **49**, 132 (1972).

1866. A. G. Hinze and D. J. Frost, *J. Catal.* **24**, 541 (1972).

1867. E. N. Frankel, *J. Am. Oil Chem. Soc.*, **48**, 248 (1971).

1868. E. N. Frankel and F. L. Thomas, *J. Am. Oil Chem. Soc.*, **49**, 10 (1972).

1869. N. F. Noskova, N. I. Marusich, and D. V. Sokol'skii, *Izv. Akad. Nauk. Kaz. SSR, Ser. Khim.*, **21**, 21 (1971); through *Chem. Abstr.*, **75**, 150890 (1971).

1870.   G. N. Sharifkanova and N. I. Marusich, *Vestn. Akad. Nauk. Kaz.*, **27**, 46 (1971); through *Chem. Abstr.*, **75**, 62999 (1971).

1871.   F. K. Shmidt, S. M. Krasnopol'skaya, and V. G. Lipovich, *Izv. Nauch.-Issled. Inst. Nefte-Uglekhim. Sin. Irkutsk, Univ.*, **12**, 24 (1970); through *Chem. Abstr.*, **75**, 53727 (1971).

1872.   D. V. Sokol'skii, G. N. Sharifkanova, N. F. Noskova, A. D. Dembitskii, and M. I. Goryaev, *Zh. Org. Khim.*, **7**, 1556 (1971).

1873.   J. C. Falk, *J. Org. Chem.*, **36**, 1445 (1971).

1874.   W. A. Butte, U.S. Pat. 3,542,898; through *Chem. Abstr.*, **74**, 87312 (1971).

1875.   W. A. Butte, U.S. Pat. 3,542,899; through *Chem. Abstr.*, **74**, 87313 (1971).

1875a. W. Morris, D. Y. Waddan, and D. Williams, Ger. Pat. 2,052,730; through *Chem. Abstr.*, **75**, 64575 (1971).

1876.   L. Farady and L. Marko, *J. Organomet. Chem.*, **28**, 159 (1971).

1877.   W. R. Kroll, U.S. Pat. 3,644,445; through *Chem. Abstr.*, **76**, 154452 (1972).

1878.   C. Lassau, J. Gaillard, and L. Sajus, *Hydrocarbon Process*, **50** (10), 97 (1971).

1879.   C. Lassau and L. Sajus, Ger. Pat. 2,032,141; through *Chem. Abstr.*, **74**, 87499 (1971).

1880.   C. Lassau, D. Maincon, and L. Sajus, Ger. Pat. 2,062,425; through *Chem. Abstr.*, **75**, 63100 (1971).

1881.   C. Lassau, R. Stern, and L. Sajus, Fr. Pat. 2,045,021; through *Chem. Abstr.*, **76**, 46686 (1972).

1882.   C. Lassau, R. Stern, and L. Sajus, Ger. Pat. 2,116,313; through *Chem. Abstr.*, **76**, 24664 (1972).

1883.   C. Lassau, R. Stern, and L. Sajus, Fr. Pat. 2,057,465; through *Chem. Abstr.*, **76**, 64008 (1972).

1884.   T. Yoshimoto, T. Kaneko, T. Marimiya, and H. Yoshii, Jap. Pat. 71 34,004; through *Chem. Abstr.*, **76**, 60477 (1972).

1885.   Y. Yoshimoto, T. Narumiya, H. Yoshii, K. Irako, and Y. Kaneko, Jap. Pat. 70 39,275; through *Chem. Abstr.*, **74**, 127211 (1971).

1886.   T. Yoshimoto, S. Kaneko, H. Yoshii, and T. Sasaki, Jap. Pat. 71 17,126; through *Chem. Abstr.*, **75**, 152761 (1971).

1887.   T. Yoshimoto, S. Kaneko, and H. Okado, Jap. Pat. 71 02,831; through *Chem. Abstr.*, **75**, 7138 (1971).

1888.   T. Yoshimoto, M. Kaneko, T. Narumiya, and H. Yoshii, Jap. Pat. 71 08,874; through *Chem. Abstr.*, **75**, 7143 (1971).

1889.   T. Yoshimoto, S. Kaneko, and H. Okado, Jap. Pat. 71 02,832; through *Chem. Abstr.*, **75**, 7137 (1971).

1890.   A. W. Shaw, Ger. Pat. 2,045,622; through *Chem. Abstr.*, **75**, 7117 (1971).

1891.   A. W. Shaw and E. T. Bishop, U.S. Pat. 3,634,549; through *Chem. Abstr.*, **76**, 128453 (1972).

1892.   M. M. Wald and M. G. Quam, U.S. Pat. 3,585,942; through *Chem. Abstr.*, **75**, 152743 (1971).

1893.   K. Bronstert, V. Ladenberger, and G. Fahrbach, Ger. Pat. 2,013,263; through *Chem. Abstr.*, **76**, 47138 (1972).

1894. R. J. A. Eckert and J. Heemskerk, Ger. Pat. 2,132,336; through *Chem. Abstr.*, **76,** 128357 (1972).

1895. W. C. Kray, Jr., Ger. Pat. 2,051,251; through *Chem. Abstr.*, **75,** 37623 (1971).

1896. L. E. De Winkler, Ger. Pat. 2,125,413; through *Chem. Abstr.*, **76,** 86893 (1972).

1897. C. W. Strobel, U.S. Pat. 3,646,142; through *Chem. Abstr.*, **76,** 154487 (1972).

1898. H. L. Hassell, Ger. Pat. 2,045,621; through *Chem. Abstr.*, **74,** 142672 (1971).

1899. E. W. Duck, J. M. Locke, and J. C. Mallison, Brit. Pat. 1,229,573; through *Chem. Abstr.*, **75,** 50244 (1971).

1900. J. C. Falk and R. J. Schlott, *Macromolecules*, **4,** 152 (1971).

1901. J. C. Falk, *J. Polym. Sci.*, **9(A1),** 2617 (1971).

1902. J. C. Falk and R. J. Schlott, *Angew. Makrol. Chem.*, **21,** 17 (1972).

1903. G. A. Shagisultanova, A. A. Karaban, A. S. Tikhonov, and S. P. Gorbunova, *Khim. Vys. Energ.*, **6,** 186 (1972); through *Chem. Abstr.*, **76,** 131995 (1972).

1904. E. A. Symons and E. Buncel, *J. Am. Chem. Soc.*, **94,** 3641 (1972).

1905. C. D. Ritchie and H. F. King, *J. Am. Chem. Soc.*, **90,** 833 (1968).

1906. F. Botter, G. Dirian, J. Pauly, and E. Roth, Ger. Pat. 2,108,944; through *Chem. Abstr.*, **74,** 119088 (1971).

1907. J. Legall, D. V. Dervartanian, E. Spilker, J. P. Lee, and H. D. Peck, Jr., *Biochim. Biophys. Acta*, **234,** 525 (1971).

1908. J. Chatt and G. J. Leigh, *Chem. Soc. Rev.*, **1,** 121 (1972).

1909. A. Shilov, N. Denisov, O. Efimov, N. Shuvalov, N. Shuvalova, and A. Shilova, *Nature*, **231,** 460 (1971).

1910. R. E. E. Hill and R. L. Richards, *Nature*, **233,** 114 (1971).

1911. G. N. Schrauzer, P. A. Doemeny, G. W. Kiefer, and R. H. Frazier, *J. Am. Chem. Soc.*, **94,** 3604 (1972).

1912. K. Watanabe, T. Kondow, M. Soma, T. Onishi, and K. Tamaru, *Chem. Commun. (Lond.),* **1972,** 39.

1913. M. Ichikawa, T. Kondo, K. Kawase, M. Sudo, T. Onishi, and K. Tamaru, *Chem. Commun. (Lond.),* **1972,** 176.

1914. M. Ichikawa, K. Kawase, and K. Tamaru, *Chem. Commun. (Lond.),* **1972,** 177.

Index

A ligand given in parentheses is optional. Thus *(hydrido)phosphines* then includes *phosphine complexes* and *hydridophosphine complexes.*

225, 248, 270, 284, 285, 311, 321,
324, 342, 344, 365, 371, 415, 416,
427, 436
see also Benzanthrone; Benzil; Benzoquinone; Dyes; Flavones; Quinones;
Reductive-amination

Lactone formation, 159, 160, 167, 170,
268, 422
Lanthanides, 389, 411
Lead complexes, metal, mixed, 323, 341,
358
salts, stearates, 384, 389
Ligand effects, survey of, dissociation, 403
electronic, 403, 404
steric, 404, 405
Lignins, hydroformylation of, 160, 163,
168
Linalool, hydrogenation of, 221, 222
Lineoleate, hydrogenation of, 221, 223, 325,
341, 344, 346-360, 443
Linolenate, hydrogenation of, 344, 348,
349, 352-354, 356-359, 362, 443
Linseed oils, hydrogenation of, 348, 350,
355
Lithium salts, 389; see also Alcoxyalanates
Lupenyl acetate, hydrogenation of, 221

Magnesium complexes, alkyls, 411
salts, 37
Manganese complexes, acyls, 63, 169, 170
alkyls, 63, 415
carbonyls, 63, 354
enols, π-, 415
hydridocarbonylphosphines, 415
hydroformylation catalysts, 63
salts, 62
stearates, 62, 384-386
Ziegler catalysts, 63, 363, 364, 371-374,
379
see also Permanganate
Markownikov (Anti-) addition of metal
hydride, 118, 119, 144, 146, 172,
173, 181-183, 185, 253, 254, 262,
266-268, 415, 422
Mercury atoms, 38, 39
Mercury complexes, 404
allyls, 118
aquo, 37-41, 407
carboxylates, 37, 39-42

cocatalysts, 321, 359
en, 39, 41
metal, mixed, 274, 301, 341
salts, 39, 158, 389, 403
Mercury, reduction of, 37-42, 301, 389, 407,
420
Metal atoms, 34, 38, 39
Metal hydrides, bond energy, 109, 289, 290,
308, 344
formulation, 6, 7, 401
Metal-metal complexes, mixed, 56, 59, 72,
274, 295, 301, 309, 319, 322, 323,
338, 341, 359, 430, 433. See also
Hydrogenolysis; Stannous chloride
Metal olefin complexes, stability, constants,
86, 211, 212, 297, 298, 300, 434
Metal salts, activity of, survey, 388, 389
Methoxide, as catalyst, 394
Molybdenum complexes, acetate, 59
alkyls, 60, 413, 414
allyls, 361
arene-carbonyls, 58
carbonylphosphines, -arsines, 413, 414
+ $SnCl_2$, 60, 361
chlorides, 60
+ $SnCl_2$, 59, 60, 402
cyclopentadienyls, 59, 413, 414
carbonyls, 58, 59, 444
hydrido phosphines, - phosphites, 414
hydroformylation catalysts, 413, 414
hydrogenase, 397, 398
metal, mixed, 56, 59
molybdate, 60
nitrogen, 414
nitrogen fixation models, 445
oxide, mixed, 411
quinone, 59
(thio) carboxylates, cationic, 59, 414
vinyls, 414
Ziegler catalysts, 60, 363, 365, 367, 383,
444
Monoenes, hydrogenation of, 44-50, 56-60,
63-66, 70-72, 78-104, 113, 116,
123-130, 132, 134, 137, 143-148,
153-164, 167-173, 181, 184, 186-
192, 195, 198, 199, 201, 203-245,
247-263, 265, 267, 269-287, 296-
312, 315-325, 327-331, 337, 338,
340-360, 363-367, 370, 371, 373-
380, 384-388, 390-392, 395, 397,